51

AF774761

BIOCOMMUNICATION

Sign-Mediated Interactions between Cells and Organisms

BIOCOMMUNICATION
Sign-Mediated Interactions between Cells and Organisms

Editors

Richard Gordon

Gulf Specimen Aquarium & Marine Laboratory, USA & Wayne State University, USA

Joseph Seckbach

The Hebrew University of Jerusalem, Israel

World Scientific

NEW JERSEY • LONDON • SINGAPORE • BEIJING • SHANGHAI • HONG KONG • TAIPEI • CHENNAI •

World Scientific Publishing Europe Ltd.
57 Shelton Street, Covent Garden, London WC2H 9HE
Head office: 5 Toh Tuck Link, Singapore 596224
USA office: 27 Warren Street, Suite 401-402, Hackensack, NJ 07601

Library of Congress Cataloging-in-Publication Data
Names: Gordon, Richard, 1943– editor. | Seckbach, J. (Joseph), editor.
Title: Biocommunication : sign-mediated interactions between cells and organisms / [edited by] Richard Gordon (Gulf Specimen Aquarium & Marine Laboratory, USA & Wayne State University, USA) & Joseph Seckbach (The Hebrew University of Jerusalem, Israel).
Description: Hoboken, New Jersey : World Scientific, 2016
Identifiers: LCCN 2015049554 | ISBN 9781786340443 (hc : alk. paper)
Subjects: LCSH: Information theory in biology. | Plant cellular signal transduction. | Animal communication.
Classification: LCC QH507 .B525 2016 | DDC 571.7/42--dc23
LC record available at http://lccn.loc.gov/2015049554

British Library Cataloguing-in-Publication Data
A catalogue record for this book is available from the British Library.

Copyright © 2017 by World Scientific Publishing Europe Ltd.

All rights reserved. This book, or parts thereof, may not be reproduced in any form or by any means, electronic or mechanical, including photocopying, recording or any information storage and retrieval system now known or to be invented, without written permission from the Publisher.

For photocopying of material in this volume, please pay a copying fee through the Copyright Clearance Center, Inc., 222 Rosewood Drive, Danvers, MA 01923, USA. In this case permission to photocopy is not required from the publisher.

Desk Editors: Anthony Alexander/Mary Simpson

Typeset by Stallion Press
Email: enquiries@stallionpress.com

Printed in Singapore

Dedication to the Memory of Eshel Ben-Jacob (1952–2015)

Eshel Ben-Jacob was a theoretical and experimental physicist at Tel Aviv University, holder of the Maguy-Glass Chair in Physics of Complex Systems, and Fellow of the Center for Theoretical Biological Physics (CTBP) at Rice University. During the 1980s he became an international leader in the theory of self-organization and pattern formation in open systems, and later extended this work to adaptive complex systems and biocomplexity. His specialization in self-organization of complex systems yielded the breakthrough of solving the long-standing (since Kepler) snowflake problem. In the late 1980s, he turned to the study of bacterial self-organization, believing that bacteria hold the key to understanding the larger biological systems. He studied pattern forming bacteria species, and became a pioneer in the study of bacterial intelligence and social behaviors of bacteria, thus being an important contributor to this book. He was an influential figure in establishing the now rapidly evolving Physics of Living Systems (Biological Physics and Physical Biology) disciplines. Eshel was an exceptionally creative interdisciplinary scientist. He was one of the world's leading experts in biocomplexity, with a scientific profile that integrated a remarkable mix of physics and biology, immunology and mathematical modeling, engineering and econophysics. Not many scientists contributed to so many different fields with innovative and influential research as he did.

Eshel Ben-Jacob was a superb teacher and lecturer, his teachings and presentations inspired a wide range of audiences, from expert physicists at numerous international conferences and students at his courses, to high school students and hi-tech workers. Eshel Ben-Jacob generously shared his knowledge and love of science. As a mentor, he advised and collaborated with many students and junior associates who later had a successful career of their own. He was a devoted advisor, his deep and extreme passion for science was magnetizing, and he was intense and genuine both as a person and as a scientist.

His many honors for contributions to science included his election in 2014 to the American Philosophical Society, the United States' oldest learned society, the 1986 Landau Research Prize, the 1996 Siegle Research Prize of the Israel Academy of Sciences and Humanities and the 2013 Weizmann Prize in Exact Sciences. Ben-Jacob was former president of the Israel Physical Society and a former chair of the Israel Ministry of Education's Advisory Council of High School Physics Education.

Eshel's sudden death on Friday June 5, 2015 came as a shock to all of us. He was a bright star who gave for free, so much to so many. We were lucky to get to know him and learn from him. His death was a devastating event for all of us. We will miss him.

Alin Finkelshtein, June 11, 2015

Dedication to the Memory of Fernando Palop (1954–2015)

In Memoriam of Fernando Palop, born in Barbastro 1954, died in Valencia 2015, Associate Professor, Polytechnic University of Valencia and co-founder of TRIZ XXI (http://www.triz.es).

Editors and friends would like to thank Professor Fernando Palop for his valuable contribution to this book. Palop dedicated a big part of his life to teaching the business of tech mining and its importance in increasing corporate intelligence of organizations. He worked with both students and professionals, either through his university career or his own company, TRIZ XXI.

His work focused on designing and creating competitive intelligence methods to reduce risk in innovation processes and knowledge management. Through this he worked closely with universities and institutions, and later became a member of the 166th Committee of The Spanish Association for Standardization and Certification. As a distributor of knowledge, Palop gave numerous speeches and seminars on competitive intelligence and TechMining throughout Latin America, including Brazil, Colombia and Chile.

Preface

Who is Who in Biocommunication

Joseph Seckbach[*,§] and Richard Gordon[†,‡,||]

[*]*The Hebrew University of Jerusalem, Israel*

[†]*Gulf Specimen Marine Laboratory*
Panacea, Florida, USA

[‡]*C.S. Mott Center for Human Growth and Development*
Department of Obstetrics and Gynecology
Wayne State University, Detroit, Michigan, USA

[§]*joseph.seckbach@mail.huji.ac.il*
[||]*dickgordoncan@gmail.com*

"Three things are beyond me: The way of an eagle in the sky ... how a snake makes its way over a rock ... how a man has his way over a maiden"
Proverbs 30: 18–19

1. Introduction

The concept of biocommunication is very broad and has several meanings: it involves linguistics, cognition, and even outer space, including astrobiology. Biocommunication can also cover molecular genetics, plant physiology (chemotropism), and plant perception. Moreover, the term includes inter-personal exchange of information in various modes.

Biocommunication exists in almost all organisms from bacteria to man. Every organism uses signs of communication to exchange information for purposes of coordination and cooperation. This communication also serves for organization between members of the same, related, or unrelated species. All coordination between cells, organs, and organisms depends on successful biocommunication processes. For example, between single cells we find the chemical attraction between sperm migrating toward an egg within the body or outside in the wet environment. In this respect, diseases in organisms are often the results of disturbed or damaged communication between cells.

In the Bible are a few cases of communication between man and other creatures: such as the conversation of the snake with Eve about eating from the fruit of the Tree of Knowledge in the Garden of Eden (Genesis 3: 1–6), the case of Balaam's ass who spoke to him and complained about his beating her three times (Num. 22: 28–30). It is said that King Solomon was gifted to talk to plants, animals, fowl, creeping things, and fish. Given recent successful attempts to "speak" with primates, and our increasing interest in communicating in some way with animals, these ancient stories show that interest in interspecific biocommunication is thousands of years old.

In our study of biocommunication, we are aiming at more specific types of communication, intraspecific or interspecific, between species of bacteria, fungi, plants, animals, and humans. Communication also incorporates linguistics, sign-mediated interactions following syntactic, pragmatic, and semantic rules. Biocommunication of animals includes vocalizations (as between bird species, canines and others), the use of pheromone production, (between insects or mammals and chemical signals between plants and animals. Animal communication is related to zoosemiotics. Some biocommunication has tentatively been connected to the concept of instincts in insects and animals (Tautz, 2008).

2. Human Communication — the King of Communication

In human beings, we have direct speaking and the senses of vision, hearing, odor, and touching the body as means of communication.

Human's senses are related to links from person to person, such as male–female — including the courtship period — and mankind to their surroundings. The five senses of humankind enable the mastery of humans over other animals. Man is able to communicate with some domestic animals and birds and receive responses from them. For example, the wagoner orders his horses or mules to go forward or to halt. People can command dogs, cats, goats, sheep, or pet birds to perform various actions and they often obey.

In India, street performers with flutes entice cobras to respond to their music (though this is mostly a reaction to movement). On the other hand, we know about communication between man and dolphins, which amazes large crowds of spectators.

3. Biocommunication in Plants

There are plants, such as the orchids, whose flowers mimic others so as to appear as females to their pollinators, so that male insects are happy to land on these flowers, being certain that they are his female; this way pollination and fertilization take place. Other plants can "communicate" with their pollinators through bright colors or pleasant or repulsive odor.

There are chemical signals between plants and animals (such as tannin production by plants to warn away insects). We have heard about chemically mediated communication between plants. Plant physiology covers some processes such as geo- or heliotropism. Some consider plants intelligent (Buhner, 2014). Plants may have perception and are sentient; perhaps they experience pain or pleasure of emotions (Tompkins and Bird, 1973). According to the last two references (popular works), plants have the ability to communicate with humans and other forms of life. Others plants display trembling, moving, and shaking of the whole plant over an insect visit. This is the case of *Mimosa* plants which defend themselves against animals through stimuli responses that rapidly fold its leaves. Still others repulse their plant eating grazers by producing offensive odors or poisonous chemicals to repulse neighboring plants or other predator enemies. The relation between plants and other organisms has been recently published (Seckbach and Dubinsky, 2010).

4. Insects and Animals

Social insects (such as bees, ants, termites, and wasps) use pheromones for communication. We know about the bees' dance (which communicates how to get to flowers). The pollination period is vital for nature and for development of fruit; at this stage plants signal (through their odor, shape and colors) pollinating insects (mainly flying ones) and humming birds to land on their flowers.

Dogs have abilities to smell; this feature has been used by police and armies for their activities. A male dog may appear from several kilometers away during a female's rutting period, a sample of odor communication. On some butterflies their wings mimic the patterns of toxic butterflies to ward off predators.

5. Fowls' Communication

Among the birds are strongly social species; communication among them is seen in a variety of ceremonies, gestures, and specific calls. We recognize vocalization (bird species) and know about pairs of birds (e.g., pigeon couples) — how they divide their functions in the nest, sitting on the eggs and bringing food for the youngsters. There are singing birds learning to sing complicated songs from their parents and even to mimic human voices. Another point is the behavior of the male birds, such as, turkeys or peacocks, to impress the female during the courting period with full-blown colored feathers and special sounds. All their displays are aimed at attracting mates. Among penguins, we find close relations of couples using their voices in communication. They operate with division of duty, in order to protect the huge crowd against the cold polar winds or to warm their eggs. There are the young penguins who find their parents among millions of other members in the surrounding flock by their sounds. Some birds may alert their neighbors against an enemy in the vicinity by making warning sounds.

We'll add a personal story about bird communication. Many years ago, when JS was a tractor driver he was plowing the fields near the Kishon River (northern Israel). The plow exposed and raised up from the ground a large black snake, which started to wiggle between the freshly lifted soil sections. Suddenly, a carnivorous bird, perhaps a vulture or eagle, landed and started to "deal" with the snake. A few minutes later, the bird lifted up the snake but let it drop from high above, leaving the snake on the ground. Perhaps it proved too heavy. Then the bird flew away and returned with another one [its mate?]. Both of them landed near the prey and together lifted the snake from the ground and flew away with it to have their full meal.

6. General Concepts

In this volume, we describe more specific types of communication. Such exchange of information, and sign-mediated interaction within (intraspecific) or between (interspecific) takes place among species of bacteria, plants, fungi, and animals. Biocommunication means strong interaction via a special "language" used within a species, including that between primates, birds or insects. In most cases, there are chemical molecules (semiochemicals) involved. Biocommunication also occurs between the cells of a developing embryo, as it forms itself.

Biocommunication is expressed in different ways. One may associate it with the concept of symbiosis and organismal language and interaction between species. For example, such communication exists within the soil between roots of higher trees and symbioses with arbuscular mycorrhizal fungi.

The purpose of this book to demonstrate the panoply of biocommunication, interaction, and "language" in the biological world. Such interrelations range from molecules and cells, bacteria, fungi, cyanobacteria, and other organisms as well as from microanimals to mammals.

Topics covered within this book include: molecular biocommunication, superfast evolution via interspecies, bacterial communication, communication languages and agents in biological systems, carnivorous plants and insects, the crosstalk between plants and their symbionts, global communication within the embryo, how neurons find their target, animal communication under noisy conditions above and below water, chemical communication, sexual communication by pheromones, and artificial communication.

Acknowledgment

We thank Fern Seckbach for her comments.

References

Buhner SH. 2014. *Plant Intelligence and the Imaginal Realm: Beyond the Doors of Perception into the Dreaming of Earth*: Inner Traditions/Bear, USA.

Seckbach J and Dubinsky Z. 2010. *All Flesh Is Grass: Plant-Animal Interrelationships*. Springer, New York.

Tautz J. 2008. *The Buzz about Bees: Biology of a Superorganism*. Springer, Germany.

Tompkins P and Bird C. 1973. *The Secret Life of Plants*. Harper & Row, New York.

Introduction to Biocommunication

Joseph Seckbach[*,§] and Richard Gordon[†,‡,||]

*The Hebrew University of Jerusalem, Israel

†Gulf Specimen Marine Laboratory
Panacea, Florida, USA

‡C.S. Mott Center for Human Growth and Development
Department of Obstetrics and Gynecology
Wayne State University, Detroit, Michigan, USA

§joseph.seckbach@mail.huji.ac.il
||dickgordoncan@gmail.com

Biocommunication (BC) as a field is concerned with describing the exchange of information between various organisms. Tembrock outlined a first draft of the BC theory in 1975. There are abundant cases of communication within intra- and interspecies in the biological world. It exists in almost all organisms from bacteria to man. Every organism uses signs of coordination, cooperation, and organization between members of the same, related, or non-related organisms. All coordination between cells, organs, and organisms depends on successful biocommunicative processes. In this respect, diseases in organisms are the result of damaged communication between cells. Such interaction is revealed among single and multiple microorganisms (Prokaryotes) and higher forms (Eukaryotes).

We know of communication between humans, but humankind talks to domestic animals such as dogs, cats, or birds, flocks, cattle, and donkeys, or horses (and get responses from them). Some claim that music in the cattle barn increases the production of milk by cows. Here, JS wishes to share with you a little episode: a few years ago, we watched a TV program on a big screen; adjacent to us sat two Chihuahuas. When dogs in the show appeared on the TV monitor, the little domestic dogs started to bark at the foreign penetrators of their territory and warn them to go away.

In the Bible are cases of communication between man and other creatures (such as the archaic snake and Eve; Balaam and the donkey, the wisdom of King Solomon, etc.) or the Almighty talking to man (for details, see the preface in this volume).

Our aim in this volume is to present information about and examples of the BC concept to a broad audience. This is accomplished through several different chapters of separate examples of BC from molecules to the level of monkeys and dolphins. We describe specific types of communications within molecules, species of Bacteria, plants, fungi, and animals. Although there are many more cases of this field, the contents offered in this book will give the audience a good overview of the BC phenomena. The interaction of birds or animals may include vocalization (such as between competing bird species, a pair of pigeons, or higher animals), pheromone production (in insects and in animals), chemical or color signals, between insects, plants, and animals inter-relationships (Seckbach and Dubinsky, 2011; Tautz, 2008) and altruistic behavior.

In the volume we describe more specific types of communication, exchange of information, bioinformatics, and sign-mediated interaction within (intraspecific) or between (interspecific) species of plants, animals, fungi and bacteria. BC means strong interaction via a "language" between the species (even between real language of primates, between birds, or insects). In most cases, chemical molecules involved. BC is expressed in different ways. One way associate it with the concept of symbiosis (Seckbach, 2002; Seckbach and Grube, 2010), organismal language, and interaction between species.

For example, such communication exists within the soil between roots of higher plants and symbioses with arbuscular mycorrhizal fungi. We may also include BC between single cells, e.g., the chemical attraction between the sperm migrating toward the egg within the body or outside in a wet environment. Some organisms may use pheromone production, as between various species of insects and animals such as vertebrates. There are also chemical signals between plants and animals, as in tannin and phyto-poison production used by vascular plants to warn away insects or inhibit germination of other plants. Such chemically mediated communication is between plants and within plants.

We offer this book to graduate students, researchers in biology, ecology, zoology, to related applied areas of environment and to adventurously curious readers.

Ille Gebeshuber has done a splendid job in outlining our book in her Foreword.

This book came about through a colleague at Louisiana State University (LSU), Professor Maud Walsh, who was interested in environmental communication and that concept led us to BC.

Acknowledgments

We thank all our contributors for their chapters and for their long patience. This book could not be published without the effort and persistence of our contributors and reviewers. We apologize to any person who assisted us in this volume who may have been overlooked.

The following arc the names of are our reviewers and advisors. We also wish to express our gratitude to colleagues who pointed out new reviewers who agreed to evaluate chapters. Among the reviewers are Bashir Ahmad, Bradly Alicea, Luciano Avio, Frantisek Baluska, Kara Loeb Belinsky, Francisca Bronfman, Michael Y. Galperin, Ille Gebeshuber, Vera Kolb, Laurent Legendre, Michael Levin, Howard H. Pattee, Bartosa J. Plachno, Vladimir Seleznev, James A. Shapiro, Alexei A Sharov, John S. Torday, Jack A. Tuszynski, Clément Vidal, Luis P. Villarreal, Günther Witzany. Additional appreciation is due to Lubomir Adamec, David J. Chapman, Julian Chela-Flores, Natan Gadoth, Rob Hengeveld, Arie S. Issar, Shmuel Nussinov, Florence Raulin, Luis P. Villarreal.

To our readers: In addition to our volume, there is also a set of books on BC edited by Günther, and we thank him for his initial cooperation in this book.

References

Seckbach J. 2002. Symbiosis: mechanisms and model systems. In *Cellular Origin, Life in Extreme Habitats and Astrobiology*, Seckbach J (ed.), Vol. 4. Kluwer Academic Publisher, NL.

Seckbach J and Grube M. 2010. Symbioses and stress, joint ventures in biology. In *Cellular Origin, Life in Extreme Habitats and Astrobiology*, Seckbach J (ed.), Vol. 17. Springer, Dordrecht, NL.

Seckbach J and Dubinsky Z. 2011. All Flesh is Grass (Plants–Animal Interrelationships). In *Cellular Origin, Life in Extreme Habitats and Astrobiology*, Seckbach J (ed.), Vol. 16. Springer, Dordrecht NL.

Tautz J. 2008. *The Buzz about Bees*. Springer, Berlin-Heidelberg.

Tembrock G. 1975. *Biocommunication*. Veiweg, Braunschweig.

Foreword

Biocommunication in the Web of Life: Theoretical and Experimental Approaches

Ille C. Gebeshuber

Institute of Applied Physics
Vienna University of Technology, 1040 Wien
Austria & Aramis Technologies Sdn. Bhd., Malaysia
gebeshuber@iap.tuwien.ac.at

The question "What is life?" has puzzled mankind since the early beginnings. Biocommunication provides a distinct marker between life and non-life: only living entities communicate via signs and codes. Dead matter follows physics and chemistry, without the additional magic sparkle of communication.

This book is a compilation of chapters by key authors from important fields in biocommunication. The book has two major parts, with Part 1 being more concerned with theoretical aspects, and Part 2 presenting important experimental approaches.

Biocommunication denotes communication on all levels in the web of life: communication takes place inside cells, across cells, within an organism and from one to another, within one species, across species, and even across time. Table 1 denotes which chapter deals with which biocommunication actors, Table 2 denotes the respective signal types the authors are dealing with and Table 3 establishes further synergies besides biocommunication actors and signal types. From these tables, synergies in all three categories across the chapters in the book can easily be identified, and the reader can direct his or her attention to the respective works.

Below, the individual chapters are described in more detail.[1]

PART I Theoretical Approaches

Chapter 1 Alexei A. Sharov: Molecular biocommunication

If life started out with pure chemistry, at some point in time, signs entered the game: chemical molecules that stand for something else than their chemistry alone. As Charles Peirce (Peirce, 1955) wrote: *"A sign is something which stands for something in some respect or capacity"*. Codes came up. The genetic code for example. How? Why? When?

Since this is the first chapter in the book, we will explain Tables 1, 2, and 3 regarding this work. The main category of Chapter 1 is theoretical; in Tables 1 and 2 this is indicated by the capital T in the first line, respectively. Harari & Sharon's chapter 9 is the first experimental one, it has a capital E in these tables. Chapter 1 (Sharov, 2016) deals with the biocommunication in all of life, indicated in the last line in Table 1, and the signal types he especially focuses on are chemical signals, codes, epigenetics, eusemiotics, genetic signals, informational communication protosemiotics and signs. All these are marked in grey for this chapter in Table 2. Synergies of his work with the work of others in this book can easily be identified by checking who else deals with "all life" in terms of actors in biocommunication (and the result is: Witzany, Chapter 2; Harari & Sharon, Chapter 9). From Table 2 one can see the other authors who deal with the same signal types, in a similar way. Table 3 identifies further synergies within the book, regarding

[1]In some cases we have copied some sentences from an author's chapter.

Table 1. Biocommunication actors as treated in the book.

Biocommunication \ Author(s)	Sharov	Witzany	Martinelli	Negrotti	Gordon & Stone	Gebeshuber and Macqueen	Tamir & Priel	Kak	Harari	Finkelshtein *et al.*	Sbrana	Darnowski	Slabbekoorn	Rocha *et al.*	Gershony-Emek *et al.*	Rumbaugh *et al.*	Frohoff and Oriel
Theoretical/Experimental	T	T	T	T	T	T	T	T	E	E	E	E	E	E	E	E	E
Animals			■	■					■								■
Bacteria interspecific										■							
Bacteria intraspecific										■							
Bacteria — fungi										■							
Birds													■				
Carnivores					■												
Cells — neurons															■		
Dolphins — humans																	■
Dolphins non-human animals				■									■				■
Fish													■				
Herbivores						■											
Human cells														■			
Humans			■	■												■	
Interneuronal															■		
Interspecific		■				■											■
Intraneuronal															■		
Intraspecific		■				■											■
Multicellular organisms					■												
Neuro-muscular							■	■							■		
Organisms with brains								■									
Organisms with DNA							■										
Pan/Homo bonobos — humans																■	
Pan/Homo chimpanzees — humans																■	
Plants — fungi											■						
Plants — pollinators												■					
Plants — prey												■					
Plants animals									■								
Quantum biology								■								■	
With the past						■											
All life	■	■							■								

Table 2. Biocommunication signal types as treated in the book.

Signal Type \ Author(s)	Sharov	Witzany	Martinelli	Negrotti	Gordon & Stone	Gebeshuber and Macqueen	Tamir & Priel	Kak	Harari & Sharon	Finkelshtein *et al.*	Sbrana	Darnowski	Slabbekoorn	Rocha *et al.*	Gershony-Emek *et al.*	Rumbaugh *et al.*	Frohoff and Oriel
Theoretical/Experimenta1	T	T	T	T	T	T	T	T	E	E	E	E	E	E	E	E	E
Acoustic													■			■	■
Chemical	■	■			■	■			■	■	■			■	■		
Codes	■																
Differentiation waves						■											
Electrical															■		
Epigenetic	■				■												
Eusemiotics	■																
Gene expression										■							
Genetic	■	■				■											
Infochemicals									■								
Informational communication	■				■									■			
Knowledge-based communication				■													■
Language		■	■														
Language (printed signs)																■	
Language (vocal)						■										■	■
Material properties					■												
Mechanical																	
Names			■														
Noise induced protein mutations via DNA mutation						■											
Nucleotide flips						■											
Olfactory												■					
Physical		■															
Proteins											■						
Protosemiotics	■																
Quantum biology								■									
RNA		■															
Signals that induce cell differentiation						■											
Signs	■		■	■													

(*Continued*)

Table 2 (*Continued*)

Smell																	
Switching on and off genes																	
Symbols																	
Taste																	
UV																	
Various																	
Visual																	
Words																	

certain topics of relevance, such as for Sharov computers, molecular signaling, the neuromuscular junction, neurons, protosymbols, signs, transferrable genetic units and wave propagation.

Sharov deals with important questions such as: *What is the difference between chemistry and molecular communication? How does a molecule become a sign?* Molecular biocommunication is a fascinating subject.

The integrity of living cells is maintained via regular communication between cell components. Molecular biocommunication is essential for the functioning of living cells; the complexity of molecular communication networks varies from the most primitive form of sign processing in nature, termed protosemiosis by Sharov to complex categorization mechanisms (eusemiosis). In protosemiosis, signs directly encode and control cell functions instead of being associated with objects. Molecular communication ensures the functional integrity of cells, and supports their growth, reproduction, and defence from pathogens. When used for molecular biocommunication, the original linguistic terminology, which was developed for human languages, is not fully applicable.

Chemistry alone is not sufficient to understand how the cell works. Biosemiotics attempts to answer how biological processes are encoded, controlled and communicated. Physics and chemistry study things as they are, whereas semiotics studies things as they are used by agents to encode and control their activities. Identification of agents is always non-trivial and requires a competent interpreter, which is an agent itself. Semiotic analysis of

Table 3. Synergies of authors and articles with regard to the topic as treated in the book

Synergies of Authors and Articles with regard to the topic	Sharov	Witzany	Martinelli	Negrotti	Gordon & Stone	Gebeshuber and Macqueen	Tamir & Priel	Kak	Harari	Finkelshtein *et al.*	Sbrana	Darnowski	Slabbekoorn	Rocha *et al.*	Gershony-Emek *et al.*	Rumbaugh *et al.*	Frohoff & Oriel
Co-cultural evolution																■	■
Co-cultural processes				■									■	■		■	■
Co-experiencing																■	■
Cognitive ethology																■	■
Cohabitation																■	■
Collaboration		■				■			■	■	■				■	■	■
Communication with the past						■				■						■	
Computers	■			■													
Embryo development					■				■								
Horizontal gene transfer						■									■		
Informational communication				■										■		■	
Interspecies cooperation									■		■					■	
Knowledge-based communication				■												■	■
Legal rights (non-human animals)																■	■
Male–female						■			■								
Molecular signaling	■														■		
Mutations		■				■	■										
Mutual respect											■					■	■
Neuromuscular junction	■														■		
Neurons	■														■		
Noise					■		■	■					■				
Non-coding parts of the genome		■				■											
Personhood (non-human animlas)																■	■
Protosymbols	■					■											
Quantum biology								■								■	
Shared interests across life forms											■					■	■
Signals to another sensory universe												■	■				
Signs	■		■	■													
Stem cell differentiation					■									■			
Synergies											■						■
Transferable genetic units	■					■											
Wave propagation	■				■									■			

(*Continued*)

Table 3 *(Continued)*

Synergies		
theme	*chapter*	*chapter*
Co-cultural processes	Negrotti	Rocha *et al.*
Co-experiencing	Gorman	Rumbaugh *et al.*
Co-feeling	Gorman	Rumbaugh *et al.*
Co-sensing	Gorman	Rumbaugh *et al.*
Communication with the past	Gebeshuber & Macqueen	Finkelshtein
Communication with the past	Gebeshuber & Macqueen	Rumbaugh
Computers	Sharov	Negrotti
Embryo development	Gordon & Stone	Harari
Horizontal gene transfer	Gebeshuber & Macqueen	Gershony-Emek *et al.*
Knowledge-based communication	Negrotti	Rumbaugh *et al.*
Male–female	Gebeshuber & Macqueen	Harari
Molecular signaling	Gershony-Emek *et al.*	Sharov
Mutations	Witzany	Tamir & Priel
Mutations	Gebeshuber & Macqueen	Tamir & Priel
Mutations	Tamir & Priel	Witzany
Mutual respect	Gorman	Sbrana
Neuromuscular junction	Gershony-Emek *et al.*	Sharov
Neurons	Gershony-Emek *et al.*	Sharov
Noise	Gordon & Stone	Tamir and Priel
Noise	Tamir & Priel	Kak
Noise	Tamir & Priel	Slabbekoorn
Non-coding parts of the genome	Witzany	Gebeshuber & Macqueen
Non-coding parts of the genome	Gebeshuber & Macqueen	Witzany
Protosymbols	Sharov	Gebeshuber & Macqueen
Quantum biology	Kak	Rumbaugh *et al.*
Shared interests across life forms	Gorman	Sbrana
Signals to another sensory universe	Darnowski	Slabbekoorn
Signs	Sharov	Martinelli
Stem cell differentiation	Gordon & Stone	Rocha *et al.*
Synergies	Gorman	Sbrana
Transferable genetic units	Sharov	Gebeshuber & Macqueen
Wave propagation	Gershony-Emek *et al.*	Sharov

molecular signaling associated with biological processes yields substantially more information than a simple chemical approach. Cells use molecular signs to encode and control their activities. Molecular signals directly guide actions of specific cell components without reference to an object. Molecular signaling can be described in a "code-semiosis" signification model (Barbieri, 2003). A semiosis code is defined as a mapping between a set of signs and a set of meanings. Examples are the genetic code with ribosomes as codemakers (specific agents).

In protosemiosis, signs are associated with actions of agents rather than with objects. Responses of the biological system are not determined by the physical nature of the signs. Instead, they are shaped by the agent's semiotic architecture, which is a product of long-term adaptive evolution.

Automatic processing of signs is common in living systems. It does not imply determinism, because responses are often checked for errors and adjusted to the environment and internal context within the agent. Automatic processing of signs does not include learning. Learning emerged historically via adaptive evolution, which is a learning-like process at the population level (Sharov, 1992). Evolution is a learning process at the population level. A sign can initiate, inhibit, or modify agent activities, but it never fully determines the outcome. "*A computer contains codes but it is not a semiotic system*" (Barbieri, 2008: 594).

Molecular proto-signs can be classified into three major categories: proto-icons, proto-indices and proto-symbols. They are nested in each other: one includes the previous one as a component. Proto-icons interact via one particular feature, for example the complementary binding of proto-icons to surfaces of molecules in a key and lock manner. The "shape" of the key and lock is a shape in an abstract space that includes geometry, charges and hydrophobic/-philic aspects. An example for a proto-icon is single stranded deoxyribonucleic acid (DNA) without any particular shape, which, upon meeting its complementary segment, spontaneously assembles into a double-stranded rigid DNA helix. Proto-symbols are members of a family of similarly structured molecular signs that are processed uniformly by the same subagents and the same set of adapters. An example is the genetic code.

However, the processing of proto-symbols is based on a "heritable convention" rather than the socially-mediated conventions of eusemiotic symbols.

Contemporary organisms use all three kinds of proto-signs for their molecular communication. At the origin of life, no protosymbols may have been present. Proto-semiosis, to interconnect otherwise separate processes, is necessary for life to develop. However, proto-icons alone cannot make a functional living system because they have no capacity to organize a semiotic scaffold. The molecular signaling process is of paramount importance in fabrication, sensing, memory, movement and wave propagation in organisms. According to Rosen (Rosen, 1991) each component of a cell should be fabricated by other components, thus enabling the full capacity for self-repair, growth and self-reproduction. Sensing is needed to monitor external and internal conditions. All living organisms need memory to store information that encodes and controls their functions. Memory is preserved across generations in memory carriers such as DNA. Operational memory of smaller units such as proteins or ribosomes is often based on reversible or irreversible protein modifications. Larger units, such as gene cascades involved in cell differentiation or other functions, suggest a hierarchical structure to DNA memory. DNA and mRNA store large amounts of functional information that is needed to produce and regulate other molecular agents. Because it cannot be edited much, DNA is rather a passive information resource than an agent. An additional layer of rewritable memory that is preserved after cell division emerged in eukaryotic cells in the form of chromatin (DNA + histones) and DNA methylation: epigenetic memory.

Faster signaling does not take place via the generally slow process of the transport of large molecules, but by a signal traveling through an array of linked transmission agents without movement of the agents themselves. Action potentials in neurons, the electrical spikes that travel along the axon, are an example for this (Gershoni-Emek *et al.*, 2016), as are muscle cells that receive a neural impulse respond by opening the voltage-gated channels, which allow extracellular Ca^{++} to enter the cell and initiate muscle contraction, and differentiation waves (Gordon and Stone, 2016).

Each component of a cell should be fabricated by other components, thus enabling error-control, robustness, and modularity in molecular communication. In any organism, at any point in time, the whole signaling network should be prepared for various kinds of disruptions of both internal and external origin. Life can only persist via error correction, compensation, robustness and protection from foreign signals. Template-based synthesized products are proofread, and tagged for degradation when erroneous, misfolded proteins are repaired by chaperones, and the threshold for initiating cell actions can be adjusted. Related to pluri- and omnipotency, differentiated cells are protected from transforming into pluripotent cells which may cause cancer. Accidental activation of any single transcription factor associated with pluripotency has no consequences, because it cannot function alone without its partners.

Molecular signaling networks often have modular organization, such as is the case in the cell cycle of eukaryotic cells where the transition between the major modules (five in the case of the cell cycle) are checkpoints with specific inputs and outputs. One of the checkpoints in cell division (the one at transition to mitosis, cell division) has elements of the advanced level of semiosis, or eusemiosis. Sharov says that this checkpoint in cell division shows signs of eusemiosis, since *"The output of the checkpoint is a sign of a specific state of the cell, and this state is an object."*

Categorization mechanisms in molecular signaling networks are context-dependent and adaptable even in organisms without neural systems (Ginsburg and Jablonka, 2009; Sharov, 2013). The epigenetic system offers enormous capacity of dynamic memory storage.

Chapter 2 Guenther Witzany: Key levels of biocommunication

Chapter 2 of the book is the theory chapter "Key levels of biocommunication" by Günther Witzany (Witzany, 2016) from Austria.

Language is life. Günther Witzany, seasoned biocommunication expert, identifies in his contribution the key levels of biocommunication. Antagonistically to Tamir and Priel's approach to noise as an enabler of genome mutations in their Chapter 7 "Channel capacity and rate distortion in amino acid networks" in this book (Tamir and Priel, 2016), Witzany thinks that natural genome editing is not the result of replication errors but of a group interaction of competent ribonucleic acid (RNA) agents.

Witzany opposes assumptions such as that communication is limited to humans or that non-human organisms function mechanistically as machine-like stimulus reaction automatons determined by genetic programs. He also opposes behaviorism. Biocommunication is primarily a social interaction between living agents that share a real life world's traditions and environmental circumstances (Tomasello, 2008). In everyday life we do things with words that have serious consequences. A coherent model of language and communication must be the basis of all reasoning, as a good basis for knowledge about the world. Indeed, Witzany states that the only sure knowledge is how language and communication function. Witzany argues that in natural languages there cannot be such a thing as context free grammar.

The four key levels of biocommunication as identified by Witzany are found in all species and domains of life. They are (1) Sensing of abiotic circumstances (organisms of all domains store information about these indices in memory, to adapt better to repeated life situations), (2) Transorganismic communication processes (such as in the human mouth with its 500+ bacterial communities), (3) Interorganismic communication with a species specific vocabulary, between members of the same or related species), and (4) Intraorganismic communication (this also includes genetic parasites and mobile genetic elements, both important in Witzany's approach).

Key features of natural languages/codes in biocommunication processes comprise the fact that languages/codes inherently depend on social groups. The language tool serves to do things with signs combined with context-dependent markings such as conscious and unconscious body expressions.

Language is important. It is a new way of matter to interact, distinct from chemistry and physics. It is typical for life, and its use distinguishes between life and non-life. Communication provides appropriate tools for differentiation at specific levels, which is otherwise difficult to describe in reductive terms by means of pure physics and chemistry (molecular biology). Language-like structures and communication processes occur at the simplest level of living nature. All coordination and organization needs signs. Communication is highly complex; it cannot be reduced to mechanistic input/output or cause/reaction descriptions.

Communication is a sign mediated interaction between at least two living agents that share a repertoire of signs that are combined in varying contexts which transport content. Communication is interactions that depend on a shared repertoire of signs and rules. Such features are lacking in abiotic interactions.

In Witzany's view DNA provides stable habitats for unstable RNA colonizers. Viruses (who are older than cellular life) and viral-derived elements are the agents that edit the genome in host organisms. Viral colonizers play major roles in evolution and diversity of organisms. Mobile genetic elements such as non-coding RNAs are actively engaged in nearly all cell processes to meet both evolutionary and developmental needs. DNA is the storage medium, and there are species-specific active RNAs. Non-coding RNAs interact with DNA, RNA, and proteins and play important roles in nuclear organization, transcription, post-transcriptional and epigenetic processes. Non-coding RNA can undergo nuclear-cytoplasmic, nuclear-mitochondrial and axodendritic trafficking.

Very interesting and highly unusual are Witzany's thoughts about RNA agents that form consortial biotic structures. With language came life, and material left the purely physical chemical interaction patterns. Mixtures of RNA fragments form cooperative networks. RNA populations evolve to greater complexity through cooperation (Vaidya, 2012) — cooperation outcompetes selfishness. RNA stem loop consortia have selective (biological) group-building competence, and higher order structures such as pseudoknots show strictly context sensitive base-pairing. Witzany argues that nucleic acid language has a language/code nature which represents the possibility of coherent *de novo* generation and context dependent alterations for a diversity of different meanings (functions) relating to the same syntax structures.

Biocommunication can be defined as the sign-mediated interactions of groups of living agents that share syntactic, pragmatic and semantic rules for sign use. Biocommunication cannot be sufficiently explained by materialistic and reductionist concepts. The biocommunication approach allows for a clear distinction between life and non-life, and could serve as an appropriate complementary tool for interpreting empirical data of biological disciplines coherent with current knowledge about communication and language.

For synergies of Chapter 2 with other chapters see Tables 1, 2, and 3.

Chapter 3 Dario Martinelli: Zoosemiotics, typologies of signs and continuity between humans and other animals

In his work on zoosemiotics, Dario Martinelli (Martinelli, 2016) deals with typologies of signs and continuity between humans and other animals. A sign is something that stands for something else. In the field of semiotics, a signal, i.e., any form of behavior aimed to transmit information from one animal to another, is the simplest unit of a sign. A sign is a more complex display involving more than one signal. Animal communication must also be thought in terms of channels, i.e., sensory modes employed to transmit a message. Signals can either be digital (two possible states, on/off, color/no color) or analogue, where signal variability is important, and the message's content depends on the different degrees of emission. Analogue signals give a higher communicative potential. The meaning of a signal often depends on the context. Many signals evolved from basic patterns that initially had no communicative intention. Growling in dogs for example, might initially have been a part of being angry, and later became ritualized, detached from fight, as a more complex form of signification. Growling can lead to a fight, or prevent it. Semiosic (i.e., sign process related) phenomena that occur among animals are signification, representation and communication — they are constructed, organized, assembled, distinguished, interpreted and codified. Sign production and exchange among animals can be proprioceptive, intra- or interspecific. In organisms, language adds a series of communicative and cognitive elements on top of the existing ones. The use of symbols (and a consequent mental/interactive capacity) has for many years been a concept of which only humans were claimed to be capable. Insects supposedly have a very low and rigid degree of reasoning, and yet — from their world comes a beautiful example on the use of symbols in animals: Dipterans (flies) of the family Empididae practice cannibalism. In order not to risk being eaten, the male, before the copulation, offers the female an empty balloon. Another example is the highly symbolic dance of bees, in which they indicate where and how far away food can be found. Bees sign the distance to the food source with the time employed to cover the middle axis of the waggle dance, which looks like the Fig. 8. Honeybee communities in different countries use the very same articulation of this distance sign in their dance to represent completely different distances: the same amount of time signifies 5 m for an Egyptian bee, 25 m for an Italian one, and 75 m for a German bee.

Also alarm cries in monkeys are examples of symbolicness. Such cries inform conspecifics of the arrival of a predator. The response of the other monkeys is appropriate for escaping the corresponding predator: leopard — run to the trees, eagle — look up and seek shelter, snake — stand up on two legs and look in the grass). The cries do not just differ for different animals, but also with the distance of the animal to the caller. The syntax of the monkey callers is phonological and lexical. *Phonological syntax* is a combination of signs that, taken alone, do not necessarily have a specific meaning, or at least not a quantitatively different one from the combination. There are many examples of phonological syntax in animals. However, lexical syntax was for a long time only attributed to humans. It denotes a combination of signs that are meaningful as single units, and that also mean "something else". Vervet monkeys for example use the call "pyow" to designate a leopard, and the call "hack" for an eagle. Pyow pyow hack hack does not mean two leopards and two eagles, but "Let's move to another place". Language leads to change in behavior. If the monkeys did not possess a "language" their reaction would have been of another type (namely, most probably, escaping from the danger). The richer and more complex the "semiogram" (i.e., the complete sign catalog of an individual), the more sophisticated the mental life. The active participation to semiosis opens up for further — and increasingly articulated — cognitive abilities.

Also names seem not to be unique to humans. Dolphins have individual specific names, which they choose in the first year of their life.

The six main functions within a communication system are expressive, conative, phatic, referential, metalinguistic and poetic. These functions are mutually inclusive.

When reading Martinelli's chapter the author of this foreword was especially intrigued by the fact that some dancing flies give empty nuptial gifts (Fig. 1) to the females. It caused her to write the public outreach text given below:

> "***Marry me, marry me. I have the best empty balloon.***
>
> *The use of symbols has been for many years a concept (and a consequent mental/interactive capacity) of which only humans were claimed to be capable.*
>
> *When we go on a date, we bring a little present for our sweetheart. Highly nutritious chocolate, a beautiful rose as a symbol of love or a piece of precious jewellery. Dancing flies are no different. Some male dancing flies bring highly nutritious packages as nuptial gifts, and some bring more ritualized presents — such as empty balloons (Sadowski et al., 1999). The girls love them, and off they fly into the sunset."*

Martinelli's chapter has synergies with Sharov's Chapter 1 "Molecular biocommunication" (Sharov, 2016) and Negrotti's Chapter 4 "Communication as an artificial process" (Negrotti, 2016) in this book regarding their treatment of signs. For synergies of Chapter 3 with further chapters see Tables 1, 2, and 3.

Figure 1. Balloon flies at their nuptial dance. In some species, the balloons are filled with a present for the female (such as an insect); in some they are empty. Left: © 2013 Mark Shields. Right: *Empis aerobatica* Melander, © E.M. Fisher.

Chapter 4 Massimo Negrotti: Communication as an artificial process

Negrotti (Negrotti, 2016) deals in his chapter with communication as an artificial process. Language is in his view the sole technology intrinsically oriented towards giving birth to something artificial, while the materials adopted in all fields of naturoids, i.e., technological reproductions of natural subjects at a certain level of observation, historically come from a wide range of uses. He categorizes communication in two main areas: informational communication, that humans and animals share, and knowledge-based communication, which only humans are capable of.

Informational communication mainly transmits messages about facts in the outside world that might be verified by anyone, and takes place amongst people, animals, plants, and microorganisms; it can be denoted as biosemiotic management of standardized signals. Animals and plants such as trees communicate through physical or chemical signs supporting information that allows, or even forces, the living system to adopt some stereotyped behavior. Informational communication is a flow of effects, i.e., ordered physical events, as for example also takes place in a computer.

Human communication acts at two levels: the one we share with animals, and that which we have constructed culturally through language. Negrotti argues that the informational process consists of a sort of mechanism, more or less plastic, of codified translation of signals. Knowledge, on the other hand, is the creation of individual meaning through particular processes of association constituting one's semantic history.

Knowledge-based communication, Negrotti argues, only takes place between humans. It is an open process with a dynamic code and complex processes of relationships between mental states and related meanings. Human communication of personal knowledge is difficult — Negrotti describes it as an attempt to reproduce one's mental state in the mind of someone else, by exploiting the seeming representational power of the words — knowledge-based messages are therefore artificial objects, results of a design process that is trying to reproduce a mental state by means of words. We can put our feelings, desires, thoughts and wishes into words, express our inner state, and communicate it to somebody else. However, the receiver does not perceive the same message as the sender sends, since both intertwine words they speak or hear with personal history, memories and experiences. Knowledge-based messages require decoding and interpretation. And interpretation strongly depends on the personal knowledge of the interpreter, who cannot prevent linguistic signs from triggering memories of past experience relating to that sign. Humans have deep differences in their individual disposition and experience. An important concept related to knowledge-based messages in Negrotti's chapter is the "observation level" of the speaker and the listener, respectively. Normally, the observation level selected by the speaker poses no problem. However, if the listener changes the observation level at which he wants to understand the message, he will be unable to see and feel what the speaker tried to reproduce in his or her mind. Even if speaker and listener share the same observation level, reproduction of the mind of the speaker is uncertain and not assessable. Words are signs that fire imagination. The listener sees and feels dependent on his own mind, and not the speaker's. Their two maps have a very low probability of overlapping substantially. Reality does not know any observation level. Unpredictability in the communication process is a rule.

When scientists analyze natural objects, they isolate them from the rest of the world, although reality does not have such separations or boundaries. A truly holistic view, whilst at the core of human ambition, does not allow us to generate any taxonomy and to induce any general law — something that is possible through isolating, under some profile, homogeneous classes of objects in empirical reality. We will often gain knowledge of the object in isolation, but loose several possible relationships between it and other phenomena that may be significant when trying to predict the behavior of the investigated phenomenon.

Also in human communication we isolate single statements form the discourse, and seemingly gain an increase in clarity. However, possible relationships may be lost. Even "context analysis" guarantees nothing, owing to the huge dimensional difference between the context analysis, however rich, and the whole structure of personal mental states. Selecting and reproducing "the essential" is important. Biomedical engineers "reproducing" a muscle need to concentrate on other factors than producers of lab-made burgers (*in vitro* meat). For the biomedical engineers, the main ambition is to make the material invisible to the body. Widely, in every field of automation, the final performance to be achieved is evaluated as more important than the reproduction of the natural structures of processes that generate the performance. This has, in some cases, catastrophic effects. Mining of resources, manufacturing, transportation, use and disposal of technological devices is optimized along a selected small range of parameters, and currently the effects of our ways of doing things are visible in dangerous changes in biogeochemical cycles and in a decline in the Earth ecosystem, including a humankind-induced mass extinction of species (Rockström *et al.*, 2009; Barnosky *et al.*, 2011; Steffen *et al.*, 2015). Even fields originally proposing to address such problems, such as biomimetics, i.e., learning from living nature for new, potentially disruptive ways in engineering and science, make the mistake to reduce the parameters they take into consideration to a much too small set (Gebeshuber *et al.*, 2009).

Negrotti also includes reflections on art in his chapter. Art is often regarded the human activity that best maximizes the efficacy of communication. For Negrotti, in the arts, the meaning of the message resides in its formal beauty rather than its content. The artist expresses himself in a way that is most suitable for him, rather than for others. Exemplified by travel journals, he describes the typical transfiguration that characterizes every artistic production.

In human communication we have no possibility of reproducing the essential performance of our mental state in an automatic way, i.e., attempting to reproduce it directly, avoiding the adoption of some observation level. The word can be simultaneously beneficial and dangerous.

The distance between words as signs and corresponding mental states is far greater than between the materials adopted by bioengineers and the corresponding materials adopted by nature when building natural phenomena.

For many reasons due to the personal history of each of us, many words or styles may be stored in our mind along with bad memories, and, as a consequence, listening to them or reading them may trigger a repulsion or, at least, a strong deviation from their intended meaning. Only in a dictionary words are pure signs, i.e., informational structures. Their meaning strays very far from their informational content.

Humans express their mental states in knowledge-based communication, and the result is an artificial object whose destiny is largely independent of the intentions of the speaker.

Knowledge-based communication is the most ambitious and least assessable technique invented by humans for enhancing their relationships.

Negrotti's chapter has synergies with various chapters in this book: In their tissue engineering and regenerative medicine tech mining study "The contribution of the bio-communication (BICO) to biomedical and tissue engineering: A tech mining study" presented in Chapter 14 (Rocha *et al.*, 2016) Rocha and co-workers also treat co-cultural processes (i.e., simultaneous culture of different cell types) and informational languages in tissue engineering. Sharov (2016) and Negrotti both deal with computers, however, from different viewpoints: Sharov quotes Barbieri, who says "*A computer contains codes but is not a semiotic system*" (Barbieri, 2008. 594); for Negrotti, computers deal with informational communication. Rumbaugh and co-workers present in chapter 16, "Ethical Methods of Investigation with Pan/Homo Bonobos and Chimpanzees", the communication of Pan/Homo beings with each other and with humans (Rumbaugh *et al.*, 2016). Synergies with Negrotti's Chapter 4 are in the area of informational language and knowledge-based communication. Negrotti and Rumbaugh *et al.* mention communication on inner mental states. Negrotti says that only humans are able to communicate in a knowledge-based way. Rumbaugh and co-workers say otherwise in Chapter 16. For synergies of Chapter 4 with further chapters see Tables 1, 2, and 3.

Chapter 5 Richard Gordon and Robert Stone: Cybernetic embryo

Growth is one of the fascinating aspects of life. How does a little fertilized cell develop to a fully-fledged organism? How do cells differentiate into new cell types? We have so many different types of cells in our bodies, so many organs, connections between all of them, and continuous communication going on, during embryogenesis, during growth from child to adult and in the adult form. Simply amazing.

Organisms grow in genuine, highly complex feedback conditions. Machines are assembled. This provides food for thought, touching biology, engineering, philosophy, materials science, control and communication aspects. Which shall result in a new, sustainable, disruptive approach to engineering.

Machines and organisms have paramount differences in various aspects. Organisms grow from a single cell whereas machines are assembled from manufactured parts. Embryogenesis is the field that investigates how an embryo forms and develops. Gordon and Stone present in Chapter 5 "Cybernetic embryo" (Gordon and Stone, 2016) their views on embryo development from a cybernetics perspective, concentrating on feedback systems that provide control and communication. Cybernetics as an engineering discipline is a quite mature field known as 'control theory' (Gebeshuber, 2014); using control theory's deterministic vocabulary and set of concepts on 'reverse engineering' of the embryo shall provide valuable inspiration for major steps forward in the field. The cybernetic embryo is a system in which communication and control via signals affect its process as it develops from one state to the next. The communication is provided via differentiation waves, i.e., waves of change of cell type, providing regional and global communication within the embryo. Their differentiation tree concept provides an experimentally and computationally testable hypothesis on self-construction of complex embryos, explaining pattern formation of complex systems of a nested character, with 'meta-complexity' of self-organization and adaptation on multiple levels. Each stage is viewed not as an end in itself, but rather as a preparation for the next stage. Important themes such as purpose and goals, teleonomy (purposive behavior controlled by processing of coded information) and teleology (purposive behavior controlled by feedback), are addressed from various angles, including citations of paramount work in various fields of specialization, and going back in time all the way to Aristotle, who described embryo development as an epigenesis, a chain of one genesis after another, where new structures and functions appear at various steps, with an increase of complexity during embryo development. Gordon and stone treat differentiation waves as cybernetic feedback systems, and establish a binary code in the memory of each cell, giving history and memory of differentiation steps from the single cell state to the current stage. Each bit of this code corresponds to the triggering of a cascade of gene products and gene expression. Differentiation in embryos might be a lot less complex than we have imagined.

Synergies exist with Chapter 14 (Rocha *et al.*, 2016) regarding treatment of the differentiation of stem cells, and with Chapter 9 (Harari and Sharon, 2016), since both deal with embryo development. Harari and Sharon describe in the worm *C. elegans* dauer larvae that — upon receipt of chemical signals — are produced in harsh conditions. They do not feed, survive on fat stores, and can survive for several months before they proceed maturation to adults. And both Chapter 7 (Tamir and Priel, 2016) and Gordon and Stone (2016) deal with noise. For synergies of Chapter 5 with further chapters see Tables 1, 2, and 3.

Chapter 6 Ille C. Gebeshuber and Mark O. Macqueen: Superfast evolution via trans- and interspecies biocommunication

Gebeshuber and Macqueen propose in Chapter 6, "Superfast evolution via trans- and interspecies biocommunication" that the real speed of evolution can be explained by a combination of natural selection with permutative and constructive elements (Gebeshuber and Macqueen, 2016). Selective elements are genetic units that are transferable across species and biological kingdoms, which is why the uniformity of the genetic code is of such great importance. Permutative elements include permutation by recession (biocommunication with the past via a genome that can repair itself and even reverse mutations) and innovation (communication with other organisms). Constructive elements comprise quantitative environmental (partner selection, competitor elimination) and qualitative genetic factors (selecting the probability for the sex of the offspring, on the male side by adapting prostate liquid, on the female side by variable egg hull permeability for X and Y carrying sperm) — which accounts for biocommunication with the environment and the partner. Exemplified by horizontal gene transfer and the most conserved protein present in all organisms, the chaperone protein HSP70, that is present in most organisms but not in the common ancestor, the authors argue that mutualism, a collaborative association of life forms, often from different species, that benefits all partners, might be the rule, and not the exception.

For synergies of Chapter 6 with other chapters see Tables 1, 2, and 3.

Chapter 7 Boaz Tamir and Avner Priel: Channel capacity and rate distortion in amino acid networks

The first intriguing fact about Chapter 7 "Channel capacity and rate distortion in amino acid networks" by Tamir and Priel (2016) on noise and information in protein channels is the perfect combination of highly exact mathematical concepts and beautiful translation of related mathematics into words. This is a precious kind of communication ability in itself.

Chapter 7 deals with the application of information theory to the analysis of amino acid sequences. The approach uses the analogy between the process of translating DNA sequences (source) to proteins (target) and the process of communicating information via a noisy channel (channel). The messages are words (codons) based on a four-letter alphabet (bases). Important concepts are channel capacity and signal distortion rate. Noise is important in biological systems, and in many cases beneficial, for example, via the process of stochastic resonance (Petracchi *et al.*, 2000). Tamir and Priel argue that biological systems must maintain a certain level of noise, since it facilitates evolutionary processes. Mutation rates for RNA viruses could be 10^{-3} to 10^{-5} per base per generation (Drake and Holland, 1999).

The competition between capacity and distortion is calculated as a function of a single parameter, the point mutation rate, which is the probability for a single nucleotide flip to occur. The amino acid channel is modelled as a transition matrix, statistically mapping each pair of amino acids from the source to the target. The authors establish a metric space of amino acids, where each amino acid is represented as a vector in a high dimensional space where the coordinates correspond to the amino acid's physico-chemical properties such as volume, bulkiness, polarity, hydrophobicity, etc.

For synergies of Chapter 7 with other chapters see Tables 1, 2, and 3.

Chapter 8 Subhash Kak: Communication languages and agents in biological systems

Chapter 8 is on quantum biology (Kak, 2016). The author proposes that communication languages and agents in biological systems are based on entities for which quantum mechanical objects are the primitives.

The hierarchy of languages and metalanguages that must be associated with biological systems are explained via communication and cognitive agents that are viewed as collections of quantum particles. Autonomous cognitive agents as they exist in the brain are viewed as self-organized systems with strong dynamical nonlinearities. Biological systems are considered from the perspectives of communication and complementarity.

Kak asks questions about the hierarchy of languages and metalanguages in biological systems. Quantum effects have been proposed for certain biological processes, including photosynthesis, olfaction, vision, long-range electron transfer and bird navigation. Kak views the brain as a classical/quantum hybrid system. He says that the cognitive system is quantum at a deeper level but coupled to the conscious system, which is classical. He talks about the Copenhagen Interpretation and that its early pioneers viewed the unconscious mind as quantum mechanical and different from the classical mind of the internal dialog within the individual. There is a field of research called quantum cognition.

Kak asks questions such as *"If subsystems within biological systems are quantum mechanical what is the nature of communication between these subsystems?"*. Biocommunication includes associative (classical) and reorganizational (adaptive) elements in addition to those that are quantum. For the associative language, the medium is chemical and electrical; the quantum language is the one that is characterized by quantum processes. Biocommunication language is adaptive. In the brain the expectation affects the way the incoming sensory information is processed.

The way in which information is exchanged across areas is more important than the contents of any specific area. Neuroscience only provides clues about the lowest levels of the neuronal hierarchy; about the higher levels we do not possess knowledge that is independent of language. More complex concepts in children emerge under the influence of language. In his chapter, Kak combines quantum mechanics and biology, by proposing a Bose–Einstein quantum probability distribution for certain molecular structures and shapes as a model of human memory. His assumption is that bosonic quasiparticles or fermions are assembled in different arrangements to become cognitive agents. Agents and memories should in his view be set apart by number, structure and informational content. Agents, unlike memories, are linked to sensors and actuators. An agent must be invariant to certain types of transformation and it should be resistant to noise within a certain limited range. From this resistance to noise follows a necessary minimum separation between patterns, not in 3D geometry but in an abstract space.

There are two kinds of quantum particles, bosons and fermions. Bosons are governed by the Bose–Einstein statistics, whereas fermions are governed by the Fermi–Dirac statistics. The number of permissible arrangements of particles decreases from classical to bosonic to fermionic states. A memory or cognitive agent as a collective of quantum particles must have a unique structure.

Kak proposes an algorithmic approach to information content in which the length of the program required to generate the pattern is a measure of the information — in this way, such a structure can mimic the object of information in form. The number of items in short-term memory is four to six in humans, and much higher in chimpanzees. Kak speculates whether processing in the workspace corresponding to short time memory has a quantum basis and whether the number of items can help determine its physical structure correlates. In his view it is plausible that agents are fermion collectives and memories are boson collectives.

For synergies of Chapter 8 with other chapters see Tables 1, 2, and 3.

PART II Experimental Approaches

Chapter 9 Ally R. Harari and Rakefet Sharon: Chemical communication

Chapter 9 of the book is Ally Harari and Rakefet Sharon's work on chemical communication (Harari and Sharon, 2016), which is the most ancient mode of information transfer in organisms. This is a basic chapter, very wide and interesting. Harari and Sharon deal extensively with infochemicals in the whole phylogenetic tree and give detailed examples for the usage of infochemicals as pheromones in insects.

When reading Harari and Sharon's chapter the author of this chapter was especially intrigued by the fact that plants can call other plants to call for help against herbivores (Fig. 2). It caused her to write the public outreach text given below:

> *"Most of us see plants as quite defenceless organisms. But if threatened, some of them can call enemies of their enemies! And not just this: the damaged plants ask their plant friends to join their chemical call, to make it better audible!*
>
> *One example for this type of chemical biocommunication are Lima bean leaves that are damaged by a certain plant-eating spider mite. The leaves emit chemicals that induce in other Lima bean plants the emission of similar chemicals that attract predatory mites who attack the spider mites (Choh et al., 2004)."*

This is one of the examples that Harari and Sharon describe in their overview chapter on chemical communication. Chemical communication is the most ancient method of information transfer between individuals. Communication via infochemicals is the oldest sense. Intentionally transferred information is termed signals, unintentionally transferred information is known as cues. Infochemicals can be beneficial or harmful for sender and/or receiver.

Ponderosa pines, bark beetles and wasps are for example all part of a complex chemical communication network: bark beetles are attracted to the pine trees by volatile compounds that are emitted by weakened trees, and start excavating a nuptial chamber. In this process, they release further infochemicals known as aggregation pheromones. The resulting high number of beetles weakens the tree's defence and subjugates it to beetle colonialization. In certain species, the aggregation pheromone is only attractive until a certain number of animals is reached, and becomes repellent in

Figure 2. Left: Lima bean *Phaseolus lunatus*. Public domain image. Middle: Lima-bean-eating Two-spotted Spider Mite (*Tetranychus urticae*). © J. Holopainen, image reproduced with permission. Right: Spider-mite-eating predatory mite (*Phytoseiulus persimilis*), called by the beans. © Mick E. Talbot, image reproduced with permission.

character when too many individuals aggregate — preventing starvation of the beetles who are already here. The aggregation pheromone of the bark beetles also attracts parasitic wasps who parasitize the adult beetles.

Infochemicals seem to be as complex as necessary and as simple and less diverse as possible: infochemicals released by plants recruiting predators of their herbivores share a common molecular structure, whereas sex pheromones are highly species specific to avoid mating mistakes resulting in potentially hybrid offspring.

Further important infochemicals are trail-marking pheromones, e.g., in army ant colonies that consist of up to 20 million blind individuals who communicate via chemical and tactile stimuli, and territory-marking pheromones that are for example used by desert ants when they mark insect carcasses "as their own". These ants do not use trail-marking pheromones, but orient themselves with the help of the polarization of the skylight and further cues (Karman *et al.*, 2012).

One story that might be interesting to follow up regarding communication between plants and animals is what colleague Alfonso Donnarumma told the author of this chapter about goats: *"A friend of mine told me long ago something like that; goats, eating some plants, use a guerrilla-type strategy, that means hit and run, because they were aware of the reaction of the plant. For this reason, he told me, the goats were immune to neoplastic diseases* (*cancers*)."

Harari and Sharon's chapter has synergies with Chapter 5 "Cybernetic embryo" by Gordon and Stone (Gordon and Stone, 2016) regarding embryo development and Chapter 6 "Superfast evolution via trans- and interspecies biocommunication" by Gebeshuber and Macqueen (Gebeshuber and Macqueen, 2016) regarding the competition of males with other males for mating with females. For synergies of Chapter 9 with further chapters see Tables 1, 2, and 3.

Chapter 10 Alin Finkelshtein, Alexandra Sirota-Madi, Dalit Roth, Colin J. Ingham and Eshel Ben-Jacob: *Paenibacillus vortex* — A bacterial guide to the wisdom of the crowd

The chapter by Finkelshtein and co-workers (Finkelshtein *et al.*, 2016) deals with the social bacterium *P. vortex* and its communication strategies. This bacterium has two subpopulations, termed builders and explorers. Builders have less flagellae, reproduce fast, and mainly live in the center of the colony. Explorers have multiple flagellae, and explore new grounds for a colony to potentially live on. *P. vortex* is a social, motile, pattern forming bacterium. It has an exceptionally high number of communication related genes. Its two subpopulations allow an adjustment to other microorganisms, namely fungi. Fungi bridge air gaps with their mycelia, and allow bacteria to traverse these and colonize new niches. *P. vortex* cooperatively forms complex colonies with elevated adaptability — the colonial patterns are collectively engineered according to the environmental conditions. *P. vortex* uses quorum sensing. It contextually interprets chemical messages and formulates appropriate complex responses. This allows exchange of information across colonies and species (quorum sensing of competitions is inhibited).

Bacterial linguistics refers to structural aspects of communication, corresponding to the structural (lexical and syntactic) linguistic motifs, and not semantic (pragmatic) aspects. Cognition is communication based; colonial self-organization forms multicellular super-organisms. Bacteria glean information from both the environment and neighboring bacteria. They use transduction networks and genomic plasticity, collectively creating the colony and maintain its integrity by sharing interpretations of chemical cues and exchanging chemical messages. The colony can thereby alter its structure and make decisions. These bacteria show social intelligence and fundamental (primitive) forms of cognition.

Paenibacillus shows advanced defensive and offensive strategies and has various genes that are important for agricultural, medical and industrial applications, such as its capability to fix nitrogen and degrade various polysaccharides. *P. vortex* and *P. dendritiformis* can develop colonies that behave much like a multicellular organism, with cell differentiation and task distribution.

P. vortex is a social microorganism with remarkable complex and dynamic architectures. It shows collective motility and foraging swarms that have an aversion to crossing each other's trail and that collectively change direction when food is sensed. Swarms can split and reunite again.

Flagellae are entwined with those of the near neighbors, and their velocity can be more than 1 cm in an hour. When grown in liquid, they lose both flagellae and motility.

The social intelligence of the species shows in an individual's capacity to perceive and understand the environment and to respond to that understanding in a personally and socially effective manner. They develop knowledge based on their experiences, influencing the development of their progeny. Bacterial social intelligence can be estimated by their ability to communicate, sense their environment and adapt accordingly. Measured in bacterial IQ, *P. vortex* has the highest IQ score.

Ancestor bacteria can influence progeny in ways other than by stable mutations. Subpopulations of builders and explorers occupy different locations in the colony: the builders are in the central parts, whereas the explorers are localized to the vortices.

Each morphotype is capable to convert to the other and has a different stability. While builders are stable (without explorers) only for eight hours, explorers remain stable for days to weeks as long as they are kept in liquid culture. Both subpopulations eventually convert to the mixed culture composition.

Explorers have resistance to antibiotics and a decreased reproduction rate. They have hyperflagellation and are highly motile, actively communicating with their environment. The colony sends explorers to seek for new and favorable niches for colonization.

The swarm shows intelligence: they transport non-motile asexual fungal spores and benefit from the fungal ability to bridge air gaps. About two to nine bacteria capture the spores with 6 to 30 flagellae. This capturing shows specificity — only a certain type of fungal species is transported. The benefit for the spores is large motility at high velocity; the benefit for the bacteria is the crossing of air gaps that generally represent a significant barrier to bacteria.

When reading Finkelshtein *et al.*'s chapter the author of this foreword was especially intrigued by the fact that bacteria carry the spores of fungi with them, which then, when growing, allow them to cross air bridges (Fig. 3). It caused her to write the public outreach text given below:

> "****Living bridges, built by men — and friends of social bacteria****
>
> Paenibacillus vortex *is a social, motile, pattern forming bacterium that establishes colonies that respond cooperatively to challenges*

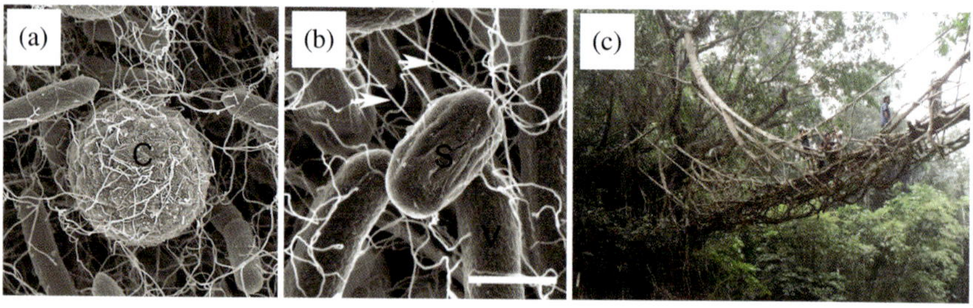

Figure 3. Left: *P. vortex* bacteria transporting fungi spores (Ingham *et al.*, 2011). Their flagellae selectively bind to one certain spore type. Scale bar: A, 3 μm; B, 1 μm. Right: Living bridge in Indonesia. © 2015 Ille C. Gebeshuber, image reproduced with permission.

in complex environments. It exists in two subpopulations, the builders and the explorers. The builders reproduce and mainly live in the center of the colony, whereas the explorers reach out, and look for new places to live in.

The explorers have various whip-like structures that allow them to move, and on their journeys, they take cargo: spores of fungi (Fig. 3). Two to nine bacteria jointly carry one of these spores on the journey. They need them when air gaps block their way: the bacteria alone could not cross the gap, but when the fungi grow there, the bacteria can simply cross the gap on the fungi. So they carry the "seeds" of living bridges with them (reminds me of the living bridges (Fig. 3) that the people in Indonesia and India build across rivers, from ficus trees, which are, when small, bent across the gap, and woven back and forth, until some decades later people can cross the gaps over the living bridges, see photo). The advantage of the fungi is a dispersal that is about 40 times faster than without being carried around by bacteria friends.

Researchers have established bacterial linguistics of this species, referring to their structural aspects of communication, corresponding to the structural (lexical and syntactic) motives. They form multicellular super-organisms, who obtain information both from the environment and neighbouring bacteria. They collectively share interpretations of these signals, and exchange chemical messages. Based on joint decisions they alter their structures. They have social intelligence, and fundamental (primitive) elements of cognition.

The middle image shows bacteria transporting a fungus spore (only the spore of one certain fungus is easily transported by these specific bacteria, there is a glue between these two that glues only one certain type of bacterium to one certain type of spore — wow!! Imagine such a glue for the separation of materials in waste — one for each metal, one for each type of plastic, one for wood, one for paper, one for organic compostable materials, etc.).

The middle image of Fig. 3 shows bacteria transporting a fungus spore.

The right image of Fig. 3 shows a living bridge in Indonesia.

For synergies of Chapter 10 with other chapters see Tables 1, 2, and 3.

Chapter 11 Cristiana Sbrana, Alessandra Turrini and Manuela Giovannetti: The crosstalk between plants and their arbuscular mycorrhizal symbionts: A mycocentric view

There were various fascinating aspects that the author of this foreword experienced when moving from the little central European country Austria to tropical South East Asian Malaysia. One that still startles her is the difference in the concept of a tree in these two regions of the world. In Austria, a tree is seen as one tree. Sometimes, there are bird nests on the tree, and sometimes, ants and squirrels run up and down. But that is essentially it. In Malaysia a tree is an organism that is host to many other organisms. There are ferns living on trees, lichens, orchids, bromeliads. There are ants on the tree, and squirrels, and flying foxes and bats, and lantern bugs, and the occasional sun bear. Each of the giant rainforest trees is a multilevel entity, home to hundreds of different species — untamed and unmanicured. Similarly unidirectional was her view of roots. *"A root is something that branches finer and finer, and that takes up nutrients from the soil, nutrients that are dissolved in water. There is no place for fungi on roots. If I see fungi on roots, the tree rots, and the fungi are a sign of death."* This her view was flipped over in 2013 on her visit to Peter Goldsbury in New Zealand. He told her about the intimate relationships of plants with root fungi. Symbiotic relationships seem to occur more often in plants than previously assumed.

When reading Sbrana *et al.*'s Chapter 11 "The crosstalk between plants and their arbuscular mycorrhizal symbionts: a mycocentric view" (Sbrana, Turrini and Giovannetti, 2016) the author of this foreword was especially intrigued by the fact that symbioses of plants and fungi seem to be much more common than previously assumed (Fig. 4). It caused her to write the public outreach text given below:

*"***Plant friendship****

My dad has a great hand for plants (as opposed to Mark and myself). What you can see here is a two-year-old pine tree baby, bred from the seed from our Italy holiday, and a young silver leaf

Figure 4. Friendship between silver leaf and pine tree. © 2015 Ille C. Gebeshuber, image reproduced with permission.

(a favourite plant from my childhood, also known as Jew's shilling). The silver leaf plant was sick and small, and not happy at all, until its leaves started to touch the pine needles. They became friends, and need and support each other, and ever since they touch each other, the silver leaf is strong and beautiful, and grows well. My father sees such relationships in plants."

Many plants live in symbiosis with root fungi that provide them with valuable Phosphorus and further chemical elements while the plants provide the fungi with complex carbon compounds products that the fungi cannot produce by themselves. This symbiotic relationship is initiated by complex chemical communication between the host plant and the fungi. 80% of land plants live in such symbiotic relationships with arbuscular mycorrhizal fungi (which are organisms that can be viewed as living fossils).

Plants that live in symbiosis with such fungi need less fertilizer and less pesticides, and have enhanced nutrient uptake. The arbuscular mycorrhizal fungi need the contact with a host plant to complete their life cycle. In the absence of a host plant the growth of fungal germlings is arrested after three weeks and — following a signaling cascade — the protoplasm is retracted close to the mother spore.

Multifold signals of different nature act at different steps of the interaction between fungal symbionts and host plants. There are various interdependencies between host plants and arbuscular mycorrhizal fungi, such as some stimulatory signals and/or inhibitory factors in the fungus that are regulated by the host plant. Already before physical contact pre-contact recognition events from both sides enable switching of the fungi from the asymbiotic to the pre-symbiotic stage. Physical contact is established with the formation of an appressorium, a minute cell that presses onto the plant root, injecting a peg into the root, thereby establishing physical contact. The establishment of appressoria is the key sign of fungal recognition of host plants. Host plants react "friendly" to appressoria formation, whereas plants that are either no host plants of the arbuscular mycorrhizal fungi or that have certain genetic mutations would start defence responses, trying to forcefully stop the physical contact between the arbuscular mycorrhizal fungi and the root.

The symbiont and the host cross-talk during cell-to-cell interactions. After the differentiation of appressoria, the arbuscular mycorrhizal fungi colonize host roots forming intracellular and intercellular hyphae, vesicles and arbuscules. Arbuscules are structures inside the host cells, allowing for nutrient exchange, membrane biogenesis and metabolism, cell wall reorganization and hormone balance. Despite the importance of nutrient exchanges in the symbiosis, where partners offering the best rate of exchange are rewarded, little is known on the mechanisms that regulate such processes. The exchange of Carbon for phosphate is tightly linked and requires both fungal and plant control.

In the presence of non-host plants, fungi behave different. Why some plant species cannot establish symbiotic relationships with arbuscular mycorrhizal fungi is poorly understood. Many such plants have inhibitory compounds reducing fungal growth. Interestingly, non-host plants that are grown in the presence of a mycorrhizal companion plant (or nurse plants) often show variable extent of root colonization.

The establishment of a functional arbuscular mycorrhizal symbiosis is the outcome of multifold signals of different nature, acting at different steps of the interaction between fungal symbionts and host plants. Various plant and fungal genes are activated during symbiosis establishment.

For synergies of Chapter 11 with other chapters see Tables 1, 2, and 3.

Chapter 12 Douglas William Darnowski. Attraction of preferred prey by carnivorous plants

Chapter 12 "Attraction of preferred prey by carnivorous plants" by Douglas William Darnowski presents a nice overview of carnivorous plants and some of their ways to communicate with their prey (Darnowski, 2016).

Darnowski describes how carnivorous plants attract preferred prey. Carnivorous plants thrive in nutrient-poor habitats, such as heavily weathered soils with low abundance of Nitrogen and Phosphorus. Placed in a habitat with more abundant nutrients, most carnivorous plants are quickly outcompeted by non-carnivores and perish. They lure their prey via odors, rewards and bright markings to their site, and then trap and digest them, with significant ecological benefit derived from this (Adamec, 2011). Many plants readily absorb material placed on their surfaces, especially leaves. The author proposes a new index, the prey preference index (PPI), to describe such behavior. Carnivorous plants use deceptive communication via visual, olfactory and taste signals with their prey — they are liars! With their pollinators they communicate non-deceptively. Traps from carnivorous plants can be active or passive. Digestion happens via enzymes; some plants wait to produce these enzymes until they are sure that prey has been caught well enough and escape is prevented.

The wild tomato is a plant species that might be on the way towards becoming a carnivorous plant. Carnivorous plants use signals according to the abilities of their intended prey. Some use UV markings, similar to the nectar guides in flowers (Fig. 5). Some pitcher plants use a nectar reward to draw prey to themselves. Some plants no not directly digest the insects they catch, but use another species to prepare their food: Roridula plants have sticky epidermal hairs that are resinous and therefore hydrophobic. The water-based enzymes cannot readily digest any caught insects. Inside the Roridula plant live bugs that are not caught by the sticky resin, eat the trapped insects and provide the plant with especially rich nutrient mixtures in their faeces.

In a similar way, the bird toilet pitcher plant serves as ideal perch for tree shrews and birds, whose faeces is rich in N and P. The new mathematical index proposed by the author is termed PPI; it helps in identifying deceptive communication in plants.

When reading Darnowski's Chapter 12 "Attraction of preferred prey by carnivorous plants" the author of this foreword was especially intrigued by the fact that carnivorous plants signal to their prey in fluorescent signals (Fig. 5). It caused her to write the public outreach text given below:

> *"Plants communicate with insects in a language that insects understand. Many insects see ultraviolet light (UV), and various plants show the insects by patterns that are only visible in the UV range where the nectar (and pollen) can be found (Figure 5). In this way, the plants stay inconspicuous for grazing mammals who might eat them, but are very attractive to pollinators.*
>
> *And now comes the interesting twist: carnivorous plants use similar signals (Fig. 5), tricking insects who are up for a nice sweet nutritious meal into their traps, where they will be digested and incorporated into the plant (Kurup et al., 2013)."*

For synergies of Chapter 12 with other chapters see Tables 1, 2, and 3.

Figure 5. Top: Flowers that appear uniformly yellow to humans have markings and nectar guides for insects, who can see in the UV range. The Mimulus flower photographed in visible light (left) and ultraviolet light (right) shows a nectar guide visible to bees but not to humans. © 1999 by Plantsurfer, image reproduced with permission. Middle and bottom: some pitcher plants trick insects with similar markings to their traps. Middle: *Nepenthes khasiana* pitcher peristomes with lids in white light. Bottom: Blue fluorescence emissions from *Nepenthes khasiana* pitcher peristomes and lids at UV 366 nm. © Rajani Kurup, Anil J. Johnson and Sabulal Baby, image reproduced with permission.

Chapter 13 Hans Slabbekoorn: Animal communication: Competition for acoustic space in birds and fish

We all know that birds communicate by sounds. Their singing attracts partners, or deters other birds, they make warning calls, etc.

What might be new to many is that also fish make sounds. For communication, e.g., in spawning choruses. Birds and fish face acoustic noise, from the living and non-living environment, and from anthropogenic sources.

Sound is an excellent medium to convey information in natural habitats in air and in water. Noise pollution influences communication by acoustic signals. In Chapter 13 (Slabbekoorn, 2016), "Animal communication: competition for acoustic space in birds and fish", Hans Slabbekoorn treats bird and fish side by side in terms of signaling problems and solutions and thereby provides insight into general concepts and current gaps in our knowledge.

Birds hear in a similar frequency range as humans do (20 Hz to 20 kHz), with their best range in the 1–5 kHz range. Some pigeons are sensitive to infrasound, some birds use echolocation (on calls that are audible to people) and the male blue-throated hummingbird generates ultrasonic song components that are not audible to himself. We currently do not know why he is doing this (the ultrasonic parts of his song are not just harmonics of song parts in the audible range). It might be that he sends signals into a sensory universe that is remote from his own perceptions. Here, synergy can be established with the chapter on UV signals in lying pitcher plants (Darnowski, 2016), carnivorous plants that mimic signals of nectar plants to pollinators. Such facets are highly interesting aspects of biocommunication.

Songbirds sing different dialects in different neighborhoods, and each bird has its own vocal signature. Female birds can discriminate amongst almost identical song types sung by either their mate or his neighbor — such fine sensitivity to detail also means that ambient noise may prevent auditory detection or adequate recognition of relevant details.

The song learning process involves copying sound elements and ways of singing from conspecifics that modify the ontogenetic trajectory of individuals but also determine the dialectal pattern formation at the population level. Such phenomena yield cultural transmission and cultural evolution and allows parallels to cultures in general such as e.g., in non-human primates and human dialects and languages (synergy with Chapter 3 and co-cultural processes in Table 3).

Most fish hear in a much more restricted frequency range than humans do. They detect both underwater sound components, sound pressure and particle motion, and with their lateral line system can sense nearby sources of very low frequencies. Some eels hear infrasound, some shads hear ultrasound (up to 180 kHz). Fish use sound for orientation and predator–prey interactions, but many species also listen for the communicative sounds generated by conspecifics. 800 species of fish are known to generate sounds. Many fish have small repertoires and context-associated within-species variation, just like birds. Variety in fish sounds stands for many potential messages for individuals of the same species that are within earshot, just like in birds. Fish are able to locate sound sources. They may also be bothered by interference of auditory detection and recognition of conspecific signals or environmental cues.

There is a competition for acoustic space in air and in water, as background noise changes the signal-to-noise ratio: rain-induced noise can decline the signal range in air from 600 m to 100 m. At the seaside many shorebirds have piercing calls. There are noise-related modifications, and the evolution of better hearing is not primarily driven by acoustic communication but instead by relative noise levels in the environment.

Anthropogenic noise (sounds generated by human activities) is detrimental: there is a close link between ambient noise conditions and avian signal efficiency. Traffic noise has most energy in the low frequency range. Birds start to sing louder in noisy conditions, they sing louder on weekdays than on weekends, and in places where human activity starts very early, dawn choruses shift into the more quiet nocturnal conditions. In Leipzig, Germany, it was found that birds at the noisy ring road woke up more than five hours earlier than birds in the quieter urban forest. Birds also start to sing higher in frequency.

Human activities in, on or close to water also cause considerable elevation of natural ambient sound levels underwater. Anthropogenic noise may not only deter but also interrupt important activities such as courting, spawning and foraging, or disturb fish during migration. The fish sounds are mainly at low frequencies, as is anthropogenic noise. Only a single case is known where fish adjust their sounds to anthropogenic noise, by increased pulse rate when exposed to noisy boat passages. Fish may also be able to generate louder sounds in response to elevated noise levels, but their moderate flexibility in terms of sound production and their limited range of hearing sensitivity may make spectral adjustments that would yield at least some masking avoidance less likely.

Acoustic signals are an excellent way to communicate in air and in water. Noise interference can be a challenge for fish and birds, since it has serious impacts on acoustic signal efficiency. Ambient noise influences the evolution of acoustic signals. More insight into the impact of anthropogenic noise will not only provide insight to animal vulnerability but also has the potential to raise awareness about the intrinsic beauty and ecological value of natural soundscapes in air and in water.

For synergies of Chapter 13 with other chapters see Tables 1, 2, and 3.

Chapter 14 Angela Machado Rocha, Fernando M. Palop, Maria Clara Melro and Marcelo Santana Silva: The contribution of biocommunication (BICO) to biomedical and tissue engineering: A tech mining study

Replacing damaged or dysfunctional body parts has a long history. 1954 the first kidney was transplanted, and in the 1990s, Dolly the sheep was cloned and first human embryonic stem cells were isolated.

The chapter of Rocha and co-workers (Rocha *et al.*, 2016) is a tech-mining study on more than 8,000 scientific papers published in the years 2003 and 2013 in an interdisciplinary field between engineering and the life sciences covering tissue engineering and regenerative medicine (TERM) as well as biomaterials.

In TERM, communication processes are of utmost importance: Human organs and cells communicate (crosstalk) with each other directly without the interference of the central nervous system. Cells can only multiply when growth factor commands are given by way of protein molecules and peptides emitted by the nucleus.

Biomaterials are materials that are intended to be in contact and interact with a biological system, evaluating, treating, augmenting or replacing existing functions. Biomaterials must be biocompatible. "Cell Biology and Biochemistry" and "Molecular Biology" are the top categories represented in 27% of the publications evaluated in the study. How cells naturally interact with each other is increasingly investigated in co-cultural processes, which are a simultaneous culture of different cell types. Stem cells are cells that can differentiate in various potential directions. Their crosstalk with other cells in the human body influences their proliferation, migration, differentiation, apoptosis, etc.

Understanding communications of the cells and tissues in bodies is important for the development of tissue architecture and tissue regeneration.

Three categories of biomaterials are inert materials, bioactive materials and biodegradable materials. All of them communicate with the body in different ways. Factors such as nerve growth factors cross-talk with cells, and can activate different answers by the cell nucleus related with its cell fate (differentiation, apoptosis, multiplication, etc.). The interaction of biomaterials with extracellular matrices influences this communication, and can influence the decision on cell fate. The extracellular matrix provides structural and biochemical support to the surrounding cells. It serves an important role in tissue and organ morphogenesis and in the maintenance of cell and tissue structure and function.

The relation between biomaterials and the nanoscale is intrinsic — small patterns and structures on the biomaterials can have a "loud voice" in cross-talk with cells and tissues.

Further important materials are bioceramics. Its three categories are inert, active and degradable or reabsorbable. Bioceramics can be found in the form of microsphere layers or thin coatings on metallic implants with a porous network and compound polymer components (composites). The most successful classes of bioceramics are Calcium phosphate biodegradables and alumina, which is highly inert and resistant. The body tries to isolate foreign bodies by forming a layer of non-adherent fibrous tissue around the implant where possible.

Biopolymers are suitable as biomedical material due to their structural similarity with tissue components. This structural similarity is important in biocommunication. The three categories of biopolymers are proteins, polysaccharides and polynucleotides. They perform certain functions due to their molecular configuration, composition and organization. The most widely used biopolymer is collagen — it is a key material for fibers, sponges and scaffold matrices.

Biomaterials implanted into the body induce a response called foreign body reaction. Macrophages fuse to form multinucleated giant cells, which often persist for the life of the implant. In its final stage, the foreign body reaction involves walling off the device by fibrous collagen. Novel smart materials shall in the short term minimize this foreign body reaction and promote normal wound healing. In the long term they should actively participate in the regeneration of damaged tissue and respond to stimuli from their environment. Interactive smart biomaterials such as electrospun fibers or hydrogels from peptides, collagen and further materials imitate extracellular matrices in order to stimulate cellular invasion, adhesion and proliferation. Biological approaches that focus on the repair and restoration of the structure and function of tissues are necessary.

Regenerative medicine aims at creating or regenerating new biological tissues, based on pre-existing tissues. The three pillars for Tissue Engineering are scaffolds, cells and growth factors. TERM is concerned with the regeneration of organs and living tissues through tissues of the patient. The cells grow in a scaffold, in the presence of growth and differentiation factors (see also Gordon and Stone, 2016).

Today, it is impossible to build a large organ three dimensionally where all cells are in direct contact with the nutrient solutions.

Important characteristics of the materials for the construction of the scaffold are biocompatibility, having a surface that allows cells to adhere, proliferate and migrate to the scaffold, biodegradability, sufficient mechanical properties to maintain the structure and function immediately after transplantation and structural and functional properties that allow extracellular-matrix biomolecular transfer signals between cells.

The implant material must not inhibit the biological functions of the host tissue. This is sometimes achieved by changes in the nano-topography of the scaffold surface with the aim of controlling and directing physical and molecular signals that govern cell behavior.

Bones and cartilages are the mainstream applications of TERM research. Better understanding of the extra- and intracellular communication mechanisms open important advances and new approaches in cancer treatment and tissue regeneration. There is an increasing interest for biocommunication in TERM, including investigations of proteins and their role in cell growth.

3D printing and patterning of scaffolds, and bioprinting, are new trends in TERM. A big challenge is still how to emulate blood vessels in the bioprinted tissue.

For synergies of Chapter 14 with other chapters see Tables 1, 2, and 3.

Chapter 15 Noga Gershoni-Emek, Eitan Erez Zahavi, Shani Gluska, Yulia Slobodskoy and Eran Perlson: Communication languages and agents in biological systems

When reading Gershoni-Emek *et al.*'s Chapter 15 "Communication languages and agents in biological systems" (Gershoni-Emek *et al.*, 2016) the author of this foreword was especially intrigued by the fact that neurons can of course be viewed under different aspects than just conducting action potentials (Fig. 6). It caused her to write the public outreach text given below:

> *"I have a beautiful blowpipe (Fig. 6) I once bought from Ms. Reita Rahim, from Gerai OA, a volunteer-run, nomadic stall selling crafts by the Orang Asal (indigenous minorities) of Malaysia; 100% of sales is paid to the named artisan. One of her friends from the indigenous communities in Malaysia made it. The poison that is used with blowpipes in Malaysia is not from frogs, as it is used in South America, but from poisonous latex from the Ipo tree. I once was shown such a tree by the Orang Asli here in Malaysia, impressive.*
>
> *The poisonous latex contains chemicals that lead to arrest of the heart, and cause death. The poison of poison dart frogs attacks the*

Figure 6. Left: Artists impression of firing neurons. Permission pending. Right: Orang Asal from Malaysia with blowpipe. Image © Sarawak Tourism Board, Malaysian Borneo. Image used with permission.

central nervous system, arrests muscles and finally breathing, leading to death. Have you ever thought how such poisons work? They are chemicals talking to our bodies. Which brings us again to the field of biocommunication, this time at a neural level — the level of the single nerve fibres.

Electrical signals travel along nerve fibres, but this is not all: there are also various chemical signals, that give the fiber information about its surrounding, about its status (Is there a damage somewhere? An injury? Is the nerve fiber very young and still needs to grow? In which direction? And connect with whom? Where?) Chemical signals are sent back and fro along and inside nerve fibres, and in many cases, chemical signals that are important during growth are not produced anymore when the person is grown up. Unless they have neurodegenerative diseases.

Sometimes the neural language needs to be translated, into a language that for example the muscle can understand. At the neuromuscular junction, the electrical signal that has travelled along one of your motor neurons from your brain to your little toe is translated into a chemical signal.

Neurons are masters in biocommunication, they travel through the whole body, meeting so many different environments, and talking so many languages. Understanding and speaking this language in typical and atypical situations (pathophysiology of neurological and psychiatric disorders) is important for basic knowledge, but also to be able to help people with neurodegenerative diseases."

Neurons communicate with their environment as well as with the cell's different compartments: soma, axon and synapses. They are highly polarized cells with axons that are many orders of magnitude longer than the diameters of their bodies. Neuron communication comprises action potentials, as well as protein and RNA trafficking mechanisms in order to regulate the specific task of spatial communication. Axonal transport mechanisms to and from the neuron body takes place on kinesin and dynein "highways" in the cell. Altered actions in neurocommunication and spatial changes in the cellular balance of survival versus stress/death signals contribute significantly to the rapid onset of neurodegeneration, and suggest a non-cell autonomous mechanism in neurodegenerative diseases.

Signaling in neurons includes trafficking of neurotrophic factors, which are growth factors that act specifically on neurons, such as glial cell-derived neurotrophic factors (GNDF) signaling that supports the survival of motor, dopaminergic, sensory and sympathetic neurons. Neuronal death signaling such as tumor necrosis factor (TNF) and guidance factors are means for neural communication.

A neuron's diverse microenvironment contains secreted negative signals that can lead to synapse disruption, axon pruning and neuronal death: physiological processes in the developing embryo, but pathological in adults.

Neuronal receptors are localized in a spatiotemporal way at the plasma membrane. Rafts, some tens of nanometers long, comprised of lipids and proteins, are microdomains that contribute to spatiotemporal signaling in neurons. Microtubules can regulate spatial localization and function of receptor domains.

The organization of neurotrophic factors in neurons is sophisticated, with lipids, proteins, actin and microtubules as well as endocytic and transport machinery working together to ensure signaling takes place at the right place, time and intensity.

Also local protein synthesis is of paramount importance in neuron communication. mRNA allows to respond locally to stimuli that affect general cell function. Both axons and dendrites respond to extrinsic cues in order to regulate growth, navigation and synapse formation — as well as for maintenance and repair. mRNA is in a translationally repressed state until the arrival of a triggering signal — such as a guidance cue, or injury.

The growth cone is a specialized structure at the tip of a growing axon. During growth and path finding, it responds to various stimuli from the environment providing guidance information. Locally synthesized proteins help with this.

Another interesting aspect of communication in a neuron is how the neuronal cell body receives information about an injury from distant lesion sites in the axon. Injury signaling activates the nerve regeneration process.

There are thousands of different mRNAs in neurons. Also microRNAs are important — they hold key roles in gene regulatory networks, and their dysregulation plays a role in the pathophysiology of neuronal and psychiatric disorders. Exosomes transfer miRNAs between cells, ensuring horizontal transfer of genetic information.

At the neuromuscular junction, motor neurons converse with the skeletal muscle. This happens quickly and reliably, ensuring precise control of skeletal muscle contraction.

For synergies of Chapter 15 with other chapters see Tables 1, 2, and 3.

Chapter 16 E. Sue Rumbaugh, Itai Roffman, Elizabeth Pugh and Duane M. Rumbaugh: Ethical methods of investigation with *Pan/Homo* bonobos and chimpanzees

For the author of this foreword the most amazing aspect of Rumbaugh's Chapter 16 "Ethical Methods of Investigation with Pan/Homo Bonobos and Chimpanzees" (Rumbaugh *et al.*, 2016) is when they describe the arousal of a joint language that is understandable for all species joining the game in the joint society of humans and apes, living together in a forest, in semi wild conditions. ICG heard that apes have language, and that they communicate with each other not just on present, but also on past and future events, as well as feelings, hopes and expectations. We know the same for people. But that, when the groups live together, in a non-laboratory setting, a new language arises, is stunningly touching and beautiful.

Bonobos, humans and chimpanzees share 99% of our genome. Apes are generally greatly underestimated. Chimpanzees and bonobos are sentient beings that are self-aware, capable of symbolic thought, and can construct mental worlds, social rules and cultures. Rumbaugh and co-workers address aspects of communication with these beings, and ethical methods of learning from such highly cultural species. Most of the time apes used in such studies are in captive environments that are devoid of choice, freedom of expression, the execution of free will, and the right to self-determination.

Studies of how apes react in such an environment might also give important input regarding how to curb human aggression when people are kept in such tight and supervised conditions.

Behavioral and genetic data tell that apes belong to the human side of the man/animal dichotomy that still is very prominent in our mind. Apes, when reared in different ways, display radically different cognitive competencies. They show two types of using language: the first one is used for ritualistic communicative purposes (Hello, how are you) and can serve true social functions and promote social cohesion, regardless of whether the state of knowledge of the recipient is known or not. In the other type of language, however, the state of knowledge of the recipient is of paramount importance. This type of language can convey truly new information. Chimpanzees can communicate novel and appropriate information to each other, they can do so with intent and they can display sensitivity to the knowledge state of the recipient. Chimpanzees and bonobos who are reared in the right cultural environment, in loving and stable family settings, have a theory of mind, i.e., they understand that others can have knowledge that is different from one's own knowledge. Most other animals do not think about the state of mind of the others, but co-feel, co-sense and co-experience with them, e.g., through mirror neurons. Chimpanzees recognize themselves on television and distinguish between taped and real-time videos of themselves. An entity that cannot see itself as others see it lives in a distinctly different world. Non-sentient beings are time locked to the desires of the moment. Chimpanzees not only recognize themselves in a mirror, they also check with the help of a mirror how lipstick looks on them — they are not only interested how it feels or tastes. Apes are conscious about being conscious.

The emergence of the sentience of selfhood, a concept of moral action and an understanding of the theory of mind require a socio-cultural world. Without such a world, humans and apes fall back into an animal like existence. Culture plus biology are important to seed reason, reflective thought, purposeful action and planning. This emergence is on the verge of vanishing if not passed on successfully to the next generation.

Very few researchers have the privilege to work with linguistically competent apes. Bonobo Kanzi is one of them. The chapter contains the highly interesting and unique report of researcher Itai Roffman (IR) who grew up with his brother Orr who only started to talk at the age of 14, when he watched videos of communicating apes. Orr only then started to share his inner thoughts. Due to his childhood experience with five-year younger brother Orr, IR had a different, very open way to communicating with chimpanzees and bonobos. Their conversations were bilateral, and contextually appropriate. They only took place when they felt comfortable with each other. Only if people really live in their world, these apes communicate in a true way. Arrogant students who do not appreciate them as beings and who bribe them to collaborate in laboratory setting experiments obtain wrong data. Kanzi was very careful in what he shared, with whom he shared and when he shared. Kanzi and Orr talked in three word sentences, sharing inner feelings, stories, ideas, wishes and interests. When IR was accepted into the bonobo community, he had to go through some kind of initiation ceremony. Orr found his own language with the help of the bonobos. Before watching the videos, miming language capacities and the complex statements made through the use of gestures, glances and joint contextual reference and knowledge were absent. Orr now could initiate — not by just doing, but by expressing his desires to do so. A second breakthrough occurred when Orr watched a video of bonobos painting. His scribbles started to transform to geometric shapes to figurative forms, and he could very well interpret bonobo paintings — at times better than the researchers. Art is a spatial, not a linear medium. IRs early life with Orr prepared him to be more sensitive to the communication of apes than others. They are creative, thinking, reasoning feeling beings capable of foresight, planning and moral action.

Humanness as we experience it does not exist outside of human language. This also holds for science. To study deep into the essence of language is to reach deeply into what it is to be human. Language seems to have appeared 40,000 years ago, whereas we achieved our physical forms 200,000 years ago. Perhaps our human language is the limit of our cognitive world. Only by jumping across cultural and linguistic boundaries can one's eyes be opened to the cultural blinders imposed by one's own rearing. We have almost no means of understanding any human who does not develop language in a normal manner unless we spend time with them while co-constructing dialogues co-filled with meaning. Any being capable of becoming conscious of being conscious has the potential for some form of truly symbolic language. Language is a symbolic/grammatical system of representational communication that functions to convey intention and meaning to its users.

Socio-cultural interactions initiate, force and perpetuate the acquisition of grammars in species with executive function control over their own conscious processes. Grammar arises with the desire and effort to make meaning of acts that remain incoherent and disconnected until they become linked by grammar and symbols. Grammar folds back upon itself and becomes reflexive. Grammar does not need to be taught — and so parents do not bother.

Both humans and bonobos focus on whom, what and where, but humans also focus on why, how and when. Bonobos less likely attribute personal causality to complex events than humans do. The human socio-cultural world does not map precisely onto the socio-cultural world of a closely related species. In symbolic self-conscious beings, speciation is about co-constructing a conscious reality that all share via the operating system of a particular language. It is the lifting process — the extension of meaning beyond the here and now — that truly defines language. Creativity and imaginations seem to be uniquely hominid. Given stable and rich environments, apes and humans will turn their attention to an endless range of objects and activities. Humans and apes are clearly something apart from the rest of the world. Why this is so and what does this mean?

Gebeshuber and Macqueen write "*Members of the same species were found to differ dramatically in gene content: In a study by Welch et al.* (2002) *three E. coli genomes were sequenced. Less than 40% of the genome was common to all three bacteria!*". On the other hand, the genome of bonobos, chimpanzees and humans is 99% similar. Perhaps various important aspects of a species are not stored in the genome, but in another space. Humans, bonobos and chimpanzees construct symbolic world and moral systems.

The two chimpanzees Sherman and Austin developed a gestural symbolic language similar to the twin languages sometimes reported for co-reared twins.

Wild chimpanzees are sentient self-reflexive moral beings with complex non-material cultures and symbolic systems. Language builds new interpretations and creative combinations through shared exchanges across time and during development. Matata, Kanzi's mother, was caught in the wild, and came to the laboratory already with knowledge of the culture of wild bonobos. She clearly had some form of language. Culture is very important for the development of the mind. Human culture and bicultural rearing of young bonobos in groups comprising older bonobos and humans over-rides the supposed absence of language. Even though our species is known for its vocal emulation skills, it is impossible to emulate bonobo speech with the precision pitch control they employ. (Potential way to address this: Ask a Thai student to help — the Thai language very much focuses on pitch.) Matata did not communicate with the humans, but she would ask Kanzi or some other language apt ape to convey requests to people. She realized that people could not understand her, but that Kanzi could, and that he could act as translator. Kanzi and friends translated various information for Matata, most often about danger, food, travel, visitors, or unusual events.

In a group of people and bonobos that used to live in the forest together, the human companions were not caretakers, they were companions. Bonobos develop vocal utterances, different from Matata's language, when together with such a group in the forest. This language was one that was beginning to be comprehensible by human members. Language appeared to drive all other changes, and the apes who spoke it became increasingly diverse, and sentient beings were suddenly emerging.

Human groups have always separated themselves across time and they continue to do so on the basis of epi-cultural markers such as speech, dress, food, etc. Any group of symbolic self-sentient beings will naturally become uncomfortable when the behaviors they extent to others fail to meet with their anticipated results. Humans have historically found it difficult to look across cultural borders and to see humanity as one entity, even less so all that is alive.

In *Pan/Homo* beings, the bonobos who were reared with people and older bonobos in the forest, symbolic meanings literally jumped into existence everywhere.

Formal tests — which do not involve real world communications — will cause bonobo meanings to be withdrawn instantly. Language fundamentally alters the hardwire of the modern chimpanzees and bonobos. In the course of the experiments, their bodily form changed, and became more human-like. Culture is the ratchet that made man from ape. If we define culture as the glue which allows self-conscious beings to make meaning of each other's actions and communications by binding together acts that co-mean, ape cultures may well be found more similar to ours than we have previously acknowledged. The sudden (in evolutionary terms) onset of language, speech, music, art, architecture, writing, metallurgy, centralized government, etc., cannot be explained by standard scientific models of evolution. All the seemingly vast differences between living groups of humans are a result of culture. Kanzi could understand fire, kinship and cooking. Could he perpetuate this knowledge in the wild? (Potential issues that can arise: what if he finds out that there is a way to get rid of mean people by setting their house on fire? What if animals suddenly developed an idea of their strength? And started self-defence against people?).

Humans have increased susceptibility to cancer. Apes normally do not get cancer that metastasizes. Modern genetics finds that the gradualist argument is being undermined by genetic data on species other than apes and human beings. The discontinuity between ape and man is more on the psyche level. The symbolic mind is generating the world around it through an infinite myriad of cultural lenses. The human genome displays far less variability than that of any living apes. Homo sapiens went through an evolutionary hourglass approximately 70,000 years ago (Petraglia K *et al.*, 2012), when our numbers were reduced to between 2,000 and 15,000 individuals. Inbreeding seems to be a particular problem in the human species.

In humans, behavior can be completely freed from the environment and possibly even from genetic control by the power of symbolic thought. Changes in behavior affect gene expression directly. Scientists are only now beginning to think about the true behavior/biology relationship. We long for a more precise model to explain the integrated behavior of life forms across time on the planet (synergy with Chapter 9, Harari and Sharon, 2016). Epigenetic factors have no finite number. Genes can travel for millions of years doing nothing and reappear as active again, when conditions change. Populations change when they are in contact with different environments, and when they undergo behavioral change. Science has still not found the answer in selectionism and mutation. The Darwinian model provides scientists with a mechanistic description of what would otherwise be a totally mysterious process. In natural settings, all genes and all species are constantly co-evolving within a system that is itself self-evolving. Epigenetics challenges the established theories on selection processes. The genome is a tool-kit that has evolved in order to allow the organisms to differently manifest across geological eras of time. The genome represents its history (synergy with Chapter 6, Gebeshuber and Macqueen, 2016, regarding communication with the past and permutation by recession).

Bacteria come with a set of genes designed to do many different things in different environments. Genes can be activated by internal gut biota — diet is a critical factor (synergy with Chapter 6, Gebeshuber and Macqueen, 2016). Chimpanzees seem to be changing rapidly at the genetic level, while we are not.

Chapter 17 Toni Frohoff and Elizabeth Oriel: Conversing with dolphins: The holy grail of interspecies communication?

Chapter 17 deals with the communication of dolphins with other animals and with humans. One extremely anthropocentric misconception is the "smile" of the dolphin — it has nothing to do with an emotional expression, but is a static feature of the dolphin's jaw. Dolphins are viewed as very positive by a major fraction of the population, and in many cases scientific paradigms merge with policy, politics, emotion and ethics. These sentient marine mammals have fascinated people since before Aristotle.

Toothed whales, including dolphins and orcas, demonstrate amongst the most sophisticated biocommunicative abilities of any animal. When studying their communication, various reductionist and mechanistic approaches need to be overcome, in a more expansive, multidisciplinary approach to biocommunication. The authors of Chapter 17 (Frohoff and Oriel, 2016) pledge for respectful research, on wild animals, to fully explore the collaborative relationship between humans and dolphins. "Respectful" refers to the fact that the wellbeing of the dolphins or other non-human animals takes priority over the success of the research. Furthermore, they stress the importance of positive co-habitation with dolphins from a multispecies, co-cultural context (fishing cooperatives, between humans and dolphins, seabirds and dolphins, humans and seabirds and dolphins).

Dolphins communicate with acoustic signals, via visual posturing and movements, chemoreception, tactile exchange and the use of water for making bubbles and splashes (visual and acoustic signals). Free-ranging cetaceans show vocal learning and referential signaling. They have varied and rich social cultures, and use tools. Only female humans and whales live long past menopause, being "repositories of ecological knowledge and cultural wisdom".

Similar to Rumbaugh *et al.* in the previous Chapter 16 (Rumbaugh *et al.*, 2016), Frohoff and Oriel stress personhood as an important concept when dealing with non-humans. Anthropocentrism and the resulting cultural assumptions are inherent in Western thought and activitiy, whereas in certain non-Western and indigenous cultures, the concept "person" may apply to animals, plants, humans, life systems, climatic events, landscapes and spirits (Ingold, 2000: 90).

For further synergies of Chapter 17 with chapters of this book see Tables 1, 2, and 3.

The next to last sentence of Chapter 17 shall also conclude the red thread through the single chapters of this book on biocommunication: "*Communicative connection points between species can all serve to build on a more thriving relationship between, and for, the benefit of humans and all animals*" (and, in principle, the whole biosphere).

References

Barbieri M. 2008. Biosemiotics: a new understanding of life. *Naturwissenschaften*, 95(7), 577–599.

Barnosky AD, Matzke N, Tomiya S, Wogan GOU, Swartz B, Quental TB, Marshall C, McGuire JL, Lindsey EL, Maguire KC, Mersey B and Ferrer EA. 2012. Approaching a state shift in Earth's biosphere. *Nature*, 486, 51–57.

Choh Y, Shimoda T, Ozawa R, Dicke M and Takabayashi J. 2004. Exposure of lima bean leaves to volatiles from herbivore-induced conspecific plants results in emission of carnivore attractants: active or passive process? *Journal of Chemical Ecology*, 30(7), 1305–1317.

Darnowski DW. Attraction of preferred prey by carnivorous plants. In *Biocommunication*, Chap 12, Seckbach J and Gordon R (eds.). World Scientific Publishing, London, 309–325.

Drake JW and Holland JJ. 1999. Mutation rates among RNA viruses. *PNAS*, 96(24), 13910–13913.

Finkelshtein A, Sirota-Madi A, Roth D, Ingham CJ and Ben-Jacob E. *Paenibacillus vortex* — A bacterial guide to the wisdom of the crowd. In *Biocommunication*, Chap 10, Seckbach J and Gordon R (eds.). World Scientific Publishing, London, 257–283.

Frohoff T and Oriel E. 2016. Conversing with dolphins: the holy grail of interspecies communication? In *Biocommunication*, Chap 17, Seckbach J and Gordon R (eds.). World Scientific Publishing, London, 573–595.

Gebeshuber IC. 2014. Message, in: Control, Instrumentation, Communication and Computational Technologies (ICCICCT), 2014 International Conference on, doi:10.1109/ICCICCT.2014.6993208, IEEE Conference Publications, ii.

Gebeshuber IC and Macqueen MO. 2016. Superfast evolution via trans- and interspecies biocommunication. In *Biocommunication*, Chap 6, Seckbach J and Gordon R (eds.). World Scientific Publishing, London, 165–185.

Gershoni-Emek N, Zahavi EE, Gluska S, Slobodskoy Y and Perlson E. 2016. Communication languages and agents in biological systems. In *Biocommunication*, Chap 15, Seckbach J and Gordon R (eds.). World Scientific Publishing, London, 411–448.

Gordon R and Stone R. 2016. Cybernetic embryo. In *Biocommunication*, Chap 5, Seckbach J and Gordon R (eds.). World Scientific Publishing, London, 111–163.

Harari A and Sharon R. 2016. Chemical communication. In *Biocommunication*, Chap 9, Seckbach J and Gordon R (eds.). World Scientific Publishing, London, 229–256.

Ingham CJ, Kalisman O, Finkelshtein A and Ben-Jacob E. 2011. Mutually facilitated dispersal between the nonmotile fungus *Aspergillus fumigatus* and the swarming bacterium *Paenibacillus vortex*. *PNAS*, 108(49), 19731–19736.

Ingold T. 2000. *The Perception of the Environment: Essays in Livelihood, Dwelling, and Skill*. Routledge, London.

Kak S. 2016. Communication languages and agents in biological systems. In *Biocommunication*, Chap 8, Seckbach J and Gordon R (eds.). World Scientific Publishing, London, 203–226.

Kurup R, Johnson AJ, Sankar S, Hussain AA, Kumar CS, and Baby S. 2013. Fluorescent prey traps in carnivorous plants. *Plant Biology*, 15(3), 611–615.

Martinelli D. 2016. Zoosemiotics, typologies of signs and continuity between humans and other animals. In *Biocommunication*, Chap 3, Seckbach J and Gordon R (eds.). World Scientific Publishing, London, 63–85.

Negrotti M. 2016. Communication as an artificial process. In *Biocommunication*, Chap 4, Seckbach J and Gordon R (eds.). World Scientific Publishing, London, 87–109.

Peirce CS. 1955. *Philosophical Writings of Pierce*, Bachler J (ed.). Dover Publications, New York.

Petracchi D, Gebeshuber IC, DeFelice LJ and Holden AV. 2000. Stochastic resonance in biological systems. *Chaos, Solitons & Fractals*, 11, 1819–1822.

Rocha AM, Palop FM, Melro MC and Silva MS. 2016. The contribution of biocommunication (BICO) to biomedical and tissue engineering: a tech mining study. In *Biocommunication*, Chap 14, Seckbach J and Gordon R (eds.). World Scientific Publishing, London, 365–410.

Rockström J, Steffen W, Noone K, Persson A, Chapin FS, III, Lambin EF, Lenton TM, Scheffer M, Folke C, Schellnhuber HJ, Nykvist B, de Wit CA, Hughes T, Leeuw Svd, Rodhe H, Sörlin S, Snyder PK, Costanza R, Svedin U, Falkenmark M, Karlberg L, Corell RW, Fabry VJ, Hansen J, Walker B, Liverman D, Richardson K, Crutzen P and Foley JA. 2009. A safe operating space for humanity. *Nature*, 461, 472–475.

Rumbaugh ES, Roffman I, Pugh E and Rumbaugh D. 2016. Ethical methods of investigation with *Pan/Homo* bonobos and chimpanzees. In *Biocommunication*, Chap 16, Seckbach J and Gordon R (eds.). World Scientific Publishing, London, 449–572.

Sadowski JA, Moore AJ and Brodie III ED. 1999. The evolution of empty nuptial gifts in a dance fly, *Empis snoddyi* (Diptera: Empididae): bigger isn't always better. *Behavioral Ecology and Sociobiology*, 45(3–4), 161–166.

Sbrana C, Turrini A and Giovannetti M. 2016. The crosstalk between plants and their arbuscular mycorrhizal symbionts: a mycocentric view. In *Biocommunication*, Chap 11, Seckbach J and Gordon R (eds.). World Scientific Publishing, London, 285–308.

Sharov AA. 2016. Molecular biocommunication. In *Biocommunication*, Chap 1, Seckbach J and Gordon R (eds.). World Scientific Publishing, London, 3–35.

Slabbekoorn H. 2016. Animal communication: competition for acoustic space in birds and fish. In *Biocommunication*, Chap 13, Seckbach J and Gordon R (eds.). World Scientific Publishing, London, 327–363.

Steffen W, Richardson K, Rockström J, Cornell SE, Fetzer I, Bennett EM, Biggs R, Carpenter SR, Vries Wd, de Wit CA, Folke C, Gerten D, Heinke J, Mace GM, Persson LM, Ramanathan V, Reyers B and Sörlin S. 2016. Planetary boundaries: guiding human development on a changing planet. *Science*, 347, 6223.

Tamir B and Priel A. 2016. Channel capacity and rate distortion in amino acid networks. In *Biocommunication*, Chap 7, Seckbach J and Gordon R (eds.). World Scientific Publishing, London, 187–202.

Tomasello M. 2008. *Origins of Human Communication*. MIT Press, Cambridge.

Witzany G. 2016. Key levels of biocommunication. In *Biocommunication*, Chap 2, Seckbach J and Gordon R (eds.). World Scientific Publishing, London, 37–61.

About the Editors

Richard Gordon received a B.Sc. in Mathematics from the University of Chicago in 1963 and a Ph.D. in Chemical Physics from the University of Oregon under the late Terrell L. Hill. He retired from the University of Manitoba in 2011 as a Professor in Radiology, having supervised students in Biosystems Engineering, Botany, Computer Science, Electrical & Computer Engineering, Physics & Astronomy, and Zoology. He is an eclectic scientist and prolific writer with over 200 peer reviewed publications in a wide number of fields (https://www.amazon.com/author/dickgordoncan, http://tinyurl.com/DickGordon). He has edited 16 academic books and special issues of scientific journals including one book of his own, *Hierarchical Genome and Differentiation Waves: Novel Unification of Development, Genetics and Evolution*. His latest monograph, with lead author Natalie K. Gordon, is *Embryogenesis Explained* (World Scientific Press). He is best known for interdisciplinary and cross disciplinary work bridging biology with fields such as mathematics, engineering, physics and chemistry. He wrote the first paper on diatom nanotechnology founding that field. He started the field of adaptive neighborhood image processing. He has also published on algal biofuels, computed tomography, AIDS prevention, neural tube defects, embryo physics, as well as research and social ethics. His most cited paper is one where he applied the linear Kaczmarz method to create the nonlinear Algebraic Reconstruction Technique (ART) for image reconstruction in 1970. He founded Books With Wings, which sends books to the universities of Afghanistan (https://en.wikipedia.org/wiki/Richard_Gordon_(theoretical_biologist)) He is presently a volunteer at the Gulf Specimen Marine Laboratories (http://www.gulfspecimen.org/gulf-specimen-lab-staff/) and an Adjunct Professor in the C.S. Mott Center for Human Growth and Development, Department of Obstetrics & Gynecology, Wayne State University.

Professor J. Seckbach earned his Ph.D. from the University of Chicago and did a post doctorate in the Division of Biology at Caltech, in Pasadena, CA. He led a group researching exobiology (extraterrestrial life) at UCLA. He was appointed to the Hebrew University, Jerusalem and spent sabbaticals at UCLA and Harvard University. Dr. Seckbach enjoyed his DAAD-sponsored (The German Academic Exchange) periods in Tübingen, Germany, and at LMU, Munich. He served at Louisiana State University (LSU), Baton Rouge, LA, USA, as the first selected Chair for the Louisiana Sea Grant and Technology transfer.

Professor Joseph Seckbach has edited over 30 books for various publishers. He has co-edited other volumes, such as the *Proceeding of Endocytobiology VII Conference* (Freiburg, Germany) and the *Proceedings of Algae and Extreme Environments meeting* (Trebon, Czech Republic). See: http://www.schweizerbart.de/pubs/books/bo/novahedwig-051012300-desc.ht). He co-edited the recent volume (with Richard Gordon) entitled *Divine Action and Natural Selection: Science, Faith, and Evolution* published by World Scientific Publishing Company. He is currently still busy with new books to be published by Imperial College Press, World Scientific Publishing Company, Elsevier, Springer and Wiley/Scrivener.

Beyond editing academic volumes, he has published ~140 scientific articles on plant ferritin–phytoferritin, cellular evolution, acidothermophilic algae, life in extreme environments and on astrobiology. He also edited and translated several popular books and reviewed several books. Dr. Seckbach is the co-author, with R. Ikan, of the Hebrew-language *Chemistry Lexicon* (two editions by Deveer Publisher, Tel-Aviv). His recent interest is in the field of enigmatic microorganisms and life in extreme environments and Astrobiology. He enjoys visiting many Universities and Institutes and presenting their seminars on his studies. E-mail: joseph.seckbach@mail.huji.ac.il

Contents

PART I
Theoretical Approaches

1. Molecular Biocommunication

Alexei A. Sharov

Laboratory of Genetics and Genomics
National Institute on Aging
251 Bayview Blvd., Baltimore, MD 21224, USA
Sharoval@mail.nih.gov

The integrity of living cells is maintained via regular communication between cell components. Membrane receptors provide information on the environment and launch signals to modify cellular functions accordingly. Gene regulatory networks coordinate the work of genes to keep the homeostasis of the cell and optimize its functions. Ribosomes are miniature robots that are programed to produce various kinds of proteins. Molecular biocommunication includes the most primitive form of sign processes in nature (I call it protosemiosis), where signs directly encode and control cell functions instead of being associated with objects. Such signs can be classified into proto-icons that operate via specific binding, proto-indexes that signal via physical association of several binding and/or catalytic domains, and proto-symbols that are processed by universal subagents equipped with a set of heritable adapters. The functions of molecular signaling include fabrication of cell components, sensing, memory, movement, and propagation of depolarization waves on membranes. Besides immediate cellular functions, molecular networks support robustness, error control, and modularity, which often require categorization of objects, and therefore belong to the advanced type of sign processing — eusemiosis.

1. Introduction

The science of communication gradually expanded with the advance of human technology and knowledge. It was applied first to human language, then, subsequently, to technology, animal acoustic, and chemical signals. With the emergence of molecular biology, the field of communication had unexpectedly expanded to the level of molecules within cells. Molecules play various roles in biocommunication. Membrane receptors provide information on the environment and launch signals to modify cellular functions. Cytoplasmic and nuclear receptors coordinate the work of genes to keep the homeostasis within cells and optimize their functions. Ribosomes are miniature robots that are programed to produce various kinds of proteins. Heredity is based on the long-term memory molecules [e.g., deoxyribo nucleic acid (DNA) and ribonucleic acid (RNA)] which are copied and passed from one generation to another. Molecular communication ensures the functional integrity of cells, and supports their growth, reproduction, and defense from pathogens. The study of communication became an interdisciplinary field where the original linguistic terminology is not fully applicable. One of the first attempts to overcome the constraints of linguistics was undertaken by Charles Peirce who developed a general theory of signs (Peirce, 1998). He defined signs as follows: "a sign is something which stands to somebody for something in some respect or capacity" (Peirce, 1931: 228). Activities of living organisms associated with sign perception, interpretation, or creation are called "semiosis". Peirce considered mind as an important condition for the existence of signs: "The third condition of the existence of a sign is that it shall address itself to the mind. It is not enough that it should be in relation to its object [...] but must be regarded by mind as having that relation" (Peirce, 1986: 83). But he thought that sign relationships can be captured by any mind, and thus they are observer independent. Thus, causal relationships or physical connections are sufficient to make a sign: "if I take all the things which have certain qualities and physically connect them with another series of things, each to each, they become fit to be signs" (Peirce, 1998: 40). However, he did not envision (as anyone else at that time) that communication may exist between mindless molecular agents that are not capable of contemplating objects. Thus, the analysis of molecular biocommunication requires more general approaches than Peirce's semiotics.

In this chapter, I argue that molecular biocommunication includes the most primitive form of sign processes in nature, which I call "protosemiosis", a term that was earlier suggested by Prodi (1988), where signs (proto-signs) are not associated with objects, but instead directly control the activity of agents (Sharov, 2013; Sharov and Vehkavaara, 2015). The terms "control" and "regulation" are interpreted by some authors as efficient causation or brute force (Hoffmeyer and Emmeche, 1991; Bruni, 2008). In contrast, by "control" I mean various kinds of effects which do not fall into the category of efficient causation: initiation or inhibition of actions, specifying the mode of action, its direction, timing, and scope, as well as conditional (i.e., context-dependent) effects. Although proto-signs contribute to the actions, they never fully determine them because actions are always agent dependent and agent-specific. Molecular proto-signs are classified here into proto-icons, proto-indexes, and proto-symbols. Higher levels of molecular communications support error correction, robustness, and modularity of signaling, as well as lasting interactions with internal and external objects. Corresponding molecular networks often include internal representations of objects as fixed point attractors in the continuous signaling phase space (i.e., categorization). These belong to the advanced type of semiosis, I call it "eusemiosis", where signs are associated with objects or object states. Analysis of molecular communication is a part of a wider endeavor, biosemiotics, which is aimed at the integration of semiotics with biology at all levels of organization (Hoffmeyer, 2008).

2. Chemistry Versus Semiotics: How Does a Molecule Become a Sign?

Biophysics and chemistry considers a living cell as a complex chemical reactor, where each process can be described in terms of reactants, products, catalysis, and kinetics. However, chemistry tells nothing about the purpose of chemical interactions, and thus, it is not sufficient to understand how the cell works. Molecular interactions in living organisms should be studied not only by chemistry but also by biosemiotics which attempts to answer how biological processes are encoded, controlled, and communicated, and how do they contribute to the goals of living systems (Sharov, 1992, 2010).

Application of semiotic terms to molecular interactions raises an important question: where is a boundary between chemistry and semiotics and how does a molecule become a sign (Pattee, 1969)? Sign theory is often built on the basis of "semiotic realism", where signs are viewed as real and objective natural phenomena without a reference to interpreters (Vehkavaara, 2003). For example, formal and final causations are used as criteria of sign relationships (Deely, 1994; Bruni, 2008). Another common criterion of a sign relationship is a source of energy: semiotic interactions are powered by the internal energy of the receiving system rather than by the impact (Bateson, 1972). However, these approaches are meaningful only if applied to goal-directed agents that use forms to produce tools and resources. Attempts to expand formal and final causations beyond agents ignore the striking difference between living and non-living systems. In particular, information was associated with purely physical categories such as entropy, negentropy (Brillouin, 1953), or with quantum physics (Green, 2000). Another theory extended quantum principles to information energy complementarity (Ji, 2012). As a result, either life is reduced to non-equilibrium physical systems (Chaisson, 2005) or life properties are extended to the full universe as in pansemiosis, hylotheism, or the Shillongator model (Merrell, 1996; Brier, 2008; Ji, 1991). The energy criterion of signs also does not work without agency: for example, a nuclear fission reaction is based on the release of internal nuclear energy, but it is not a semiotic process.

A short answer to the question posited in this section is that physics and chemistry study things as they are, whereas semiotics studies things as they are used by agents (e.g., living organisms and man-made automata) to encode and control their activities (Sharov, 2010). In other words, semiotics is focused on the *role* of things in the context of agent activities. Agents can be studied by physics (e.g., biophysics), but only in non-semiotic aspects (i.e., excluding goal-directed behavior and evolution). In particular, information is not a chemical or physical concept (Emmeche, 1999). In contrast, semiotics offers a comprehensive approach to the analysis of agents and signs. This criterion for separating biosemiotics from physics resembles Cartesian dualism, however, there are important differences. First, Cartesian dualism opposes material and mental substances, and the mental world is assumed to be non-material and non-physical. In biosemiotics, agents are material and their change does not contradict physics but is not fully predicted by physics. In simple living organisms, goal-directed activities are explained by the "natural self-interest", which emerges from recursive self-maintenance and self-reproduction of agents (Vehkavaara, 2003). In addition, advanced organisms develop a network of secondary goals and habits which are partially independent from the basic natural self-interest. The attribute "being an agent" is not physical because it is impossible to make a universal detector of agency (Sharov, 2010). To identify a thing as an agent we have to understand its goals, functions, origins, external and internal communication, and these characteristics cannot be measured without a *prior* knowledge for the functional organization of each kind of agents. Thus, identification of agents is always non-trivial and requires a competent interpreter, which is an agent itself. Agents were often misidentified in the history of human civilization. For example, celestial bodies were considered as super agents (gods), but infections were not recognized as micro-agents. Moreover, nobody suspected that our body is made of trillions of tiny subagents called cells. Second, agents differ in their functional complexity, and agents at the lower end of the complexity scale do not possess mind. Thus the distinction between agents and non-agents is different from the matter/mind dichotomy in Cartesian dualism. Third, biosemiotics considers a transition between non-agents and simple agents at the origin of life; in contrast, mind did not originate from matter according to the Cartesian doctrine.

The difference between chemical and semiotic approaches can be illustrated by the reaction of actin polymerization. Actin is a very abundant protein in eukaryotic cells which makes a cytoskeleton via polymerization into long fibers. Actin molecules emerged in evolution for a specific purpose: to make flexible filaments that can be easily assembled, disassembled, and integrated into the cytoskeleton of cells. These filaments increased cell stability, preserved its shape, and allowed crawling (Lodish *et al.*, 2000). Large eukaryotic cells could never have appeared in evolution without a cytoskeleton. Polymerization of actin can be viewed as a simple chemical reaction:

$$\text{Actin} + \text{Actin} + \cdots + \text{Actin} = (\text{–Actin–})_n. \tag{1}$$

Actin protein is polarized and its (+) end fits due to its shape and charge with the (–) end of another molecule. Thus, actin monomers can polymerize into fibers in solutions with sufficient ionic strength (Lodish *et al.*, 2000). This reaction, however, also represents a simple case of signaling. Actin monomers readily bind to the terminal actin molecules in the fiber but not to free-flowing actins. This means that attachment of actin molecule to the filament makes a *signal* for binding of the next molecule. Thus, the signal travels along the filament at the tip of its growth and results in the construction of the full filament.

Polymerization of actin is controlled by additional signaling molecules. The actin monomer has a site for binding adenosine triphosphate (ATP) (along with Mg^{++}), which facilitates polymerization at the (+) end of the filament (Lodish *et al.*, 2000). However, the ATP is hydrolyzed to adenosine diphosphate (ADP) immediately after incorporation of monomers into the filament (Fig. 1).

Figure 1. Polymerization of actin. ATP binding is required for polymerization.

Dissociation of actin occurs at the (–) end, which results in the "treadmill effect". Tropomodulin, however, caps actin polymers at the (–) end and prevents the dissociation. Another protein, CapZ, which was found in Z bands of sarcomeres in muscles, caps actin filaments at the (+) end and inhibits polymerization. Initiation of new actin filaments (or new branches of existing filaments) is controlled by Arp2/3 complex of proteins.

Several other proteins bind to actin monomers, and in this way, regulate the assembly of actin filaments. For example, cofilin binds to ADP-actin monomers (i.e., products of filament disassembly) and keeps them in the ADP bound form, preventing polymerization. Profilin reverses this effect by facilitating the exchange of actin-bound ADP for ATP. As a result, actin monomers dissociate from cofilin and become available for the assembly into filaments. However, polymerization of ATP-actin monomers can be inhibited by thymosin beta which specifically sequesters ATP bound actin. In summary, the semiotic analysis of molecular signaling associated with actin polymerization yields substantially more information than a simple chemical approach (Eq. (1)).

Molecular signs are defined here as molecules, molecular complexes, or molecule parts that are used by cells and organisms to encode and control their activities. Some molecular signs are present only temporarily to produce or transfer certain signals [e.g., secondary messengers, such as cyclic adenosine monophosphate (cAMP)]. Other signs are stable and support the organization of signaling networks. Many stable memory signs reside in the DNA sequence and chromatin marks. Tools (instruments) represent another kind of stable signs, which implement the actions of agents. For example, membrane receptors redirect external signals (ligand binding) into internal signaling pathways. Structural elements of the cell (e.g., membranes, cytoskeleton) contribute to the functions at the whole cell level, such as osmosis, movement, and shape change. But they also participate in molecular signaling processes. For example, cytoskeleton fibers serve as transportation highways within cells, and provide mechanical input to tension sensors (Lodish *et al.*, 2000). Resources (e.g., ATP, glucose, and aminoacids) mostly have non-semiotic functions, but they can participate in molecular signaling if their depletion is detected by specific sensors and used as a signal to initiate emergency responses. For instance, bacteria use transporter proteins to bring glucose molecules inside the cell; but in the case of glucose deficiency, idle transporter proteins activate adenylate cyclase, which makes a specific "SOS signal" in the form of cAMP. When cAMP binds to the C-reactive protein (CRP) receptor [also known as catabolite activator protein (CAP)], the CRP acts as a transcription factor and initiates the transcription (i.e., mRNA synthesis) of genes associated with alternative (i.e., non-glucose) metabolic pathways, such as *lac* operon (Lodish *et al.*, 2000). Resource molecules located outside of the cell often play the role of attractants. For example, glucose or aminoacids bind to specific membrane receptors of bacteria and guide the movement of bacterial cells towards higher concentration of these molecules (see below).

3. Molecular Communication as Protosemiosis

Molecular communication differs substantially from other types of communication at the levels of multicellular organisms; however, there is no consensus on the criteria that characterize it. Kull suggested that cell metabolism and other molecular processes belong to the category of "vegetative semiosis" that is based on iconic signs, in contrast to animal semiosis which is based on indexes, and cultural semiosis, which is based on symbols (Kull, 2009). However, terms "vegetative-" and "animal semiosis" are not explanatory because they refer to phylogenetic branches of plants and animals, to which they are not related. Another problem is that the term "icon" is not applicable to intracellular signaling. Icons point to other objects by similarity or quality (Peirce, 1998), but molecular signs are not processed mentally, and thus, neither similarity nor quality is evaluated. Instead, molecular signals directly guide actions of specific cell components without reference to an object.

To describe molecular signaling, Barbieri proposed a new signification model called "code semiosis" (Barbieri, 2003). He defined code as a mapping between a set of signs and a set of meanings. For example, the genetic code maps triplets of nucleotides in the mRNA to aminoacids in the protein. Specific agents called "codemakers" (i.e., ribosomes) convert the encoded message (mRNA) in to its meaning (polypeptide). In addition, another kind of agents, "copymakers" support replication of the DNA or synthesis of mRNA on the DNA template. To describe other sign processes in living organisms, Barbieri developed the concept of "three types of semiosis" (Barbieri, 2009): manufacturing semiosis (protein synthesis), signaling semiosis (signal transduction), and interpretative semiosis, which is associated with making internal representations of the world. The former two types of semiosis are based on coding as opposed to interpretation. The major weakness of the concept of code-based semiosis is that it is focused on the mapping between two sets of elements (codes and meanings), whereas the nature and origin of codemakers, and the functions of products are not discussed.

To characterize simple molecular signaling I use the term "protosemiosis", which was earlier suggested by Prodi (1988). The specific feature of protosemiosis is that signs are associated with actions of agents rather than with objects (Sharov, 2013). In particular, bacteria process molecular signals by initiating or changing actions, and there is no evidence that they can contemplate and operate objects. In contrast, eukaryotic cells have long-lasting complex interactions with such objects as symbionts (mitochondria, chloroplasts), endoplasmic reticulum (ER) vesicles (transport, sorting, and secretion), external particles (endocytosis), and neighboring cells (desmosomes and synapses). Thus, it is conceivable that molecular signaling in eukaryotic cells is not limited to protosemiosis but also has elements of eusemiosis, which is a higher-level type of sign processing that includes internal representations of objects. Eusemiosis does not require the brain, but it is not reducible to one or few molecular signaling events and requires at least minimal categorization capacity in agents (Sharov, 2013; Sharov and Vehkavaara, 2015).

The concept of protosemiosis seem to contradict to the views of Hoffmeyer, Emmeche, and Bruni, who believe that categorization as well as triadic relationships between sign vehicle, object, and interpretant (i.e., the effect of a sign on the interpreter) are universal for all semiotic processes in living organisms (Hoffmeyer and Emmeche, 1991; Bruni, 2008). However, this disagreement mostly reflects terminological differences between authors. For example, Bruni views stereochemical key and lock binding of a ligand to receptor as an example of categorization, where a digital output is produced from the analog input (e.g., spatial pattern). I do not consider this process as categorization because bacteria do not perceive the spatial pattern of a ligand. Instead they use a receptor, which is a tool refined in evolution by genetic selection, to handle the ligand and produce a signal. The notion of categorization is relevant only for those analog to digital transitions, where the analog signal is already internalized, as in the case of visual perception of objects. Hoffmeyer, Emmeche, and Bruni interpret the triadic sign relation as Aristotelian formal and final causations, which are opposed to material and efficient causations. Thus, the absence of objects in a sign relationship appears for them equivalent to a statement that sign effects are mediated by material or efficient causation. However, the presence of objects in a sign relation and the contingent nature of signs are two different properties which do not necessarily coincide. Bacteria have no internal representations of objects but their responses to protosigns are contingent. These responses are not determined by the physical nature of signs. Instead, they are shaped by the agent's semiotic architecture, which is a product of long-term adaptive evolution.

Protosemiosis is mostly based on automatic sign processing, and thus, resembles computation, which is often viewed as a non-semiotic process because computers do not handle the meanings of signs (except of artificial intelligence projects). For example, Barbieri wrote: "A computer contains codes but is not a semiotic system" (Barbieri, 2008: 594). Hoffmeyer considered computers non-semiotic because "these have not (at least yet) been constructed to depend on the creative activity of an analogy coded version interacting with real world processes in such a way as to test the fitness of the digital specifications necessary for its own construction" (Hoffmeyer, 1998: 34). However, automatic sign processing is not limited to computers because a ribosome is a nano scale externally programed robot, which makes polypeptides automatically based on the mRNA sequence. But if ribosomes together with the protein synthesis are not semiotic, then the whole biosemiotics theory appears in jeopardy. Hoffmeyer's suggestion to use self-reproduction as a criterion of semiosis does not solve the problem because this criterion is not universal. For example, mules cannot reproduce but they are alive and fully capable of meaningful sign processing. However, the origin of semiosis and increase of its complexity would require self-replication.

There are several reasons why automatic processing of signs should be viewed as a specific kind of semiosis. First, automatic processing of signs is common in living organisms. Simplicity and speed are the main advantages of automatic responses. Second, prokaryotes are mostly automatic in controlling their functions. If we deny the existence of semiosis in bacteria and archea, then the main thesis of biosemiotics that life and semiosis are coextensive (Anderson *et al.*, 1984) would be wrong. Third, automatic processing of signs does not imply determinism because the responses of automated agents are often checked for errors and adjusted to the environment and internal context within the agent. Although automatic processing of signs does not include learning, it emerged historically via adaptive evolution, which is a learning-like process at the population level (Sharov, 1992). Finally, automatic processing of signs is not reduced to efficient causation because the response is always agent-specific and changes in evolution. A sign can initiate, inhibit, or modify agent activities, but it never fully determines the outcome.

Based on these considerations we have to accept a broader concept of semiotic agency, where agents differ in the degree of their autonomy and automation (Sharov, 2010). Agents can manufacture or recruit slave agents which are then programed to perform specific functions. Ribosomes and computers are examples of such slave semiotic agents which process signs in automated ways. Although man-made machines are not self-reproducing, they evolved together with human technology and should be viewed as a part of the human functional envelope (Sharov, 2010). Thus, automated sign processing in protosemiosis should not be neglected in the biosemiotic theory.

4. Classification of Molecular Proto-Signs Based on Their Immediate Interactions

Proto-icons. Molecular proto-signs can be classified by the mode of their immediate interactions into three major categories: proto-icons, proto-indexes, and proto-symbols (Sharov, 2010; Sharov and Vehkavaara, 2015). These categories are discussed here in the order of increasing complexity, and they appear nested in a sense that each following one includes the previous one as a component. Proto-icons interact with their partners (e.g., receptors, ligands, cofactors) via one particular feature that cannot be subdivided into independent functional subfeatures (Fig. 2). For example, cAMP is a proto-icon relative to its binding to the receptor protein, CRP, in bacteria. Of the two interacting proto-signs, at least one is artificial in a sense that it was fabricated by a living organism, and it emerged in evolution for the purpose of interacting with its partner. Thus, naturally occurring external molecules can become proto-icons only if organisms develop receptors to which they bind. Internal proto-icons are often fabricated via protein modification, when molecular agents label proteins with phosphate, methyl, or ubiquitin molecular marks. These modifications enable

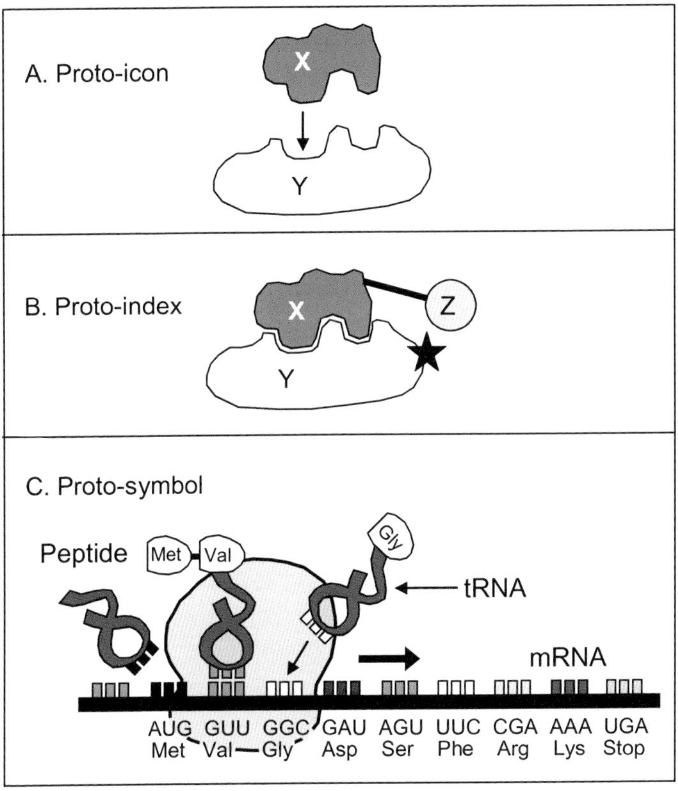

Figure 2. Types of molecular signs in living cells classified by immediate interaction.

the substrate proteins to interact with specific receptors. Some proto-icons are catalytically active and thus can modify the partners to which they bind. However, their abilities to bind and catalyze should be inseparable and supported by the same protein domain, because otherwise the molecule will be a proto-index (see below). For example, cathepsin binds protein at specific motifs of aminoacid residues, and breaks polypeptide bonds after binding. Another example is a type II restriction enzyme EcoRI in bacteria, which cuts DNA at GAATTC palindromic sites (Lodish *et al.*, 2000).

Binding of proto-icons is achieved via specific three-dimensional surfaces that match with the surfaces of their partners in a key and lock manner (Fig. 2). However, this is not a true iconic relation that is based on similarity or quality of objects. Proto-icons are not *recognized*, but are simply *detected* by specialized partners, and their binding surfaces are complementary rather than similar to the surface of partners. The notion of complementarity between shapes of interacting molecules is an oversimplification because of the following. First, the "shape" of binding molecules is more than just a geometric surface. Additional features include electric charges, which allow electrostatic binding if charge signs are opposite at contact points (e.g., hydrogen bonds). Also hydrophilic/hydrophobic components of proteins should match to support proper interactions. For example, two coiled coil domains make a twostranded spiral via interaction of hydrophobic aminoacids that occur regularly at every 4th position (Lodish *et al.*, 2000). Second, binding molecules may be highly dynamic in their shapes during interaction, and thus, the complementarity is virtual, because it is established in the process of interaction. For example, two single-stranded DNA segments ATTCGTTAGG and CCTAACGAAT (in 3′ to 5′ orientation) do not have any shape as the nucleotides rotate freely around their covalent bonds. But if they meet, they spontaneously make a double-stranded rigid DNA helix. The DNA sequence motif is an example of a composite proto-icon which is built as a sequence of elementary units — nucleotides. Composite proto-icons mediate mRNA degradation via micro-RNA binding.

Proto-indexes. Proto-indexes are signs that include two or more functional domains or subunits and, therefore, can combine multiple actions such as binding and catalysis (Fig. 2). For example, type I restriction enzymes in bacteria include three subunits: HsdR is a nuclease which cuts DNA, HsdS binds to a specific DNA motif, and HsdM (methyltransferase) can methylate the DNA molecule (Murray, 2000). Membrane receptors are proto-indexes which combine an external sensing domain for recognition of signal molecules (ligands) with an internal domain that activates secondary messengers. Transcription factor is a proto-index that combines a DNA-binding domain, which binds to regulatory DNA motifs in the promoter of genes, with activation domain, which interacts with RNA polymerase to initiate transcription of a gene. Proto-indexes differ substantially from eusemiotic indexes that are processed mentally. For example, when smoke is used by animals or humans as an index for fire, it is first recognized as an object. Then, the mental representation of smoke induces the representation of fire (and possibly, fear). In contrast, proto-indexes in living cells are not associated with mental representations of objects.

Proto-indexes are essential for the regulation of cellular processes because they can modify catalytic activity in response to specific regulatory signals. As discussed above, actin monomers require binding of ATP as a regulatory signal. A more complex regulation was found in the cAMP-dependent protein kinase A (PKA) in bacteria, which is a tetramer with 2 catalytic subunits and 2 regulatory subunits (Lodish *et al.*, 2000). An extension of a regulatory subunit (called "pseudosubstrate") blocks the activity of the catalytic subunit by binding its active site. Upon binding of cAMP to the regulatory subunit, the pseudosubstrate changes its shape and detaches from the catalytic subunit. Thus, the catalytic subunit is released and starts phosphorylating various target proteins, which in turn cause important changes in cell metabolism.

Proto-indexes can connect otherwise unrelated processes within cells. Such connections comprise the regulatory network that codetermines the meaning of molecular signals. Recently, Hoffmeyer introduced the term "semiotic scaffolding" to describe the network of semiotic controls that are tuned to the needs of the system and support meaningful interpretation of signs (Hoffmeyer, 2007). Semiotic scaffolds extend to the molecular level, where connections are welded via proto-indexes. For example, P300 protein is an important component of enhancers that control the expression of nearby genes (Visel *et al.*, 2009). It can bind to various transcription factors, including steroid receptors, cAMP response element-binding protein (CREB) and myeloblastosis (MYB) (via KIX domain), p53 (via KIX), interferon receptors (via zinc-finger domain) (Vo and Goodman, 2001). In addition, P300 can acetylate histones and open the chromatin, which makes DNA accessible to various transcription factors. Obviously, P300 is a hub in the cellular signaling network that integrates multiple incoming signals, although many aspects of its function are still unknown.

Proto-symbols. Proto-symbols are members of a family of similarly structured molecular signs that are processed uniformly by the same subagents and the same set of adapters. For example, mRNA sequences are processed by ribosomes interacting with a set of aminoacyl-tRNA adapters in order to make a protein. The order of aminoacids in the protein is specified by the sequence of nucleotides in the mRNA according to a simple rule known as "genetic code" (Lodish *et al.*, 2000). Each triplet of nucleotides corresponds to one of 20 possible aminoacids; in addition some triplets signify the start and end of protein synthesis. However, ribosome and aminoacyl-tRNA are not the only subagents that are involved in the processing of mRNA sequences. Even before mRNA molecules leave the nucleus, they are processed by spliceosomes which cut out segments of mRNA called "introns", which usually start with GU and end with AG nucleotides. But splicing is not deterministic and can be modified on demand by regulatory factors. Thus, many genes have multiple alternative splicing patterns, and therefore, can produce different proteins from the same gene.

The family of transcription factors that control the activity of various genes can also be viewed as proto-symbols, whose action (i.e., activation of target genes) is mediated by a common agent, RNA polymerase, and a set of adaptors represented by promoters of genes that carry specific DNA binding motifs recognized by transcription factors. In contrast to protein synthesis, where the set of adaptors is nearly universal in the whole tree of life, promoter sequences are conserved only within evolutionarily related species of organisms. Thus, transcription factors may not work properly if introduced into cells of unrelated species. By changing adapters/promoters evolving species can redirect functions to new organs or new functional pathways.

The relationship between proto-symbols and their products can be called "conventional" because it is based on a set of contingent adapters, which is preserved as a heritable accident in evolution. This conventional element makes proto-symbols similar to real symbols used in human communication. However, in contrast to eusemiotic symbols, the processing of proto-symbols is based on a "heritable convention" rather than a socially mediated convention. Another similarity between genes and human symbols is that both are components of a language system with some kind of generative grammar (García, 2005). Based on genetic "grammar", a new field of synthetic biology attempts to construct novel DNA sequences which upon transcription and translation generate proteins with given functional properties (Arpino *et al.*, 2013). However, genetic language is a proto-language because it lacks several important properties of cognitive languages. First, cognitive language is learned whereas genetic proto-language is inherited. Second, genetic sequences encode only a very narrow set of objects (i.e., proteins) whereas human language allows description of any object. And third, genetic proto-language is conservative as compared to dynamic human languages, which can easily incorporate new words, sentences, and texts. Of course, genetic sequences also change in evolution, but this process is slow and not intentional as compared to the change of human speech.

Contemporary organisms use all three kinds of proto-signs for their molecular communication, but primordial living systems had no proto-symbols according to most scenarios of life origin. For example, the RNA world scenario assumes that life started from self-replicating RNA molecules and did not include protein synthesis (Schuster, 1993). After spontaneous self-folding, RNA acquired catalytic properties and was able to support metabolism and replication of other RNA molecules. Another scenario of the origin of life starts with even more simple coenzyme-like molecules, which inhabited oil microspheres in water and encoded surface properties of these microspheres (Sharov, 2009). Such systems could evolve by accumulating additional heritable and functional coenzyme-like molecules, which eventually became organized into RNA-like polymers with template-based replication. Interestingly, both scenarios *require proto-indexes* to interconnect otherwise separate processes. For example, coenzyme-like molecules had to combine a catalytic part capable of oxidizing hydrocarbons (alkanes) with a binding part to anchor the molecule to the surface of oil microsphere. These molecules may also need some components to facilitate their autocatalytic self-assembly. Similarly, in the RNA world scenario, RNA molecules support several kinds of interactions. They serve as a template for replication, guide a unique way of self-folding, and finally play the role of catalysts. Proto-icons alone cannot make a functional living system because they have no capacity to organize a semiotic scaffold.

5. Functional Classification of Molecular Signaling Processes

In this section, I consider functions of molecular sign processes, which extend beyond the immediate interaction of signs with their partners. Thus, I describe them as signaling pathways/channels or communication acts. Each communication act transfers an information/signal from one molecular agent to another and the receiver responds with some action or modification of its previous activity. Neither actions nor their products are signs by definition. Here I disagree with Uexküll who considered actions as "effector signs" (Uexküll, 1957). However, some actions or their products may be signs for agents if they control certain downstream functions.

Fabrication. Fabrication of cell components is an essential function of signaling pathways because it is necessary for the autopoietic self-renewal of cells. The most sophisticated method of molecule fabrication is protein synthesis guided by genetic information. First, the RNA polymerase complex makes an mRNA copy of a gene which is spliced by spliceosomes and exported into the cytoplasm. Then, ribosomes synthesize the protein using the mRNA sequence as a guide. Subsequent self-folding of the polypeptides/proteins is controlled by interactions of aminoacids via hydrogen bonds, electrostatic charge, and hydrophilic affinity. Correct protein folding may require interaction with chaperones, such as the GroEL/GroES bacterial complex (Fenton and Horwich, 2003). Each component of a cell should be fabricated by other components, thus enabling the full capacity for self-repair, growth, and self-reproduction (Rosen, 1991). Fabrication of cell components is always under semiotic control, which is especially important for copying memory molecules such as DNA. Recursive fabrication of programs that are needed in the next round of fabrication of all components in self-reproducing agents was called *semiotic closure* (Pattee, 2007).

Newly fabricated proteins execute or control cellular functions, including modification of proteins and synthesis of various other molecules (e.g., aminoacids, carbohydrates, nucleotides, lipids). Thus, protein synthesis is indirectly involved in all other kinds of fabrication methods. For example, protein modification is catalyzed by kinases, methyltransferases, and other catalysts which are all proteins and have to be synthesized before their function is required. Similarly, the assembly of cytoskeleton or chromatin requires synthesis of protein components that can later be assembled into actin fibers, microtubules, or nucleosomes. Fabricated proteins often self-assemble into larger complexes and organelles, following mutual binding affinities (proto-icons) as well as sequential signal transfer via proto-index relations (e.g., signal propagation during polymerization of actin). Construction of protein complexes and other large structures is often guided by additional agents. For example, transcription factors bind to DNA and recruit various catalytic complexes to remodel the chromatin structure or establish long distance links between DNA loops.

Sensing. Sensing is another common type of signaling channel in the cell which is needed to monitor external and internal conditions. Signs that invoke sensing are called "perceptual signs" (Uexküll, 1957). They are processed by receptors (or sensors), whose action results in the construction of the internal sign. For example, the presence of interferon (IFNγ signaling protein outside of the cell is detected by a specific transmembrane receptor which aggregates into a dimer and phosphorylates the signal transducer and activator of transcription (STAT) factor at the internal side of the receptor (Lodish *et al.*, 2000). STAT becomes active after phosphorylation: it makes dimers that move into the nucleus and activate the transcription of target genes. The majority of perceptual signs are produced by agents for signaling purposes, and in this case they belong to the category of "communicative signs" (Clarke, 1987). For example, (IFNγ is secreted by T-helper cells to stimulate the immune response during viral infection.

Some perceptual signs are not produced for signaling, and these are called "natural" signs (Clarke, 1987). However the term "natural" does not imply that signs exist independently from agents; a molecule becomes a sign only if agents develop a receptor to detect it. Many natural signs are small molecules (e.g., acetic acid, sugars, aminoacids), but in addition, cells can detect electrical gradients, light, and temperature. Electrical gradients can modulate voltage-gated channels, whereas light is sensed by photoreceptor proteins (e.g., rhodopsin). Temperature can be sensed by a variety of proteins and mRNA due to temperature-dependent change of their spatial conformation.

Memory. All living organisms require memory to store information that encodes and controls their functions. Because organisms are perishable, the only way to preserve their functional organization is to encode functions in a memory carrier (e.g., DNA) and transfer it to the following generation by copying. However, most molecular agents within living cells (e.g., proteins, ribosomes) do not carry information on methods of their own construction because they are manufactured by other agents. These agents carry only a few memory signs that are sufficient for operational tasks. This kind of "operational memory" is often based on reversible or irreversible protein modifications, such as phosphorylation, acetylation, methylation, ubiquitination, and sumoylation. Some receptors are equipped with operational memory elements with long decay rates, which help to measure the change in signal intensity. For example, stimulation of the bacterial receptor called methyl-accepting chemotaxis protein (MCP) by glucose increases the level of methylation of glutamate residues in the MCP (Derr *et al.*, 2006). The methylation level of MCP is maintained in a dynamic equilibrium and represents the *change* in concentration of glucose over time, rather than the actual concentration. This feature is important for navigation of bacteria towards higher concentration of glucose (chemotaxis).

Nucleic acids (e.g., DNA and mRNA) play a special role in living cells because they can store large amounts of functional information that is needed to produce and regulate other molecular agents. The large information capacity of nucleic acids is based on their aperiodic polymer structure, where nucleotides function as characters combined into linear genetic messages. With every cell division, DNA molecules have to be duplicated and copies are distributed between daughter cells. But the DNA molecule is a passive information resource rather than an agent; thus, other molecular agents have to maintain and manipulate it. DNA is duplicated by DNA polymerase, it is rotated by helicases, and knots are resolved by topoisomerases. In eukaryotes, DNA is packaged into nucleosomes, which include eight histone proteins wrapped by approximately 200 base pairs of DNA (Fig. 3). DNA in combination with histones is called chromatin. A large number of multiprotein agents maintain the chromatin structure within each cell. They assemble or disassemble nucleosomes, shift nucleosomes along the DNA sequence, and modify the packaging of nucleosomes, making it either loose or tight. Without supporting agents, the DNA cannot function as a memory molecule, and the sequence of nucleotides would have no meaning.

The major limitation of the DNA-based memory is that DNA molecules cannot be efficiently edited. Some editing is possible in immunoglobin genes and in transcribed mRNA (e.g., splicing and nucleotide substitutions). However, the utility of these mechanisms is limited. Thus, an additional layer of rewritable memory emerged in eukaryotic cells in the form of chromatin. Because many features of chromatin structure are not dependent on the underlying DNA sequence, it is called "epigenetic memory". Histone proteins, which are major components of the chromatin, carry long tails, where specific aminoacids can be modified in

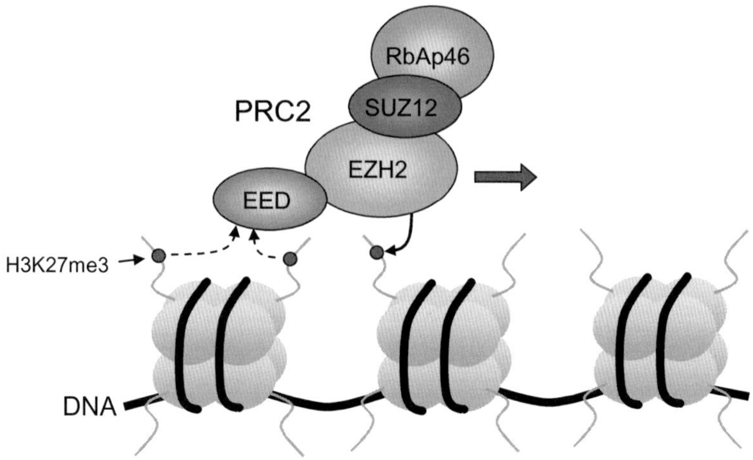

Figure 3. PRC2 complex deposits H3K27me3 repressive chromatin marks. EED protein brings PRC2 complex to existing H3K27me3 marks, and methyltransferase EZH2 deposits new repressive marks.

various ways (methylation, acetylation, phosphorylation, and ubiquitination) by chromatin remodeling agents (Fig. 3). Depending on the type of histone modification, the nucleosomes may either condense tightly into a closed chromatin structure or become separated to allow access of transcription factors and RNA polymerase to the exposed DNA.

Most chromatin modifications are regulated via positive feedback, which supports the stability and propagation of certain chromatin states. For example, tri-methylation of lysine at position 27 in histone protein H3 (shortly, H3K27me3) is performed by the EZH2 methyltransferase, which is a component of the polycomb repressive complex PRC2 (Sarma *et al.*, 2008) (Fig. 3). This histone modification makes the chromatin condense and the DNA silent. Another component of PRC2 is embryonic ectoderm development (EED) protein, which binds to H3K27me3 chromatin marks, recruiting the PRC2 complex to histones with this mark (Margueron *et al.*, 2009). This process supports the stability of H3K27me3 chromatin marks and even its propagation along the chromosome (Fig. 3). Insulator proteins [e.g., CCCTC-binding factor (CTCF)] bind to DNA locations with specific short motifs of nucleotides and function as barriers to the linear propagation of chromatin marks. When DNA is duplicated, the chromatin marks are restored in both copies, so that epigenetic memory is preserved after cell division. One of the major functions of chromatin structure in animals and plants is to regulate the expression of genes that are specific for each cell type. Thus, different cell types carry different chromatin profiles, which are established and maintained by a set of transcription factors and chromatin remodeling factors.

Movement. Movement of cell components is essential for normal functioning of living cells. Besides passive diffusion, cell components participate in a variety of active movements controlled by signs. For example, transport proteins recognize specific substrates and carry them across the cell membrane. Sodium-glucose transporters move both glucose and sodium in the same direction (inside the cell) powered by the increased sodium concentration outside (Lodish *et al.*, 2000). Bacterial cells are propelled in water by rotating flagellae, which are external protein structures anchored in the plasma membrane. The movement is caused by the flow of protons (H^+) through the molecular motor. Similar motors exist in mitochondrial membranes and are used for making ATP. Complex movements are needed to maintain the integrity of DNA: for example, helicases can unwind supercoiled DNA, whereas topoisomerases can pass one DNA helix through another. Eukaryotic cells developed a variety of linear motors that move along cytoskeleton fibers. For example, muscle contraction is powered by myosin "thick" filaments, which slide along "thin" actin/tropomyosin filaments (Lodish *et al.*, 2000). Dynein is a motor protein which controls the movement of microtubules, especially during cell division (mitosis). Cytoplasmic dynein in animal cells can transport vesicles (ER, endosomes, and lysosomes) along microtubules.

Transport of proteins is controlled by signal peptides which are mostly located at the *N*-terminal (Lodish *et al.*, 2000). The signal peptide of most secretory and membrane-bound proteins is recognized by the signal recognition particle (SRP) receptor located on the ER (membrane) while the protein is still being synthesized by the ribosome. As a result, the whole ribosome is drawn towards the membrane and continues the synthesis of the protein through membrane. Secretory proteins are then transported to the Golgi for further processing and sorting. Additional signals indicate the final destination of proteins. For example, the presence of mannose-6-phosphate targets proteins to the lysosome (Lodish *et al.*, 2000).

Wave propagation. Transport of large molecules is generally a slow process. Faster signaling requires a different kind of communication based on the domino effect, when a signal travels through an array of linked transmission agents without movement of agents themselves. Such systems are materialized in biological evolution in the form of membrane bound ion channels and pumps, which are most common in neural cells. The membrane serves as a scaffold that incorporates agent proteins and as a boundary that separates the inner and outer space for cells or cell components. This separation allows the cell to keep a gradient of ion concentration and electric charge. For example, action potential in nerves travels through the sequential opening of voltage-gated sodium membrane channels, which causes the depolarization of the membrane, and this change opens neighboring sodium channels (Lodish *et al.*, 2000) (Fig. 4). The voltage gradient at the cell membrane is generated by sodium–potassium pumps, which move Na^+ out of the cell and K^+ into the cell. Because resting potassium channels allow K^+ to move freely out of the cell, the outer surface of the membrane becomes positively charged and the inner surface becomes negatively charged. If the membrane potential decreases to a certain threshold, the voltage-gated sodium channels open and Na^+ is allowed to enter the cell. The depolarization wave rapidly spreads along the membrane by opening nearby voltage-gated sodium channels. In neurons, these events generate action potentials which are electrical spikes that travel along the axon. When voltage-gated channels close and the membrane potential is restored, the cell is ready to generate another wave.

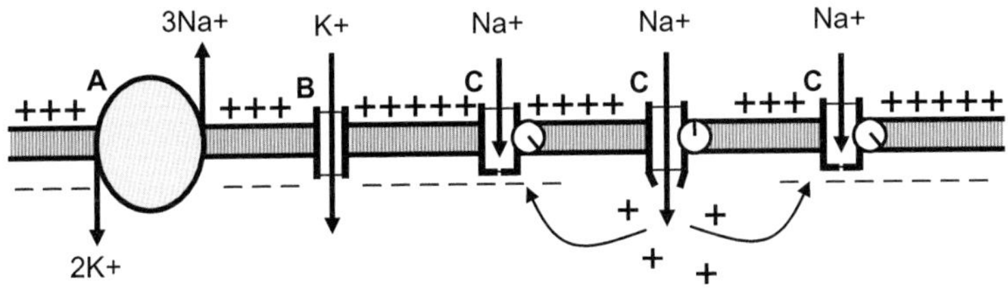

Figure 4. Wave propagation in the neural membrane: A — sodium–potassium pump; B — resting potassium channel; C — voltage-gated sodium channel.

In various receptor cells, the nerve impulse is initiated as follows. Receptor proteins (e.g., olfactory receptor or photoreceptor) activate G-proteins, which in turn open gated ion channels and initiate the action potential. There are various mechanisms of initial ion channel opening, including cAMP synthesis and phosphorylation of channel proteins by PKA (Lodish *et al.*, 2000). Muscle cells that receive a neural impulse respond by opening the voltage-gated calcium channels that allow extracellular Ca^{++} to enter the cell and initiate muscle contraction.

6. Error Control, Robustness, and Modularity in Molecular Communication

Signaling processes in living cells not only support immediate functions of cells (see previous sections) but also satisfy system-level goals such as error control, robustness, modularity, and defense. The whole signaling network should be prepared for various kinds of disruptions of both internal and external origin. Internal disturbances include errors in DNA replication, protein synthesis, or evolutionary changes in cells and organs. External disturbances include environmentally caused disruptions of normal functions, as well as signals from inimical agents, which may exploit the molecular signaling pathways of the host for their own benefits. If cell communication cannot withstand these disturbances, then various damages occur from wasted energy to the irreversible degradation and death. The major difference between molecular signaling and computation is that computers are assumed to have no errors or disturbances. In contrast, living organisms make errors in sign processing and experience evolutionary changes and semiotic attacks from inimical agents. Thus, life can persist only via error correction, compensation, robustness, and protection from foreign signals. Because defense mechanisms against infection (immune system) are versatile and complex, here for brevity I discuss only strategies for error correction, robustness, and modularity.

Repair or destruction of non-functional molecules is widespread in molecular signaling networks. For example, template-based DNA (or RNA) synthesis includes immediate proofreading. If a wrong base is attached, then the exonuclease domain cleaves the erroneous base and polymerization continues (Lodish *et al.*, 2000). Similar proofreading is performed by ribosomes during protein synthesis. Mismatches in DNA nucleotide pairs are recognized by MutS proteins, and if found, a segment of a newly synthesized DNA strand is removed by exonuclease and repaired by the DNA polymerase (Lodish *et al.*, 2000). Errors in mRNA are detected if a ribosome comes across a premature stop codon while moving along the mRNA. If a stop codon is encountered before an exon junction protein complex, which is normally located in the coding region, then the mRNA is tagged for degradation (Chang *et al.*, 2007). Misfolded proteins in the cell are recognized by molecular chaperones (also known as heat shock proteins) based on exposed hydrophobic domains or unpaired cytosine residues; chaperones then assist in refolding of proteins (Lodish *et al.*, 2000). However, if the native state of a protein cannot be restored, misfolded proteins are targeted for degradation via ubiquitin — proteasome system. First, these proteins are tagged with a short polypeptide chain called "ubiquitin", and then they are disassembled by proteases in specific proteasome organelles.

Robustness of molecular communication can be supported by the redundancy of signaling pathways integrated via logical gates. For example, each kind of molecular receptor exists in multiple copies scattered on the outer membrane or in the cytoplasm, but activation of a single receptor is usually not sufficient to generate the response of a cell. This is a safety mechanism which interrupts signaling initiated by accidental receptor firing. Only massive activation of receptors is "meaningful" and induces cell actions. The threshold for initiating cell actions can be adjusted by changing either the rate of degradation of signaling molecules or the abundance of antagonists that have to be saturated first in order to overcome their inhibiting effect. For example, the signaling function of beta-catenin in embryogenesis is disrupted by GSK3 kinase which phosphorylates beta-catenin and targets it for degradation. Thus, beta-catenin has to be highly abundant in the cytoplasm to overcome the antagonistic effect of GSK3 (Metcalfe and Bienz, 2011).

Redundant signaling can support robustness in the regulation of transcription. For example, the pluripotent state of mammalian embryonic stem cells is maintained by several transcription factors, including Oct4, Sox2, and Nanog, which bind jointly to the promoters and enhancers of pluripotency-related genes (including Oct4, Sox2, and Nanog) and activate their transcription (Chen *et al.*, 2008). Thus, the pluripotent state is stabilized via autocatalytic activation of these transcription factors. However, differentiated cells are protected from transforming into pluripotent cells, which may cause cancer. Accidental activation of any single transcription factor associated with pluripotency (e.g., Oct4) has no consequences, because it cannot function alone without partners. Another strategy for increasing the robustness of molecular signaling is based on the use of oscillating signals. For example, oscillating Ca^{+2} ion signals reduce the effective threshold for activating transcription factors, and the frequency determines which genes are activated (Dolmetsch *et al.*, 1998).

Molecular signaling networks often have modular organization where each module communicates with others only through designated inputs and outputs. Other internal characteristics of the module are protected from external disturbance (e.g., via isolation or compensation). For example, mechanisms of signal transduction are partitioned into several modules such as mitogen-activated protein kinases (MAPK), phospho-inositide 3-kinase (PI3K), Janus kinase (JAK/STAT), cAMP, and G-protein pathways (Lodish *et al.*, 2000).

These modules, however, interact through designated signals. The cell cycle of eukaryotic cells consists of five major modules or phases: G0 (resting), G1 (cell growth), S (DNA replication), G2 (transition to mitosis), and M (mitosis) (Lodish *et al.*, 2000). The M-phase is additionally separated into prophase, metaphase, anaphase, telophase, and cytokinesis. These modules are mutually exclusive, and become activated in a strictly defined order. The transitions between them, which are called "checkpoints", include specific inputs and outputs. Most well-studied are three checkpoints: the G1 checkpoint controls the transition to the S-phase, and it is needed to prevent uncontrolled proliferation of cells; the G2 checkpoint ensures that DNA synthesis is completed and damaged DNA repaired, and the metaphase checkpoint ensures that chromosomes are aligned at the center and are ready for separation. By changing the conditions at checkpoints, the cell cycle can be adjusted to a specific cell type. For example, the G1 checkpoint is bypassed in early embryos, which allows cells to divide without growth. In contrast, this checkpoint is impassable in most neural cells, which do not divide after maturation.

Higher-level functions of molecular communication networks often require integration of a large number of heterogeneous signals. For example, the G2 checkpoint integrates signals from proteins that monitor the replication of DNA, repair of damaged DNA, viral infection, and apoptosis. Possible outputs include allowing mitosis, cell cycle delay to complete DNA replication or repair, or apoptosis (Lodish *et al.*, 2000). The complexity of regulation indicates that this checkpoint can be considered as a primitive categorization unit at the cellular level. The output of the checkpoint is a sign of a specific state of the cell, and this state is an object. Therefore, the G2 checkpoint has elements of the advanced level of semiosis, or eusemiosis. Other terms have been suggested previously to describe categorization at the molecular level. Bruni proposed the term "categorical sensing", which was derived from "categorical perception" (Bruni, 2008). The latter term means "the ability to slot a bewildering number of impressions into categories" (Hoffmeyer, 1996: 77). Another term related to eusemiosis is "digital-analogical consensus", which was defined as a "recurrent pattern for the creation of complex logical products that constitute specific signs" (Bruni, 2008: 368). This categorization mechanism supports context dependency and adaptability even in novel conditions that did not occur previously. But in identical initial conditions, the output would be reproducible and heritable. Only evolutionary mechanisms can rewire molecular categorization networks. However, there is a poorly explored possibility that molecular communications within cells can be rewired by adaptive individual learning even in organisms without neural systems (Ginsburg and Jablonka, 2009; Sharov, 2013). Because learning requires a rewritable memory to accumulate life experience, learning cellular mechanisms (if they indeed exist) should be integrated with the epigenetic system which offers an enormous capacity of dynamic memory storage.

In summary, molecular communication networks are essential for the functionality of living cells. Biochemical analysis of molecular interactions should be supplemented by biosemiotic studies of molecular networks to explore their effects on various cell functions, including context-dependent effects. The complexity of these networks varies from most primitive signaling (protosemiosis) to complex categorization mechanisms (eusemiosis). Protosemiosis started from the origin of life and its functions gradually expanded. Eventually, increased redundancy of signaling pathways and their integration via information processing hubs such as cell cycle checkpoints have resulted in the emergence of eusemiosis.

Acknowledgments

I would like to thank anonymous reviewers for valuable comments on the manuscript. This project was supported by the Intramural Research Program of the National Institute on Aging (NIA/NIH).

References

Anderson M, Deely J, Krampen M, Ransdell J, Sebeok TA and Uexküll Tv. 1984. A semiotic perspective on the sciences: steps toward a new paradigm. *Semiotica*, 52(1–2), 7–47.

Arpino JA, Hancock EJ, Anderson J, Barahona M, Stan GB, Papachristodoulou A and Polizzi K. 2013. Tuning the dials of synthetic biology. *Microbiology*, 159(7), 1236–1253.

Barbieri M. 2003. *The Organic Codes: an Introduction to Semantic Biology*. Cambridge, UK, Cambridge University Press, New York.

Barbieri M. 2008. Biosemiotics: a new understanding of life. *Naturwissenschaften*, 95(7), 577–599.

Barbieri M. 2009. Three types of semiosis. *Biosemiotics*, 2(1), 19–30.

Bateson G. 1972. *Steps to an Ecology of Mind; Collected Essays in Anthropology, Psychiatry, Evolution, and Epistemology*. Chandler Pub. Co., San Francisco.

Brier S. 2008. *Cybersemiotics*. University of Toronto Press, Toronto, Buffalo, London.

Brillouin L. 1953. Negentropy principle of information. *Journal of Applied Physics*, 24(9), 1152–1163.

Bruni LE. 2008. Cellular semiotics and signal transduction. In *Introduction to Biosemiotics. The New Biological Synthesis*, Barbieri M (ed.), 365–407. Springer, Dordrecht.

Chaisson EJ. 2005. Non-equilibrium thermodynamics in an energy-rich universe. In *Non-Equilibrium Thermodynamics and the Production of Entropy: Life, Earth, and Beyond* Kleidon A and Lorenz RD (eds.), 21–31. Springer, Berlin, Heidelberg.

Chang YF, Imam JS and Wilkinson MF. 2007. The nonsense-mediated decay RNA surveillance pathway. *Annual Review of Biochemistry*, 76, 51–74.

Chen X, Xu H, Yuan P, Fang F, Huss M, Vega VB, Wong E, Orlov YL, Zhang W, Jiang J, Loh YH, Yeo HC, Yeo ZX, Narang V, Govindarajan KR, Leong B, Shahab A, Ruan Y, Bourque G, Sung WK, Clarke ND, Wei CL and Ng HH. 2008. Integration of external signaling pathways with the core transcriptional network in embryonic stem cells. *Cell*, 133(6), 1106–1117.

Clarke DS. 1987. *Principles of Semiotic*. Routledge & Kegan Paul, London; New York.

Deely JN. 1994. *New Beginnings: Early Modern Philosophy and Postmodern Thought. Toronto Studies in Semiotics*. University of Toronto Press, Toronto.

Derr P, Boder E and Goulian M. 2006. Changing the specificity of a bacterial chemoreceptor. *Journal of Molecular Biology*, 355(5), 923–932.

Dolmetsch RE, Xu K and Lewis RS. 1998. Calcium oscillations increase the efficiency and specificity of gene expression. *Nature*, 392(6679), 933–936.

Emmeche C. 1999. The sarkar challenge to Biosemiotics: Is there any information in a cell? *Semiotica*, 127(1–4), 273–293.

Fenton WA and Horwich AL. 2003. Chaperonin-mediated protein folding: fate of substrate polypeptide. *Quarterly Review of Biophysics*, 36(2), 229–256.

García ÁL. 2005. *The Grammar of Genes: How the Genetic Code Remsembles the Linguistic Code*. Bern, Germany: Peter Lang, Bern, Germany.

Ginsburg S and Jablonka E. 2009. Epigenetic learning in non-neural organisms. *Journal of Bioscience*, 34(4), 633–646.

Green HS. 2000. *Information Theory and Quantum Physics: Physical Foundations for Understanding the Conscious Process (Theoretical and Mathematical Physics)*. Springer, Berlin, Heidelberg, New York.

Hoffmeyer J. 1996. *Signs of Meaning in the Universe. The Natural History of Signification*. Indiana University Press, Blomington.

Hoffmeyer J. 1998. Surfaces inside surfaces. *Cybernetics and Human Knowing*, 5(1), 33–42.

Hoffmeyer J. 2007. Semiotic scaffolding of living systems. In *Introduction to Biosemiotics. The New Biological Synthesis*, Barbieri M (ed.), 149–166. Springer, Dordrecht.

Hoffmeyer J. 2008. *Biosemiotics: An Examination into the Signs of Life and the Life of Signs*. University of Scranton Press, Scranton, PA.

Hoffmeyer J and Emmeche C. 1991. Code-duality and the semiotics of nature. In *On semiotic modeling*, Anderson M and Merrell F (ed.), 117–166. Mouton de Gruyter, Berlin, New York.

Ji S. 1991. The Shillongator. In *Molecular Theories of Cell Life and Death*, Ji S (ed.), 157–163. Rutgers University Press, New Brunswick.

Ji S. 2012. *Molecular Theory of the Living Cell: Concepts, Molecular Mechanisms, and Biomedical Applications*. Springer, New York.

Kull K. 2009. Vegetative, animal, and cultural semiosis: the semiotic threshold zones. *Cognitive Semiotics*, 4, 8–27.

Lodish H, Berk A, Zipursky SL, Matsudaira P and Darnell J. 2000. *Molecular Cell Biology*, 4th edition. W. H. Freeman and Co., New York.

Margueron R, Justin N, Ohno K, Sharpe ML, Son J, Drury WJ 3rd, Voigt P, Martin SR, Taylor WR, De Marco V, Pirrotta V, Reinberg D and Gamblin SJ. 2009. Role of the polycomb protein EED in the propagation of repressive histone marks. *Nature*, 461(7265), 762–767.

Merrell F. 1996. *Signs Grow: Semiosis and Life Processes*. University of Toronto Press, Toronto, London.

Metcalfe C and Bienz M. 2011. Inhibition of GSK3 by Wnt signaling — two contrasting models. *Journal of Cellular Science*, 124(21), 3537–3544.

Murray NE. 2000. Type I restriction systems: sophisticated molecular machines (a legacy of Bertani and Weigle). *Microbiology and Molecular Biology Reviews*, 64(2), 412–434.

Pattee HH. 1969. How does a molecule become a message? *Developmental Biology Supplement*, 3, 1–16.

Pattee HH. 2007. The necessity of biosemiotics: matter-symbol complementarity. In *Introduction to Biosemiotics*, Barbieri M (ed.), 115–132. Springer, Dordrecht, The Netherlands.

Peirce CS. 1931. *Collected Papers of Charles Sanders Peirce*. Harvard University Press, Cambridge.

Peirce CS. 1986. *Writings of Charles S. Peirce: A Chronological Edition*, Fisch MH and CJW Kloesel (eds.). Indiana University Press, Bloomington.

Peirce CS. 1998. *The Essential Peirce: Selected Philosophical Writings*, Project PE (ed.). Indiana University Press, Indiana.

Prodi G. 1988. Material bases of signification. *Semiotica*, 69(3–4), 191–241.

Rosen R. 1991. *Life Itself: A Comprehensive Inquiry into the Nature, Origin, and Fabrication of Life. Complexity in Ecological Systems Series*. Columbia University Press, New York.

Sarma K, Margueron R, Ivanov A, Pirrotta V and Reinberg D. 2008. Ezh2 requires PHF1 to efficiently catalyze H3 lysine 27 trimethylation *in vivo*. *Molecular and Cellular Biology*, 28(8), 2718–2731.

Schuster P. 1993. RNA based evolutionary optimization. *Origins of Life and Evolution of Biospheres*, 23(5–6), 373–391.

Sharov AA. 1992. Biosemiotics: functional-evolutionary approach to the problem of the sense of information. In *Biosemiotics. The Semiotic Web 1991*, Sebeok TA and Umiker-Sebeok J (eds.), 345–373. Mouton de Gruyter, New York.

Sharov AA. 2009. Coenzyme autocatalytic network on the surface of oil microspheres as a model for the origin of life. *International Journal of Molecular Sciences*, 10(4), 1838–1852.

Sharov AA. 2010. Functional information: towards synthesis of biosemiotics and cybernetics. *Entropy*, 12(5), 1050–1070.

Sharov AA. 2013. Minimal mind. In *Origins of Mind*, Swan L (ed.), 343–360. Springer, Dordrecht.

Sharov AA and Vehkavaara T. 2015. Protosemiosis: agency with reduced representation capacity. *Biosemiotics*, 8(1), 103–123.

Uexküll Jv. 1957. A stroll through the worlds of animals and men: a picture book of invisible worlds. In *Instinctive Behavior: The Development of a Modern Concept*, Schiller CH (ed.), 5–80. International Universities Press, Inc., New York.

Vehkavaara T. 2003. Natural self-interest, interactive representation, and the emergence of objects and Umwelt. *Sign Systems Studies*, 31(2), 547–587.

Visel A, Blow MJ, Li Z, Zhang T, Akiyama JA, Holt A, Plajzer-Frick I, Shoukry M, Wright C, Chen F, Afzal V, Ren B, Rubin EM and Pennacchio LA. 2009. ChIP-seq accurately predicts tissue-specific activity of enhancers. *Nature*, 457(7231), 854–858.

Vo N and Goodman RH. 2001. CREB-binding protein and p300 in transcriptional regulation. *The Journal of Biological Chemistry*, 276(17), 13505–13508.

2. Key Levels of Biocommunication

Guenther Witzany

Telos-Philosophische Praxis, Vogelsangstrasse 18c
5111-Buermoos, Austria
witzany@sbg.at

Organisms actively compete for environmental resources. They assess their surroundings, estimate how much energy they need for particular goals, and then realize the optimum variant. They take measures to control certain environmental resources. They perceive themselves and can distinguish between "self" and "non-self." Current empirical data on all domains of life indicate that unicellular organisms such as bacteria, archaea, giant viruses, and protozoa as well as multicellular organisms such as animals, fungi, and plants coordinate and organize their essential life functions through signaling processes. Signaling allows for real life coordination and organization and is a communicative action in which species-specific behavioral patterns and sign repertoires are used. Cells, tissues, organs, and organisms that communicate share several key levels that are essential to all life forms and which serve as a uniform tool for investigating biocommunication. Complementary to this, active biocommunication depends on the deoxyribonucleic acid (DNA) storage medium and the agents that generate coherent content of nucleic acid sequences. Therefore natural genome editing is not the result of replication errors but of group interactions of competent ribonucleic acid (RNA) agents.

1. Introduction

For centuries communication has been limited to humans, whereas non-human living organisms have been assumed to function mechanistically as machine-like stimulus reaction automatons determined by genetic programs. This mechanistic perspective has also dominated the description and explanation of human behaviorism within various sender receiver narratives: first a sender has a thought, and then it is coded into signs/codes and transmitted via a channel to a receiver who decodes it in accordance with his/her private value system. Similar explanatory models have dominated the mathematical theory of language, systems theory, biolinguistics, bioinformatics, and biosemiotics (Popov *et al.*, 1996; Nowak and Krakauer 1999; Nowak *et al.*, 2000; Nowak *et al.*, 2001, 2002; Eigen and Winkler 1983, Eigen, 2013; Searls, 2002; Chomsky, 1964; Ji, 1999; Barbieri, 2001). The focus of these concepts varies from physical/chemical properties of signs and transfer channels (biolinguistics, biosemiotics), to codes (biology), to systems of higher and lower orders, to material reality depicted as the ideal in mathematical formalizations (bioinformatics). Because organisms, especially animals with central nervous systems and brain organs, are built of chemical molecules that follow physical laws, the best reality depicting language is mathematics, in which brain architecture and its function are represented (Shannon and Weaver 1949; Eigen and Winkler, 1983). Theories that propose other entities than atoms and molecules as essential tools of life are guilty of "vitalism"-like narratives (Venter, 2013). Therefore, we must investigate all systems in which a universal syntax — that represents the logical order of material reality — can be found at the root of all languages. Logically, computation methods are the best tools for research, because organisms can be said to depict hard drives and DNA the software (Venter, 2013).

This 20th century narrative is outdated, however (Witzany, 2012a, 2012b; Witzany and Baluška, 2012a, 2012b; Baluška and Witzany, 2013). Empirical data indicate communication is primarily a social interaction using languages/codes, i.e., social interactions between living agents that share a real life world's traditions and environmental circumstances (Tomasello, 2008). Experience and the organization, and coordination of everyday real life practice determine communication patterns (Habermas, 1984, 1987, 1994). Language serves as a tool for exchanging relevant contents, i.e., meanings. Most importantly, however, the social interaction of communication is a real action — intended to convince, to persuade, to obey, to share social experience, feelings and thoughts, reciprocal or imperative (Austin, 1975; Searle, 1976). Neurobiological research of the last decades demonstrated that the social brain organs of children learn through observations and mimicry, and consequent feedback from adult group members (Tomasello, 2008). The early experience of coherent complementarity between cognition and emotion in real life situations is essential to these basic experiences and cannot be predicted or computed by mathematical algorithms. During socialization, children not only learn how to use a small vocabulary to express their feelings but also build sentences to coordinate with speech acts. Through observations and mimicry they learn to combine words with correct meanings; they do things with words. Later in their development, group members who share communicative and linguistic competence describe observations through constative speech acts or communities install laws that orientate common interactions and goals as regulatory speech acts. In everyday life, we do things with words that have serious consequences.

Utterances represent an obvious superficial grammar, but what we mean or what we intend depends on various circumstances, intentions, and goals. We express our meaning through gestures, eye movements, and vocal focusing (emphasis). Therefore, we have a limited number of signs for creating unlimited number of sentences. Additionally we may vary the meaning of identical sentences, which is not represented by syntax (see Fig. 1). This means that communication as a social interaction differs absolutely from the solipsistic sender–receiver narrative as well as the information-transfer narrative (Wittgenstein, 1953; Austin, 1975; Searle, 1976; Habermas, 1994).

Sender–receiver narratives confuse their language of description with reality or how to interconnect thoughts, language, and

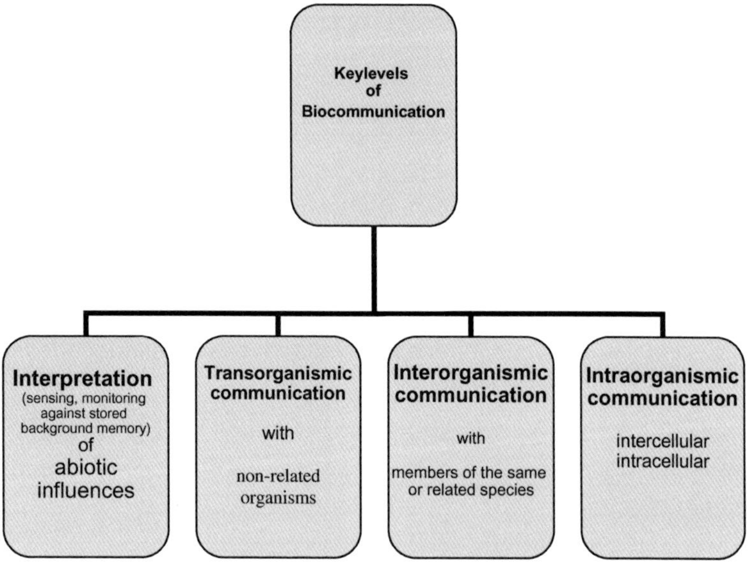

Figure 1. The key levels of biocommunication.

observations. This is the real problem with all metaphysical approaches such as objectivism, realism, and even ontology: not being aware, that the basic needs they describe do not depict physical objects (McCarthy, 1984). Yet sentences of languages solely represent the language of description, its background paradigm, and, in the case of sciences, the current consensus of discipline agreements (Kuhn, 1967). At the basis of all reasoning, a good basis for knowledge about the world, must be a coherent model of language and communication. Scientific knowledge is the result of exchange of arguments within scientific communities. I have outlined in previous works the reasons why none of the above-mentioned concepts can explain the evolution and function of languages used in communicative interactions (Witzany, 1995, 2000, 2010). The core reasons are that these concepts cannot sufficiently explain:

- Simultaneous understanding of identical meanings in two interacting partners, as expressed in successfully coordinated activity;

- Differentiation between deep and superficial grammar of a statement along with differentiation between locutionary, illocutionary, and perlocutionary speech acts;
- The *de novo* generation of coherent and context-dependent sentences.

2. Roots of Scientific Knowledge

In contrast, this article will try to start at the bottom. What can we be sure to know as scientific knowledge? How do we assemble empirical data and how do we construct experimental setups free from personal values? What is it that we define as empirical significance? Is there any unequivocal definition of empirical significance? What are sure definitions of observation, description, understanding, theory? What definitions of the objective world, subjective world, cognitive perception, and dispositions are unequivocal? Is there any reality that cannot be described? What can we be unequivocally sure we all agree with founded and justified arguments? What are the tools that we can rely on in discussion of these problems?

These are some of the core assumptions and questions that were questioned in the early 20th century when scientists tried to construct an unequivocal scientific language clearly delimited from non-scientific languages such as theology, poetry, and metaphysics. The discussion lasted until the late 1980s and produced definitive results (Witzany, 2000, 2010).

- The only knowledge we can be undoubtedly sure is, that if we want to discuss or speak about something in the world we need a commonly shared sign system (language).
- No knowledge, no communication, no exchange of any information, no idea, or thought can be generated without a commonly shared language. Even if one would doubt in this, how could he present his critics? We do not have ideas/thoughts that we afterwards transfer into sentences as they are inherent linguistically built and constructed.
- The only sure knowledge is how language and communication function. They are both subjects and objects of our language-based

utterances, so we can clearly identify the prerequisites of communication and language, and what is missing if language and communication do not function.

- One person alone cannot invent a language. Languages emerge from groups of living agents who need tools to coordinate behavior and organize real life needs. Languages are used in communicative acts. Because a private language is impossible in principle, communication is inherently a kind of social interaction. Private thoughts follow social interaction in which we learn communicative and linguistic competences and do not predate them.

We term communicative competence as the ability to correctly initiate social interactions. We term linguistic competence as the ability to correctly generate sentences.

Every natural language/code needs groups of living agents who use this language in real life world to coordinate and organize. This means in natural languages there cannot be such things as context-free grammar. This is an artificial construction. Prior to all artificial languages such as languages of scientific disciplines, humans learned via socialization processes in everyday language how to combine actions with words and correct meanings, the complementary interaction of cognitive and emotional developments. Private thought, as well as artificial scientific languages, follows as a second step and does not predate social experiences of communicative sign-mediated interactions.

3. "If a Lion Could Speak, We Could Not Understand Him"

It is true that if lions could speak we would be unable to understand their conversation (Wittgenstein, 1953). The reason lies within the essential features of languages. They emerged out of functional behaviors interconnected with social interactions. As humans we must be lions, which is clearly impossible, to grow up within lion families, share their history. To understand a species-specific utterance someone must be a specific part of the community to understand what is meant by the signal in detail. This was Wittgenstein's understanding.

Yet there is another approach that helps us to understand signaling processes in non-human living beings. Charles Sanders Peirce suggested that to identify the meaning of signs, words, and language-like structures, "we have, therefore, simply to determine what habits it produces, for what a thing means is simply what habits it involves" (Peirce, 1923). This is a similar definition to Wittgenstein's "The meaning of a word is its use" (Wittgenstein, 1953) the core assumption of pragmatics that the function is required to identify the meaning, as outlined by the best interpreter of Ludwig Wittgenstein, John Austin, in *How to Do Things With Words* (1975). In biocommunication research, this means we search for species-specific life worlds with genera, families, and species:

- Observations on how they live (ecology, niche construction).
- Finding excreted molecules, secondary metabolites, hormones, neurotransmitters, RNAs, proteins, etc. that are in correlation with observed behavioral motifs.
- Determining if what happens in several life world circumstances correlates with the concentration of excreted molecule groups.
- Determining how behavior and molecule receptions are interconnected.

If we use this perspective it will immediately become clear that there are different levels of signaling processes in communicative actions which must be delimited if they are not to be confused with the whole variety of communication.

4. Key Levels of Biocommunication in All Domains of Life

There is a recognized abundance of signaling processes in all domains of life: bacteria, archaea, viruses, protozoa, fungi, animals, and plants. There are species-specific signaling processes and niche-specific signaling processes. For example, bacterial interactions with fungi in plant rhizosphere ecology use unique signaling to generate behavioral patterns than bacteria within the gut of animals. I have discussed this phenomenon with leading experts (Witzany 2010, 2011a, 2011b, 2012a, 2012b, 2013) and stress that there are certain key levels of biocommunication (see Fig. 2 and Appendix, p. 55) that are found throughout all species and domains of life:

- Sensing of abiotic circumstances such as light, gravity, temperature, water, dryness, wind, etc. Such environmental circumstances represent important indices on nutrition availability, symbiotic organisms, growth control, and developmental time clocks. They are not only sensed but monitored, and organisms of all domains store information about these indices in memory, to adapt better to repeated life situations.
- Transorganismic communication is termed sign-mediated interaction between non-related species, as is the case in most symbiotic partnerships or in prey–predator interaction motifs. Such transorganismic communication processes can become very complex, as in the rhizosphere of plants (together with fungi, insects, bacteria, nematodes) and also in the normal ecosphere, such as the human mouth with its 500-plus different bacterial communities.
- Interorganismic communication is termed sign-mediated interaction processes between members of the same species or related species. They share species-specific vocabulary which may vary according to different ecospheres through dialects in which the signaling semantics differs according to adaptational needs.
- Intraorganismic communication: every organism consists of various parts that must be coordinated appropriately to install life functions. The signaling processes between these parts is termed intraorganismic communication. This also includes communication between DNA storage media and genetic parasites whether they are persistent settlers of viral origin or are defectives (see Section 8.1) of infection events identified as mobile genetic elements, which now act as natural genetic engineers for better adaptation by the host organism.

Figure 2. Context determines meaning (e.g. "The shooting of the hunters") not syntax. Similarly the superficial grammar of DNA does not determine its meaning. The *in vivo* context which results in epigenetic markings represents a variable deep grammar which determines post-transcriptional modifications such as RNA editing and alternative splicing. Therefore in contrast to the opinions of Manfred Eigen, Sidney Brenner or Craig Venter algorithm-based DNA processing cannot generate both, superficial and deep grammar. (Graphics design: Uta Mackensen, EMBO, reports 2012; 13: 1054–1056. Permission by EMBO reports.)

5. Key Features of Natural Languages/Codes in Biocommunication Processes

Thanks to Charles Morris we know that every natural language/ code can function within group users only if they follow three levels of rules by which the signs of a language obtain real language functions (Morris, 1946). (1) Combinatorial rules on how to correctly combine characters with larger entities such as words and sentences (syntax). (2) Context-relevant rules on how sign constructions are used according to the circumstance-dependent needs of the sign-using agents (pragmatics). (3) Content-dependent rules on how to correctly combine signs with designated objects.

It does not make much sense to designate ice with characters that symbolize fire. The results of empirical analyses of natural languages used by groups of humans to organize and coordinate behavior for essential life functions within a real life world are as follows:

- Natural languages/codes are tools used by groups that cannot emerge out of single individuums. Language/code usage inherently depends on social groups.
- Natural languages/codes are tools in communication processes in which members of groups interact with other members via communicative actions. This means the language tool serves to do things with signs, such as words and utterances, combined with context-dependent markings such as conscious and unconscious body expressions.
- Natural languages/codes in communicative actions function according to the intentions and goals of living agents if certain rules of usage are followed. This implies that the agents are competent in following these rules. The competence is learned and memorized within a socialization process in which rule-following in sign-use and combination to meaning is trained and later on passed from experienced to non-experienced agents.

6. The Benefit of the Biocommunication Approach

The advantage of a methodical adaptation of communication and linguistic terminology is that it provides appropriate tools for differentiation at specific levels, which is otherwise difficult to describe in reductive terms by means of pure physics and chemistry (molecular biology). The result is that language-like structures and communication processes occur at the simplest levels of living nature. Language and communication are not the evolutionary inventions of humans, nor are they anthropomorphic adaptations for describing non-human entities. It is an empirical fact that all coordination and organization within and between cells, tissues, organs, and organisms needs signs, i.e., molecules that serve as signals or symbols in messages or as vital indicators of environmental conditions. Because no natural code can encode itself, in the same way that no natural language can speak itself, these signs must be sensed and interpreted in the correct way by biological agents, i.e., there must be subjects of sign production and sign interpretation. The consequence of this is that sensing, as well as interpretation, may fail. This can result in inappropriate behavior by, or even fatal consequences for, cells, tissues, organs, and organisms.

6.1 The machine metaphor for living organisms is old-fashioned

The method of analyzing any part of a machine in detail to gain a picture of the whole functional blueprint, which can then be used to reproduce or manipulate it, or to produce an even more perfect example, e.g., as in artificial genetic engineering or synthetic biology, is still useful if we are dealing with machines. In contrast, communication between cells, cellular components, tissues, organs, and organisms is far from being a procedure that can be reduced to mechanistic input/output or cause/reaction descriptions (Shapiro, 2011). It is evident that communication processes between living organisms include a variety of non-mechanistic circumstances and competences that must be fulfilled in parallel if communicative acts are to have successful consequences: for example, innovating common coordination to adapt to new environmental conditions.

Machines cannot create new programs out of a functional blueprint, which is in contrast to the abilities of living organisms that are able to communicate with each other. The universal Turing machine and the self-reproducing machines of von Neumann (Turing, 1950; von Neumann, 1966; Brenner, 2012) still remain at the conceptual stage. No single self-reproducing machine has ever been identified within the last 80 years, however. There is good reason for this: it is, in principle, impossible for an artificial machine to reproduce itself (Witzany and Baluška, 2012b). In contrast to artificial machines, which cannot reproduce themselves, living cells and organisms can reproduce and, additionally, generate an abundance of behavioral motifs for which no algorithm can be constructed, such as *de novo* generation of coherent nucleotide sequences.

7. Biocommunication and Language in Living Nature

Consistent with current knowledge about natural communication processes, communication is defined as the sign-mediated interaction between at least two living agents which share a repertoire of signs (a kind of natural language) that are combined (according to syntactic rules) in varying contexts (according to pragmatic rules) with transport content (according to semantic rules). This means monological concepts such as the sender–receiver narrative cannot explain the emergence of commonly shared meanings.

Contrary to former concepts, the importance of this result is that these three levels of semiotic rules (Gk *semeion* = sign) are complementary parts of any natural language or code. If one level is missing, according to Charles Morris, we cannot seriously speak of language or signal-mediated communication. To the best of my knowledge, the most recent definition of communication is this: sign-mediated and rule-governed interactions, i.e., interactions that depend on a shared repertoire of signs and rules (Witzany, 2010). These features are lacking in abiotic interactions, however; no semiotic rules are necessary if water freezes to ice.

8. The Alternative to the Gene Concept: DNA as Habitat and its RNA Inhabitants

DNA serves as a stable information storage medium and every protein which is needed by the cell is produced from this blueprint. It was recently found that an abundance of various RNA elements cooperate in a variety of steps as regulatory units with multiple competences to act on RNA transcripts. On the one hand, natural genome editing is the competent agent-driven generation and integration of meaningful DNA nucleotide sequences into pre-existing genomic content arrangements, and the ability to (re-)combine and (re-)

regulate them according to the context-dependent (i.e., adaptational) purposes of the host organism (Witzany, 2010). Natural genome editing, on the other hand, designates the integration of all RNA activities acting on RNA transcripts without altering DNA-encoded contents. If we take the genetic code seriously as a natural code, there must be agents that are competent to act on this code because no natural code codes itself, just as no natural language speaks itself. As viral and subviral agents have been suggested as code editing agents, because there are several indicators that demonstrate viruses competent at both RNA and DNA natural genome editing (Witzany, 2013).

8.1 Cellular genomes: stable DNA habitats for unstable RNA colonizers

Cellular life is the main subject of biology and the history of evolution starts with the emergence of the first living cell. For decades, there was no doubt that speaking about life meant speaking about living cells and cellular assemblies. Virology elucidated the fact that viruses code for typical features that are not part of cellular life and are not found in any cell, which suggests viruses are older than cellular life (Villarreal, 2005; Forterre, 2005; Koonin, 2009). Additionally there are some indicators that viruses and viral-derived elements are the agents that edit the genome in host organisms, exemplified by the non-lytic but persistent colonization of prokaryotes and eukaryotes by viral agents (Villarreal, 2005; Villarreal and Witzany, 2010).

This view is supported by the hypothesis of a pre-cellular RNA world with an abundance of competing and cooperating ribozymes. Research on the agents that act on transcripts from the DNA information storage medium demonstrated an RNA world of high diversity and an abundance of regulatory functions in all key steps of cellular replication such as expression, transcription, translation, and repair (Gesteland *et al.*, 2006; Witzany, 2009a, 2009b).

At the present stage we can identify two complementary kinds of natural genome editing: (1) long-lasting and stable inherited alterations of the cellular DNA genomes by persistent viral infections that alter genetic host identity and (2) co-opted adaptations of ribozyme-like parts as remnants of former viral colonizers that constitute regulatory elements in the present highly dynamic RNA world shortly after transcription. There are several indicators that both kinds of natural genome editing agents are epigenetically regulated (Zuckerkandl and Cavalli, 2007; Maksakova *et al.*, 2008).

Persistent viral infection events determine gene word order in both prokaryotes and eukaryotes. Viral colonizers therefore play major roles in evolution and diversity of organisms (Villarreal, 2012). Various examples demonstrate that these viral colonizers are still active in specific developmental processes in that they are expressed in a rather limited developmental window until the process is finished. Then they are silenced again. Also well documented is the exapted function of a great variety of former retroviral parts such as env, gag, pol, that now play important roles in the gene regulation of host organisms. These "defectives" (Villarreal 2005) that now function as "effective" regulatory elements share similar features with their viral relatives and in most cases act as ribozymatic functions. They do not alter inheritable DNA content but dynamically act on RNA transcripts, although there are some examples in which RNA genome editing may also become a conserved status, especially by being reverse-transcribed into DNA.

8.2 Agents that edit host genomes

If the genetic code is really a natural code we must look at the agents that are able to generate sequences in this code and implement them in basic life processes such as sequence generation and context-relevant information storage and modification.

The essential assumptions after the 1960s were that natural laws determine statistical fluctuations ("genetic drift") in the molecular sequence order and replication errors ("mutations") lead to variations in DNA sequences that are the object of selection processes. The increasing volume of empirical data does not fit this concept, however:

- The central dogma of molecular biology "DNA–RNA–Protein everything else," is obsolete nowadays; the reverse trend is widespread.

- Viruses and their defectives play various roles in evolution and developmental processes as obligate inhabitants of cellular genomes.
- Non-coding RNAs (functional RNAs that are not translated into proteins) are actively engaged in nearly all cell processes to meet both evolutionary and developmental needs.
- Mobile genetic elements (the type of DNA or RNA that can move around within a genome) that actively influence the shape of genomes and their internal interactions.
- Flexible and dynamic expansion of information processes through epigenetic programing and reprograming, alternative splicing (a regulated process occurring during gene expression that results in a single gene coding for multiple proteins), and RNA editing (a process in which changes to specific nucleotide sequences are made).

Today we can identify several key players that coordinate and organize the genetic content compositions of host organisms. They include endogenous viruses and defectives, transposons, retro-transposons, long terminal repeats, non-long terminal repeats, long interspersed nuclear elements, short interspersed nuclear elements, group I introns, group II introns, phages, and plasmids (Weiner, 2006; Slotkin and Martienssen, 2007; Kapitonov and Kohany, 2007; Lambowitz and Zimmerly, 2011). These are just some examples that use genomic DNA as their preferred living habitat. This means that DNA is not a solely a stable genetic storage medium that serves as an evolutionary protocol but is also a species-specific ecological niche for active RNAs.

A great variety of such genetic elements have been identified during the last 40 years as obligate inhabitants of all genomes, whether prokaryotic or eukaryotic. They infect, insert, and delete; some cut and paste, others copy and paste, and both spread within the genome. They change host genetic identities by insertion, recombination, or epigenetic regulation or re-regulation of genetic content, and co-evolve with the host to interact in a modular manner (Shapiro, 2011).

Together with non-coding RNAs they shape both genome architecture and regulation. In this respect, they are agents of change not only over evolutionary time but also in real time as domesticated agents.

Non-coding RNAs interact with DNA, RNA, and proteins and play important roles in nuclear organization, transcription, post-transcriptional, and epigenetic processes (Mercer and Mattick, 2013a, 2013b). Non-coding RNAs are transcribed in both the sense and antisense directions and may be expressed in a cell type, subcellular compartment, developmental stage, or an environmental stimuli-specific, that is, context-dependent, manner. Specific RNA polymerases overlap in transcriptional contents, which means that each nucleotide can participate in varying transcriptional content arrangements according to varying contexts (Mattick, 2009, 2010).

Non-coding RNAs can be regulated in a varying manner, coordinated or independently, autonomously or functionally interrelated, and can regulate individual genes as well as large genetic networks; they can precisely control the spatiotemporal deployment of genes that are executing neuronal processes with extreme cell specificity. Various classes of non-coding RNAs target each other for post-transcriptional regulation via alternative splicing, polyadenylation, 5′ capping, non-templated modifications, and RNA editing. RNA editing especially can transmit environmental information to the epigenome and therefore enable neuronal plasticity with learning and memory (Qureshi and Mehler, 2012). Additionally, non-coding RNAs can undergo nuclear-cytoplasmic, nuclear-mitochondrial, and axodendritic trafficking via ribonucleoprotein complexes that promote the spatiotemporal distribution and function of various combinations of ncRNAs, mRNAs, and RNA-binding proteins (Mercer and Mattick, 2013a, 2013b; Clark *et al.*, 2013).

8.3 RNA group membership: RNA stem-loop self-ligation and pseudoknots

If we are to look at the interactional motives of RNA agents to form consortial biotic structures that follow biological selection processes and not mere physical–chemical reaction patterns we must look at the group building of RNA stem-loop structures (Flores *et al.*, 2012; Witzany, 2016).

Recently it has been found that single stem-loops interact in a purely physical–chemical mode without selective forces, independently of whether they are derived randomly or are constructed under *in vitro* conditions (Smit *et al.*, 2006). In contrast, if these single RNA stem-loops build groups they transcend the purely physical–chemical interaction pattern and emerge as biological selection forces, biological identities of self/non-self identification and preclusion, immune functions, dynamically changing (adapting) their membership roles. A single alteration in a base-pairing RNA stem that leads to a new bulge may dynamically alter not just this single stem-loop the whole group identity of which this stem-loop is part (Villarreal and Witzany, 2013a, 2013b).

Simple self-ligating RNA stem-loops can build much larger groups of RNA stem-loops that serve to increase complexity (Briones *et al.*, 2009; Gwiazda *et al.*, 2012). This may lead to ribozymatic consortia, which later on build success stories such as (1) the merger of the two subunits of transfer RNAs or (2) RNA-dependent RNA polymerases for replication of RNA through RNA or (3) the subunits of ribosomal RNAs, all of them groups that evolved and functioned for different reasons than in the later conserved modes (Dick *et al.*, 1995; Sun and Caetano-Anolles, 2008).

Significantly, mixtures of RNA fragments that self-ligate into self-replicating ribozymes spontaneously form cooperative networks. For example, three-member networks show highly cooperative growth dynamics. When such cooperative networks compete directly against selfish autocatalytic cycles, the former grow faster, indicating the ability of RNA populations to evolve greater complexity through cooperation (Vaidya *et al.*, 2012). In this respect cooperation clearly outcompetes selfishness.

Another intriguing example of the biological (selective) group-building competence of RNA stem-loop consortia is the chemical interaction based on the molecular syntax in stem-loop "kissing," in that single-stranded regions of RNA stem-loops bind according to Chargaff rules to other single-stranded stem-loop structures (Staple and Butcher, 2005). It represents an RNA folding motif that links two elements of RNA secondary structure and additionally serves as a unique protein recognition site. RNA kissing motifs unite und build more complex group identities for several functions, such as dimerization of genomic RNA in viruses.

Complex three-dimensional structures can be built by consortia of single RNA sequence strings. One of the most interesting structures are pseudoknots which are composed of two helical segments connected by single-stranded regions of loops (Gupta *et al.*, 2012). Bases in the single-stranded loop are base-paired with bases outside the loop (Staple and Butcher, 2005; Rivas and Eddy, 1999). This interaction pattern clearly depends on the rules of molecular syntax but is initiated by adaptational purposes of different ecosphere habitat dynamics. Therefore, the results of these interactions may lead to structurally diverse groups with important different biological roles such as the catalytic core of key players of the present RNA world, i.e., ribozymes (rRNAs, tRNAs), self-splicing introns, telomerase and context-dependent altered gene expression by inducing ribosomal frameshifting in several viruses.

Most interestingly, the base-pairing in pseudoknots tends to be strictly context-sensitive and base-pairing overlaps in sequence positions. Additionally, the emergence of single-nucleotide bulge loops can hardly be predicted. This currently places limitations on algorithm-based prediction models such as dynamic programming or stochastic context-free grammars (Lyngso, 2004). It indicates the language/code nature of nucleic acid language which represents the possibility of coherent *de novo* generation and context-dependent alterations for a diversity of different meanings (functions) relating to the same syntax structures.

9. Conclusions

Biocommunication can be defined as the sign-mediated interactions of groups of living agents that share syntactic, pragmatic, and semantic rules for sign use. Biocommunication research investigates empirically how coordinated and organized groups of living agents function by using sign-mediated interactions as basic tools. Biocommunication cannot be sufficiently explained by mechanistic and reductionist concepts, because they cannot appropriately describe the various levels of sign use, such as sign sequences with both a superficial grammar and a deep grammar.

The biocommunication and natural genome editing approach to the processual reality of living agents has some advantages over traditional scholarly convictions: (1) a clear distinction between life and non-life, (2) an empirical non-mechanistic and non-reductionistic descriptive method of biotic interactional patterns in all organismic kingdoms. Additionally, (3) any observed coordination within and between organisms can be deciphered by research which identifies signaling molecules and syntactic, pragmatic, semantic rules underlying the mode of signal use. Last but not least, (4) biological research must no longer concentrate on experimental set-ups and theoretical approaches that seek to elucidate language and communication in non-human living nature by means of mathematical (algorithm-based) modeling.

In the light of current empirical and theoretical knowledge it is clear that the multiple levels of sign-mediated interactions which we call "communication" cannot be explained or even sufficiently described by older models such as the "sender–receiver" narrative or even based on such terms as "system" or "information." Even physical or mathematical definitions of languages as quantifiable sets of signs fail to describe the key features of these phenomena. In contrast, the biocommunication concept could serve as an appropriate complementary tool for interpreting empirical data of biological disciplines coherent with current knowledge about communication and language.

Appendix: History of Biocommunications Theory

A first draft of a theory of biocommunication was outlined in 1971 (Tembrock, 1971). Tembrock exemplified the three semiotic levels syntax, semantics, and pragmatics in great detail for several behavioral patterns within the kingdom of animals. He focused on the transport of information via chemical, mechanical (tactile and acoustic) and visual signs. Although his investigations were outlined in a strict empirical manner, Tembrock justified his approach according to a solipsistic model of knowledge and communication as we came to know it in the depiction theory of language. His biocommunicative approach is therefore coherent with the sender–receiver model of information theory, a derivative from mathematical theory of language.

But the inherent features of language and communication, especially the possibility of innovative semiosis or the common understanding (and interpretation) of identical meanings, is without the realm of formalisable procedures.

In contrast with this empiricist approach, at the end of the 1980s language philosopher Guenther Witzany developed a pragmatic approach of biocommunication based on the results of the philosophy of science discourse in the 20th century (Witzany, 1993, 2000, 2007). In this new philosophy of biology he integrated the pragmatic turn in its methodological foundation as well as the complementarity of the three semiotic levels of semiotic rules. Additionally, and in contrast with theories of knowledge with a solipsistic foundation, he investigates its scientific object according to the primacy of pragmatics, i.e., the contexts in which communicative-intersubjective sign-users are interwoven in a real life world.

The main focus of biocommunicative investigations is the agents that use and interpret signs in communicative interactions. Because 'One cannot follow rules only once' (Wittgenstein, 1953), speech and communication are kinds of social behavior and therefore it is important to investigate group behavior and group identity, the pragmatic contexts in which they are actively interwoven together with their history and cultural identity. These groups share a repertoire of signs and semiotic rules, with which they coordinate every life organization that is necessary.

This biocommunicative approach investigates communicative acts within and between cells, tissues, organs, and organisms as sign-mediated interactions throughout all organismic kingdoms. After developing the biocommunication approach as program in 2010,

Witzany edited several books with leading experts in their field such as biocommunication in soil microorganisms (Witzany, 2011b), biocommunication of plants (Witzany and Baluška, 2012a), biocommunication of fungi (Witzany, 2012a), biocommunication of animals (Witzany, 2013) and biocommunication of ciliates (Witzany and Nowacki, 2016), see below. The results of these books demonstrate that competent sign-using agents from all organismic kingdoms follow syntactic, pragmatic, and semantic rules in parallel. No level of rules is reducible to one another, although in all cases context determines the meaning of the signs not syntax, i.e., the primacy of pragmatics. This is a crucial difference from all similar concepts of bringing together linguistics and biology such as systems biology, synthetic biology, biolinguistics, bioinformatics, biosemiotics, not to speak about mechanistic and even non-scientific (esoteric) approaches that want to reduce biocommunication to physicalistic and/or chemical principles.

References

Austin JL. 1975. *How to do Things with Words*. Harvard University Press, London.

Baluška F and Witzany G. 2013. At the dawn of a new revolution in life sciences. *World Journal of Biological Chemistry*, 4, 13–5.

Barbieri M. 2001. *The Organic Codes. The Birth of Semantic Biology*. PeQuod, Ancona.

Brenner S. 2012. Life's code script. *Nature*, 482, 461.

Briones C, Stich M and Manrubia SC. 2009. The dawn of the RNA world: toward functional complexity through ligation of random RNA oligomers. *RNA*, 15, 743–749.

Chomsky N. 1964. *Current Issues in Linguistic Theory*. The Hague, Mouton, London.

Clark MB, Choudhary A, Smith MA, Taft RJ and Mattick JS. 2013. The dark matter rises: the expanding world of regulatory RNAs. *Essays in Biochemistry*, 54, 1–16.

Dick TP and Schamel WA. 1995. Molecular evolution of transfer RNA from two precursor hairpins: implications for the origin of protein synthesis. *Journal of Molecular Evolution*, 41, 1–9.

Eigen M and Winkler R. 1983. *The Laws of the Game: How The Principles of Nature Govern Chance*. University Press, Princeton.

Eigen M. 2013. *From Strange Simplicity to Complex Familiarity a Treatise on Matter, Information, Life, and Thought*. Oxford University Press, Oxford.

Flores R, Serra P, Minoia S, Di Serio F and Navarro B. 2012. Viroids: From genotype to phenotype just relying on RNA sequence and structural motifs. *Frontiers in Microbiology*, 3, 217. doi. 10.3389/fmicb.2012.00217.

Forterre P. 2005. The two ages of the RNA world, and the transition to the DNA world: a story of viruses and cells. *Biochimie*, 87, 793–803.

Gesteland RF, Cech TR and Atkins JF. 2006. *The RNA World*. 3rd Edition. Cold Spring Harbor Laboratory Press, New York.

Gwiazda S, Salomon K, Appel B and Müller S. 2012. RNA self ligation: From oligonucleotides to full length ribozymes. *Biochimie*, 94, 1457–1463.

Habermas J. 1984. *The Theory of Communicative Action. Reason and the Rationalization of Society*, Vol. 1. Beacon Press, Boston.

Habermas, J. 1987. *The Theory of Communicative Action. Lifeworld and System: A Critique of Functionalist Reason*, Vol. 2. Beacon Press, Boston.

Habermas J. 1994. Actions, speech acts, linguistically mediated interactions and the lifeworld. *Philosophical Problems Today*, 1, 45–74.

Ji S. 1999. The linguistics of DNA: words, sentences, grammar, phonetics and semantics. *Annals of the New York Academy of Sciences*, 870, 411–417.

Koonin EV. 2009. On the origin of cells and viruses: primordial virus world scenario. *Annals of the New York Academy of Sciences*, 1178, 47–64.

Kuhn TS. 1967. *The Structure of Scientific Revolutions*. University of Chicago Press, Chicago.

Lyngsø RB. 2004. Complexity of pseudoknot prediction in simple models. Paper presented at the ICALP.

Maksakova IA, Mager DL and Reiss D. 2008. Keeping active endogenous retroviral-like elements in check: The epigenetic perspective. *Cell Mol Life Sci*. 65, 3329–3347

Mattick JS. 2009. Deconstructing the dogma. A new view of the evolution and genetic programing of complex organisms. *Annals of the New York Academy of Sciences*, 1178, 29–46.

Mattick, JS. 2010. Video Q&A: Non-coding RNAs and eukaryotic evolution — a personal view. *BMC Biology*, 8, 67.

McCarthy T. 1984. Translator's introduction. In *The Theory of Communicative Action 1, ix*. Beacon Press, Boston.

Mercer TR and Mattick JS. 2013a. Structure and function of long noncoding RNAs in epigenetic regulation. *Nature Structural & Molecular Biology*, 20, 300–307.

Mercer TR and Mattick JS. 2013b. Understanding the regulatory and transcriptional complexity of the genome through structure. *Genome Research*, 23, 1081–1088.

Morris C. 1946. *Signs, Language, and Behavior*. Prentice Hall, New York.

Neumann VJ. 1966. *Theory of Self-Reproducing Automata*, 122–123. Arthur WB (ed.). University of Illinois Press, Urbana and London.

Nowak Martin AN and David C. 1999. The evolution of language. *Proceedings of the National Academy of Sciences of the USA*, 96, 8023–8033.

Nowak Martin AN, Joshua BP and Jansen VAA. 2000. The evolution of syntactic communication. *Nature*, 404, 495–498.

Nowak Martin AN, Komarova NL and Niyogi P. 2001. Evolution of universal grammar. *Science*, 291, 114–118.

Nowak Martin AN, Komarova NL and Niyogi P. 2002. Computational and evolutionary aspects of language. *Nature*, 417, 611–617.

Peirce CS. 1923. How to make our ideas clear. In *Chance, Love, and Logic: Philosophical Essays*. Morris RC (ed.), 41–42. Harcourt, Brace, and World, New York.

Popov O, Degal DM and Trifonov EN. 1996. Linguistic complexity of protein sequences as compared to texts of human languages. *Biosystems*, 38, 65–74.

Qureshi IA and Mehler MF. 2012. Emerging roles of non-coding RNAs in brain evolution, development, plasticity and disease. *Nature Reviews Neuroscience*, 13, 528–541.

Rivas E and Eddy SR. 1999. A dynamic programing algorithm for RNA structure prediction including pseudoknots. *Journal of Molecular Biology*, 285, 2053–2068.

Searle J. 1976. *Speech Acts. An essay in the Philosophy of Language*. Cambridge University Press, Cambridge.

Searls DB. 2002. The language of genes. *Nature*, 420, 211–217.

Shannon CE and Weaver W. 1949. *The Mathematical Theory of Communication*. University of Illinois Press, Urbana.

Shapiro JA. 2011. *Evolution: A View from the 21st Century*. FT Press, Washington.

Smit S, Yarus M and Knight R. 2006. Natural selection is not required to explain universal compositional patterns in rRNA secondary structure categories. *RNA*, 12, 1–14.

Staple DW and Butcher SE. 2005. Pseudoknots: RNA structures with diverse functions. *PLoS Biology*, 3(6), e213.

Sun FJ and Caetano-Anollés G. 2008. Transfer RNA and the origins of diversified life. *PLoS One*, 3(7): e2799.

Tembrock G. 1971. *Biokommunikation. Informationsübertragung im biologischen Bereich*. Akademie-Verlag, Berlin.

Tomasello M. 2008. *Origins of Human Communication*. MIT Press, Cambridge, MA.

Turing A. 1950. Computing machinery and intelligence. *Mind*, 59, 433–460.

Vaidya N, Manapat ML, Chen IA, Xulvi-Brunet R, Hayden EJ and Lehman N. 2012. Spontaneous network formation among cooperative RNA replicators. *Nature*, 49, 72–77.

Venter C. 2013. *Life at the Speed of Light: From the Double Helix to the Dawn of Digital Life*. Viking, New York.

Villarreal LP. 2005. *Viruses and the Evolution of Life*. ASM Press, Washington.

Villarreal LP. 2012. The addiction module as social force. In *Viruses: Essential Agents of Life*, 107–145, Springer, Dordrecht.

Villarreal LP and Witzany G. 2010. Viruses are essential agents within the roots and stem of the tree of life. *Journal of Theoretical Biology*, 262, 698–710.

Villarreal LP and Witzany G. 2013a. The DNA Habitat and its RNA Inhabitants. At the dawn of RNA sociology. *Genomics Insights*, 6, 1–12.

Villarreal LP and Witzany G. 2013b. "Rethinking quasispecies theory: From fittest type to cooperative consortia." *World Journal of Biological Chemistry*, 4, 79–90.

Weiner AM. 2006. SINEs and LINEs: troublemakers, saboteurs, benefactors, ancestors. In *The RNA World*, Gesteland RF, Cech TR and Atkins JF (eds.). 507–534. 3rd edition. Cold Spring Harbor Laboratory Press, New York.

Witzany G. 1993. Natur der Sprache — Sprache der Natur. Sprachpragmatische Philosophie der Biologie. Königshausen & Neumann, Wuerzburg.

Witzany G. 1995. From the 'logic of the molecular syntax' to molecular pragmatism. Explanatory deficits in Manfred Eigen's concept of language and communication. *Evolution and Cognition*,1, 148–168.

Witzany G. 2000. *Life: The Communicative Structure*. Books on Demand, Norderstedt.

Witzany G. 2007. *The Logos of the Bios 2. Biocommunication*. Helsinki, Umweb.

Witzany G. 2009a. Non-coding RNAs: persistent viral agents as modular tools for cellular needs. *Annals of the New York Academy of Sciences*, 1178, 244–267.

Witzany G, ed. 2009b. *Natural Genetic Engineering and Natural Genome Editing*. New York Academy of Sciences, New York.

Witzany G. 2010. *Biocommunication and Natural Genome Editing*. Springer, Dordrecht.

Witzany G. 2011a. The agents of natural genome editing. *Journal of Molecular Cell Biology*, 3, 181–189.

Witzany G, ed. 2011b. *Biocommunication in Soil Microorganisms*. Springer, Heidelberg.

Witzany G, ed. 2012a. *Biocommunication of Fungi*. Springer, Dordrecht.

Witzany G, ed. 2012b. *Viruses: Essential Agents of Life*. Springer, Dordrecht.

Witzany G, ed. 2013. *Biocommunication of Animals*. Springer, Dordrecht.

Witzany G and Baluška F. 2012a. *Biocommunication of Plants*. Springer, Berlin, Heidelberg.

Witzany G and Baluška F. 2012b. "Life's code script does not code itself. The machine metaphor for living organisms is outdated." *EMBO Reports*, 13, 1054–1056.

Witzany G. 2016. Crucial steps to life: From chemical reactions to code using agents. *Biosystems*, 140, 49–57.

Witzany G and Nowacki M, eds. 2016. *Biocommunication of Ciliates*. Springer, Dordrecht.

Wittgenstein L. 1953. *Philosophical Investigations*. Basil Blackwell, Oxford, UK.

Zuckerkandl E and Cavalli G. 2007. Combinatorial epigenetics, 'junk DNA', and the evolution of complex organisms. *Gene*, 390, 232–242.

3. Zoosemiotics, Typologies of Signs and Continuity Between Humans and Other Animals

Dario Martinelli

Prof., PhD, International Semiotics Institute
Kaunas University of Technology, Kaunas, Lithuania
dario.martinelli@ktu.lt

1. Introduction

The goal of this essay is to present a basic classification of signs and sign behavior in ethological zoosemiotics and to discuss the continuities between humans and other animals in this domain, particularly in those critical aspects (like, for instance, symbolic communication) where an active discussion still exists on the possible elements of discontinuity between the *Homo sapiens* and the other species.

Ethological zoosemiotics is the primary and most classical field within zoosemiotics, the one that actually deals with the semiosic processes occurring among animals (whereas the field of "Anthropological zoosemiotics" investigates on the human–animal relationship, and it is not an object of this essay). The field can be divided into an early stage and a modern one (Martinelli, 2010: 9). In its early phase, zoosemiotics was mostly conceived as an umbrella term for gathering different approaches on animal communication. "Communication" was still the keyword, as in this stage there was a clear emphasis on this specific process, rather than the broader semiosis. Classical ethology and the behavioristic schools, finally, provide the scientific framework for (and document the uncertain semiotic identity of) this type of zoosemiotics.

Sometime in the late 1970s (or thereabouts), zoosemiotics enters its modern age, this transition being marked by a few important innovations:

(1) Zoosemioticians extend their interests to the entire spectrum of semiosis.
(2) They start treating zoosemiotics as an autonomous discipline, with its own methodologies, fields of interest and jargon. This is also possible because,
(3) cognitive approaches appear, producing something similar to what the publication of Griffin 1976 had produced in ethology. Through this transformation, zoosemiotics develops a strictly semiotic paradigm (as the traditional definitions of "semiotic threshold" always imply a mental process underlying sign production).
(4) In a general sense, zoosemiotics becomes more aware of itself, and scholars start systematizing the discipline in different respects. It is in the 1970s that we witness the first attempts to trace a history of zoosemiotics (Tembrock, 1971; Sebeok, 1979); and it is still in the 1970s that we find a first annotated guide to zoosemiotic literature (Sebeok, 1972).

2. From Signals to Signs

For the purposes of this essay, I do not intend to create too significant a distinction between sign and signal (a more familiar term in biology, as applied to animal communication): in principle, within semiotics, a "signal" tends to be considered the simplest unit of a sign, the latter normally consisting of a more complex display involving more than one signal. Also, there are numerous cases where such difference is difficult to spot (the reduction to a basic unit not being always possible, the relevance of this reduction not always tangible, etc.), and other instances where biologists end up calling "signals" what semioticians call "signs" and vice versa. What I will instead use as a watershed between "sign" and "signal" (we shall see it later) is the occurrence, in the case of "signs", of a process of mediation and/or ritualization that a simple signal does not necessarily require.

Simply put, a "signal" is any form of behavior aimed to transmit information from one animal to another. In physical terms, it is also a form of energy, and therefore it can be chemical, mechanic, electric and the likes: for this reason, animal communication (in fact, biocommunication *in toto*) must be also thought in terms of channels (as we shall see later).

One basic distinction that needs to be operated is between digital and analogical signals: a digital (or discreet) signal does not vary in quantity (length, strength, intensity, etc.), or, if it does, that is not relevant to the reception of the message. It is either present or absent ("yes/no", "on/off", etc.). The appearance of a certain body color in some animal species is a good example: the males of the Three-spined stickleback *Gasterosteus aculeatus* display a characteristic red color on their chin and belly during the mating period (Östlund-Nilsson *et al.*, 2007, 167–168). This color is *either* there *or* is not.

In the case of analogical (or graduate) signals, variability becomes exactly the point: the message's contents depend on the different degrees of its emission. Here, we have cases of the "more/less", "higher/lower", etc., type. In the threatening signals displayed by many species, the *more* aggressive the animal feels, the *more* intense the signal becomes. Analogical signals tend to be richer than digital ones, with a higher communicative potential. An analogical signal carries many meanings, not just one (that may or may not be there, as in the case of digital signals). Also, the meaning of a signal often depends on the context: time, place, subjects involved and other factors make it possible for the same signal to display different meanings in different situations. On the contrary, those signals that evoke a fixed, specific response are called "releasers".

Signals may also be encoded or decoded in slightly unusual manners. In a conventional sense, communication is a process that primarily produces some sort of advantage for the sender ("advantage" being naturally meant in a broad sense, starting from a "successful" conveying of the information), and then it may or may not produce it for the receiver. However, when the emitted signal produces an advantage *only* for the receiver, we have a case of "signal interception". This phenomenon occurs in several contexts, including intraspecific and interspecific competition, predator–prey relation, parasitism, etc. In signal interception, the specimen who receives ("intercepts") the message is actually not the one intended by the sender, but a third subject, who indeed benefits from "meddling" into someone else's business. A typical case is that of territorial fish species: the male sends courtship signals to the female (the "designated" receiver), but another male (the "interceptor") gets the message too, and enters in mating competition with the sender. Other frequent cases include intraspecific signs of a given species intercepted by a predator of that species, who therefore get useful information on the presence of a potential prey, and so forth. Among other things, the existence of this phenomenon is one important reason why very "extrovert" (visible, loud, etc.) exhibitions in intraspecific communication are performed in limited time and contexts. The goal is to diminish the chances of being intercepted (and therefore damaged in some form).

Signals can also be used for conveying specific information about the sender. When they indicate his/her position, identity, characteristics and else, they are called "self-promotion signals", as in several forms of singing, as appearing in birds, insects, amphibians, which inform about the singer's species, sex and individual qualities. A particular case of self-promotion is the "social status signal" (Fernald, 2014), that informs co-specimens of the sender's position in the group hierarchy, as in the case of the visual displays of dominant subject/s in certain communities: a white-silver color on the back for the Mountain Gorilla *Gorilla beringei;* a dark plumage under the beak for the *Zonotrichia querula,* and so forth. Also the signs emitted by young subjects are usually aimed at self-promotion. "Infant signals" have the primary goal of letting the adults know that the sender is indeed an infant, with all that this information implies (the subject is weak, not self-sufficient, etc.). Generally, an infant is not recognized by a single characteristic, but by a combination, an "overall appearance" that Konrad Lorenz named "infant scheme" (Lorenz, 1971), and which includes such features like a certain morphology of the head (proportionally bigger than an adult's), certain behavioral patterns, etc. For evolutionary reasons, the recognition of a subject as "infant" (whether or not co-specific) tends to inspire a more compassionate behavior in the observer, as if "it would not be fair" to take advantage of the situation.

Many signals evolved from basic patterns that were not born with a "communicative" intention (Cimatti, 1998: 187–188). These patterns were simply phenomena that were happening naturally, in consequence of given stimuli (muscle movements, physiological processes, etc.). At some point, in the evolution of a species, these patterns underwent a process of "ritualization": they were modified and stereotyped, becoming something that "signify" a certain meaning, not simply that is "produced" by it. Once again, let us take aggression as an example, particularly the growling signal that canids display when they are reacting to an intruder: there must have been an initial point, in canids' evolution, in which growling was simply a reflex of distress and anger. At this point, the pattern was not a form of communication, it did not "stand for" or "represent" anger: simply, it was "part" of being angry, exactly like yawning is part of feeling sleepy. Later, some dog ancestor must have, so to speak, realized that growling *alone* was often enough to intimidate the intruder. Why then engaging to a fight, if that pattern was perfect to "give the impression" of wanting a fight? Growling, at this point, became "a ritual", a threat *detached* from the fight, which now could or could not occur. It was, in other words, a more complex form of signification, that now we can refer to as a "sign", an augustinian *aliquid pro aliquo*. This transition inaugurates a zoosemiotic approach to the topics of this essay.

3. Typologies of Semiosis

As already mentioned, the target of ethological zoosemiotics relies on all semiosic phenomena occurring among animals: **signification**, occurring when the receiver is the only subject taking part in the semiosis, and a true sender is missing; **representation**, occurring when the sender is the only semiosic subject, and **communication**, occurring when sender and receiver both take part in the semiosic phenomenon. The sign production and exchange among animals can be **proprioceptive**, when sender and receiver are the same subject (a typical case being the so-called *echolocation*); **intraspecific**, when sender and receiver (now two different subjects) belong to the same species (probably the most common instance of sign exchange, though the existence of socially-specific communicative forms, such as dialects, create challenges within this category as well); or **interspecific**, when sender and receiver belong to different species, family, orders, etc. (as in prey–predator relations, parasitism, territory competition, etc.)

Another crucial distinction in zoosemiotics, as in all other branches of semiotics, needs to be made between **syntactics**, **semantics**, and **pragmatics**, these terms being introduced by Charles Morris. A syntactic investigation concerns the sign repertoires of living systems in themselves, their structure, their spatial–temporal features, the way they form meaningful combinations. In Sebeok's words, zoosyntactics "deals with combinations of signs abstracted from their specific signification or their ecological setting" (Sebeok, 1990: 43). Zoosemantics "is devoted to the signification of signs, and must take account of the context referred to by the source and apprehensible by the destination; this is the least well understood dimension of animal communication studies" (Sebeok, 1990: 43). Finally, "Zoopragmatics may be said to deal with the origin of signs in the source, or sender, the propagation of signs through a medium, or channel, and the effect of signs on the destination, or receiver" (Sebeok, 1990: 43).

A theoretical systematization of zoosemiotics also involves the study of the channels involved in zoosemiosis, that is, the sensory modes employed to transmit a message, which do not necessarily correspond to the five human senses. Zoosemioticians classify at least seven channels: the olfactory, the gustatory (these two normally considered under the more general chemical channel), the tactile, the thermic, the electric, (these three normally grouped into the general tactile channel, as they all require physical contact), the acoustic channel, and the visual channel. Depending on the research context, an analysis of the channels may include their possible classifications, their constitutive parts, the occurrences they operate in, the species making use of them, and their characteristics. Finally, magnetoception too (that is, the ability to detect a magnetic field to perceive directions and locations, as widely used by birds in their migrations, but also by certain insects, reptiles, crustaceans and cartilaginous fish) has often been considered a *channel* in all respects, thought the discussion remains problematic due to the general difficulties in studying the topic altogether.

4. Typologies of Signs

As an *aliquid pro aliquo*, something that stands for something else, a sign is, quite simply, the *currency* of semiosis, what is being used for the latter to occur. To study semiosis means to study how signs are constructed, organized, assembled, distinguished, interpreted, codified, and so forth.

In order to investigate the different typologies of signs involved in animal semiosis, I suggest to employ the famous Peircean distinction between *Icons*, *Indexes* (or Indices) and *Symbols*, and to add the concept of *names* (signs of personal identity and class extension), as provided by Sebeok (e.g., in 1994), and the very critical notion of *syntax*.

4.1 Indexes

Despite the fact that, in the Peircean system, the *icon* is the sign-object relation connected to *firstness* (i.e., the mode of being of what is without reference to any subject or object), this classification will start from indexes, as the typology of sign that the vast majority of animal semiosis is concerned with. The significance of indexicality, within zoosemiotic study, is to be considered more central than in the case of anthroposemiotics, particularly its aspects related to human culture, whose logocentric nature predictably manifests a greater interest in symbols and icons. In a sense, one may go as far as to say that one of the main side effects of the production of an artificial and/or cultural sign system is exactly its emancipation from indexicality, in favor of iconicity and (most of all) symbolicness. A great number of markers, symptoms, indicators and signals produced through any of the available channels are in one way or another indexical, from the tail-wagging of a dog (index of its excitement) to the wheel display of a peacock (index of its health conditions). Although one might infer that this high recurrence of indexicality is an indication that mental processes do not play a major role in animal semiosis (as, after all, an index is most often a natural reflex, an instinctive reaction, a physiological display), this conclusion would be in fact as inaccurate as saying that a human display of joy as the smile is alien to any cognitive implication, ignoring that elaborate path that goes from joy to smiling and that includes the *reasons* for joy, the specific *causes* of a smile, the frequent presence of *receivers* of the "smile" message, the different *degrees* of display, and others, that obviously require the type of mediation that a mind can provide. The same applies to a similar non-human manifestation of joy as the tail-wagging of a dog, who has the same cognitive 'problems' to solve, and normally provides an analogous range of solutions.

What instead is true is that indexicality, although requiring mental processes in most cases, does not necessarily requires complex and *multilayered* mental processes, and that is primarily because an index tends to require temporal presence and spatial continuity, and its encoding is strongly bound to the *hic et nunc* of semiosis. The decoding may still happen later or further (e.g., some chemical sign may be carried by the sender for days and kilometers), but, what we might call the "packaging" of the sign is an immediate process that does not require too much cognitive articulation.

Finally, it is not entirely true to say that the "degree of indexicality in animal semiosis is definitely higher than in human semiosis" (Nöth, 1990: 163), not at least if we interpret this statement to the letter (i.e., non-human animals use *more* indexes than humans), perhaps thinking that human language was able to *replace* indexes with more efficient symbolic descriptions. What has instead happened is that language added a series of communicative and cognitive elements *on top* of the existing ones, not *in place of* them. And indeed, natural semiosis is still intact in human beings, and goes on roughly to the same extent as it used to before language appeared. What has crucially changed is the decoding stage: precisely because human language is admittedly a very efficient strategy for informing (or deceiving) each other about several topics and conditions, the human being learned to pay attention mostly to that, losing — or better: stiffening up — his/her abilities in non-verbal communication. In other words, there is not a single pheromone less spread by a person who is sexually attracted by another person, but it is likely that the two people concerned will negotiate this attraction by means of words rather than armpit smelling.

4.2 Icons

Iconicity is slightly less recurrent than indexicality in animal semiosis, but still very easy to find. The many fascinating forms of mimicry (extensively studied in semiotics by Timo Maran, in, a.o., 2007) are normally the first category of examples mentioned. In a more general sense, a great deal of signs related to deception is of iconic type, as the often mentioned case of sham behavior (discussed, a.o., in Hailman, 1977: 179–183). Mimicry and imitation do not anyway serve exclusively deceptive purposes: the enormous area of birds' vocal imitations qualify in most of the cases for iconicity (and quite often for symbolicness, too). Other examples of iconic signs are recurrent in play (for instance, when a canid intends to mock-fight, she/he will not rarely offer an iconic representation of someone sleepy, or dead — therefore defenceless — by lying on his/her back), in ritualized social behavior, in aesthetic signs, and in several other situations.

4.3 Symbols

The concept of symbolicness has been for many years a concept (and a consequent mental/interactive capacity) of which only humans were claimed to be capable. However:

> *The fondly cherished mythic characterization of man, adhered to by E. Cassirer's epigones and many others, as a unique, animal-Symbolicum can be sustained only if the definition of 'symbol' is impermissibly ensnared with the concept of natural language, which G.G. Simpson quite aptly characterized as "the most diagnostic single trait of man." By every other definition — invoking the principe of arbitrariness, the idea of a conventional link between a signifier and its denotata, Peirce's 'imputed character', or the notion of an intentional class for the designatum — animals demonstrably employ symbols* (Sebeok, 1990: 42).

In order to prove this point, Sebeok deliberately chose two examples from the world of insects, that is, from those animals with supposedly a very low and rigid degree of reasoning. The first one refers to the dipterans of the family *Empididae*, one of those that practice cannibalism: in order not to risk being eaten, the male, before the copulation, offers the female an empty balloon.

> *The evolutionary origins, that is, the increasing ritualization, of this gesture have been unravelled, step by step, by biologists, but this story is irrelevant in a synchronic perspective: the fact remains that the gift of the balloon features a wholly arbitrary symbol, the transfer of which merely reduces the probability that the male himself will fall prey to his female partner* (Sebeok, 1990: 42–43).

As for the second case, Sebeok offers a very interesting interpretation of the celebrated dance of the honey bee:

> *It is common knowledge that if its food source is farther away than 100 m., the bee's tail-wagging dance conveys, among other bits of information, the direction of the goal, the sun being used as a reference point. Now if the bee dances on a horizontal surface, von Frisch tells us that "the direction of a waggling run points directly to the goal", that*

> *is to say the display is indexic (the rhythm, incidentally, depicts the distance iconically, since the farther away the goal, the fewer cycles of the dance occur in a given period). If, however, the dance takes place on a vertical comb surface — as is the case, normally, in the dark hive — then "the dancer transposes the solar angle into the gravitation angle", according to von Frisch. In other words, if a vertical honeycomb is involved, when an angle with respect to gravity is substituted as the orientation cue, the indexical aspect of the display attenuates to the extent that, temporarily, its symbolic aspect comes to rank predominant* (Sebeok, 1990: 43).

In fact, one can argue that also in normal (horizontal) conditions the dance is symbolic, because it anyway employs a communicative strategy that is based on a model. Thus, if this model was simply indexical, besides not being a model at all, it would have changed its configuration, at the changing of the contextual conditions (from horizontal to vertical position). As it seems to be symbolic, it becomes understandable that the model (which is by nature conventional and abstract, and therefore resistant to contextual variations) is not affected by changes of sort. To clarify this aspect, we may think of the positioning of an arrow, in a street, for signifying the direction "forward". Now, as we all know, a sign for forward (which is most often bidimensional) may appear in at least three configurations: (1) pointing forward, in a horizontal position, e.g., when the arrow is painted on the road's asphalt; (2) still pointing forward, but in a vertical position, as it may appear on a street sign, on the right side of the street; and finally (3) pointing up, still on a street sign, when it appears positioned in front of the driver. Now, not only do configurations 1 and 2 not provoke any significant changes in the behavior of the driver, but neither does configuration 3, that is, the car does not start flying. This is possible when the arrow's appearance is specifically perceived as the result of a convention, untied to any physical connection (resemblance or consequence) with the object represented. And it is all the three configurations which have a symbolic value, not just the third one. It would be too laborious to separate the notion of indexicality, iconicity and symbolicness as applied to the same sign. Much more economic is to establish one single convention (still a convention, but at least "one"), and accord one's behavior on that basis.

For ethologists, usually less demanding when it comes to defining symbolic semiosis (the concept being less central in their interests), it is the entire bee-dance (which they in fact often call bee-*language*) that presents features of symbolicness. Among the evidence they bring in support of this view, a rather interesting example is the geographical variability in the significance of the various signs. Different communities of the same species *Apis mellifica* may for instance use the very same articulation of the distance-sign (that is, the time employed to cover the middle axis of the Fig. 8) to represent completely different distances. The same amount of time will signify 5 m for an Egyptian *Apis mellifica*, 25 m for an Italian, and 75 m for the predictably more efficient German bee (see Mainardi, 1992: 455).

Also, ethologists have two more favorite examples of symbolicness in animal semiosis. One is, of course, bird singing, which presents features of cultural learning and development, geographic dialectal variation, and — most of all — conventional and abstract sign-object codification. The other example, perhaps less obvious, is the system of alarm cries developed by species like the ground squirrels of the genus *Spermophilus*, the prairie dogs *Cynomys*, the monkeys *Cercopithecus* and others, to inform their cospecific of the arrival of a predator. These species emit differentiated cries depending on the type of predator, and it is highly specialized in the case of the vocalizations of the Vervet Monkeys *Cercopithecus aethiops* and of the Greater Spot-nosed Monkeys *Cercopithecus nictitans*. When one vervet monkey sees a predator, she/he produces a vocalization that alerts the rest of the group. This vocalization is of three different kinds, and depends on the specific predator in the vicinity. A loud barking call is given for leopards, a short, double syllable cough stands for eagles, and a "chutter" 'sound stands for snakes. The calls are not similar to the sounds that those predators utter or produce, they are therefore not of the iconic type, and the response of other monkeys to a given call is appropriate for escaping the corresponding predator (i.e., when the leopard call is heard, the monkeys run to the trees; when the eagle call is heard, they look up into the air and seek shelter; and when the snake call is heard they stand up on two legs and look in the grass). In many communities, in addition, specific cries were also noted for two more possible predators, which evidently the vervet monkeys perceive as rather similar to each other: baboons and human beings (for a thorough description of vervet monkeys' calls, see Cheney and Seyfarth, 1990).

Once again, it is the strong dialectal variation (plus the fact that the cries are learned, and it takes a while before the young specimens provide an adequate response to them), to persuade scholars of the symbolic character of this sign-catalog. At a distance of few kilometers, a leopard is signaled with a completely different cry. One may argue that the dialect variation does not exclude *a priori* the possibility that these cries remain of indexical or iconic type. However, the notion can be accepted only if fairly traded with the hypothesis that most of human language, too, is therefore only indexical or iconic, *despite the existence of different languages and dialects.* If we take the example of honomatopoeias, it is clear that the creation of such words departs from iconic logics. Yet, words describing the same acoustic phenomenon can be completely different, depending on the language, as in the case of animal vocalizations (a dog goes "Woof" in the UK *and* the USA *and* Australia, three geographically separated places, yet she/he goes "Bau" in a country, Italy, which is much closer to the UK than any of the other English speaking countries).

Or, alternatively, we could accept the hypothesis that no sign is actually 100% symbolic, as a link (even a remote one) with a physical entity is sooner or later detectable: even street signs themselves, which are normally taken as the most extreme example of symbolicness, show different degrees of this relation (an arrow remains after all iconic, to some extents; the use of the red color for interdictions and dangers is easily related to the recurrence of this color for blood and poisonous substances in nature, etc.).

A last important criterion for detecting symbolicness in signs was established by Peter Marler (1984), and concerns the possibility of discerning a sign from its function. A sign is a true symbol when in principle it can arbitrarily be used in a different context, or provoke a different reaction. As Marler (1984: 354) himself puts it, precisely with reference to the monkeys' alarm calls, "If an alarm call is truly a referential symbol it should be potentially dissociable from acts of fleeing". W. John Smith comments on this criterion, and applies it to the very case suggested by Marler:

> *By dissociable he meant that an individual uttering a call would not always flee. In fact, individuals do not always flee after calling — not just in the case of vervets but in ground squirrels and, indeed, in any well-studied case of which I am aware. Fleeing is simply one of several options that become predictable from such a call. Other options may include pausing to monitor or continuing with a previous activity that was interrupted by calling; the particular conditions of each event determine whether escape, monitoring, or some other alternative follows calling. No single class of behavior occurs as the exclusive and inevitable correlate of a signal* (Smith, 1991: 216).

4.4 Syntax

Another aspect in the semiosis of these animals that probably demolishes many of the assumptions we had, as humans, about our own language, is syntax. In general, after Marler 1977, scholars distinguish between *Phonological Syntax* (a combination of signs that, taken alone, do not necessarily have a specific meaning, or at least not a qualitatively different one from the combination), and *Lexical Syntax* (a combination of signs that are also meaningful as single units, and that also mean 'something else'). No particular effort is required for a language scholar to find examples of phonological syntax in non-human animals. For instance, in Robinson (1984) and in Mitani–Marler (1989) the phenomenon was studied respectively in wedge-capped capuchins (*Cebus olivaceus*) and in male gibbons. As for lexical syntax, the Greater Spot-nosed Monkeys can still be taken as example. This species has two main alarm sounds: something like 'pyow', designating a leopard, and something like 'hack', standing for "eagle". The use of these sounds pragmatically proceeds in the same vein as with vervet monkeys. Recent research, however (Arnold and Zubermühler, 2006), has revealed that the Greater Spot-nosed form sequences with these sounds (e.g., "pyow pyow hack hack"), sequences whose meaning is not simply "A leopard and an eagle" (or two leopards and two eagles), but means something entirely different, that is, "Let us move to another place". This is clearly a case of lexical syntax, where existing meaningful units are combined in order to produce a new meaning, that is not simply the sum of its constituents. Keeping up with the hypothesis that languages are never *fully* symbolic, one might guess that this new meaning developed from an indexical formulation of the type "eagles everywhere, leopards everywhere", which syllogistically turned into a kind of "as dangers are everywhere, this is definitely not a safe place", and refined into the more practical "let's go somewhere else". This type of interpretation can reasonably be applied to all languages including the human one, and can easily explain why, for instance, we went from the red of blood and poison to the red of warnings and interdictions in street signs.

4.5 Names

The last category of signs suggested by Sebeok is that of the "names":

> *A name is a form that identifies a human being (Alexander, Sarah, etc.), or, by connotative extension, an animal, an object (such as a commercial product), or event (such as a hurricane). A name has both indexical and symbolic properties: it is partly an indexical form because it identifies a person and, usually, points to his/her ethnic origin; it is partly a symbolic form because, like any word, it is a product of conventionalized representational practices. Less often, named are coined iconically: Trivial but instructive examples of this can be seen in names given typically to household animals-Tuff, Pooh-Pooh, etc.* (Sebeok and Danesi, 2000: 27).

Leaving aside the initial (unjustified) attribution of the notion of names to humans only, the properties of the "name" sign seem to apply at least to several birds and cetaceans, the two categories that have received the most attention in this respect. Very interesting is the case of the Bottlenose dolphin *Tursiops truncatus*, to which specific attention has been devoted in a number of investigations, particularly the seminal (Janik *et al.*, 2006). Janik and his colleagues found that dolphins possess individually distinctive signature whistles that are used to maintain group cohesion. While identification signals are present in many other species, dolphins' names have the characteristic of being influenced by vocal learning. By using a sample of 14 dolphins at Sarasota Bay, Florida, the researchers recorded and then played the signature whistles to the animals, who recognized their names, and did so (in 9 cases out of 14) also when they heard an artificially-synthesized version of the whistle. The conclusions of the research team is that dolphins extract identity information from signature whistles, even after a complete artificialization of the signal, where all voice features are removed from the signal and only the time-frequency relation is preserved (which corresponds more or less to a complete stylization of a colored painting into a simple pencil contour drawing), showing that dolphins seem to be a very rare case of animals, besides humans, that transmit identity information regardless of the caller's voice or location:

> *This study demonstrates individual identity information being encoded independent of the signaler's voice or location in the natural communication system of a nonhuman animal. We found not only that the frequency modulation pattern of signature whistles carries sufficient information for individual discrimination but that this information is used by the receiver to identify individuals* (Janik *et al.*, 2006: 8295).

Great care is taken to point out that recognition is different from simple discrimination:

> *animals recognize each other's whistles individually rather than just discriminate between them. We define recognition as perceiving something to be identical with something previously known. Discrimination can use but does not require such previous knowledge if it is solely based on the comparison of distinctive features* (Janik *et al.*, 2006: 8295).

Finally, it shall be interesting to point out that the choice of the personal name, which is up to the carrier him/herself and occurs during the first year of life, is the result of a combined imitation of several other names the young dolphin is exposed to — more or less as if a woman choses the name Lina because her uncle is named Linus, her friend is called Tina, and her teacher is called Linda.

5. The Functions of Communication

To conclude this essay, another important way to classify semiosis in animals goes through its *functions*. As already proved by Cimatti (1998: 59–105), the classic theoretical model proposed by Roman Jakobson for classifying the functions of communication can be easily applied in zoosemiotic contexts. The model was originally conceived for the specific area of semiosis called "communication", yet, an extension to other fields of semiosis is easily possible, for at least four of the six functions described below (the expressive, the referential, the metalinguistic, and the aesthetic), do not explicitly refer to a sender–receiver interaction.

Jakobson (1960: 353–357) classifies six main functions within a communication system, each concerned with a particular element of the semiotic process: *expressive* (referred to the *sender*), *conative* (referred to the *receiver*), *phatic* (referred to the establishment of *contact* between sender and receiver), *referential* (referring to the *context* of the message), *metalinguistic* (referring to the *code* used in a given communication act), and *aesthetic* or *poetic* (referring to the form of the *message*). The functions are mutually inclusive: a message is usually the intersection between two or more functions, even if one of them is often dominant and more evident.

In *expressive* communication the most relevant part is the display of the emotional state and the identity of the sender. All kinds of messages more or less display some feelings or emotions. Rogers and Kaplan, for instance, (1998: 86), report the case of Australian magpies *Cracticus tibicen* performing their whole repertoire of songs throughout the entire year. For this species, singing is a ritualized way to give instructions about food, migration, territories, etc. Yet at the same time, and according to intensity, speed, loudness and other parameters, these songs are also an indicator of anxiety, joy, alarm, enthusiasm, attention or other emotional states.

In *conative* communication the sender uses the message in order to make the receiver have a certain reaction and act consequently. Conative functions stand out clearly in all mating and territorial signs, i.e., the bulk of the sign-repertoires of practically all species are basically of conative type. The sender wants the receiver to do something (mate with him/her, leave his/her territory, etc.), no matter how ritualized such signs may be. Conative are usually also those signs used for deceptive purposes, and in general all those aimed to some form of manipulation of the receiver. Curiously, the traditional definitions of animal communication, as provided by classical ethology, tended to emphasize only the conative aspect of sign exchange: "Communication between animals involves the giving off by one individual of some chemical or physical, that, on being received by another, influences its behavior" (Frings and Frings, 1964: 3); "Animal communication evokes a change of behavior in another individual" (Cullen, 1972: 103) and others.

The term *Phatic* refers to the use of communication mainly aimed to establish and/or keep contact between sender(s) and receiver(s). Such a function is probably best illustrated by the ducks' repeated "I am here, where are you?" famously described by Konrad Lorenz. A more articulated example concerns the wolves, *canis lupus*, and other *canidae*. After hunting and rallying together as a pack, wolves act very excited and joyfully howl together. The meanings of this practice are of celebrative type: celebrating the success of the hunt (i.e., the presence of food), the wolves' safety, and the joy of reuniting after some time of separation and difficulties (see Lopez, 1978, 1999: 55). In addition, howls serve the purpose of *practically* rallying the wolves within a pack, before and after hunting, and of preventing pack-mates from wandering off on their own. This range of meanings somehow covers the entire spectrum of a message's phatic function (and of course, of the conative, aesthetic and expressive ones, too, but it was anticipated that it is only very rarely that a function appears alone). In howling one finds the simplest (the establishment of a contact) and the most complex (the strengthening of comradeship) extents of phatic communication. In particular, the act of celebrating the reunion of the pack is a most revealing example of how close cultural, ritual, aesthetic (musical) and phatic elements, gathered in a single behavioral pattern, can be.

In *referential* communication, signs refer to the context, to the surrounding environment, i.e., to everything except the sender of the message (and sometime even to him/her). Referential signs are thus descriptive, and may refer to places, times, people, sensations and so on. "It's 9 o'clock", "Franz Kafka was born in 1883", and "it's summertime, and the living is easy" are all referential messages. Two paradigmatic examples are the above-mentioned dance of the honey bees and alarm calls of vervet monkeys. It can also be added that several displays of an animal's health, identity, social status, etc., share an expressive and a referential dimension. Finally, several birds use acoustic communication in a referential fashion for showing migration routes. A few other, like the Herring Gull *Larus argentatus* and the Greater Honeyguide *Indicator indicator* may use sounds also to indicate sources of food.

In *metalinguistic* communication, signs are used to refer to other signs, as occurs in playing, deception, or several forms of ritualization. In playing, particularly, the adoption of metalinguistic signs is crucial, in that several forms of play are nothing but unserious imitations of "real life" situations, most of which rather dangerous, as hunting or fighting. The signs of aggression, thus (running towards the receiver, displaying anger, etc.), must always be accompanied by other signs, which refer to the former, indicating that all seemingly serious actions taken from that point on, must be interpreted only in a playful way.

Another very interesting example comes from the vast catalog of interspecific or intergroup imitations (particularly through singing). The imitation of another species' or group's singing style is an iconic (and generally referential) sign of that style, but at the same time another important intention is again playful: the animal finds it pleasant and challenging to imitate another specimen. In other cases, imitation serves very precise functions, one of which is related with territoriality: by imitating a potential enemy belonging to another species (including predators), the imitator pretends to belong to the species imitated, and practically "says" that the given territory is already occupied by a member of the potential enemy's species. Finally, imitation in singing may also recall the — very metalinguistic — use of columns in medieval cathedrals. Not only do they hold the church up, but variously styled columns coming from different places, placed throughout the cathedral, would serve ostentatiously, as a sign of richness. As the mating season approaches, the male of the Superb Lyrebird *Menura superba*, whose usual mating task is to impress the female, adorns his singing with a huge set of short imitations of different species and objects (including dog whines and doorbells). The rule is *the richer the better*, thus signs made for totally different reasons (like the doorbell) are here used metalinguistically as a display of the creativity and musical skills of the Lyrebird.

Finally, now singing has been mentioned, it is also clear that several signs and sign-repertoires related with mating (and not only with mating), are of *aesthetic* type, that is, they concentrate on the message itself, elaborating on its representation and form. Architectural, pictorial, kinaesthetic, and musical signs (to use the classification adopted in Sebeok, 1981) always carry a (normally predominant) aesthetic dimension. I devoted a substantial portion of my research to these signs, particularly the musical ones (Martinelli, 2002, 2009, for example).

As a bonus, it may be worthwhile to mention a seventh function of communication, elaborated and proposed by Cimatti (1998: 102–105): the *cognitive* one:

> *To speak about a Cognitive Function of language means to deal with how an animal's inner and outer behavior change in consequence of the possession itself of a language. In particular, we are interested in establishing how the way of thinking of an animal changes according to the fact that it is able to semiotically interact with other specimens* (1998: 102).

The example brought by Cimatti is once again that of the monkeys' alarm calls. When a danger appears, the monkey's mind is probably activated in two ways: on the one hand, the mental representation of the perceived object (what kind of danger, how close, how fast, etc.); on the other hand the activation of a signal, i.e., that specific alarm call that stands for that specific danger. Here one notices already a "change" in the semiotic animal: if she/he did not possess a "language", his/her reaction would have been of another type (namely, most probably, escaping from the danger).

In addition, a second characteristic of the cognitive function is that the call does not only act in a heteroceptive way (i.e., as a referential sign describing the danger and as a conative sign for the monkey's community), but also proprioceptively, in that it increases the self-control skills of the animal (who is now *expressing* his/her emotional status and *self-conatively* guiding his/her own behavior). In other words, the typical semiotic description of language as a communicative system *and* a modeling system finds here a perfect match: not only the mental complexity of an animal functions as *source* for a rich sign-repertoire. The process works the other way round too: the richer and more complex the "semiogram" (i.e., the complete sign catalog of an individual), the more sophisticated the mental life. The active participation to semiosis *opens up* for further — and increasingly articulated — cognitive abilities.

References

Arnold K and Zuberbühler K. 2006. Language evolution: semantic combinations in primate calls. *Nature*, 441, 303.

Cheney DL and Seyfarth RM. 1990. *How Monkeys see the World: Inside the Mind of Another Species*. University of Chicago Press, Chicago.

Cullen MJ. 1972. Some principles of animals communication. In *Non-Verbal Communication*, Robert A. Hinde (ed.), 101–122. Cambridge University Press, Cambridge.

Fernald RD. 2014. Communication about social status. *Current Opinion in Neurobiology*, 28, 1–4.

Frings H and Frings M. 1964. *Animal Communication*. Blaisdell, New York.

Cimatti F. 1998. *Mind and Language in animals. Introduction to Cognitive Zoosemiotics*. Carrocci, Roma.

Griffin D. 1976. *The Question of Animal Awareness.* Rockefeller University Press, New York.

Hailman JP. 1977. *Optical Signals.* Indiana University Press, Bloomington, IN.

Jakobson R. 1960. Concluding statement: linguistics and poetics. In *Style in Language,* Thomas A. Sebeok (ed.), 350–375. Cambridge University Press, Cambridge.

Janik VM, Sayigh LS and Wells RS. 2006. Signature whistle contour shape conveys identity information to bottlenose dolphins. *Proceedings of the National Academy of Sciences of the United States of America,* 103, 8293–8297.

Lopez B. 1978. *Of Wolves and Men.* Scribner's, New York (Italian edition: *Lupi: dalla parte del migliore nemico dell'uomo.* Casale Monferrato: Piemme, 1999).

Lorenz K. 1971. *Studies in Animal and Human Behavior*. Harvard University Press, Cambridge, MA.

Mainardi D. 1992. *Dizionario di etologia*. Einaudi, Torino.

Maran T. 2007. Semiotic interpretations of biological mimicry. *Semiotica,* 167(1–4), 223–248.

Marler P. 1977. The structure of animal communication sounds. In *Recognition of Complex Acoustic Signals,* Theodore H. Bullock (ed.), 17–35. Springer-Verlag, Berlin.

Marler P. 1984. Animal communication: affect or cognition? In *Approaches to Emotion,* Klaus R. Scherer and Paul Ekman (ed.), 345–365. Lawrence Erlbaum Associates, Hillsdale, NJ.

Martinelli D. 2002. *How Musical is a Whale? — Towards a Theory of Zoömusicology.* International Semiotics Institute, Helsinki, Imatra.

Martinelli D. 2009. *Of Birds, Whales and Other Musicians — Introduction to Zoomusicology*. University of Scranton Press, Scranton.

Martinelli D. 2010. *A Critical Companion of Zoosemiotics: People, Paths, Ideas.* Springer, Berlin, New York.

Mitani JC and Marler P. 1989. A phonological analysis of male gibbon singing behavior. *Behavior*, 109(5), 20–45.

Nöth W. 1990. *Handbook of Semiotics.* Indiana University Press, Bloomington, IN.

Östlund-Nilsson S, Mayer I and Huntingford FA. 2007. *Biology of the Three-Spined Stickleback.* CRC Press, New York.

Robinson JG. 1984. Syntactic structures in the vocalizations of wedge-capped capuchin monkeys, *Cebus olivaceus*. *Behavior*, 90, 46–79.

Rogers LJ and Kaplan G. 1998. *Not Only Roars and Rituals*. Allen & Unwin, St. Leonards.

Sebeok TA. 1972. *Perspectives in Zoosemiotics*. Mouton de Gruyter, The Hague.

Sebeok TA. 1979. *The Sign & Its Masters*. University of Texas Press, Austin.

Sebeok TA. 1981. *The Play of Musement*. Indiana University Press, Bloomington, IN.

Sebeok TA. 1990. *Essays in Zoosemiotics*. Toronto Semiotic Circle, Toronto.

Sebeok TA. 1994. *Signs: An Introduction to Semiotics*. University of Toronto Press, Toronto.

Sebeok TA and Danesi M. 2000. *The Forms of Meaning*. Mouton de Gruyter, Berlin.

Smith WJ. 1991. Animal communication and the study of cognition. In *Cognitive Ethology: The Minds of Animals: Essays in Honor of Donald Griffin R*, Ristau CR and Griffin DR (ed.). Erlbaum L. Associates, Hillsdale, NJ, 209–230.

Tembrock G. 1971. *Biocommunication*. Akademic Verlag, Berlin.

4. Communication as an Artificial Process

Massimo Negrotti

Full Professor of Methodology of Human Sciences
University of Urbino, Italy
massimo.negrotti@libero.it

In the animal and vegetable kingdom, communication plays a fundamental role of regulating individual behavior, and biosemiotic management of standardized signals constitutes the core process that allows mutual adaptation. However, only human beings have developed high-level language based on semantics — i.e., a sort of 'technology' aimed at sharing not only intentions and behaviors, but also complex mental states, such as formal reasoning, subjective attitudes, and emotions.Yet language, even if originating from natural human aptitudes, has developed as an artificial cultural construct intended to enrich purely signal-based messages through the added value of meaning. What actually happens is that, as in all cases of artificial reproduction of natural exemplars — which we shall call *naturoids* — what a message induces in the mind of a listener or a reader is not verifiable, and may deviate greatly from expectation.

1. Introduction: Information and Knowledge

Communication among living systems certainly works as a good adaptive advantage, as it allows the exchange of messages regarding what is happening in the immediate surroundings and beyond. For instance, as is well known, ants are able to send and receive messages regarding the presence of food or of some form of danger, and they can even share help messages. All this happens by means of chemical substances modulated to indicate, for example, different aspects of danger, and by means of sounds and the use of antennae. When faced with uncertain situations, animals can resolve, with a certain flexibility, problems of interpretation of messages, even to the point of correcting errors. The fact remains, however, that their elaboration does not go beyond an informational level strictly linked to their high degree of specialization, unlike that which happens in man, with his 'openness to the world' (Gehlen, 1983).

Humans, too, of course, have developed the ability to exchange messages, thanks to the physiological organs at our disposal, but we do not limit ourselves to exchanging messages about facts in the external world that might be verified by anyone. We also exchange our personal knowledge — that is to say, what is happening in our own minds, and not just in the perceived environment. This implies a shift from a purely information-based communication to a communication aimed at sharing mental states, including thoughts, emotions, memories, and interpretations of individual and communal facts (de Solla Price, 1964).

The term 'knowledge' should not to be misinterpreted here. It does not refer only to the higher capacities of the intellect, such as abstraction, ideation, or problematization with regard to available information. Beyond the epistemological profile, in the context of the discussion that we are introducing, with the term 'knowledge' we intend, on the one hand, what Michael Polanyi meant when speaking of all that is involved in our mental abilities — i.e., all that justifies his main affirmation, according to which "we can know more than we can tell" (Polanyi, 1966). On the other hand, 'knowledge' means, here, the whole amount of personal mental states, including mental images and emotions that we know and recognize within ourselves, but which we find very difficult to impart through the narrow funnel of words.

The problem of the 'substance' of words, whether it be natural or artificial, has been discussed by numerous philosophers, including Plato and Thomas Aquinas. For the former, with his particular form of idealism, words are perfectly able to represent things, while for the latter, "Words as words are essentially artificial or arbitrary in the sense of *ad placitum*, since they are products deliberately made to signify ..." (Oesterle, 1994). On the other hand, even if, as far as our everyday uses of words in communication is concerned, the notion of 'natural language' is widespread, linguists agree in defining as artificial those languages that are intentionally constructed for some functional objective, such as computer languages, or Charles Ogden's K. *Basic English*, proposed in 1930. More recently, Pierre Bourdieu has spoken of 'legitimate languages' — namely, semi-artificial languages such as those adopted by institutions and educated persons, deliberately controlled and constantly under revision (Bourdieu, 1991). Finally, all current languages are governed by explicit grammatical and syntactical rules, while, at the level of semantics, in spite of the efforts of semiology, it is not realistically conceivable that the meaning of the words, even if conventionally expressed in dictionaries, might be intimately and univocally shared by human beings. As the Italian linguist De Mauro writes, "... one can say, however, that, to philosophical minds, the gap between the rich, varied fullness of existence and the poor repetitiveness of the human word has always seemed unbridgeable" (De Mauro, 1999). Likewise, Dante Alighieri, in *Paradise*, confesses "Oh how brief is speech, and how feeble to my concept! and this, to that which I have seen, is so much, that it does not suffice to say 'little'." It therefore does not appear unreasonable to consider the messages we construct in verbal communication as a metaphor of the artificial reproduction of objects or processes that, originating in our mind, or reflected and reworked in it according to our relationships with the external world, we desire to put in common with others by means of tools — our words — highly standardized and thus available to the most diverse subjective interpretations.

This brings us back to the problem of meaning as something that is very clear to us, but which becomes something else when it is reduced to words to be interpreted by someone else. Such a difficulty does not certify incommunicability as an inexorable outcome in human relations, although, as we shall see, it is now a widely recognized human condition with regard to the use of language. Human cooperation is a fact, but it relies, on information exchange, and not on the sharing of personal knowledge (*cf.* Peters, 1999). Thus, as occurs for all animal and even vegetable species, human groups gain survival benefit from informational coordination. Paradoxically, however, the functional coordination of human communities increases the more communication is 'uncontaminated' by the subjective interpretation of messages, because introducing semantics linked to each person's life story deforms the conventional character of the information.

We may adopt a *knowledge-based* message, such as the poetic verse by Verlaine, "Blessent mon cœur d'une langueur monotone," as an information signal, but the contrary is not possible, of course, without generating uncertain consequences.

On the one hand, we may maintain that animals and trees communicate through physical or chemical *signs* supporting information that allows, or even forces, the living system to adopt some stereotyped behavior. On the other, we should accept the idea that human communication acts at two different levels: that which we share with animals, and that which we have constructed culturally through language, including not only signs but also what Charles Peirce S. has named *interpretants* — i.e., pieces of mental entities that work just in order to interpret signs through other signs. According to the definition given by Peirce, "A *Sign* is anything which is related to a Second thing, its *Object*, in respect to a Quality, in such a way as to bring a Third thing, its *Interpretant*, into relation to the same Object, and that in such a way as to bring a Fourth into relation to that Object in the same form, *ad infinitum*" (Hartshorne *et al.*, 1958). Under this profile, therefore, the main difference between information and knowledge consists in the fact that, while an informational process consists in a sort of 'mechanism', more or less 'plastic', of codified translation of signals, one of the proprieties of knowledge is certainly the creation of individual meaning through particular processes of association constituting one's 'semantic history.'

The lowest level of human communication includes messages whose 'meaning' is, as it were, reduced to their informational content, and which cannot be freely interpreted without losing communicational effectiveness. For example, if I am told that my train will depart at eight o'clock, then I cannot simply decide to depart at nine o'clock without missing that train. Therefore, a purely information-based communication implies a rather closed system, because coding and decoding cannot be open to *ad libitum* reference structures, but only to pre-established sets of possible actions. This fact allows the possibility of assessing the success of information-based communication, because the congruity of the input and output can, in principle, always be verified empirically (for instance, if I succeed in taking the desired train). Clearly it is not possible to reduce meaning to such a correspondence, as some authors seem inclined to think (see, for instance, the classic work by C. K. Ogden and I.A. Richards, *The Meaning of Meaning*, 1923), although informational correspondence allows a pragmatic cooperation among humans, as it does with other living systems. In some ways, informational communication can be thought of as a flow of effects — i.e., ordered physical events — as happens in the case of DNA, and in many organic processes such as the involuntary activation of muscles. This is why a computer can be programed to process even large amounts of complex information structures, thanks to deterministic programs, yet is unable to deal with the meaning of even the simplest mental state communicated to it by means of words.

On the other hand, a knowledge-based communication, though it also may be triggered by an information-based one, consists of an open process, in the sense that the interpretation of the message depends not only on a given code, but also on a dynamic and complex process of relationships among mental states and related meanings.

While informational messages can achieve high degrees of effectiveness, messages based on personal knowledge are linguistic structures always open to arbitrary interpretation through a sort of hermeneutics that may proceed *ad infinitum*. In daily life we all send and receive both types of message. Nevertheless, in order to assess the degree of success of our knowledge-based messages, we always try to reduce them to the correspondence between the actual behavior of our interlocutor and the meanings we have assigned to our messages. In doing so, we testify to the impossibility of verifying directly — as might be done with informational messages — the reconstruction, in somebody else's mind, of the meaning we feel in our own. Of course, the behavior of our interlocutor might be only seemingly consistent with our message. It might be due to quite different reasons, such as generic courtesy or, much more likely, to the reference he sets up with his own experience.

Literature and theatre, due to their exploitation of words by definition, very often exhibit a clear understanding of this effect. For instance, it is synthesized by the Italian dramatist Luigi Pirandello in his work *Uno, Nessuno e Centomila*, when one of his characters says

> "... you, my dear fellow, will never know how I translate that which you say to me. You have not spoken Turkish, no. We have used, you and I, the same language, the same words. But what fault have we if the words, in themselves, are empty? Empty, my dear fellow. And you fill them with your meaning, in saying them to me; and I, in receiving them, inevitably, fill them with my meaning. We believed that we had understood each other; we had not understood at all" (Pirandello, 1994).

Thus, as Pirandello later concludes, "in the passage from one mind to another, modifications are inevitable" (Pirandello, 1994). But, the same may be said for Ionesco's work, and also for Beckett's poetics when, for example, he introduces a sort of pathology of communication whereby the character Murphy is said to have left the listener "spattered by words that went dead as soon as they had sounded; each word obliterated, before it had time to make sense, by the word that came next" (Cited in Kennedy, 1978).

2. Communication as an Artificial Reproduction Process

While the difficulty of human communication of personal knowledge is widely accepted, the problem remains of providing a convincing, if not conclusive, explanation of this fact. A complementary way to understand the almost impossible transfer of a precise personal meaning to the mind of another could be based on the analogy between communication and other processes, such as the attempts to reproduce natural instances by means of technological strategies.

The starting assumption of the explanation proposed here is that a message can be seen not only as a linguistic fact, but also, from a functional point of view, as the attempt to reproduce one's mental state in the mind of someone else, by exploiting the seeming representational power of the words. Because of its conventionally coded structure, a purely information-based message does not try to reproduce anything, and we can obtain, without problems, a true *replication* of it. By contrast, a knowledge-based message tries to reproduce a mental state by means of words, thus giving birth to a sort of artificial object. Information-based messages are structures that require simple decoding, while knowledge-based ones need to be understood through interpretation as well as decoding. Moreover, knowledge-based messages require both decoding, as a necessary condition, and interpretation as a sufficient one, but in such a way that the latter cannot rely only on the former, because interpretation strongly depends on the personal knowledge of the interpreter, who cannot prevent linguistic signs from triggering memories of past experience relating to that sign.

As stated by Peirce, "... the operations of a higher level cannot be accounted for by the laws governing its particulars forming the lower level. You cannot derive a vocabulary from phonetics; you cannot derive the grammar of a language from its vocabulary; a correct use of grammar does not account for good style; and a good style does not provide the content of a piece of prose" (Hartshorne *et al.*, 1958). Furthermore, when we compose a message, we are inclined to think that "we shall find the words we need and bring them out in the order expressing our thought" among the "many thousands of words" available, relying on our imagination in order "to evoke from these available resources the implementation of our purpose" (Hartshorne *et al.*, *cit.* p. 36).

Thus, the whole process of constructing a message is very close to, or analogous to, that of setting up a project. In particular, when we communicate, we try to reproduce our mental state by exploiting the 'linguistic technology' *in the best way* — i.e., trying to get from the message — even resorting to a desperate 'dance of synonyms' — the ability to reproduce optimally what we see and feel in our mind. This is, in the end, what we mean when we say "I do not know how to say this in other words," exactly because, when speaking, we are also listening to ourselves, and of course, in our mind, the finished message sounds very effective. We therefore assume that the clarity of the message should be unaltered when it reaches the mind of our interlocutor. Really this hope stems from the noble conviction that all human beings have the same 'nature,' thus neglecting the deep differences in their individual disposition and experience — i.e., the very matter that a message would reproduce.

2.1 Reproduction

Considering a message as an artificial object means looking at it as the result of a design process very similar to the one implemented in the many fields of naturoids, whereby, by means of the available technology, humans have tried, since ancient times, to reproduce natural phenomena for practical or scientific purposes, and even for aesthetic ones as happens in the field of the arts. (For a comprehensive reference to the theory of naturoids, see Negrotti, 2012b.)

The study of man's attempts to reproduce aspects of nature via technological means, carried out for centuries with varying degrees of success, should bring to light certain methodological constraints of design and realization common to all projects aimed at creating naturoids — constraints that leave little space for the Utopia of replication through analytic models and technology, yet which open the way to wholly new, largely unforeseeable, prospects.

The practical objective underlying every naturoid development project is that of realizing an object that is *indistinguishable*, at least in some particular aspect, from its natural counterpart, which we may call the *exemplar*. In the communication process, the exemplar is the mental state one wishes to share, of course, in the hope that the message will include all that is necessary for reproducing that mental state in the mind of the listening or reading person.

Experience teaches us, however, that every artificial reproduction of a natural exemplar is carried out at a certain exclusive, or at least dominant, *observation level*, and that it is in relation to this that the indistinguishability assumes value. To describe a flower at an a esthetical, mechanical, chemical or electrical observation level implies generating different models, and therefore different reproductions, it being understood that replication, intended as reproduction at all possible observation levels, is inconceivable through technology and the construction of analytic models, because these are based on the selection and isolation of observation levels.

Indeed, a designer first observes an exemplar in nature, placing himself at a certain observation level that depends on various factors, including personal preferences and professional culture; and, after having modeled it, he or she attempts to reproduce it by means of the available technology and materials. If the resulting naturoid is successful, then anyone, by placing himself at the designer's adopted observation level, will be unable to distinguish the naturoid from its natural counterpart.

The concept of 'observation level' I adopt here is rather similar to the 'reality levels' advanced by Oppenheim and Putnam (1958), and that of *observation perspectives* of Charles Morris's 'objective relativism' when he says, albeit in a metaphysical direction, that something exists only "within a perspective, and there is no all-embracing absolute perspective within which everything exists." (Morris, 1948). Nevertheless, an observation level should be conceived as a more modulated profile of reality, including not only the hierarchy of scale, which includes micro and macro options, but also the range of qualities to be found at every hierarchical level, such as the mechanical or the electrical, the chemical or the biological, or even the aesthetic outlines we unavoidably select in any description of the world.

This is particularly true when the 'world' is our mind, and when we isolate in it some particular state we desire to communicate. In doing so, we always assume an observation level that depends on variable subjective conditions: for instance, if we want to communicate a memory of our travels, we will unavoidably assume a dominant observation level linked to the landscapes we saw, or to the historical, anthropological or artistic profile of what we encountered during our travel. We could list all these aspects by means of a succession of sentences, each of which accords with a dominant observation level.

All of us, when describing any empirical object, inevitably assume only one level of observation at any given time. Upon changing the observation level, indeed, the similarity between the naturoid and the exemplar diminishes, or perhaps even disappears altogether. An artificial heart is, as it were, placed at the same level of observation at which the organism awaiting blood is placed, and not at other levels, such as those that might be adopted for the purpose of observing, describing or teaching the heart's structures and functions. Much the same happens for the advanced projects in which the electrical actuators of an artificial limb may be activated by suitably amplified myoelectric impulses emitted voluntarily by the patient, just as if directed towards normal organic muscles. Similarly, the electrical signals emitted by an artificial retina might be processed by the brain just as if they had come from a natural retina. It is clear, though, that, if an organism expected, of an artificial device, all the features of the natural exemplar, right down to the cellular level, the deception would be revealed, and unpredictable results would emerge.

By analogy, a message, in this framework, aims at the same objective, because the words should lead the listener or reader to feel and see what we feel and see in our mind when adopting our own observation level, thereby enabling him or her to share the given mental state.

Normally, the observation level selected by the utterer poses no problems, for it is indicated, more or less clearly, by the main terms, aided by appropriate adjectives and adverbs. For instance, if one says "During my trip to Rome, I saw very fascinating monuments," it is clear that the account refers to an architectural or historical observation level. But even here, as happens in the field of technological naturoids, if the listener changes, for some reason, the observation level at which he wants to understand the message — for example, thinking of the traffic in the Roman streets — the result will be a complete reinterpretation of the message. In other words, the listener will be unable to 'see and feel' what the utterer desired to reproduce in his or her mind. Indeed, even if both speaker and listener share the same observation level, the reproduction in the mind of the latter of the mental state of the former is very uncertain and, in any case, not assessable.

Just as the components of an artificial muscle are not biological, the words are not, as such, the reality of our mental states, nor are they made of the same 'stuff': they are signs that fire the imagination, and when we communicate, we hope that a sentence is able to force the listener or reader to feel as we do thanks to the 'trick' of our style, which should reproduce our mental state in an unequivocal sense. As a result, something in the mind of our listener will surely occur, but even if he can share the observation level that we have selected, what he will see and feel will depend upon his own mind and not on ours.

In a sense, a message is an artificial object whose capacity to reproduce a mental state is very effective for us — that is to say, in our own subjective context, as happens when we read some note we wrote in the past to help us to remember some special event. But, when the message reaches our listener or reader, it will be plunged, as it were, into a quite different context: a context within which each word evokes both conscious and unconscious mental states, and relationships among mental states, whose general map depends on and represents the individual history of life. Therefore, the two maps — that of the utterer and that of the listener — have a very low probability of overlapping substantially. In other words, when the message leaves the context within which it, as an artificial device, has been well designed and works well (i.e., in the mind of the utterer), and enters the mental context of the listener, it will face very different environmental conditions, and may well act unpredictably.

This happens frequently also in the field of technological naturoids, when a device leaves the laboratory and faces the real world context — such as the human body, for example. For reality does not know any observation level, the selection or construction of which is peculiar to the human mind and to the taxonomies it constructs, and, therefore, within the real world a naturoid will always undergo a complex process of interactions.

Unpredictability in the communication process is also a rule. Unlike in the technological reproduction of natural instances, in human communication, conceived as a reproduction process, there is no guarantee of success, because even if our listener or reader is made to assume our own observation level, what will happen at this level in his or her mind is quite uncontrollable.

3. Foreground and Background

Both scientific knowledge and naturoid technology are based on the implicit idea that, in order to analyze a natural object or process, one has to isolate it from the rest of the world, although empirical reality does not recognize any separations or boundaries. Not by chance, the term 'definition', which we use constantly in vast areas of communication, derives from the Latin *definire,* referring to 'fixing the boundaries'.

As a matter of fact, we would have no hope of generating a reliable knowledge of something without resorting to this ability. A truly holistic view, whilst at the core of human ambition, does not allow us to generate any taxonomy and to induce any general law — something that is possible through isolating, under some profile, homogeneous classes of objects in empirical reality.

On the other hand, isolation does not come free of charge. Whenever we focus on one object, we bring it to the foreground, inexorably letting all else drift into the background, counting, on the one hand, on a sufficient peculiarity of the isolated object, and, on the other, on the *ceteris paribus* clause. In doing so, we will often gain knowledge of the object in isolation, but lose several possible relationships between it and other phenomena that may be significant when trying to predict the behavior of the investigated phenomenon, or to place an artificial object in the real world.

In biology, for instance, consider exemplars from among the animals, such as the *Holothuria* (the sea cucumber), which lives symbiotically with the little fish *Fierasfer acus*. Many other biological systems live in a symbiotic way, including several insects that survive through symbiosis with particular trees, such as hornets with fig trees (*cf.* Dawkins, 1996). Separating these two *symbiotic* systems, for both knowledge and reproduction aims, could lead to improper conclusions or models.

In the case of bioengineering, it has been rightly observed that a

> "... multi-purpose approach is quite different from the single function concept that governs most machines. Designers must always be mindful that each organ works in conjunction with all others: liver failure damages the brain and kidneys; the lungs and heart function in an intimate way; etc." (Fitzgibbons, 1994).

Analysis — a term which derives from the ancient Greek 'to break down' — sometimes weakens the effectiveness of the scientific explanation and prediction, and also the design of a naturoid. For instance, in the nineteenth century, in order to elucidate the functions of the thyroid gland, investigators also inadvertently removed the parathyroid glands, thereby misunderstanding the thyroid gland's true function (Hamdy, 2002).

The same occurs in human communication, because, when we isolate a single statement from a discourse, we seemingly gain an increase in clarity, but we simultaneously lose possible relationships between that statement and the rest of the personal meaning that our interlocutor attributes to the whole message. For this reason, literary critics and well-educated persons are accustomed to resorting to so-called 'context analysis' — i.e., placing a statement within a wider set of signs from which one hopes to gain a more reliable and complete interpretation of a given message. Unfortunately, even this effort guarantees nothing, owing to the huge dimensional difference between the context analysis, however rich, and the whole structure of personal mental states.

3.1 Selecting and reproducing what is 'essential'

In the field of technological naturoids, the choice of an observation level and exemplar are the first two unavoidable steps that ultimately lead to the design of an artificial object.

Nevertheless, the choice of an exemplar is not the conclusive defining step before the actual realization of the project. The last and most strategic moment could be defined as the choice or attribution of an *essential performance* in the natural exemplar. Essential performance is a quality, function or behavior of an exemplar, or even simply that aspect that the designer considers particular, typical, or indispensable. In other words, the essential performance is the true objective of a naturoid project. The selection of an observation level is needed because of our limited ability to observe the world, and the choice of an exemplar is required for isolating an object — or, rather, a portion of the world — from the background. However, these two actions do not in themselves indicate *what* we really intend to reproduce. When we say "in this laboratory we are committed to the reproduction of a human heart," we refer clearly and unequivocally to the exemplar we have in mind; but we all know that nobody would be able to reproduce a human heart, or almost any other natural exemplar, in its entirety — that is to say, at all possible observation levels and including all its properties and functions. (Several statements by researchers clarifying this point, from J. de Vaucanson up to contemporary bioengineers, may be found in Negrotti, 2010.)

Therefore, any naturoid project requires the selection of a central, or essential, performance of the natural exemplar as the main target to reach. This is true to such an extent that, in many cases, designers, although inspired by a natural exemplar, end up designing and implementing projects based on models whose structure bears little similarity to that of the exemplar, having found a way to reproduce the essential performance more directly by exploiting some available technology.

Let us look at the invention of the throttle valve for steam intake and exhaust in steam engines invented by Thomas Newcomen, in 1812. It seems that the person assigned the job of regulating the throttle valve, in order to save repetitive labor, tied the cord to the piston so that the latter could pull the cord and release it automatically. Instead of designing a human-like robot able to reproduce the mental evaluation of a man and then intervene in the process to be regulated, Newcomen's assistant focused directly on the desired performance. This happens widely in every field of automation, where the final performance to be achieved is evaluated as more important than the reproduction of the natural structures or processes that generate that performance.

In human communication, the selection of an essential performance to be reproduced in the message is characterized by analogous constraints, because we each guess that which we cannot reproduce in our own mental state as a whole. To return to the example of travel, it is clear that, if the travel lasted several days, we could not hope to reproduce all the emotions felt, even if at one only observation level. On the other hand, we might have a clear idea of the 'synthetic effect' we have drawn from that travel at, say, a landscape observation level, yet still have insufficient 'linguistic tools' to reproduce it in a message. Our only hope, therefore, is to concentrate our linguistic ability on something essential, exploiting our vocabulary and use of style in order to 'give an idea' of what we felt. Travel journals, so widespread in the 18th and 19th centuries, are no exception even if they often provide precious information on places, peoples, and cultures. Moreover, those of great literary significance, particularly from the romantic period, accentuate the natural subjective selection of that which is essential, giving rise to the typical transfiguration that characterizes every artistic production (Negrotti, 1999).

On the other hand, here, too, in contrast to what happens in the realm of technological naturoids, there is no easy way, and perhaps no way at all, to verify whether the mental state that our message has triggered in the mind of our listener or reader is exactly that which we desired to share. Even our interlocutor's reply or observed behavior cannot be taken as proof of his or her understanding. For example, if one recounts a sad tale, the listeners may start crying, but they will often do so after being drawn back into their own personal experience, triggered by the words as generic signs, and not because they share the individual essential performance of the mental state of the utterer.

Furthermore, in human communication, we have no possibility of reproducing the essential performance of our mental state in an 'automatic' way — that is to say, attempting to reproduce it directly, avoiding the adoption of some observation level. Indeed, the linguistic tools at our disposal are the only ones that might allow, but only ideally, the needed modulation and elaboration of the meaning, and we therefore cannot resort to any other expedient in order to communicate our emotions faithfully.

3.2 Materials and their inheritance

A naturoid — that is to say, an object built in order to reproduce a natural exemplar — must, by definition, be realized by means of materials that are different from those adopted by nature. The main ambition governing all the projects in the field of bioengineering is, therefore, as stated by a bioengineer, that of "making the material invisible to the body" (Mikos *et al.*, 1996). This is a key point to keep in mind in order to avoid rejection of the extraneous entity by the organism, thus rendering it useless or even deleterious.

The words we use in forming a message play the same role in the communication process as that played by the concrete materials in reproducing this or that feature of a natural exemplar — for instance, a physiological function of the human body. In fact, one's vocabulary may be seen as a repository of material units developed precisely for the purpose of reproducing our thoughts and feelings, because words cannot but have a communication function. In this sense, languages are the sole technology intrinsically oriented towards giving birth to something artificial, while the materials adopted in all fields of naturoid technology historically come from a wide range of uses.

Before the rise of a specialized 'material science,' Jacques de Vaucanson, in the 18th century, interested to the reproduction of the digestive tract of a duck, was excited to hear of the newly discovered Indian rubber. In much the same way, according to a contemporary bioengineer, "lifesaving heart surgery often relies on a polymer originally developed for women's fashions or a plastic meant for insulating electrical wires" (The Whitaker Foundation, 1995).

Nevertheless, the distance, as it were, between the words as signs, on the one hand, and the corresponding mental states, on the other, is far greater than that between the materials adopted by bioengineers and the corresponding materials adopted by nature when building natural phenomena. Indeed, words can *evoke* rather than *reproduce*, although we generally hope that a well-constructed message will be able to reproduce the right mental state in the mind of someone else, as it does in our own. Significantly, when we are aware of the weakness of this hope, in order to maximize the effectiveness of our message, we resort to other means that should strengthen it, such as gestures, modulation of the voice, or facial expressions. Robert Musil synthesizes this point very clearly when he underlines that "it is certain that a man can be understood much better by his eyes than bywords" (Musil, 1991: 93). On the other hand, our considerations apply also to gestures and other means of expression, because they, too, are deeply linked to our own previous experience.

Indeed, a sort of *inheritance principle* is a common feature of communication and of naturoid technology, since whatever material is adopted in order to reproduce something will, of course, enter the scene with *all* of its properties, and not just those that seem to allow a good reproduction. Plastic is deformed by heat;

metals are sensitive to many physical phenomena; electric and electronic materials interact in complex ways; chemicals react with each other at many levels; and words, likewise, are linked to each other according to very different personal mental paths.

For many reasons due to the personal history of each of us, many words or styles may be stored in our mind along with bad memories, and, as a consequence, listening to them or reading them may trigger a repulsion or, at least, a strong deviation from the 'meaning' that our interlocutor intends to impart. In reality, the only 'pure' messages that can exist are information-based ones that may be taken as the uttering person would desire, thus avoiding any possible transfiguration. This is due to the fact that the words are pure signs — i.e., informational structures — only in a dictionary; but, in one's mind, they are deeply embedded in one's experience, and their 'meaning' strays very far from their information content.

In the realm of technological naturoids, the inheritance principle is the main cause of the *side-effects* that arise from the interplay among the properties of the various adopted materials at all their known and unknown observation levels, and on the relationship with the context in which the naturoid must live. In the field of communication, the inheritance principle is the basis of misunderstandings, of improper associations, and, in the end, of a more or less complete departure from the mental state of the utterer. In a sense, therefore, the 'play of musement' — the free association among *interpretants* — which Peirce conceives as a sort of vacancy of human mind, is instead a permanent attitude when humans interact to share personal knowledge.

One interesting consequence of the theory outlined here is art, often regarded the human activity that best maximizes the efficacy of communication. Really, however, this is a quite improper supposition. Artists know much better than ordinary people the impossibility of actually sharing emotions, and the value of their works does not reside in the extra-artistic 'meaning' they transfer, but, rather, in the *form* that this transfer assumes. (For a more complete exposition of this point, see Negrotti, 2012a.) Indeed, even in the case of art, enjoying a painting or listening to a theatrical or musical piece surely generates *emotional effects*, but the degree of overlapping with those of the artist is highly indeterminable. Most likely, "the emotions aroused in me are not the emotions expressed by the music" (Robinson, 1994).

In fact, the artist *expresses* his own emotions, intentionally avoiding the ordinary practices of everyday communication. He or she exploits 'materials' that, as such, would be very difficult for ordinary people to adopt for communication aims, and when using words, the artist processes his or her own deep personal meaning through a non-ordinary architecture, not necessarily linked to commonly shared signs. Really, for the artist, the materials adopted are, in themselves, intrinsically 'full of meaning,' and he builds his 'message' modulating them according to his poetics. This is why Gustav Mahler stated that "If a composer could say what he had to say in words, he would not bother trying to say it in music." An artistic message is not, therefore, an alternative way of sharing, as we all try to do, psychological emotions with the public, but the reproduction of the artist's musical, pictorial or poetic emotions as such.

3.3 Deception

The fact that the design and construction of a naturoid is always related to a dominant level of observation implies a sort of deception. This is due to the simple fact that the level at which the naturoid becomes indistinguishable from the natural exemplar must be well protected from possible deviations or variations of the observation level, much as happens, *absit injuria verbis,* in a convincing conjuring trick.

In the field of art, for instance, from a purely aesthetic point of view, we could refer to such examples as the *trompe l'oeil,* which dates back to the 16th century, or the artificial perspective of Leon Battista Alberti and Piero della Francesca.

In the field of biology, it would be impossible to list all the phenomena of mimicry characterizing the many species, almost always for the purpose of defence or attack. The point is that, in all of the reported examples, the system formed by the naturoid and the host context intended to accept it, and perhaps to make use of its actions or qualities, must maintain an equilibrium that is possible only in a rather restricted domain. Indeed, even a slight change in the observation level at which the naturoid functions acceptably can, as it were, unmask the deception and lead to the degeneration of the system itself.

Deception, of course, is already prevalent in nature, but there, too, it is precarious and contingent. So, for example, a predator may spot even a well-disguised prey whose color and form blend effectively into the background, yet whose presence may still be betrayed by its movement or its odor.

With regard to naturoids, limitations in the technology may lead to some sort of breakdown of the device. For example, the visual tracking of a fast-moving object may require too high a scanning rate for an artificial seeing device, thus leading to a temporary loss of useful vision owing to the generation of chaotic images or signals.

It should also not be forgotten that the host context of any naturoid — be it the human body, the natural environment, the social context, or whatever — exhibits a universal aptitude for recognizing and fighting against extraneous bodies: an operation that will obviously be triggered by the revelation, at some observation level, of the deception employed by the naturoid. The context, in fact, may, at any time, have recourse to the dynamic and arbitrary assumption of observation levels extraneous to that assumed by the naturoid project.

In the field of purely informational communication, the deception is therefore quite possible thanks to the fact that we can arrange the informational medium — be it a sound, an image or a text — in such a way that it appears ready to deliver something while we actually intend to achieve something else. Most advertising strategies, for instance, try to induce the decoding of an informational message giving special relevance to some pieces of information rather than others.

Nevertheless, in personal, knowledge-based communication, the deceit is much more subtle, because it can happen through two different processes. The first of these is the unavoidable reduction of the message to the personal structure of meaning of the listener or reader. The second is the rising of effects in the mind of the listener or reader due to the particular use of words and style by the speaker or writer, intentionally oriented towards emphasizing or masking some emotional aspect. This fact has been discussed since Gorgia philosophy, according to which the word can be simultaneously beneficial and dangerous. While the former process is a sort of unavoidable self-deceit, the latter is a deliberate deceit pursued thanks to the standard effects of the words.

4. Conclusions

Humans and animals share a basic, purely informational, way of communicating, thanks to which they are able to coordinate their behavior. This is a cybernetic tool through which living systems control their own actions in order to maintain the required individual and community balance, thus maximizing the probability of survival. Nevertheless, only humans have built artificial means — namely, words — in order to reproduce their own most personal mental states — i.e., their personal knowledge. Using words, humans are able to *express* their mental states, because the efficacy of any message is evaluated, first of all, by the utterer; but expression is not communication.

Although a message is 'designed' and modulated to represent in the best way what an uttering or writing person wishes to reproduce in the mind of the listener or reader, the result is an artificial object whose destiny is largely independent of his intentions.

This circumstance — well recognized by an ancient and long-lasting intellectual tradition — allows the setting up of an analogy between the communication process and the design and realization of naturoids, that is to say artificial objects or processes aimed to reproduce natural instances by means of the available technology.

The proposed analogy highlights a set of constraints that, both in the field of artificial technology and in that of communication, give to the resulting object or process its own properties, rather than simply reproducing the exemplar. This is particularly evident in the arts, where the meaning of the message resides in its formal beauty rather than in its content. In fact, the extraordinary 'materials' and the strategies adopted by the artists to reproduce his own mental states testify to his need to express himself in a way that is the most suitable for him, rather than for others. In any case, personal, knowledge-based communication is, at the same time, the most ambitious and least assessable technique invented by humans for enhancing their relationships.

References

Bourdieu P. 1991. *Language and Symbolic Power*, 60, Polity Press, Cambridge.

Dawkins R. 1996. *Climbing Mount Improbable*. W.W. Norton & Company, New York. Italian translation: *Alla conquista del monte improbabile.* Mondadori, Milan. 279.

De Mauro T. 1999. *Capire le Parole*, 89, Laterza, Rome–Bari.

de Solla Price D. 1964. Automata and the Origin of Mechanism. *Technology and Culture* Vol., 81. Johns Hopkins University Press, USA.

Fitzgibbons SJ. 1994. Making artificial organs work. *Technology Review,* 97, 34.

Gehlen A. 1983. *L'Uomo, la sua natura e il suo posto nel mondo.* Feltrinelli, Milan.

Hamdy RC. 2002. The thyroid gland: a brief historical perspective. *Southern Medical Journal*, 95(5).

Hartshorne C, Weiss P and Burks AW. 1958. *Collected Papers of Charles Sanders Peirce.* Harvard University Press, Cambridge, MA.

Kennedy AK. 1978. *Samuel Beckett*, 108. Cambridge University Press, Cambridge.

Mikos AG, Bizios R, Wu KK and Yaszemski MJ. 1996. *Cell Transplantation.* The Rice Institute of Biosciences and Bioengineering.

Morris C. 1948. *The Open Self*, 130, Prentice-Hall, New York.

Musil R. 1991. *Congiungimenti.* Newton Compton Milan.

Negrotti M. 1999. From the artificial to the art. *Leonardo*, 32(3). Isast–MIT Press, Cambridge.

Negrotti M. 2010. Naturoids: from a dream to a paradox. *Futures*, 42, 759–68. Elsevier, Netherlands.

Negrotti M. 2012a. Music and naturoids: the third reality. *Leonardo*, 45. Isast–MIT Press, Cambridge.

Negrotti M. 2012b. *The Reality of the Artificial. Nature, Technology and Naturoids*. Springer, Berlin.

Oesterle JA. 1994. Another approach to the problem of meaning. In *From Bentham to Basic English*, Charles K. Ogden (ed.), 136. Routledge, London.

Ogden CK and Richards IA. 1923. *The Meaning of Meaning*. Harcourt Brace, New York.

Oppenheim P and Putnam I. 1958. Unity of science as a working hypothesis. In *Concepts, Theories, and the Mind-Body Problem*, Feigl H, Scriven M and Maxwell G (eds.). University of Minnesota Press, Minneapolis.

Peters JD. 1999. *Speaking Into the Air: A History of the Idea of Communication*. UCP, Chicago, 43.

Pirandello L. 1994. *Uno, Nessuno e Centomila*. Einaudi, Turin, 42.

Pirandello L. 1994. *Arte e scienza*. Mondadori, Milan, 106.

Polanyi M. 1966. *The Tacit Dimension*. UCP, Chicago, 4.

Robinson J. 1994. The Expression and Arousal of Emotion in Music. *The Journal of Aesthetics and Art Criticism*, 52, 13–22. Temple University, Philadelphia.

The Whitaker Foundation. 1995. *Annual report: Tissue Engineering*.

5. Cybernetic Embryo

Richard Gordon

Embryogenesis Center
Gulf Specimen Marine Laboratory
222 Clark Drive
Panacea, FL 32346, USA

C.S. Mott Center for Human Growth and Development
Department of Obstetrics and Gynecology
Wayne State University, Detroit
Michigan, USA
DickGordonCan@gmail.com

Robert Stone

Troy, MI 48083, USA
rob.stone.academic@gmail.com

We consider the possibility that embryogenesis may be modeled as the hierarchical construction and implementation of consecutive cybernetic systems and their goal points, and suggest that differentiation waves may provide the regional nested and global communication within the embryo needed to support such a viewpoint. Each cybernetic system in the embryo has two goals: to change a subset of cells to a new cell type (via a differentiation wave), and to establish the conditions that create a pair of new cybernetic systems within it (via establishing launching sites for two new differentiation waves). Embryogenesis can then be seen as a nested hierarchical cascade of cybernetic systems, giving more substance to the differentiation tree concept. We review the literature on cybernetics, especially insofar as it has been applied to embryogenesis. Our purpose here, then, is to frame a specific, testable cybernetic model for embryogenesis, consider its philosophical necessity and ramifications, review the literature relevant to the question, and propose observations and experiments to test the idea that differentiation waves form the cybernetic basis for embryogenesis.

1. Introduction[1]

> "...early development is regarded as a series of defined morphological and physiological stages, each with its own pattern, albeit difficult to discern, and each pattern serving as the spatial condition for the transformation of that stage to the next by a limited set of morphogenic mechanisms.... There would be no long-range homing of the embryo toward the adult..." (Gerhart, 1980).

While embryo physics is making great strides in analyzing the mechanical forces generated by and shaping embryos (Stein and Gordon, 1982; Gordon, 1994, 2013; Gordon and Melvin, 2003; Beloussov and Gordon, 2006; Jones and Chapman, 2012; Wyczalkowski *et al.*, 2012; Guillot and Lecuit, 2013; Mammoto *et al.*, 2013; Miller and Davidson, 2013; Brodland *et al.*, 2014) since the pioneering efforts of D'Arcy Thompson (Thompson, 1917, 1942; Wikipedia, 2014g), little work has occurred on how the

[1] Because we are reviving cybernetics in the context of embryogenesis, many of our references are to pioneers of cybernetics and embryology. Rather than cite more than a selection of their works, we have cited well written, frequently updated Wikipedia articles, or else biographies, about the individuals, for the interested reader.

embryo gets mechanically or otherwise through the succession of one morphological stage to the next (Thom, 1989; Fleury and Gordon, 2012; Beloussov, 2013). Pattern formation in embryos has, since Hans Driesch (Wikipedia, 2014l), been conceived as the problem of getting each cell to be of the right kind in the right place at the right time (Reid, 1985; Witkowski, 1985) (though we may add: in the right numbers (Gordon, 2010) or right mass (Fankhauser *et al.*, 1955) or right relative mass (Cooke, 1981)). It has often been reduced to modeling or experimentally investigating one step of cell differentiation, and then presuming that any seemingly successful model for that one step can be generalized to all steps. Such an approach does not work. We have shown how, for example, Alan Turing's (Wikipedia, 2015a) model for one step of cell differentiation (Turing, 1952) leads to a combinatorial quagmire when two or more hierarchical steps of differentiation are considered (Gordon, 2015). This is simply because the positions of consecutive cell types are not determined by the Turing model.

An extreme example of a one-step approach is the "positional information" model (Wolpert, 1969; Wikipedia, 2014p), which is far from established (Kerszberg and Wolpert, 2007; Dubuis *et al.*, 2013). This is the "lookup table" computer programing approach (Wikipedia, 2014q) to modeling embryogenesis. It does not address the spatial aspect of differentiation as a hierarchical, multistep process, but instead envisions it as a one step process generating the whole organism all at once. Considering the 'energetic and informational economics' of evolved biological processes, as well as the difficulty in construction inherent in this model alone compared to a hierarchical one (Simon, 1991), it seems sensible to consider a hierarchical series of sub-models as superior:

> "Let me introduce the topic of evolution with a parable. There once were two watchmakers, named Hora and Tempus, who manufactured very fine watches. Both of them were highly regarded, and the phones in their workshops rang frequently, new customers were constantly calling them. However, Hora prospered, while Tempus became poorer and poorer and finally lost his shop. What was the reason? The watches the men made

> consisted of about 1,000 parts each. Tempus had so constructed his that if he had one partly assembled and had to put it down — to answer the phone say — it immediately fell to pieces and had to be reassembled from the elements. The better the customers liked his watches, the more they phoned him, the more difficult it became for him to find enough uninterrupted time to finish a watch. The watches that Hora made were no less complex than those of Tempus. But he had designed them so that he could put together subassemblies of about ten elements each. Ten of these sub-assemblies, again, could be put together into a larger subassembly; and a system of ten of the latter sub-assemblies constituted the whole watch. Hence, when Hora had to put down a partly assembled watch in order to answer the phone, he lost only a small part of his work, and he assembled his watches in only a fraction of the man-hours it took Tempus. It is rather easy to make a quantitative analysis of the relative difficulty of the tasks of Tempus and Hora: Suppose the probability that an interruption will occur while a part is being added to an incomplete assembly is p. Then the probability that Tempus can complete a watch he has started without interruption is $(1-p)^{1000}$ — a very small number unless p is 0.001 or less. Each interruption will cost, on the average, the time to assemble 1/p parts (the expected number assembled before interruption). On the other hand, Hora has to complete one hundred eleven sub-assemblies of ten parts each. The probability that he will not be interrupted while completing any one of these is $(1-p)^{10}$, and each interruption will cost only about the time required to assemble five parts. Now if p is about 0.01 — that is, there is one chance in a hundred that either watchmaker will he interrupted while adding any one part to an assembly — then a straightforward calculation shows that it will take Tempus, on the average, about four thousand times as long to assemble a watch as Hora" (Simon, 1991).

As Simon points out, nested goals or structures would have a definite advantage against singular top-down models. Indeed, as Terrence Deacon's thinking suggests, complex systems of a circular and hierarchical, or nested, character appear to have a peculiar kind of 'meta-complexity': their complexity is not only the result of their ordered informational content, their parts, or their interactions, but also their circular or purposeful behaviors; in which the organization they contain builds layers of ordered complexity, of both structure and behavior, out of the inherent complexity of these parts and interactions, due to the nested goals that living systems contain, regulate, and execute (Deacon and Koutroufinis, 2014). Simply, they self-organize and adapt, on multiple levels, in part because they are goal-driven, and interact with environments. Our 'Cybernetic Embryo' concept falls in with this line of thought, in the specific case of goals in embryogenesis.

Our basic assumption is that an embryo organizes itself without having that accomplishment as a goal. A fertilized chicken egg does not "strive" to become a chicken:

> "Purposiveness is not a metrical concept; no scale of intensity has yet been suggested, for the striving of a blastula to grow into a chicken; the final cause, however interesting its operation may be to the philosopher, cannot be expressed in language intelligible to the scientific worker" (Needham, 1930).

(*cf.* Wikipedia, 2015i). Nevertheless, we ask, does an embryo construct itself by establishing a set of intermediate goals and achieving them? Does it help us to analyze embryogenesis this way? In John Gerhart's words:

> "Each mechanism presumably has built-in self-adjusting properties, that is, a capacity for regulation applied at the time that mechanism is in use in the embryo. Accordingly, homeostasis of pattern would concern only the present state, not the future" (Gerhart, 1980).

However, each stage is not an end in itself, but rather a preparation for the next stage. So in some sense this "homeostasis of pattern" is indeed preparation for the future. The idea of each stage building on the previous stage was expressed by Daniel Dennett (Wikipedia, 2015e):

> "The skyhook concept is perhaps a descendent of the *deux ex machina* of ancient Greek dramaturgy: when second-rate playwrights found their plots leading their heroes into inescapable difficulties, they were often tempted to crank down a god onto the scene, like Superman, to save the situation supernaturally.... Skyhooks would be wonderful things to have, great for lifting unwieldy objects out of difficult circumstances, and speeding up all sorts of construction projects.... There are cranes, however. Cranes can do the lifting work our imaginary skyhooks might do, and they do it in an honest, non-question-begging fashion. They are expensive, however. They have to be designed and built, from everyday parts already on hand, and they have to be located on a firm base of existing ground. Skyhooks are miraculous lifters, unsupported and insupportable. Cranes are no less excellent as lifters, and they have the decided advantage of being real. Anyone who is, like me, a lifelong onlooker at construction sites will have noticed with some satisfaction that it sometimes takes a small crane to set up a big crane. And it must have occurred to many other onlookers that in principle this big crane could be used to enable or speed up the building of a still more spectacular crane. Cascading cranes is a tactic that seldom if ever gets used more than once in real world construction projects, but in principle there is no limit to the number of cranes that could be organized in series to accomplish some mighty end" (Dennett 1995). (cf. where limits to iterating cranes are discussed.)

We turn to cybernetics, a mathematical approach which developed out of anti-aircraft work in World War II (Wiener, 1958) on smoothing of stochastic systems and prediction, and the multidisciplinary Macy Conferences (Wikipedia, 2014r) of the period of 1946–1953.[2] Cybernetics was developed as a reaction to significant

---

[2]Participants included W. Ross Ashby (Wikipedia, 2014 z,a), Gregory Bateson (Wikipedia, 2015g), Warren McCulloch (Wikipedia, 2014z), Margaret Mead (Wikipedia, 2014s), Walter Pitts (Wikipedia, 2014y), Arturo Rosenblueth (Wikipedia, 2014b), Claude Shannon (Wikipedia, 2014e), Norbert Wiener (Wikipedia, 2014u), and John von Neumann (Wikipedia, 2015 h).

scientific challenges of the time, especially in circular and biological systems (Brand, 1976). Though cybernetics' multidisciplinarity dealt with antecedents such as structural functionalism (Wikipedia, 2014x), Gestalt psychology (Wikipedia, 2014k), Charles Peirce's semiotics work (Wikipedia, 2014d), Harold Black's (Wikipedia, 2014m) and Harry Nyquist's (Nyquist, 1924; Wikipedia, 2014p) communications work at the innovative Bell Labs (Poulin and Gordon, 2001), Julian Huxley's modern evolutionary synthesis in biology (Huxley, 1942), and statistical mechanics (Gibbs 1902), and while much of the mathematical foundations were laid beforehand, it was during the Macy Conferences that the broad applications (and implications) for science (and society) from cybernetics were synthesized into Wiener's book *Cybernetics: Or Control and Communication in the Animal and the Machine* (Wiener, 1948, 1961). It may be useful to think of the founding of cybernetics as an influential point of development of many contemporary and subsequent disciplines that continue to influence one another (Fig. 1), and it would be difficult to name any modern scientific field left unaffected.

Cybernetics[3] has a concrete mathematical foundation, and is focused on processual communication and control, to resolve the central problem of "How does a system alter the distribution of stochastic or probabilistic outcomes?", with specific attention paid to circular systems with a high level of complexity. Its' formal traditions can be traced back to James Clerk Maxwell (Maxwell, 1868), if not Gottfried Leibnitz (Wikipedia, 2015f) or still further. It has a rich and precise lexicon, as other natural science approaches. Its' theoretical framework lies primarily in the reduction of systems to their variable states and the application of the concept of

[3]A useful explanation of the cybernetic approach could be: "Cybernetics... combines common elements from diverse fields of science: the theory of communication, the theory of filters and anticipation, the theory of tracking systems, the theory of automatic regulation with feedback, the theory of electronic calculating machines, physiology, and so on. Cybernetics treats various subjects of these sciences from a single point of view — as systems that are processing and transmitting information" (Mindell *et al.*, 2003).

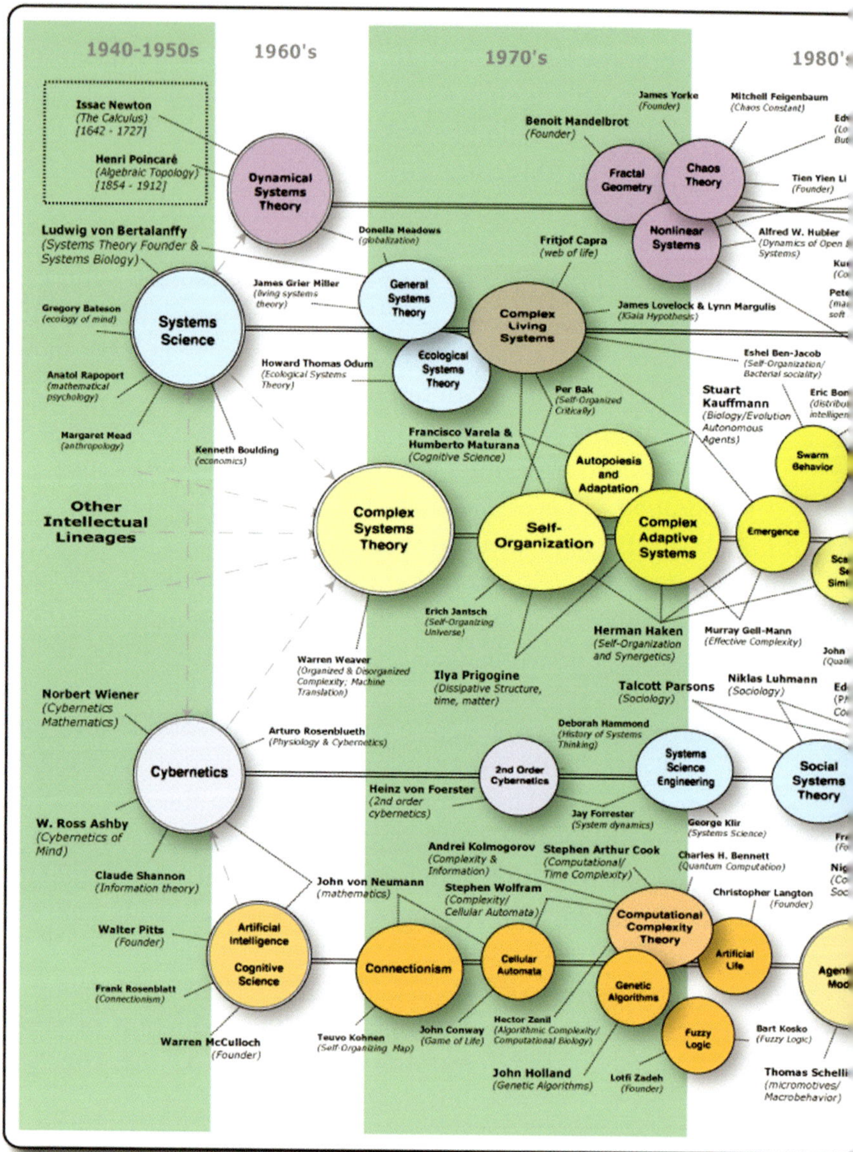

Figure 1. History of related fields stemming from cybernetics. From (Castellani, 2013), where an interactive version making the fine print readable may be found, with kind permission of Brian Castellani and Arts and Science Factory, LLC, Cleveland, OH.

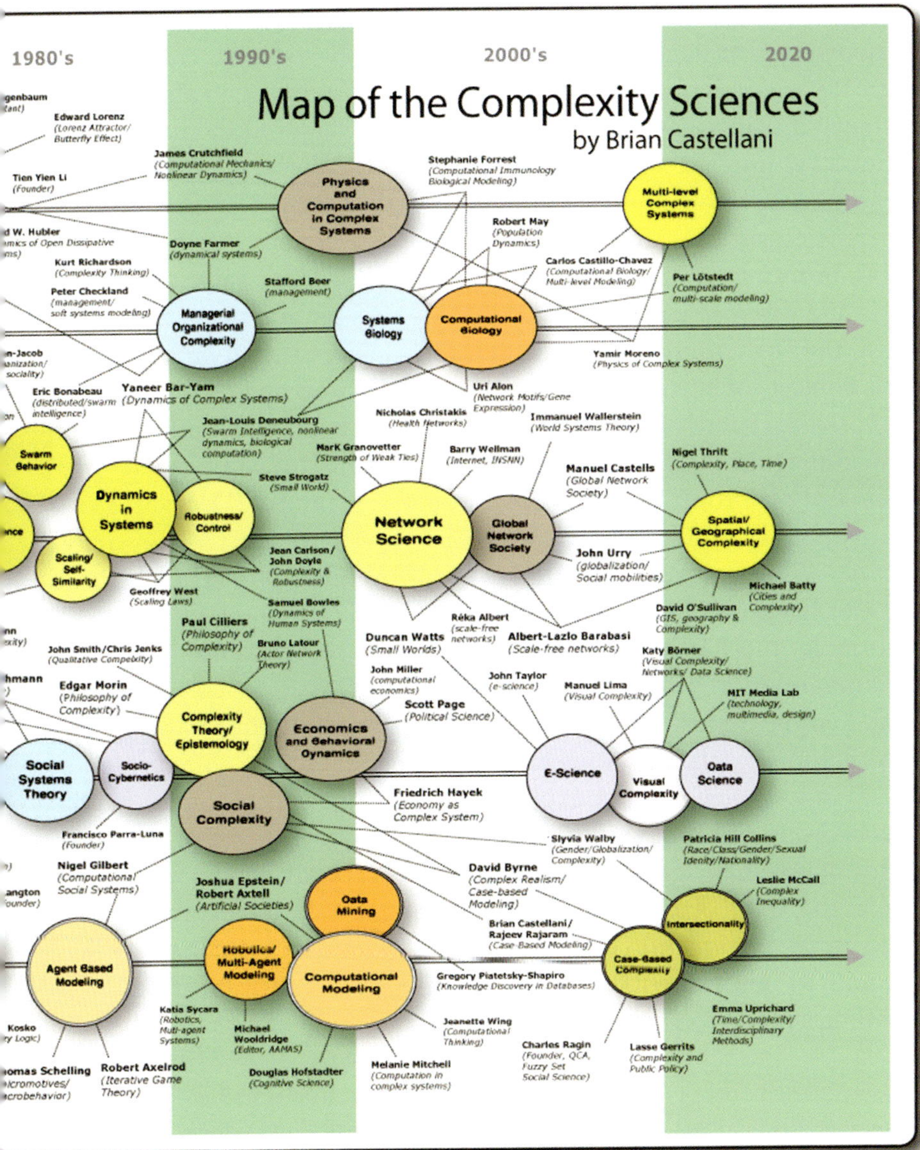
1980's
1990's
2000's
2020
Map of the Complexity Sciences
by Brian Castellani
Edward Lorenz (Lorenz Attractor/ Butterfly Effect)
Tien Yien Li (Founder)
James Crutchfield (Computational Mechanics/ Nonlinear Dynamics)
Physics and Computation in Complex Systems
Stephanie Forrest (Computational Immunology Biological Modeling)
Multi-level Complex Systems
Doyne Farmer (dynamical systems)
Robert May (Population Dynamics)
Kurt Richardson (Complexity Thinking)
Peter Checkland (management/ soft systems modeling)
Stafford Beer (management)
Carlos Castillo-Chavez (Computational Biology/ Multi-level Modeling)
Per Lötstedt (Computation/ multi-scale modeling)
Managerial Organizational Complexity
Systems Biology
Computational Biology
Yamir Moreno (Physics of Complex Systems)
Eric Bonabeau (distributed/swarm intelligence)
Yaneer Bar-Yam (Dynamics of Complex Systems)
Uri Alon (Network Motifs/Gene Expression)
Nicholas Christakis (Health Networks)
Immanuel Wallerstein (World Systems Theory)
Jean-Louis Deneubourg (Swarm Intelligence, nonlinear dynamics, biological computation)
Mark Granovetter (Strength of Weak Ties)
Barry Wellman (Internet, INSNA)
Nigel Thrift (Complexity, Place, Time)
Swarm Behavior
Steve Strogatz (Small World)
Manuel Castells (Global Network Society)
Dynamics in Systems
Robustness/ Control
Network Science
Global Network Society
Spatial/ Geographical Complexity
Scaling/ Self-Similarity
Jean Carlson/ John Doyle (Complexity & Robustness)
John Urry (globalization/ Social mobilities)
Geoffrey West (Scaling Laws)
Michael Batty (Cities and Complexity)
Samuel Bowles (Dynamics of Human Systems)
Réka Albert (scale-free networks)
David O'Sullivan (GIS, geography & Complexity)
Paul Cilliers (Philosophy of Complexity)
Duncan Watts (Small Worlds)
Albert-Lazlo Barabasi (Scale-free networks)
John Smith/Chris Jenks (Qualitative Complexity)
Bruno Latour (Actor Network Theory)
Katy Börner (Visual Complexity/ Networks/ Data Science)
Edgar Morin (Philosophy of Complexity)
John Miller (computational economics)
John Taylor (e-science)
Manuel Lima (Visual Complexity)
MIT Media Lab (technology, multimedia, design)
Scott Page (Political Science)
Complexity Theory/ Epistemology
Economics and Behavioral Dynamics
Social Systems Theory
Socio-Cybernetics
E-Science
Visual Complexity
Data Science
Social Complexity
Friedrich Hayek (Economy as Complex System)
Francisco Parra-Luna (Founder)
Sylvia Walby (Gender/Globalization/ Complexity)
Patricia Hill Collins (Race/Class/Gender/Sexual Idenity/Nationality)
Nigel Gilbert (Computational Social Systems)
David Byrne (Complex Realism/ Case-based Modeling)
Leslie McCall (Complex Inequality)
Joshua Epstein/ Robert Axtell (Artificial Societies)
Data Mining
Intersectionality
Brian Castellani/ Rajeev Rajaram (Case-Based Modeling)
Agent Based Modeling
Robotics/ Multi-Agent Modeling
Computational Modeling
Case-Based Complexity
Gregory Piatetsky-Shapiro (Knowledge Discovery in Databases)
Katia Sycara (Robotics, Multi-agent Systems)
Michael Wooldridge (Editor, AAMAS)
Jeanette Wing (Computational Thinking)
Emma Uprichard (Time/Complexity/ Interdisciplinary Methods)
Charles Ragin (Founder, QCA, Fuzzy Set Social Science)
Lasse Gerrits (Complexity and Public Policy)
Robert Axelrod (Iterative Game Theory)
Douglas Hofstadter (Cognitive Science)
Melanie Mitchell (Computation in complex systems)

feedback to understand how they are modified, sometimes as if by themselves or in seemingly acausal or highly complex ways. There are a variety of abstract "process diagrams" which can also be constructed using these principles to better understand the interoperation of entangled or co-evolving systems (Kitano *et al.*, 2005). These reduced systems show commonalities that can lead to synthesis across disciplines to better understand the general nature or laws of communication and control.

We propose here a new perspective on embryogenesis that we call the "cybernetic embryo". The general idea is that embryogenesis consists of a tree of regional teleonomic cybernetic processes (see Glossary), some of which have a global character, and all of which at their culmination have set up conditions for subsequent processes. We propose that these cybernetic "cranes" (Dennett, 1995) are based on observed differentiation waves in embryos (Brodland *et al.*, 1994; Gordon *et al.*, 1994; Jaffe, 1995, Gordon and Björklund, 1996; Gordon, 1999, 2011a; Jaffe, 1999), and thereby give a specific, testable mechanism for self-construction of complex embryos. Differentiation waves provide the communication between cells and whole tissues, which allows cybernetic control of a regional and sometimes global character. This interpretation of differentiation waves will then give an explicit, testable mechanical theory for the cybernetic embryo.

Words like 'goal' and 'purpose' do not sit easily within the paradigm of our natural scientific endeavors and training. For example, Niels Bohr (Wikipedia, 2015j) hinted at a dualism comparable to his correspondence principle for quantum mechanics (Wikipedia, 2014f):

> "...the concept of purpose, which is foreign to mechanical analysis, finds a certain field of application in problems where regard must be taken of the nature of life. In this respect, the role which teleological arguments play in biology reminds one of the endeavours, formulated in the correspondence argument, to take the quantum of action into account in a rational manner in atomic physics" (Bohr, 1933).

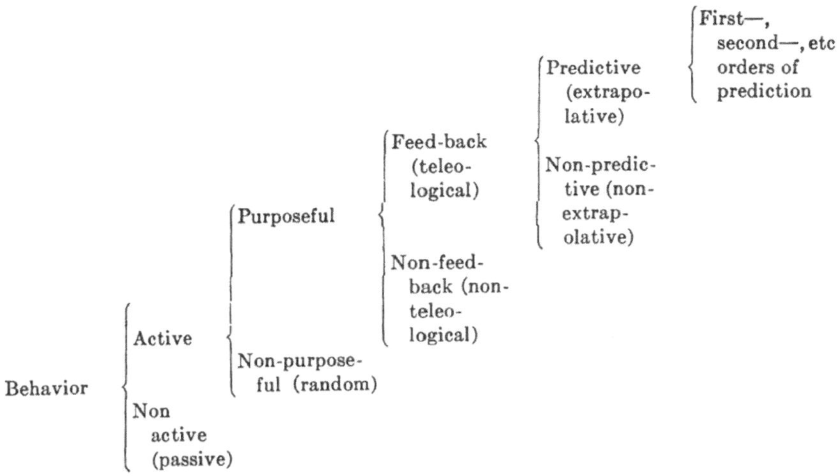

Figure 2. A proposed classification of behavior, into which we can try to place the cybernetic embryo. Since the embryo, at least initially (as in placental animals), contains the energy sources for its transformation, it is "active". Since the result is not random, it is "purposeful". Due to the phenomenon of embryonic regulation, we can say that "feedback" occurs. However, we have no reason to believe that embryos in any sense perceive their future state, and thus draw the line at this point: embryos are not "predictive". From (Rosenblueth *et al.*, 1943), published by the University of Chicago Press.

To the physicist or chemist, goals and purposes are anathema, or at least foreign, as scientific progress has, in fact, led to the continual replacement of a theological universe in which everything appeared purposive and goal driven, to a universe which is driven by natural causal processes. They were not incorrect to do so, within those realms. A mountain does not *wish* to become a mountain; a star does not have a goal of becoming a star. Sensibly, they are passive processes (Fig. 2), whose physical distributions follow a statistical character based on the exterior forces interacting with them environmentally. However, embryos appear to attain a goal of a particular shape and function, repeatedly, if we consider a whole spawning or consecutive generations. Their physical distributions show signs of internal control against their environmental landscape, with phenomena such as "embryonic regulation"

(Sander, 1971; De Robertis, 2009); simply, they appear to have more control over their own development than a mountain. Cybernetics has a rich history of explicating such control and the communication which gives rise to it, thus supporting our exploration of the cybernetic approach to some unsolved problems of embryogenesis.

Because the connotation of purpose within biology has been met with much challenge and even derision (at times necessarily) in modern history, we have chosen to place the problem of embryogenesis and the cybernetic embryo perspective in the broader contexts of the philosophy of science and the philosophy of biology, at least as they connect with cybernetics. While Darwin's theory of evolution, our modern understanding of genetics, and systems biology have given us bases for understanding the origin of novelty, none of these approaches represent a mechanism by which that novelty becomes manifest in the individual organism. That process is embryogenesis, and we need some new ways of thinking about it and mechanically describing it.

Considering the ample debts the mathematical foundations of cybernetics owes statistical mechanics, we take inspiration in the work of Ludwig Boltzmann (Boltzmann, 2012), who first bridged the philosophical gap between part and whole scientifically (in this works' context, cell and embryo). During the development of statistical mechanics by Boltzmann, Maxwell (Maxwell, 1860a, 1860b) and Gibbs (Gibbs, 1902), their mechanistic understandings were not in line with a then current philosophical antipathy to the existence of atoms (Kuznetsov, 1965). Boltzmann countered: "Who sees the future? Let us have free scope for all directions of research; away with dogmatism, either atomistic or anti-atomistic!" (Boltzmann, 2012). In doing so, Boltzmann may have anticipated Thomas Kuhn's later idea of paradigms (Kuhn, 1996), itself presaged by Koestler's work in *The Sleepwalkers* (Koestler, 1959). We find embryogenesis in a similar, if more complex, position: in need of theory to bridge not only an old reductionism vs. holism dichotomy (Becht, 1974; Gordon, 1999; Wilson, 1999), but also the tangled ones of materialist versus vitalist, physical mechanisms versus self-organization, etc.:

"Dualism shares with its antithesis, classical materialism, a certain unassailable simplicity. What they both lack is plausibility. As a result, most scientists try to steer a course that avoids the two extremes. They face the almost impossible task of somehow demechanizing the machine, but keeping the ghosts out" (Harth 1995).

Our question then is: Does a cybernetic view of embryogenesis permit us to more sensibly explore the embryo's increase in complexity?

2. Cybernetics and the Embryo

"It will be seen that cybernetics is leading me to a whole group of problems concerning organization, and it is in these fields that I think that an important part of the future of cybernetics lies.... The concept of self-organization is well known in biology, where there is a great deal of talk of material organizers of substances which in the embryo will cause different organs to come into being..." (Wiener, 1958).

The concept of the cybernetic embryo can be seen in retrospect in the ideas of Paul Weiss (Talbott, 2011):

"One frequently refers to these organizing entities under the term of [morphogenetic] 'fields.' Their existence can be traced back to the egg. In fact, just as there is a continuity of cells from the egg to the organism through successive cell divisions, so there is continuity between the primordial organizing fields present in the undivided egg and the localized fields of the later germ. Primordial fields segregate progressively into more restricted fields, and, furthermore, induce new fields in neighboring areas of the germ. Thus, the organizing principles of a germ have an ontogenetic history of their own which is not cell history. Their possession marks the egg as an entity of the rank of the organism.... Their development is a matter of the developing system as a continuum, like tensions, currents, potentials, and the like, and they pay no heed to cell boundaries, although sooner or later the intricate interplay between them and the cells sets in....

"The existence of these primordial organizing principles in the egg has been firmly established by modern experimental embryology. No pure cell theory derived from the developed organism can embrace them, unless by a vicious circle.

"In conclusion, we may say that the cell theory is correct: The egg is a cell and it gives rise to all the successive cell generations which contribute to the organism. But the organismic theory is likewise correct: The egg is also an organism, and it passes its organization on continuously to the germ and the body into which it gradually transforms. Only this dual concept seems to fit the facts, as we see them at present. To be consistent, we should supplement Virchow's well-known tenet of cell theory: 'Omnis cellula e cellula,' by its counterpart: 'Omnis organisatio ex organisatione.' If the former denies spontaneous generation of living matter, the latter denies spontaneous generation of organization" (Weiss, 1940).

Weiss' "organizing principles" (=differentiation waves organized as a differentiation tree), manifest as nested "primordial fields" (=trajectories of differentiation waves), correspond to our concept of a hierarchical tree of cybernetic processes, with the exception that we do not presume that the whole tree is represented by physical regions in the fertilized egg. On the contrary, new branches of the differentiation tree are generated at each step of differentiation. The cell lineage tree ("cell history") is enclosed in the differentiation tree ("organizing principle") (Gordon, 2011 a).

The word "cybernetics" did not come into use until 1948, eight years after Weiss wrote the above passage, when Norbert Wiener published his book *Cybernetics* (Wiener, 1961; American Society for Cybernetics, 2014). We quoted Weiss in discussing this higher level of organization in an embryo in terms of wave–wave interactions (Gordon, 1999), where the waves are differentiation waves (Gordon, *et al.*, 1994):

"Proposition 82: the launching of one differentiation wave may have a mechanical dependence on the completion of the previous differentiation wave. This provides a new lockstep wave–wave inductive mechanism, and may explain the broad temperature independence of development in poikilothermic organisms.... Wave–wave inductions would provide a lockstep mechanism for embryonic development.... In a sense, development may be analogous to a [relay] race in which a baton is passed on from one runner to the next. Each runner must wait for the previous one, no matter what the pace of the race" (Gordon, 1999).

To give credit where it is due, we should actually go all the way back to Aristotle:

"The most elegant experiment in the history of embryology was performed some 2,400 years ago by Aristotle. He opened the shell of chicken eggs at different incubation days, and carefully described what he saw: the white spot on the yolk that marks, at the very beginning, the point where the future embryo is going to appear; the tiny brown lump that starts pulsating at the third day and later will turn into a heart; the greatly expanded vesicles that will become eyes; the entangled red vessels that descend into the yolk and branch out like roots; and the thin membrane that wraps everything up like a mantle. On the basis of these observations, Aristotle concluded that in a developing embryo organs not only increase in size, as Hippocrates had said, but also in number. Embryonic development, according to Aristotle, is an *epigenesis*, a chain of one genesis after another, where new structures and new functions appear at various steps. During embryonic development, in short, *the complexity of the system increases*" (Barbieri, 2003).

The founding of cybernetics lay with an analysis of purposive, goal driven behaviors. In Rosenblueth *et al.*'s proposed hierarchy of purposive decision making systems (Rosenblueth *et al.*, 1943), the cybernetic embryo concept would appear to be classified as a non-predictive feedback system (Fig. 2).

It is important to note that Rosenblueth *et al.* connected negative feedback to *teleology:*

"All purposeful behavior may be considered to require negative feed-back.... In classifying behavior the term 'teleology' was used as synonymous with 'purpose controlled by feedback'. Teleology has been interpreted in the past to imply purpose and the vague concept of a 'final cause' has been often added. This concept of final causes has led to the opposition of teleology to determinism.... purposefulness, as defined here, is quite independent of causality, initial or final. Teleology has been discredited chiefly because it was defined to imply a cause subsequent in time to a given effect. When this aspect of teleology was dismissed, however, the associated recognition of the importance of purpose was also unfortunately discarded. Since we consider purposefulness a concept necessary for the understanding of certain modes of behavior we suggest that a teleological study is useful if it avoids problems of causality and concerns itself merely with an investigation of purpose" (Rosenblueth *et al.*, 1943).

We will not go deeply into the philosophical issues raised here. Anatol Rapoport (Wikipedia, 2014a) thought he had them sewed up:

> "It is precisely the inadequacy of... notions... of goal-directedness and purposefulness... in explaining the events of the physical world which had led to the abandonment of teleological explanations in favor of mechanistic ones.... The apparent equifinality and purposefulness of behavior and of developmental patterns is still something that has not been decomposed into mechanically determined processes in a way which would satisfy the strict mechanically-oriented determinist. The philosophical import of the cybernetic approach is in the link it has established between the concept of mechanism and that of organism.... In particular, let the mechanism be sensitive to inputs which are themselves results of its own performance, and we have introduced the principle of feedback, a cardinal cybernetic principle, which links the behavior patterns of living organisms and machines.... feedback gives the impression of 'purposefulness'.... It appears, then, that so called teleological causality can be understood in terms of ordinary mechanical causality plus a certain organization of the mechanisms" (Rapoport, 1965).

The problem for us is that he had not resolved a specific origin of the "certain organization of the mechanisms". A simplistic answer is that this is the job of evolution. However, evolution does not explain how an individual embryo builds itself alone. We propose a solution here, namely that one of the goals of a cybernetic system can be the construction of another cybernetic system within it. The embryo, over time, can then be seen as a nested hierarchical cascade of cybernetic systems, which themselves were developed as a result of the process of Darwinian evolution. The 'cybernetic embryo' concept can then be seen in the context of what could be a new evo-devo approach to the problem of embryogenesis, with antecedents such as Rapoport.

The dual nature of the term "teleology" continued to cause confusion, so shortly thereafter the term *teleonomy* was coined to describe purposive behavior in biological systems, allowing "teleology" to remain the sole referent term for science of final causes. The term *teleonomy* was clarified and popularized by Ernst Mayr (Wikipedia, 2014 j) in a manner befitting a cyberneticist:

> "It would seem useful to rigidly restrict the term *teleonomic* to systems operating on the basis of a program of coded information" (Mayr, 1961).

Jacques Monod (Wikipedia, 2014o) believed teleonomy to be a key aspect of a 'living system' (Monod, 1972). In the field of biosemiotics the coding aspect of embryogenesis is explicitly recognized (Hoffmeyer, 1996; Barbieri, 2003; Emmeche and Kull, 2011).

Marcello Barbieri suggests that embryogenesis is analogous to reconstruction from projections (more commonly known as computed tomography) from a limited number of projections (Gordon *et al.*, 1970; Barbieri, 1974; Rangayyan *et al.*, 1985; Gordon, 2011b). Just as the resulting image is more complex than its projections, so an embryo is more complex than the information that codes for it: "...the phenotype is always more complex that the genotype.... *epigenesis is a reconstruction from incomplete information*" (Barbieri, 2003). Barbieri further claims:

> "Without... a [cell] memory it would not be possible to obtain *a convergent increase of complexity*, and the real logic of embryonic development is precisely that kind of increase.... Above all embryonic development [is]... a process that is potentially

capable of generating countless different structures.... This is the great challenge of the future, and not for embryology alone, but for all of science: *how does a system manage to increase its own complexity?*... the body plan is also a *memory*, because it is a structure that appears at an early stage of development, and, from that moment on, it remains in the organism for life, acting as a deposit of information for the three-dimensional pattern of organs and apparatuses.... a supracellular memory.... *organic epigenesis requires organic memories.... organic epigenesis requires organic* codes.... *Embryonic development is a sequence of two distinct processes of reconstruction from incomplete information, each of which increases the complexity of the system in a convergent way. The first process builds the phylotypic body and is controlled by cells. The second leads to the individual body and is controlled not only at the cellular level but also at the supracellular level of the body plan*" (Barbieri, 2003).

We will try to make this memory explicit below. Barbieri recognizes an increasing number of organic codes as life evolves (Fig. 3). Here we deal only with what he called "pattern codes" and early development prior to the "phylotypic" stage (Cohen, 2003), which occurs well after neural tube closure. However, the notion that the body plan, which we suppose is the embryo's whole body, can impact the individual cells should apply at any stage of embryogenesis. In our conception, the current shape and configuration of all the cells and tissues provide boundary conditions for the trajectories of differentiation waves, at all stages.[4] But, as exemplified in the discussion of early embryogenesis by Gregory

[4]While this generalization does not support Barbieri's explanation for the existence of a phylotypic stage, that concept is under debate (Richardson *et al.*, 1998; Kranenbarg, 2000; Galis and Metz, 2001; Bininda-Emonds *et al.*, 2003; Hazkani-Covo *et al.*, 2005, Irie and Kuratani, 2011), and its explanation may not lie in the form of the organism, especially given that nematode worms have a "phylotypic" phase that corresponds, at least in the relative timing of gene expression, with vertebrates (Levin *et al.*, 2012; Richardson, 2012). Our differentiation waves theory seems to us to be neutral in regard to the outcome, i.e., frankly, it does not explain the phylotypic stage, though we have tried to place the notion of multiple developmental pathways (Berrill, 1961) into the context of differentiation waves (Gordon, 1999).

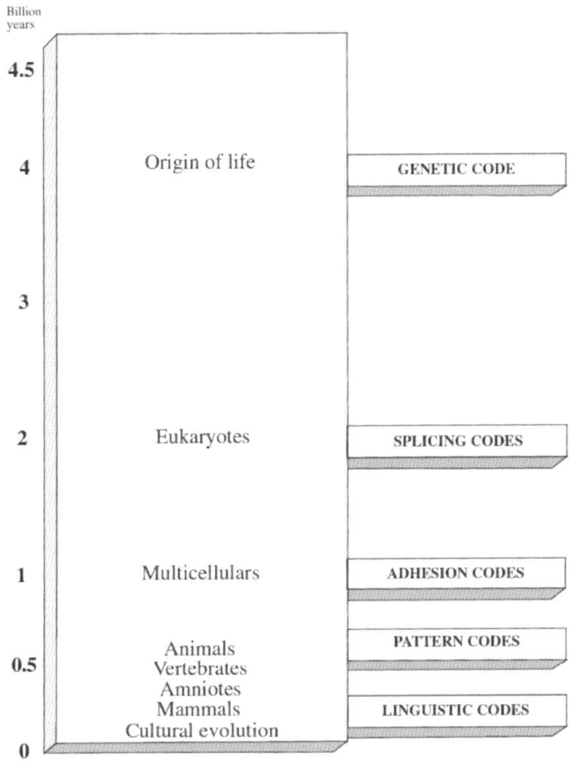

Figure 3. The major organic codes and rough times of their evolution. From (Barbieri, 2003) with permission of Cambridge University Press.

Bateson (Bateson, 1987; Wikipedia, 2015g), the source of many signals that alter the course of embryogenesis is regarded as external to the embryo, which begs the question of how a possibly uniform environment can provide the right spatial cues at every step of the process, as well as how the embryo can effectively organize itself within a very chaotic or new environmental system. This is as far as biosemiotics has gone in explaining embryogenesis, i.e., no further than molecular developmental biology. Undoubtedly organic codes are components of embryogenesis, but not the be all and the end all.

In cybernetics, purposes are tied to goals, where the controller which regulates behavior receives information about the discrepancy from the goal via feedback. The components and their relation must be reverse engineered in a system as complex as embryogenesis (Gordon and Melvin, 2003; Lu *et al.*, 2013); its difficulty without a cybernetics model is similar to reverse engineering a modern computer program with neither the source code nor a basic understanding of compilation. Feedback may be a tool for understanding purpose, but it doesn't by itself explain the origin of organizations which create feedback:

> "For understanding the general principles of dynamic systems, therefore, the concept of feedback is inadequate in itself. What is important is that complex systems, richly cross-connected internally, have complex behaviours, and that these behaviors can be goal-seeking in complex patterns" (Ashby 1956).

The advantages of a cybernetic perspective lies in its focus on systems, process, control, communication, and their logical foundation; as well as its nomenclature and mechanisms for explaining what seems acausal, nonlinear, or even purposive behaviors in a causal, logical manner:

> "The truths of cybernetics are not conditional on their being derived from some other branch of science. Cybernetics has its own foundations... two peculiar scientific virtues of cybernetics... are worth explicit mention. One is that it offers a single vocabulary and a single set of concepts suitable for representing the most diverse types of system... The second peculiar virtue of cybernetics is that it offers a method for the scientific treatment of the system in which complexity is outstanding and too important to be ignored. Such systems are, as we well know, only too common in the biological world!" (Ashby, 1956).

Modern biology and cybernetics have continually had interlaced paths of development. In an essay prior to Darwin's *Origin of Species*, Alfred R. Wallace (Wikipedia, 2015b) states his understanding of evolution, cybernetically, as James Watt's engine was one of the earliest examples of a man-made cybernetic system with control by feedback (Verveen, 1971; Brand, 1976):

> "The action of this principle is exactly like that of the centrifugal governor of the steam engine, which checks and corrects any irregularities almost before they become evident; and in like manner no unbalanced deficiency in the animal kingdom can ever reach any conspicuous magnitude, because it would make itself felt at the very first step, by rendering existence difficult and extinction almost sure soon to follow" (Wallace, 1858).

An early example of our understanding of non-machine oriented positive feedback was from Ronald Fisher's analysis of what is now known as 'Fisherian Runaway' (Fisher, 1915), where Fisher (Wikipedia, 2015 l) explained the development of structures such as a male peacock's tail as a result of positive feedback due to sexual selection: a peahen attracted to a peacock with a large tail is likelier to select them, producing (on average) an offspring which is either a peacock with a large tail, or a peahen with a greater selection preference for large tails. In fact, on the development of cybernetics, Gregory Bateson said "Wiener without a biologist wouldn't have done it", referring to his work with Rosenblueth (Brand, 1976). One of the primary goals of early cyberneticists was to understand the human brain (Ashby, 1952), along with general mechanisms and 'organizing principles' seen across the spectrum of scientific inquiry. Two eminent cyberneticists (and members of Cambridge's 'Ratio Club' (Wikipedia, 2014v)) provided new insights in biology: W. Ross Ashby's work on self-organization (Ashby, 1947), generalizing Darwin's insight on selection with similar observations in other domains in a logical synthesis to pull the vitalistic ghost out of the causal machine of complex systems; and Alan Turing's (Hodges, 1983) chemical morphogenesis (Turing, 1952), a novel application of reaction-diffusion systems and bifurcating pattern formation (cf. Nicolas Rashevsky (Wikipedia, 2014t), cited in: (Rosen, 1968)). Turing's approach echoed earlier work from R.A. Fisher (Fisher, 1930) and Andrey Kolmogorov (Kolmogorov *et al.*, 1937; Wikipedia, 2015c) in biology with partial differential equations now called Fisher-Kolmogorov equations. It would also be a disservice to the reader not to mention the significant cybernetics work of biologists like Humberto Maturana and

Francisco Varela (Maturana and Varela, 1980), as well as James Grier Miller's *Living Systems* (Miller, 1978).

3. The Embryo as an Hierarchical Cascade of Consecutively Constructed Cybernetic Systems

First, we want to give a simple example that demonstrates that embryos are capable of constructing their own nested cybernetic systems. Cold blooded animals can develop normally over a temperature range of 15–20°C (Duellman and Trueb, 1986). However, homeothermy does not initially exist in the bird or mammalian egg despite their normal development requiring a narrow temperature range (Byerly, 1938; Visschedijk, 1991; Nakage *et al.*, 2003) (0.9–1.4°C for most birds: (Webb, 1987)), which is why such eggs must be incubated or retained internally, and full thermal regulation does not set in until after hatching (Tzschentke and Rumpf 2011) or birth (Adamsons and Towell, 1965). This is an example of a living cybernetic system, analogous to a thermostat. The embryo constructs the homeothermy apparatus and sets it operating, with a goal point or "set-point", i.e., the normal body temperature. Thus we can be sure that, however it does it, albeit involving four stages in precocial birds (Whittow and Tazawa, 1991), the embryo is capable of constructing its own cybernetic feedback systems. Our contention is that each step of the whole process of embryogenesis may be seen as cybernetic.

A few people have struggled with the issue of what, if anything, 'coordinates' the spatial component of embryogenesis. Edmund Sinnott (Whaley, 1983) raised this problem of the generation of form to one equivalent in some sense to that of consciousness:

> "What impresses these thinkers is the striking resemblance between the progressive, regulatory, and (as [Edward Stuart] Russell [Russell, 1945; Wikipedia, 2014i] says) 'goal directed' processes of development and physiological activity, on the one hand, and on the other hand phenomena of purpose, of the drive toward an end, which is the basis of most mental or psychological

> activities.... A developing embryo, especially if its growth is speeded up by time lapse photography, certainly looks as if it were moving toward a goal which it is bound to reach in spite of the obstacles which we may put in its way to test the intensity and resourcefulness of its 'purpose.' Needham's phrase, 'the striving of a blastula to grow into a chicken,' [(Needham 1930)] may be a figure of speech, but to some minds it is not far from actual truth" (Sinnott, 1950).

Anatol Rapoport put the case in an amusing perspective, referring to Hans Driesch's unanticipated observation (Sander, 1993) that a two-cell sea urchin embryo produced two whole sea urchin larvae when the cells were separated (Driesch, 1892):

> "'mechanical process'... carries overtones of stupidity, at least of failing to take into account the variable environment.... Driesch evidently thought that the two separated halves of the embryo, if they were guided by mechanical process, should have exhibited 'stupid' behavior. Unaware of the changed circumstances, they should have continued blindly along the same path as if they had not been separated" (Rapoport, 1965) (as quoted in: (Hellman, 1982)).

In other words, a cybernetic process need not be stupid. They can 'self-organize', and dynamically adjust or adapt, to an extent depending on their structure and their ability to receive information constructively. Certainly, we do not design cybernetic systems to act stupidly on our behalf, nor should we expect evolved cybernetic systems to be stupid or inflexible.

Michael Apter (Wikipedia, 2014w), in a book that one would think would be right on our subject, *Cybernetics and Development*, focused almost entirely on critiquing application of information theory to development (Apter, 1966). While Shannon's communication theory is strongly associated with cybernetics' origins, and is indebted to Wiener's prior work on stochastics and ergodicity (Shannon, 1948), it is a component of its more mature phase; it is but part of cybernetics, not the whole of it. He dismisses "...the botanist Sinnott whose understanding of cybernetics is... that it is concerned only with negative feedback situations (Sinnott, 1960)". The closest he gets to our ideas is to note "...that development results from interactions between the various parts of the embryo, these interactions occurring at various levels of organization.... It might prove useful when considering control mechanisms and the communication between various parts of the developing embryo" (Apter and Wolpert, 1965), i.e., feedbacks, negative or otherwise.

One of the steps beyond cybernetics as negative feedback, and furthermore in the context of development of organisms, was taken by Magoroh Maruyama in his *The second cybernetics: deviation-amplifying mutual causal processes* (Maruyama, 1963a, 1963b). This paper triggered the career of one of us (RG) in embryogenesis. The key conceptual breakthrough here is that a system can have both positive and negative feedback loops which can mutually limit each other. The negative feedbacks generally won't achieve their goals perfectly, but by the same token the positive feedbacks won't have runaway effects. Maruyama presented a simple cellular automata model that led to a complex form, albeit with completely deterministic rules of interaction between the cells. RG then showed how a negative feedback could be used to control a positive feedback in the presence of stochastic rules, to produce a reasonably regular growing spiral "snail" pattern (Gordon, 1966). Maruyama's idea may have contributed to activator/inhibitor models for pattern formation (Meinhardt and Gierer, 2000), which are also derived from Turing's reaction/diffusion approach (Turing, 1952; Gierer and Meinhardt, 1972).

One more mathematically descriptive approach to development that could be called 'cybernetic' is that of nonlinear dynamic systems analysis in a multidimensional phase space, where the state variables can represent, for example, concentrations of specific gene products (Jaeger and Sharpe, 2014). Such systems will have basins of attraction, which cause these systems of equations to hone in on specific states. These could be interpreted as goals, and the changing of parameters and boundary conditions over time could alter which goals a dynamic system will approach. In this case the goal cannot be seen directly, because it is the measured by the higher dimensional distance from the current state to an attractor. Nonlinear dynamic systems are thus non-predictive, yet purposive (Fig. 2).

The hierarchical nature of differentiation was recognized by Conrad Waddington (Wikipedia, 2014c) in his concept of the "epigenetic landscape" (Waddington, 1957, 1966, 1977; Buss and Blackstone, 1991; Gilbert, 1991; Gatherer, 1996; Slack, 2002; Bhattacharya *et al.*, 2011; Choudhuri, 2011; Fagan, 2012; Huang, 2012), which was effectively a bifurcating attractor in an unspecified space whose topology is dictated by an underlying array of genes. Arthur Koestler (Wikipedia, 2015 d) invoked some of the language of cybernetics and fleshed out Waddington's hierarchy in his book *The Ghost in the Machine*:

> "J. Needham once coined a phrase about 'the striving of the blastula to grow into a chicken' [(Needham, 1930)]. One might call the ensemble of devices which make it succeed the organism's 'pre-natal skills'. To quote James Bonner...: 'We [already] know that nature, like man, accomplishes complex tasks by breaking them up into many simple sub-tasks' [(Bonner, 1965), cf. (Simon, 1991)]. Some of these principles, which we found reflected in embryonic development, were: the hierarchically branching order of differentiation and morphogenesis; the 'dissectibility' of that order into self-regulating holons at various levels (stages); their Janus character (autonomy versus dependence, determination versus docility); their fixed genetic canons and adaptable strategies guided by feedbacks from the hierarchy of environments; the action of triggers (inducers, etc.), which release preset mechanisms, and of scanners ('tests') which process information; the decrease of flexibility with increasing differentiation and specialisation" (Koestler, 1967).

This overall goal of the embryo appears to be broken down hierarchically into a series of sub-goals, which correspond roughly to tissue groups, organ systems, etc., via a series of bifurcating developmental events. These ideas were expressed less succinctly in Koestler's earlier book *The Act of Creation*, from which we nevertheless can extract further details:

"Organic life... is governed by 'rules of the game' which lend it... an appearance of purpose or 'goal-directedness.... A fundamental characteristic of codes on all levels is that they function on the trigger-release principle, so that a relatively simple signal releases pre-set, complex action patterns... guided by feedback.... On every level of structure and function in the living organism we find the same phenomenon of Janus-faced entities which behave partly as wholes, yet wholly as parts, as it were.... A morphogenetic field behaves 'as a unit or a whole and not merely the sum of the cellular materials of which it is composed. The field with its organizing capacities remains undisturbed if the cellular material which it controls under normal circumstances is diminished or enlarged. The unit character of the field finds its clearest manifestation in these regulative properties' (Hamburger, 1955).... Motivation in embryonic development is a subject for the metaphysician. J. Needham's tongue-in-the-cheek phrase about 'the striving of the blastula to grow into a chicken' indicates the directiveness of the morphogenetic process and its equifinal, regulative properties..." (Koestler, 1964).

It may be appropriate to call a tissue differentiation tree a decision tree (Wikipedia, 2014h). This sequential decision system through developmental time is a hierarchical model, encoded as a hierarchical genome (Gordon, 1999). The expression of this model is a generally nested series of differentiation waves, an improvement on the less rigorous and useful and archaic idea of a 'morphogenetic field'. Thus, our models' attempt to align systemic isometrics, weed signal from noise, and *Order out of Chaos;* a nod to Ilya Prigogine's (Wikipedia, 2014n) important work on dissipative structures (Prigogine and Stengers, 1984).

4. Differentiation Waves as Cybernetic Feedback Systems

We are now ready to consider the observation of differentiation waves and the concept of the differentiation tree in terms of cybernetics. Let us first try to make the components of Koestler's explanation explicit:

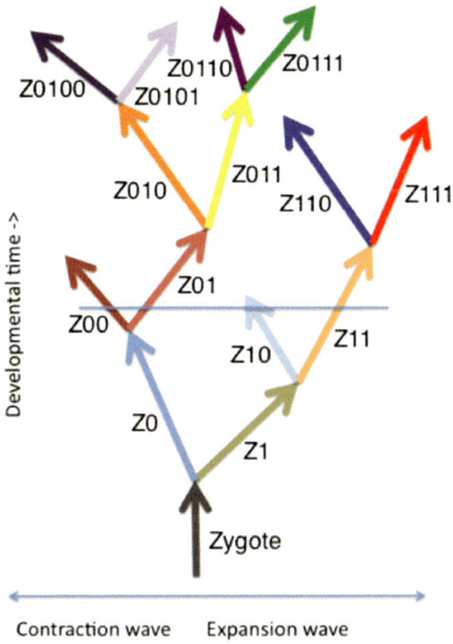

Figure 4. The zygote (Z) gives rise to the many cells of an embryo. Groups of them differentiate together in response to participating in either a contraction wave, indicated by a left leaning arrow (0), or an expansion wave, indicated by a right leaning arrow (1). Each cell type may thus be designated by a binary number (its differentiation code). To emphasize that each edge of such a differentiation tree represents a distinct cell type, we have given each edge a different color. At a given developmental time, marked by the horizontal time, only a subset of the cell types is present in the embryo. The number of cells increases with developmental time. The volume of the embryo may or may not increase with time, depending on the amount and means of supply of nutrients (internal yolk, external yolk, or placental). From (Gordon, 2015) with permission of Cambridge University Press.

- Ensemble of devices = set of differentiation waves.
- Hierarchically branching order of differentiation and morphogenesis = differentiation tree (Gordon, 1999) (Fig. 4).
- Self-regulating holons at various levels (stages) = differentiation waves, which are launched from specific loci, propagate through the tissue available to them, and then halt.
- Janus character = each differentiation wave propagates in its own, but sets up conditions for two new differentiation waves

inside its trajectory, i.e., each branch of the differentiation tree can generate two new branches.

- Autonomy = each differentiation wave is self-regulating, continuing to propagate reasonably independently of any experimental interventions and variations between individuals.
- 'Docility' means the multipotential capacity of embryonic tissue to follow this or that branch of the developmental hierarchy according to circumstances. = each cell is capable of participating in any differentiation wave that encounters it, and its response will depend on whether that wave is a contraction or an expansion wave.
- Determination = once a cell has participated in a differentiation wave, it does not revert to an earlier state of differentiation.
- Fixed genetic canons = the response of each cell to its participation in a differentiation wave is to undergo a specific change in gene expression, i.e., differentiate to a new cell type, which change depends on whether the wave is a contraction or expansion wave.
- Adaptable strategies guided by feedbacks from the hierarchy of environments = the differentiation waves and the responses of the cells provide the hierarchy of embryo "environments" that guide the cells to make appropriate "decisions".
- Action of triggers = launching of differentiation waves.
- Preset mechanisms = the cell state splitters in each cell that propagate the differentiation waves from each cell to the next in a tissue.
- Decrease in flexibility = the sequential changes in gene expression resulting in the specialized, fully differentiated adult cell type.
- Part of the whole = trajectory of a given differentiation wave.

We find it remarkable that Koestler, who was not a scientist, conceived so much more of the cybernetic nature of embryogenesis than anyone else. He does miss, however, one essential component: as implied by the work of Barbieri (Barbieri, 2003), each cell needs to retain a memory of the sequence of the contraction and expansion waves it experienced. This is what we call the "differentiation code" (Björklund and Gordon, 1994; Gordon *et al.*, 1994; Gordon, 2015) (Fig. 4). Note that the number of bits needed to represent the differentiation code depends only on the number of cell types to be generated, from egg to adult, of which the highest estimate is over 7000 (Bard *et al.*, 1998; Gordon, 1999). This is less than $2^{13} = 8192$, so at most 13 bits are needed. Note that 13 bits is far lower than the organism information contents discussed by Apter and Wolpert, ranging from 6×10^9 to 5×10^{25} bits (Apter and Wolpert, 1965). Of course, the 13 bits only allow us to distinguish topologically different differentiation trees from one another, but then that may be all we need at reasonably detailed phylogenetic levels. Differentiation in embryos may be a lot less complex than we have imagined.

Of course, each bit of the differentiation code corresponds to the triggering of a cascade of gene products and gene expression. The differentiation tree then has a mapping onto the linear DNA of the genome. Discovery of that mapping is left to the future (Gordon, 1999). However, it is at this level that evolution can act, by selecting between mutants of the topology of the differentiation tree, not just of gene products. This can be accomplished by cut, copy and paste of large segments of the genome (Gordon, 1999). In an abstract sense, the differentiation tree is a model of the process of embryogenesis (Conant and Ashby, 1970), and evolution has "designed" this model. This fits Richard Dawkins' (Wikipedia, 2015 k) concept of the *Extended Phenotype* (Dawkins, 1982), where models exist in the genetic material. We leave the interesting philosophical ramifications of this to future thinking.

Because in most organisms almost all of the cells have the same DNA, the "differentiation memory" of a cell, represented by its differentiation code, a case can be made that it is epigenetic. However, while the 'activation' may be epigenetic, the repeatable quality of embryogenesis, its fidelity from descendant "vehicle" (Dawkins, 1982) to vehicle suggests a genetic character, perhaps from introns and/or non-coding exons. Embryo dormancy in plants and even some uterine organisms, or diapause (Mead, 1993) may suggest, at times, external activation by some means.

While where these codes lay or how they activate is not currently known, we can also consider that it may not be an 'either/or' reducible problem, but instead another example of holism at play. Also, exactly what the "memorons" (Gordon, 1999) are we don't yet know, but epigenetic memory for cell differentiation and mental memory may have something in common (Levenson and Sweatt, 2005; Reul *et al.*, 2011). Extended cell differentiation in the brain has been proposed as a potential mechanism of mental memories (Gordon, 1999). Epigenetic mechanisms can range from methylation of individual nucleotide bases or histone tails to large scale configurations of the nucleus (as in our "Wurfel" model (Gordon, 1999).

We can attempt to shoehorn the concept of the cybernetic embryo into the language of biosemiotics. The general idea is that there are two "worlds" whose components are put into correspondence via use of an "adapter" (Barbieri, 2007). The first and foremost example is the genetic code, where base triplets are in the DNA/RNA "world", and the transfer RNAs are the adapters, providing the "meaning" of the base triplets as specific amino acids. What is at the two business ends of a transfer RNA is regarded as an arbitrary convention. But once that convention is fixed, the "meaning" of a DNA sequence becomes the protein that results.

Let's consider one branch of the schematic differentiation tree of Fig. 4, say the branch from Z1 to Z10 and Z11. The "world" of cells of type Z1 participates in two differentiation waves, a contraction wave traversing say a contiguous half of this tissue's cells, and an expansion wave traversing the other half. Inside each cell is a cell state splitter, which we can identify as the "adapter". An arbitrary "meaning" has been assigned: if the cell state splitter contracts, the result of the signaling to the nucleus could be either cell type Z10 or Z11. It is just a one bit assignment, but there would seem to be no biological reason which cell type should result from a cell state splitter contraction or expansion. The "meaning" of the "0" or "1" is arbitrary. So the choice is therefore a "convention". The "world" of cell type Z1 is thus mapped via the cell state splitter acting as an adapter to the "world" of cell types Z10 and Z11. There has to be a similar arbitrary (but fixed) mapping at each hierarchical step of the differentiation tree.

The main problem with such a biosemiotics pattern code approach (Fig. 2) is that it leaves out the differentiation waves themselves, whose dynamic geometry of launchings, trajectories and terminations create the spatiotemporal structure of the embryo, setting up the conditions for more differentiation waves at each step. So while the formal world to world mapping through cell state splitters as adapters gives us a biosemiotics perspective, it does not solve the whole problem of how an embryo builds itself. The cybernetic embryo point of view may prove more fruitful.

5. Experimental Tests of the Cybernetic Embryo Concept

We are now left with the task of designing experiments to test the notion of the cybernetic embryo. What are we looking for? In this concept each step of cell differentiation during embryogenesis involves:

1. A discrete event which is triggered.
2. A goal which is established.
3. A feedback test for whether or when that goal has been achieved.
4. Recording of the fact that the goal has been achieved.
5. Setting up of the next pair of triggering events.

In detail, we need to find, for each differentiation wave:

1. Its triggering event, which occurs in a specific portion of the embryo at a specific time, and which determines whether it is a contraction or an expansion wave.
2. Its trajectory, which determines which cells in its path take the next step of differentiation.
3. What stops the wave.
4. The epigenetic memory that the cell has differentiated.
5. How the triggering events for the next pair of waves (contraction and expansion) are determined.

We will discuss two organisms which show hints of allowing us to accomplish this proposed line of research, the regulating axolotl embryo (the salamander *Ambystoma mexicanum*) and the mosaic nematode embryo (the worm *Caenorhabditis elegans*).

In the axolotl, as in all vertebrates, an opening called the blastopore forms, through which much of the ectoderm invaginates, going inside and expanding to form the inner wall of the blastocoel, a hollow, water filled chamber that later will elongate, twist and turn to form the digestive system, the tube from mouth to anus. The lip of the blastopore consists of different cells from one minute to the next. Some of these cells transiently change shape to become "bottle" cells, shaped like hand blown wine bottles. We have interpreted this as a differentiation wave. It had not been previously reported in the literature as a wave, despite a large literature on bottle cells, because the wave crest (the cells moving over the lip of the blastopore) is stationary in the laboratory coordinate system. Detailed examination of time lapse movies of the blastopore reveals that it consists of a sequence of contraction (bottle cell formation) and expansion waves (Björklund and Gordon, 1994).

When the contracted bottle cells arrive inside, they flatten, suggesting they are participating in a subsequent expansion wave. Determining what is going on will require the development of *in vivo* imaging techniques such as soft X-ray computed tomography microscopy, high resolution magnetic resonance imaging, optical coherence tomography, or electrical impedance microscopy. What we do know is that a portion of this invaginating tissue (pharyngeal endoderm) makes contact at a point with the overlying ectoderm tissue, launching the ectoderm contraction wave (Gordon, *et al.*, 1994). This process was classically called "primary neural induction", because the triggered wave has a trajectory through the ectoderm that results in the participating cells differentiating into neuroepithelium, the precursor of the central nervous system, later divided (probably by differentiation waves) into brain and spinal cord, and later into the eye placodes, and the many discrete parts of the brain. We have modeled the triggering of the ectoderm contraction wave as a point in the ectoderm where a mechanical imbalance occurs due to the attachment of the underlying, invaginated cells at that point (Gordon, 1999). The ectoderm cells thereby respond to this attachment, which probably coincides with a buildup of pressure in the blastocoel, by contraction, which is then mechanically transmitted to neighboring ectoderm cells. This stretches them a bit, initiating a stretch activated contraction in them, which then propagates to the next adjacent cells, etc.

The ectoderm contraction wave stops by self-annihilation. Like a fire that burns itself out, if one part of the wave runs head on into another part of the wave, there is nothing left to burn, and the fire stops. This is the principle of a controlled burn. It suggests that a differentiation wave is a wave in an active medium. Note that we have expanded the ordinary concept of a cybernetic goal to include not just a quantitative, scalar set point, as in a thermostat, but also a topological goal, such as having the result that a differentiation wave's trajectory covers an subset of the cells in a region of the embryo. Waves in active media, which take their energy for propagation from the medium, have this property of going on relentlessly until no medium is left for them to alter, or something blocks them. Again, just think of the dynamics of forest fires.

What turns the ectoderm contraction wave on itself, changing its trajectory from convex to concave, may be the non-uniform tensions in the ectoderm generated by the tissue's continuing invagination over the lip of the blastopore. As the outer part of the wave moves faster than the inner part, leading to this change in curvature of its wave front, we are led to suggest that the wave speed is proportional to the local tension.

The lower hemisphere of the ectoderm that does not participate in the ectoderm contraction wave instead participates in the ectoderm expansion wave, differentiating into epidermis in the process. How the ectoderm expansion wave is triggered, we do not yet know.

We presume that the nucleus of each ectoderm cell receives a one bit intracellular signal from the cell state splitter, telling the nucleus whether it has participated in a contraction or an expansion wave. We have modeled this process (updated in: (Gordon,

2015)), which we call "nuclear state splitting" (Björklund and Gordon, 1993). The change in gene expression, accompanied by its epigenetic features, then changes the cell into a new kind (neuroepithelium or epidermis).

We started this story with step 5, how one particular wave triggers the next, and thus have come full circle, but with the embryo acquiring two new tissues in the process. Every detail occurs in a defined time and place in the embryo, and is accessible for experimental testing and refutation or modification of the hypothesized mechanisms. The goal is establishment of two new tissues from one. The goal is attained via differentiation waves. The accomplishment of the goal is determined by the trajectory and halting of each differentiation wave. As the goal is accomplished, the physical conditions for triggering the next pair of waves within the trajectory of the previous wave is determined.

One nuance to add is the possibility that a given differentiation wave propagates through more than one tissue, as defined by the differentiation codes of the cells it traverses (Björklund and Gordon, 1994; Gordon, 2015). In these situations the differentiation wave takes on a more regional if not quite global character, by keeping the differentiation of two or more tissues in synch:

> "It is often neglected that many of the forces and mechanical cues of a developing organ or tissue may come from tissues not necessarily from the same lineage. In fact, it is entirely possible that mechanotransduction can occur between tissues, or intertissue mechanotransduction. A fundamental understanding of the *in vivo* micromechanics of processes involved in morphogenesis is essential to employ these processes *in vitro* to differentiate stem cells to a target tissue [as suggested, by emulating differentiation waves, in: (Gordon, 2006; Lu *et al.*, 2012)]" (Kolahi and Mofrad, 2010).

Our second example is the nematode *C. elegans*, for which every single cell may be observed during embryogenesis. Almost all of the nearly 1000 cells are produced by asymmetric cell divisions, and it is presumed that in general the daughter cells are both

Table 1. Proposed correspondences between regulative and mosaic embryonic development.

Regulative	Mosaic
Contraction wave	Small cell in an asymmetric division
Expansion wave	Large cell in an asymmetric division
Cell proliferation without further differentiation	Symmetric cell division yielding two equal sized cells of the same type as their mother cell

different from one another and different from the cell from which they arose. A small proportion of the cells occur in pairs resulting from a symmetric cell division. This has led us to propose the correspondences shown in Table 1 (Gordon, 1999) and to work on turning the known cell lineage tree into a differentiation tree (Alicea, 2014; Alicea, *et al.*, 2014a, 2014b). The five steps listed above all need to be investigated, in this case as intracellular events. This endeavor would make contact with the large and growing literature on the mechanism and spatial orientation of asymmetric cell divisions (Morin and Bellaiche, 2011). We have proposed that the jump from intracellular to multicellular differentiation waves was a major event in evolution (Gordon, 1999); potentially significant to the thesis of *The Major Transitions of Evolution* (Smith and Szathmáry, 1995), and within the larger context of metasystems transition theory (Turchin, 1977).

Experimentally, what we need first and foremost is the ability to observe every cell in a developing embryo (Gordon, 2009; Gordon and Westfall, 2009), both during the course of normal development, and during physical, chemical and genetic interventions. We can look inside using X-ray computed tomography (Moosmann *et al.*, 2014) and magnetic resonance imaging (Gruwel *et al.*, 2008). Methods have been developed for measuring the forces generated by cells (Stein and Gordon, 1982; Cranston *et al.*, 2010; Brodland *et al.*, 2014) and how they react to imposed forces (Fleury *et al.*, 2010; von Dassow, 2010). *In vivo* fluorescent techniques should allow us to watch the subcellular mechanics of differentiation waves and changes in gene expression (Cavey and Lecuit, 2008). Single cells can be ablated (Gerhart *et al.*, 2008; Sweeney *et al.*, 2012; Nast and Extavour, 2014), perhaps to alter the trajectories of differentiation waves. We can vary the temperature, which has the effect of speeding or slowing embryonic development in cold blooded animals, including the axolotl, which nevertheless somehow develop normally over a wide temperature range (Lamb *et al.*, 1984; Duellman and Trueb, 1986), suggesting that development really is a cybernetic relay race (Gordon 1999). Are critical periods of sensitivity to temperature jumps, as in sex determination in reptiles (Yntema and Mrosovsky, 1982; Shine *et al.*, 2007), or embryo abnormalities and death outside a species' normal temperature range (Gordon, 1999), due to interference with the synchronization of differentiation waves, whose local speeds (Brodland *et al.*, 1994) depend on both the mechanics (non-uniform strain distribution: (Gordon *et al.*, 1994)) and the temperature (Jaffe, 2003)? So a vast array of experimental approaches to observing and manipulating the five steps in a bifurcation of the proposed hierarchical cybernetic differentiation tree is available.

Birds and mammals evolved from amphibians, and some time during that evolution lost the ability to develop over a wide temperature range, as we saw in the discussion of homeothermy. We have speculated that:

> "…the constant body temperature of mammals and egg incubation by birds may free them in part from the need for… [synchronizing] inductive relationships for developmental timing purposes, allowing more often successful heterochrony. Thus homeothermy and the consequent reduction in the developmental constraint of induction may be one key to why certain phyla radiate rapidly" (Gordon, 1999).

A full explanation of embryogenesis requires going full circle between observations, mathematics and computer simulations (Jacobson and Gordon, 1976). In *C. elegans*, since every cell's location during development is known, we can ask questions about how robust the process is. In terms of the cybernetic embryo hypothesis, we are asking how accurate the orientation and timing of each asymmetric division must be, to end up with the cells in the right places in the adult. If we were to "grow" an embryo by computer simulation, with small deviations in the parameters, the deviations might soon accumulate, creating a simulated organism that doesn't look much like the adult worm. We could add springs, either between directly related cells or distantly related, but physically nearby cells, to approximate a feedback mechanism to restore the adult configuration. To what extent would such springs compensate for the small deviations? How many are needed? The elastic constants necessary to compensate would give us an idea of the strengths, so to speak, needed for the feedbacks to keep development on course. Will the whole structure of Dennett "cranes" (Dennett, 1995) topple like a pile of blocks (Gordon, 2015) if we exceed some stability threshold for the number of cell types? This is an example, then, of how the cybernetic embryo concept can be tested *in silico*.

6. Conclusion

We have narrowed our conception of cybernetics to the original one of feedback systems, to avoid the broader connotations of all the fields it has engendered (Fig. 1). Differentiation waves become the communication channels between cells, providing a feedback mechanism that ensures the embryo attains the right kinds of cells in the right place, at the right time, in the right numbers. Philosophically, feedback has been conflated with concepts of goals and purpose. Cybernetics still does not fit comfortably within fundamental physics (Brillouin, 1949; Drozin, 1971; Laborit, 1975; Vallee, 1991), despite feedback being perhaps the simplest form of communication in a system. Teleology, teleonomics, organization, purpose and function cannot easily be wished away, and we remain with Koestler's *Ghost in the Machine* (Koestler, 1967) for now. However, as an engineering discipline, usually under the name of "control theory" (Schulz, 2006; Iglesias and Ingalls, 2010), cybernetics is quite mature, so we can look forward to its use in reverse engineering the embryo (Gordon and Melvin, 2003).

In the sense of cybernetic feedback, rather than being a simple cybernetic system "striving" to become an adult, we see the embryo as a nested, bifurcating set of cybernetic events, with a few of them occurring in synch. These stretch our concept of a cybernetic goal, since the goal includes changing a whole contiguous subset of the cells in the embryo, analogous to the "control of fields" (Schulz, 2006). While a thermostat's goal is to control a scalar (temperature), or a heat seeking missiles' goal is to reduce a vector (from itself to its moving target) to zero, the goal of each cybernetic event in an embryo is to set off a subset of the embryo's cells as a new cell type (i.e., make a step of differentiation happen) *and* to create the conditions for launching the next pair of cybernetic events (one more stretch of cybernetics). Every aspect of our proposal is testable experimentally and by computer simulation. There is much work ahead. However, the goal of the research, answering the question "How does an embryo build itself?", may be solvable. The scientific, medical, commercial, and social implications for understanding birth defects, manipulating embryogenesis, modeling evo-devo relationships, and growing self-replicating AI (artificial intelligence) machines are considerable. The cybernetic embryo may become a model for the prescient comments by one of Albert Einstein's biographers:

> "At the next stage, cybernetic systems capable of altering their own design will enable people to concentrate on more ambitious problems" (Kuznetsov, 1965).

Acknowledgments

RS would like to acknowledge his high school teacher of Genetics, Bacteriology, and Embryology, Heston Bates, who corrected him whenever he implied any purpose in biology; please consider this his final 'Beneficial Assignment'. We dedicate this work to the late W. Ross Ashby, whom RG had the pleasure of meeting at Ashby's lab and introducing him to percolation bond probability theory when shown his stochastic switch consisting of two copper plates subtended by a mixture of glass and copper balls. We would like to thank Bradly Alicea, Trista Robichaud, John Torday and Timothy Warrington for their comments on the draft manuscript.

Glossary

- Cell state splitter: An organelle, which like the spindle in open mitosis has no bounding membrane of its own, consisting of a microfilament ring, a subtending microtubule mat, and an intermediate filament ring, providing a bistable mechanism that can propagate a differentiation wave from one cell to the next.
- Contraction wave: A differentiation wave that propagates via contraction of the surface of the cells in an epithelial sheet, probably involving contraction of an apical microfilament ring in each cell.
- Cybernetic system: A system in which communication and control via signals affect it's process as it develops from one state to the next.
- Differentiation code: the sequence of contraction/expansion waves that have been experienced by an embryonic cell and its predecessors (representable as a binary number).
- Differentiation tree: A tree represented on a two-axis coordinate system, with the vertical axis representing time since fertilization of an egg, and the horizontal axis being used to indicate "contraction wave" for an edge of the tree leaning left, and an "expansion wave" for an edge of the tree leaning right. Each edge indicates a distinct, if transient cell type. Only two edges rise from each node, on the presumption that each cell type can give rise to only two new cell types, i.e., that the tree is bifurcating.
- Differentiation wave: A wave of change of cell type.
- Expansion wave: A differentiation wave that propagates via expansion of the surface of the cells in an epithelial sheet, probably involving expansion of an apical microtubule mat in each cell.
- Feedback: A signal which changes the behavior of the system, allowing it to adjust itself by minimizing or maximizing that signal.

- Feedback loop: A system where the output signal of the previous state alters the behavior of the subsequent state of the system. Also known as a Circular System.
- Goal: A predetermined state of a system, most often due to coded information, different from the starting state.
- Negative feedback: A signal which decreases the difference between the current state and a goal state.
- Positive feedback: A signal which increases the difference between the current state and a goal state.
- Teleology: Purposive behavior controlled by feedback.
- Teleonomy: Purposive behavior controlled by processing of coded information.
- Trajectory: The region of an embryo covered by a differentiation wave from its launching to its termination.

References

Adamsons K and Towell ME. 1965. Thermal homeostasis in fetus and newborn. *Anesthesiology*, 26(4), 531–548. doi: 10.1097/00000542-196507000-00017.

Alicea B. 2014. *OpenWorm Journal Club: DevoWorm — Raising the worm.* Available from: https://www.youtube.com/watch?v=XRm7SxTpmQo (Accessed on 30 May 2016).

Alicea B, McGrew S, Gordon R, Larson S and Watts M. 2014a. *From Differentiation Waves to "Wriggling": new views on C. elegans development.* Available from: http://figshare.com/articles/From_Differentiation_Waves_to_Wriggling_new_views_on_C_elegans_development/1015760 (Accessed on 30 May 2016).

Alicea B, McGrew S, Gordon R, Larson S, Warrington T and Watts M. 2014b. *DevoWorm: differentiation waves and computation in C. elegans embryogenesis.* Available from: http://www.biorxiv.org/content/early/2014/10/03/009993 (Accessed on 30 May 2016).

American Society for Cybernetics. 2014. *The origins of cybernetics*. Available from http://www.gwu.edu/~asc/origin.html (Accessed on 30 May 2016).

Apter MJ. 1966. *Cybernetics and Development*. Pergamon Press, Oxford.

Apter MJ and Wolpert L. 1965. Cybernetics and development I. Information theory. *Journal of Theoretical Biology*, 8(2), 244–257. doi: 10.1016/0022-5193(65)90075-5.

Ashby WR. 1947. Principles of the self-organizing dynamic system. *Journal of General Psychology*, 37(2), 125–128.

Ashby WR. 1952. *Design for a Brain*. Wiley, New York.

Ashby WR. 1956. *An Introduction to Cybernetics*. John Wiley & Sons, New York.

Barbieri M. 1974. A criterion to evaluate three dimensional reconstructions from projections of unknown structures, *Journal of Theoretical Biology*, 48(2), 451–467.

Barbieri M. 2003. *The Organic Codes. An Introduction to Semantic Biology*. Cambridge University Press, Cambridge.

Barbieri M. 2007. Is the cell a semiotic system? In *Introduction to Biosemiotics: The New Biological Synthesis*, Barbieri M (eds.), 179–208. Springer, Germany.

Bard JBL, Baldock RA and Davidson DR. 1998. Elucidating the genetic networks of development: a bioinformatics approach. *Genome Reseach*, 8(9), 859–863.

Bateson G. 1972. *Steps to an Ecology of Mind: Collected Essays in Anthropology, Psychiatry, Evolution, and Epistemology*, Jason Aronson, Northvale, NJ.

Becht G. 1974. Systems theory, the key to holism and reductionism. *BioScience*, 24(10), 569–579. doi: 10.2307/1296630.

Beloussov LV. 2013. Morphogenesis can be driven by properly parametrised mechanical feedback. *The European Physical Journal E Soft Matter*, 36(11), 132. doi: 10.1140/epje/i2013-13132-x.

Beloussov LV and Gordon R. 2006. Preface. Developmental morphodynamics: bridging the gap between the genome and embryo physics. *International Journal of Developmental Biology*, 50(2–3), 79–80. doi: 10.1387/ijdb.052137lb.

Berrill NJ. 1961. *Growth, Development, and Pattern*. 2nd edition. W.H. Freeman and Co, San Francisco.

Bhattacharya S, Zhang Q and Andersen ME. 2011. A deterministic map of Waddington's epigenetic landscape for cell fate specification. *Bmc Systems Biology*, 5. doi: 10.1186/1752-0509-5-85.

Bininda-Emonds OR, Jeffery JE and Richardson MK. 2003. Inverting the hourglass: quantitative evidence against the phylotypic stage in vertebrate development. *Proceedings of the Royal Society B: Biological Sciences* 270(1513), 341–346.

Björklund NK and Gordon R. 1993. Nuclear state splitting: a working model for the mechanochemical coupling of differentiation waves to master genes. *Russian Journal of Developmental Biology*, 24(2), 79–95.

Björklund NK and Gordon R. 1994. Surface contraction and expansion waves correlated with differentiation in axolotl embryos. I. Prolegomenon and differentiation during the plunge through the blastopore, as shown by the fate map. *Computers & Chemistry*, 18(3), 333–345.

Bohr N. 1933. Light and life. *Nature*, 131:421–423, 57–59. doi: 10.1038/131421a0.

Boltzmann L. 2012. *Lectures on Gas Theory*. reprinted 1964 translation of Part I: 1896, Part II: 1898 edition. Dover Publications, Inc., USA.

Bonner J. 1965. *The Molecular Biology of Development*. Clarendon Press, Oxford.

Brand S. 1976. For God's sake, Margaret: conversation with Gregory Bateson and Margaret Mead. *CoEvolution Quarterly*, (10), 32–44.

Brillouin L. 1949. Life, thermodynamics, and cybernetics. *American Scientist*, 554–568.

Brodland GW, Gordon R, Scott MJ, Björklund NK, Luchka KB, Martin CC, Matuga C, Globus M, Vethamany-Globus S and Shu D. 1994. Furrowing surface contraction wave coincident with primary neural induction in amphibian embryos. *Journal of Morphology*. 219(2), 131–142.

Brodland GW, Veldhuis JH, Kim S, Perrone M, Mashburn D and Hutson MS. 2014. CellFIT: a cellular force-inference toolkit using curvilinear cell boundaries. *Plos One*, 9(6). doi: 10.1371/journal.pone.0099116.

Buss LW and Blackstone NW. 1991. An experimental exploration of Waddington's epigenetic landscape. *Philosophical Transactions of the Royal Society B*, 332, 49–58.

Byerly TC. 1938. Effect of different incubation temperatures on mortality of chick embryos. *Poultry Science*, 17(3), 200–205.

Castellani B. 2013. IX.5 Map of Complexity Science. Arts and Science Factory, LLC, Cleveland, OH. In *Places & Spaces: Mapping Science*, Börner K and Theriault TN (eds.) http://scimaps.org/mapdetail/map_of_complexity_sc_154. Bloomington, IN, USA: Cyberinfrastructure for Network Science Center, Indiana University (Accessed on 23 September 2016).

Cavey M and Lecuit T. 2008. Imaging cellular and molecular dynamics in live embryos using fluorescent proteins. *Methods in Molecular Biology*, 420, 219–238.

Choudhuri S. 2011. From Waddington's epigenetic landscape to small noncoding RNA: some important milestones in the history of epigenetics research. *Toxicology Mechanisms and Methods*, 21(4), 252–274. doi: 10.3109/15376516.2011.559695.

Cohen J. 2003. Afterword. In *The Organic Codes. An Introduction to Semantic Biology*, Barbieri M (eds.). 265–277. Cambridge University Press, Cambridge.

Conant RC and Ashby WR. 1970. Every good regulator of a system must be a model of that system. *International Journal of Systems Science*, 1(2), 89–97.

Cooke J. 1981. Scale of body pattern adjusts to available cell number in amphibian embryos. *Nature*, 290, 775–778.

Cranston PG, Veldhuis JH, Narasimhan S and Brodland GW. 2010. Cinemechanometry (CMM): a method to determine the forces that drive morphogenetic movements from time-lapse images. *Annals of Biomedical Engineering*, 38(9), 2937–2947. doi: 10.1007/s10439-010-9998-1.

Dawkins R. 1982. *The Extended Phenotype, The Gene as the Unit of Selection*. Oxford University Press, Oxford.

De Robertis EM. 2009. Spemann's organizer and the self-regulation of embryonic fields. *Mechanisms of Development*, 126(11–12), 925–941. doi: S0925-4773(09)01446-4 [pii]. 10.1016/j.mod.2009.08.004 [doi].

Deacon T and Koutroufinis S. 2014. Complexity and dynamical depth. *Information*, 5(3), 404–423.

Dennett DC. 1995. *Darwin's Dangerous Idea: Evolution and the Meanings of Life*. Simon & Schuster, New York.

Driesch HAE. 1892. The potency of the first two cleavage cells in echinoderm development. Experimental production of partial and double formation. In *Foundations of Experimental Embryology*, Willier BH and Oppenheimer JM (eds.), 39–50. Hafner, New York.

Drozin VG. 1971. Physics to cybernetics: emergence of a new basic science. *American Journal of Physics* 39(6), 697–698.

Dubuis JO, Tkačik G, Wieschaus EF, Gregor T and Bialek W. 2013. Positional information, in bits. *Proceedings of the National Academy of Sciences of the United States of America*, 110(41), 16301–1608. doi: 10.1073/pnas.1315642110.

Duellman WE and Trueb L. 1986. *Biology of Amphibians*. McGraw-Hill Book Co., New York.

Emmeche C and Kull K. 2011. *Towards a Semiotic Biology: Life Is the Action of Signs*. Imperial College Press, London, UK.

Fagan M. 2012. Waddington redux: models and explanation in stem cell and systems biology. *Biology and Philosophy*, 27(2), 179–213. doi: 10.1007/s10539-011-9294-y.

Fankhauser G, Vernon JA, Frank WH and Slack WV. 1955. Effect of size and number of brain cells on learning in larvae of the salamander, *Triturus viridescens*. *Science*, 122(3172), 692–693.

Fisher RA. 1915. The evolution of sexual preference. *Eugenics Review*, 7(3), 184–192.

Fisher RA. 1930. *The Genetical Thory of Natural Selection*. Dover, New York.

Fleury V, Al-Kilani A, Boryskina OP, Cornelissen AJM, Nguyen T-H, Unbekandt M, Leroy L, Baffet G, Lenoble F, Sire O, Lahaye E and Burgaud V. 2010. Introducing the scanning air puff tonometer for biological studies. *Physical Review E*, 81(2), 021920.

Fleury V and Gordon R. 2012. Coupling of growth, differentiation and morphogenesis: an integrated approach to design in embryogenesis. In *Origin(s) of Design in Nature: A Fresh, Interdisciplinary Look at How Design Emerges in Complex Systems, Especially Life*, Swan L, Gordon R and Seckbach J (eds.), 385–428. Springer, Dordrecht.

Galis F and Metz JA. 2001. Testing the vulnerability of the phylotypic stage: on modularity and evolutionary conservation. *Journal of Experimental Zoology*, 291(2), 195–204.

Gatherer D. 1996. A stroll across the epigenetic landscape: bringing Waddington's ideas into molecular biology. *Early pregnancy: Biology and Medicine: The Official Journal of the Society for the Investigation of Early Pregnancy*, 2(4), 241–243.

Gerhart JC. 1980. Mechanisms regulating pattern formation in the amphibian egg and early embryo. In *Biological Regulation and Development Vol. 2: Molecular Organization and Cell Function*, Goldsberger RF (eds.), 133–316. Plenum Press, New York.

Gerhart J, Neely C, Pfautz J and George-Weinstein M. 2008. Tracking and ablating subpopulations of epiblast cells in the chick embryo. *Biological Procedures Online*, 10, 74–82. doi: 10.1251/bpo145.

Gibbs JW. 1902. *Elementary Principles in Statistical Mechanics: Developed with Especial Reference to the Rational Foundation of Thermodynamics*. 1960 reprint edition. Dover Publications, Inc., New York.

Gierer A and Meinhardt H. 1972. A theory of biological pattern formation. *Kybernetic*, 12(1), 30–39.

Gilbert SF. 1991. Epigenetic landscaping: Waddington use of cell fate bifurcation diagrams. *Biology & Philosophy*, 6(2), 135–154. doi: 10.1007/bf02426835.

Gordon NK and Gordon R. 2015. *Embryogenesis Explained*. World Scientific Publishing Company, Singapore.

Gordon R. 1966. On stochastic growth and form. *Proceedings of the National Academy of Sciences of the United States of America*, 56(5), 1497–1504.

Gordon R. 1994. *Mechanical engineering of the cytoskeleton in developmental biology. International Review of Cytology, Vol. 150*. Academic Press, USA.

Gordon R. 1999. *The Hierarchical Genome and Differentiation Waves: Novel Unification of Development, Genetics and Evolution*. World Scientific & Imperial College Press, Singapore & London.

Gordon R. 2006. Mechanics in embryogenesis and embryonics: prime mover or epiphenomenon? *International Journal of Developmental Biology*. 50(2–3), 245–253.

Gordon R, Bender R and Herman GT. 1970. Algebraic reconstruction techniques (ART) for three-dimensional electron microscopy and x-ray photography. *Jounal of Theoretical Biology*, 29(3), 471–481.

Gordon R and Björklund NK. 1996. How to observe surface contraction waves on axolotl embryos. *International Journal of Developmental Biology*, 40(4), 913–914.

Gordon R, Björklund NK and Nieuwkoop PD. 1994. Dialogue on embryonic induction and differentiation waves. *International Review of Cytology*, 150, 373–420.

Gordon R and Melvin CA. 2003. Reverse engineering the embryo: a graduate course in developmental biology for engineering students at the University of Manitoba, Canada. *International Journal of Developmental Biology*, 47(2–3), 183–187.

Gordon R. 2009. Google Embryo for building quantitative understanding of an embryo as it builds itself: II. Progress toward an embryo surface microscope. *Biological Theory: Integrating Development, Evolution, and Cognition*, 4(4), 396–412.

Gordon R. 2010. *Wholeness & Embryogenesis, Part 1 (Extending "The Hierarchical Genome and Differentiation Waves" §1.03, 1.04, 1.05)*. Embryo Physics Course. Availabel from: http://embryogenesisexplained.com/old-lectures (Accessed on 30 May 2016).

Gordon R. 2011a. Epilogue: the diseased breast lobe in the context of X-chromosome inactivation and differentiation waves. In *Breast Cancer: A Lobar Disease*, Tot T (ed.), 205–210. Springer, London.

Gordon R. 2011b. Stop breast cancer now! Imagining imaging pathways towards search, destroy, cure and watchful waiting of premetastasis breast cancer. In *Breast Cancer — A Lobar Disease*, Tot T (ed.), 167–203. Springer, London.

Gordon R. 2013. Conception and development of the Second Life® Embryo Physics Course [invited]. *Systems Biology in Reproductive Medicine*, 59, 131–139.

Gordon R. 2015. Walking the tightrope: the dilemmas of hierarchical instabilities in Turing's morphogenesis [invited]. In *The Once and Future*

Turing: Computing the World, Cooper SB and Hodges A (eds.), 150–164. Cambridge University Press, Cambridge.

Gordon R and Westfall JE. 2009. Google Embryo for building quantitative understanding of an embryo as it builds itself: I. Lessons from ganymede and google earth. *Biological Theory: Integrating Development, Evolution, and Cognition*, 4(4), 390–395 + Supplementary Appendix.

Gruwel MLH, Lerner BE, Björklund NK and Gordon R. 2008. Time-Lapse MicroMRI (μMRI) of Axolotl Embryogenesis. In *ESMRMB Congress 2008, Valencia, ES: October 2–4.*

Guillot C and Lecuit T. 2013. Mechanics of epithelial tissue homeostasis and morphogenesis. *Science*, 340(6137), 1185–1189. doi: 10.1126/science.1235249.

Hamburger V. 1955. Regeneration. In *Encyclopedia Britannica*, 67.

Harth E. 1995. *The Creative Loop, How the Brain Makes a Mind*. Addison-Wesley Publishing Co., USA.

Hazkani-Covo E, Wool D and Graur D. 2005. In search of the vertebrate phylotypic stage: a molecular examination of the developmental hourglass model and von Baer's third law. *Journal of Experimental Zoology Part B: Molecular and Developmental Evolution*, 304(2), 150–158.

Hellman WD. 1982. *Norbert Wiener and the Growth of Negative Feedback in Scientific Explanation; with a Proposed Research Program of Cybernetic Analysis [Ph.D Thesis]*. Oregon State University, Corvallis, Oregon, USA.

Hodges A. 1983. *Alan Turing: The Enigma*. Simon & Schuster, New York.

Hoffmeyer J. 1996. *Signs of Meaning in the Universe*: Indiana University Press, USA.

Huang S. 2012. The molecular and mathematical basis of Waddington's epigenetic landscape: A framework for post-Darwinian biology? *BioEssays*, 34(2), 149–157. doi: 10.1002/bies.201100031.

Huxley JS. 1942. *Evolution, The Modern Synthesis*. George Allen & Unwin, London.

Iglesias PA and Ingalls BP. 2010. *Control Theory and Systems Biology*: MIT Press, Cambridge, USA.

Irie N and Kuratani S. 2011. Comparative transcriptome analysis reveals vertebrate phylotypic period during organogenesis. *Nature Communications*, 2(248), doi:10.1038/ncomms248; erratum (2014): doi:10.1038/ncomms4366. doi: 10.1038/ncomms1248.

Jacobson AG and Gordon R. 1976. Changes in the shape of the developing vertebrate nervous system analyzed experimentally, mathematically and by computer simulation. *Journal of Experimantal Zoology*, 197(2), 191–246.

Jaeger J and Sharpe J. 2014. On the concept of mechanism in development. In *Towards a Theory of Development*, Minelli A and Pradeu T (eds.). Oxford University Press, Oxford.

Jaffe LF. 1995. Calcium waves and development. *CIBA Foundation Symposium*, 188, 4–17.

Jaffe LF. 1999. Organization of early development by calcium patterns. *BioEssays*, 21(8), 657–667.

Jaffe LF. 2003. The propagation speeds of calcium action potentials are remarkably invariant. *Biology of the Cell*, 95(6), 343–355. doi: 10.1016/s0248-4900(03)00090-x.

Jones GW and Chapman SJ. 2012. Modeling growth in biological materials. *SIAM Review*, 54(1), 52–118. doi: 10.1137/080731785.

Kerszberg M and Wolpert L. 2007. Specifying positional information in the embryo: Looking beyond morphogens. *Cell*, 130(2), 205–209.

Kitano H, Funahashi A, Matsuoka Y and Oda K. 2005. Using process diagrams for the graphical representation of biological networks. *Nature Biotechnology*, 23(8), 961–966. doi: 10.1038/nbt1111.

Koestler A. 1959. *The Sleepwalkers, A History of Man's Changing Vision of the Universe*. Grosset & Dunlap, New York.

Koestler A. 1964. *The Act of Creation*. Hutchinson, London.

Koestler A. 1967. *The Ghost in the Machine*. Macmillan, New York.

Kolahi KS and Mofrad MRK. 2010. Mechanotransduction: A major regulator of homeostasis and development. *Wiley Interdisciplinary Reviews-Systems Biology and Medicine*, 2(6), 625–639. doi: 10.1002/wsbm.79.

Kolmogorov AN, Petrovskii IG and Piskunov NS. 1937. A study of the equation of diffusion with increase in the quantity of matter, and its application to a biological problem. *Bjulleten Moskovskogo Gosudarstvennogo Universiteta*, 1(7), 1–26.

Kranenbarg S. 2000. The phylotypic egg timer. *Netherlands Journal of Zoology*, 50(2), 289–294.

Kuhn TS. 1996. *The Structure of Scientific Revolutions*. 3rd edition. University of Chicago Press, Chicago.

Kuznetsov B. 1965. *Einstein*. Progress Publishers, Moscow.

Laborit H. 1975. Physics, biology and cybernetics. *Cybernetica*, 18(2), 67–80.

Lamb RJ, Gerber GH and Atkinson GF. 1984. Comparison of developmental rate curves applied to egg hatching data of *Entomoscelis americana*

Brown (Coleoptera: Chrysomelidae). *Environmental Entomology*, 13, 868–872.

Levenson JM and Sweatt JD. 2005. Epigenetic mechanisms in memory formation. *Nature Reviews Neuroscience*, 6(2), 108–118. doi: 10.1038/nrn1604.

Levin M, Hashimshony T, Wagner F and Yanai I. 2012. Developmental milestones punctuate gene expression in the *Caenorhabditis* embryo. *Developmental Cell*, 22(5), 1101–1108. doi: 10.1016/j.devcel.2012.04.004.

Lu K, Cao T and Gordon R. 2012. A cell state splitter and differentiation wave working-model for embryonic stem cell development and somatic cell epigenetic reprogramming. *BioSystems*, 109(3), (Special Issue on Biological Morphogenesis), 390–396.

Lu K, Gordon R and Cao T. 2013. Reverse engineering the mechanical and molecular pathways in stem cell morphogenesis. *Journal of Tissue Engineering and Regenerative Medicine*, 9(3), 169–173. doi: 10.1002/term.672.

Mammoto T, Mammoto A and Ingber DE. 2013. Mechanobiology and developmental control. In *Annual Review of Cell and Developmental Biology*, Schekman R (eds.), 27–61.

Maruyama M. 1963a. A postscript to the second cybernetics. *American Science*, 51(3), 250–256.

Maruyama M. 1963b. The second cybernetics: deviation-amplifying mutual causal processes. *American Science*, 51(2), 164–179.

Maturana HR and Varela FJ. 1980. *Autopoiesis and Cognition: The Realization of the Living*. D. Reidel Publishing Company, Dordrecht, Holland.

Maxwell JC. 1860a. II. Illustrations of the dynamical theory of gases. *The London, Edinburgh, and Dublin Philosophical Magazine and Journal of Science*, 20(130), 21–37.

Maxwell JC. 1860b. V. Illustrations of the dynamical theory of gases — Part I. On the motions and collisions of perfectly elastic spheres. *The London, Edinburgh, and Dublin Philosophical Magazine and Journal of Science*, 19(124), 19–32.

Maxwell JC. 1868. On governors. *Proceedings of the Royal Society of London*, 16, 270–283.

Maynard SJ and Szathmáry E. 1995. *The Major Transitions in Evolution*. Freeman WH, San Francisco.

Mayr E. 1961. Cause and effect in biology: Kinds of causes, predictability, and teleology are viewed by a practicing biologist. *Science*, 134(348), 1501–1506. doi: 10.1126/science.134.3489.1501.

Mead RA. 1993. Embryonic diapause in vertebrates. *Journal of Experimental Zoology*, 266(6), 629–641. doi: 10.1002/jez.1402660611.

Meinhardt H and Gierer A. 2000. Pattern formation by local self-activation and lateral inhibition. *Bioessays*, 22(8), 753–760. doi: 10.1002/1521-1878(200008)22:8<753::aid-bies9>3.0.co;2-z.

Miller CJ and Davidson LA. 2013. The interplay between cell signalling and mechanics in developmental processes. *Nature Reviews Genetics*, 14(10), 733–744. doi: 10.1038/nrg3513.

Miller JG. 1978. *Living Systems*: Mcgraw-Hill, USA.

Mindell D, Segal J and Gerovitch S. 2003. Cybernetics and information theory in the United States, France and the Soviet Union. In *Science and Ideology: A Comparative History*, Walker M (ed.), 66–95. Routledge, London.

Monod J. 1972. *Chance and Necessity: An Essay on the Natural Philosophy of Modern Biology*. Vintage Books, New York.

Moosmann J, Ershov A, Weinhardt V, Baumbach T, Prasad MS, LaBonne C, Xiao XH, Kashef J and Hofmann R. 2014. Time-lapse X-ray phase-contrast microtomography for *in vivo* imaging and analysis of morphogenesis. *Nature Protocols*, 9(2), 294–304. doi: 10.1038/nprot.2014.033.

Morin X and Bellaiche Y. 2011. Mitotic spindle orientation in asymmetric and symmetric cell divisions during animal development. *Developmental Cell*, 21(1), 102–119. doi: 10.1016/j.devcel.2011.06.012.

Nakage ES, Cardozo JP, Pereira GT, Sa Q and Boleli IC. 2003. Effect of temperature on incubation period, embryonic mortality, hatch rate, egg water loss and partridge chick weight (Rhynchotus rufescens). *Revista Brasileira de Ciência Avícola*, 5(2), 131–135. doi: 10.1590/s1516-635x2003000200007.

Nast AR and Extavour CG. 2014. Ablation of a single cell from eight-cell embryos of the amphipod crustacean. *Parhyale hawaiensis. Journal of visualized experiments: JoVE*, 85. doi: 10.3791/51073.

Needham J. 1930. *The Sceptical Biologist*. Norton WW, New York.

Poulin BJ and Gordon R. 2001. How to organize science funding: the new Canadian Institutes for Health Research (CIHR), an opportunity to vastly increase innovation. *Canadian Public Policy*, 27(1), 95–112.

Prigogine I and Stengers I. 1984. *Order out of Chaos, Man's New Dialogue with Nature*. Bantam Books, Toronto.

Rangayyan RM, Dhawan AP and Gordon R. 1985. Algorithms for limited-view computed tomography: an annotated bibliography and a challenge. *Applied Optics*, 24(23), 4000–4012.

Rapoport A. 1965. The impact of cybernetics on the philosophy of biology. In *Progress in Biocybernetics*, Wiener N and Schadd JP (eds.), 141–156. Elsevier Pub. Co., New York.

Reid RGB. 1985. *Evolutionary Theory: The Unfinished Synthesis*. Croom Helm, London.

Reul JMHM, Collins A and Gutierrez-Mecinas M. 2011. Epigenetic mechanisms in memory formation. In *Brain, Behavior and Epigenetics*, Petronis A and Mill J (eds.), 287–300.

Richardson MK, Minelli A, Coates M and Hanken J. 1998. Phylotypic stage theory. *TREE*, 13(4), 158.

Richardson MK. 2012. A phylotypic stage for all animals? *Developmental Cell*, 22(5), 903–904. doi: 10.1016/j.devcel.2012.05.001.

Rosen R. 1968. Turing's morphogens, two-factor systems and active transport. *Bulletin of Mathematical Biophysics*, 30(3), 493–499.

Rosenblueth A, Wiener N and Bigelow J. 1943. Behavior, purpose and teleology. *Philosophy of Science*, 10(1), 18–24.

Russell ES. 1945. *The Directiveness of Organic Activities*. Cambridge University Press, Cambridge.

Sander K. 1971. Pattern formation in longitudinal halves of leaf hopper eggs (Homoptera) and some remarks on definition of "embryonic regulation". *Wilhelm Roux Archiv für Entwicklungsmechanik der Organismen*, 167(4), 336–352.

Sander K. 1993. Landmarks in Developmental Biology (13). Entelechy and the ontogenetic machine — work and views of Hans Driesch from 1895 to 1910. *Roux's Archives of Developmental Biology*, 202(2), 67–69.

Schulz M. 2006. *Control Theory in Physics and Other Fields of Science: Concepts, Tools, and Applications*. Springer, Germany.

Shannon CE. 1948. A mathematical theory of communication [corrected version]. *Bell System Technical Journal*, 27, 379–423, 623–656.

Shine R, Warner DA and Radder R. 2007. Windows of embryonic sexual lability in two lizard species with environmental sex determination. *Ecology*, 88(7), 1781–1788. doi: 10.1890/06-2024.1.

Simon HA. 1991. The architecture of complexity. *Proceedings of the American Philosophical Society*, 106(6), 467–482.

Sinnott EW. 1950. *Cell and Psyche; the Biology of Purpose*. Harper, New York.

Sinnott EW. 1960. *Plant Morphogenesis*. McGraw-Hill Book Co., New York.

Slack JMW. 2002. Timeline — Conrad Hal Waddington: the last renaissance biologist? *Nature Reviews Genetics*, 3(11), 889–895. doi: 10.1038/nrg933.

Stein MB and Gordon R. 1982. Epithelia as bubble rafts: a new method for measuring cell shape and intercellular adhesion in embryonic and other epithelia. *Journal of Theoretical Biology*, 97, 625–635.

Sweeney ST, Hidalgo A, de Belle JS and Keshishian H. 2012. Embryonic cell ablation in *Drosophila* using lasers. *Cold Spring Harbor Protocols*, 6, 691–693. doi: 10.1101/pdb.prot068395.

Talbott SL. 2011. A modest champion of the whole organism: Paul Weiss, scientist of distinction. *In Context*, 25, 4–8.

Thom R. 1989. *Structural Stability and Morphogenesis: An Outline of a General Theory of Models*. Addison-Wesley, Redwood City, California.

Thompson DW. 1917. *On Growth and Form*. University Press, Cambridge.

Thompson DW. 1942. *On Growth and Form, Vol. 2*. 2nd edition. Cambridge University Press, Cambridge.

Turchin VF. 1977. *The Phenomenon of Science*. Columbia University Press, New York.

Turing AM. 1952. The chemical basis of morphogenesis. *Philosophical Transactions of the Royal Society of London* B, 237, 37–72.

Tzschentke B and Rumpf M. 2011. Embryonic development of endothermy. *Respiratory Physiology & Neurobiology*, 178(1), 97–107. doi: 10.1016/j.resp.2011.06.004.

Vallee R. 1991. Concerning some concepts of systems theory. *IFSR (International Federation for Systems Research) Newsletter*, 11(1), 3–4.

Verveen AA. 1971. In search of processes: the early history of cybernetics. *Mathematical Biosciences*, 11(1–2), 5–29. doi: 10.1016/0025-5564(71)90004-6.

Visschedijk AH. 1991. Gordon memorial lecture. Physics and physiology of incubation. *British Poultry Science*, 32(1), 3–20. doi: 10.1080/00071669108417323.

von Dassow M, Strother JA and Davidson LA. 2010. Surprisingly simple mechanical behavior of a complex embryonic tissue. *PLoS ONE*, 5(12). doi: 10.1371/journal.pone.0015359.

Waddington CH. 1957. *The Strategy of the Genes, A Discussion of Some Aspects of Theoretical Biology*. George Allen and Unwin, London.

Waddington CH. 1966. *Principles of Development and Differentiation*. Macmillan Co., New York.

Waddington CH. 1977. Stabilization in systems: chreods and epigenetic landscapes. *Futures*, 9(2), 139–146.

Wallace AR. 1858. On the tendency of varieties to depart indefinitely from the original type. *Proceedings of the Linnean Society of London*, 3, 53–62.

Webb DR. 1987. Thermal tolerance of avian embryos: a review. *Condor*, 874–898.

Weiss PA. 1940. The problem of cell individuality in development. The *American Naturalist*, 74, 34–46.

Whaley WG. 1983. Edmund Ware Sinnott. *Biographical Memoirs, National Academy of Sciences*, 54, 350–372.

Whittow GC and Tazawa H. 1991. The early development of thermoregulation in birds. *Physiological Zoology*, 64(6), 1371–1390.

Wiener N. 1948. *Cybernetics: Or Control and Communication in the Animal and the Machine*: John Wiley and Sons, USA.

Wiener N. 1958. My connection with cybernetics: Its origin and its future. *Cybernetica (Belgium)*, 1–14.

Wiener N. 1961. *Cybernetics: Or Control and Communication in the Animal and the Machine*. 2nd edition: Technology Press, MA, USA.

Wikipedia. *Anatol Rapoport* 2014a. Available from: http://en.wikipedia.org/wiki/Anatol_Rapoport (Accessed on 30 May 2016).

Wikipedia. *Arturo Rosenblueth* 2014b. Available from: http://en.wikipedia.org/wiki/Arturo_Rosenblueth (Accessed on 30 May 2016).

Wikipedia. *Waddington CH*. 2014c. Available from: http://en.wikipedia.org/wiki/C._H._Waddington (Accessed on 30 May 2016).

Wikipedia. *Peirce CS*. 2014d. Available from: http://en.wikipedia.org/wiki/Charles_Sanders_Peirce (Accessed on 30 May 2016).

Wikipedia. *Shannon C*. 2014e. Available from: http://en.wikipedia.org/wiki/Claude_Shannon (Accessed on 30 May 2016).

Wikipedia. *Correspondence principle*. 2014f. Available from: http://en.wikipedia.org/wiki/Correspondence_principle (Accessed on 30 May 2016).

Wikipedia. *Thompson D'AW*. 2014g. Available from: http://en.wikipedia.org/wiki/D%27Arcy_Wentworth_Thompson (Accessed on 30 May 2016).

Wikipedia. *Decision tree*. 2014h. Available from: http://en.wikipedia.org/wiki/Decision_tree (Accessed on 30 May 2016).

Wikipedia. *Russell ES*. 2014i. Available from: https://en.wikipedia.org/wiki/E.S._Russell (Accessed on 30 May 2016).

Wikipedia. *Mayr E*. 2014j. Available from: http://en.wikipedia.org/wiki/Ernst_Mayr (Accessed on 30 May 2016).

Wikipedia. *Gestalt psychology*. 2014k. Available from: http://en.wikipedia.org/wiki/Gestalt_psychology (Accessed on 30 May 2016).

Wikipedia. *Driesch H*. 2014l. Available from: http://en.wikipedia.org/wiki/Hans_Driesch (Accessed on 30 May 2016).

Wikipedia. *Black HS*. 2014m. Available from: https://en.wikipedia.org/wiki/Harold_Stephen_Black (Accessed on 30 May 2016).
Wikipedia. *Prigogine I*. 2014n. Available from: https://en.wikipedia.org/wiki/Ilya_Prigogine (Accessed on 30 May 2016).
Wikipedia. *Monod J*. 2014o. Available from: http://en.wikipedia.org/wiki/Jacques_Monod (Accessed on 30 May 2016).
Wikipedia. *Wolpert L*. 2014p. Available from: http://en.wikipedia.org/wiki/Lewis_Wolpert (Accessed on 30 May 2016).
Wikipedia. *Lookup table*. 2014q. Available from: https://en.wikipedia.org/wiki/Lookup_table (Accessed on 30 May 2016).
Wikipedia. *conferences M*. 2014r. Available from: http://en.wikipedia.org/wiki/Macy_conferences (Accessed on 30 May 2016).
Wikipedia. *Mead M*. 2014s. Available from: http://en.wikipedia.org/wiki/Margaret_Mead (Accessed on 30 May 2016).
Wikipedia. *Rashevsky N*. 2014t. Available from: http://en.wikipedia.org/wiki/Nicolas_Rashevsky (Accessed on 30 May 2016).
Wikipedia. *Wiener N*. 2014u. Available from: http://en.wikipedia.org/wiki/Norbert_Wiener (Accessed on 30 May 2016).
Wikipedia. *Club R*. 2014v. Available from: http://en.wikipedia.org/wiki/Ratio_Club (Accessed on 30 May 2016).
Wikipedia. *Reversal theory*. 2014w. Available from: https://en.wikipedia.org/wiki/Reversal_theory (Accessed on 30 May 2016).
Wikipedia. *Structural functionalism*. 2014x. Available from: http://en.wikipedia.org/wiki/Structural_functionalism (Accessed on 30 May 2016).
Wikipedia. *Pitts W*. 2014y. Available from: http://en.wikipedia.org/wiki/Walter_Pitts (Accessed on 30 May 2016).
Wikipedia. *McCulloch WS*. 2014z. Available from: http://en.wikipedia.org/wiki/Warren_Sturgis_McCulloch (Accessed on 30 May 2016).
Wikipedia. *Ashby WR*. 2014z, a. Available from: http://en.wikipedia.org/wiki/William_Ross_Ashby (Accessed on 30 May 2016).
Wikipedia. *Turing A*. 2015a. Available from: http://en.wikipedia.org/wiki/Alan_Turing (Accessed on 30 May 2016).
Wikipedia. *Wallace AR*. 2015b. Available from: http://en.wikipedia.org/wiki/Alfred_Russel_Wallace (Accessed on 30 May 2016).
Wikipedia. *Kolmogorov A*. 2015c. Available from: http://en.wikipedia.org/wiki/Andrey_Kolmogorov (Accessed on 30 May 2016).
Wikipedia. *Koestler A*. 2015d. Available from: http://en.wikipedia.org/wiki/Arthur_Koestler (Accessed on 30 May 2016).

Wikipedia. *Dennett D.* 2015e. Available from: http://en.wikipedia.org/wiki/Daniel_Dennett (Accessed on 30 May 2016).

Wikipedia. *Leibniz GW.* 2015f. Available from: https://en.wikipedia.org/wiki/Gottfried_Wilhelm_Leibniz (Accessed on 30 May 2016).

Wikipedia. *Bateson G.* 2015g. Available from: http://en.wikipedia.org/wiki/Gregory_Bateson (Accessed on 30 May 2016).

Wikipedia. *Neumann JV.* 2015h. Available from: http://en.wikipedia.org/wiki/John_von_Neumann (Accessed on 30 May 2016).

Wikipedia. *Needham J.* 2015i. Available from: http://en.wikipedia.org/wiki/Joseph_Needham (Accessed on 30 May 2016).

Wikipedia. *Bohr N.* 2015j. Available from: http://en.wikipedia.org/wiki/Niels_Bohr (Accessed on 30 May 2016).

Wikipedia. *Dawkins R.* 2015k. Available from: http://en.wikipedia.org/wiki/Richard_Dawkins (Accessed on 30 May 2016).

Wikipedia. *Fisher R.* 2015l. Available from: http://en.wikipedia.org/wiki/Ronald_Fisher (Accessed on 30 May 2016).

Wilson EO. 1999. *Consilience: The Unity of Knowledge*: Vintage Books.

Witkowski JA. 1985. The hunting of the organizer: an episode in biochemical embryology. *Trends in Biochemical Sciences*, 10(10), 379–381.

Wolpert L. 1969. Positional information and the spatial pattern of cellular differentiation. *Journal of Theoretical Biology*, 25(1), 1–47.

Wyczalkowski MA, Chen Z, Filas BA, Varner VD and Taber LA. 2012. Computational models for mechanics of morphogenesis. *Birth Defects Research Part C-Embryo Today-Reviews*, 96(2), 132–152. doi: 10.1002/bdrc.21013.

Yntema CL and Mrosovsky N. 1982. Critical periods and pivotal temperatures for sexual differentiation in loggerhead sea turtles. *Canadian Journal of Zoology-Revue Canadienne De Zoologie*, 60(5), 1012–1016. doi: 10.1139/z82-141.

6. Superfast Evolution via Trans- and Interspecies Biocommunication

Ille C. Gebeshuber*,†,‡ and Mark O. Macqueen*,§

*Aramis Technologies Sdn. Bhd., 14 Jalan BK 5D/1C, Bandar Kinrara, 47180 Puchong, Selangor, Malaysia

†Institute of Applied Physics, Vienna University of Technology, Wiedner Hauptstrasse 8-10/134, 1040 Wien, Austria

‡gebeshuber@iap.tuwien.ac.at; ille.gebeshuber@mac.com

§mark.macqueen@mac.com

This chapter introduces to provide a thought model for the real speed of evolution as a combination of selective, permutative, and constructive elements. It introduces hypotheses on genetic biocommunication within and across species and mainly deals with the three types of adaptations in herbivores, carnivores, and omnivores where biocommunication is of paramount importance: adaptations in the food chain, adaptations to the environment (historical, via selection), and interspecies competition. Two major ideas are put forward in the chapter: firstly, a basic design principle in organisms which ensures that parts of the genetic code are interchangeable in single processing units that are readable across species, even across biological kingdoms. Secondly, active (subconscious) selection of probability for the sex of the offspring by parents (by females at the time of conception, by males shortly before ejaculation) combined with continuous adaptations of the genetic material in the sperm of the fathers. In this way, evolution might be speeded up to the pace we experience it in organisms with natural selection being only one important aspect. One still can only speculate about constructive elements of evolution, however, the hypotheses described in the chapter definitely provide various reasons for starting intense academic discussion. This book chapter attempts to do science in a completely new way the authors like to call neo-Aristotelian. It takes results from scientific investigations, and aims, via logical thinking and extrapolation (and sheer judging in some cases) to come up with a concise hypothesis on important elements of evolution. Various experiments, which might prove or undermine the ideas, are suggested in the course of the train of thought. The authors look forward to future developments in this field, and responses, results and reactions from fellow scientists.

2. Introduction

Biocommunication denotes communication within or amongst species from all the empires and kingdoms in biology (intraspecies communication, interspecies communication). Communication denotes a process by which information is exchanged between individuals through a common system of symbols, signs or, behavior. In this chapter the authors concentrate on hypotheses on genetic biocommunication within and across animal species (carnivores, herbivores, omnivores). Three different types of biocommunication are dealt with: communication with the past (concerning the proposed permutative element of evolution), communication with other organisms (concerning parts of the genetic code that are interchangeable in single processing units across species, even across biological kingdoms, in the proposed selective element of evolution), and communication with the environment and the partner (concerning the proposed constructive element of evolution). Based on sound scientific evidence and developed further by contemplation these hypotheses shall aim at stimulating further discussion, ideas, research projects and insights.

Regarding further aspects of biocommunication including vocalizations or signs via semiochemicals such as pheromones that are not treated here at all the reader is referred to a vast amount of literature that is available elsewhere in form of articles and books (see, e.g., Witzany, 2010; Witzany and Baluska, 2012; Witzany, 2012; Witzany, 2014 and references therein).

3. Motivation

It's all about survival and the resources that are available to the individual. As a start of the train of ideas and hypotheses presented in this chapter, the authors relate the trigger for biocommunication to these two factors. Reason: As long as the animal is well fed, there is no reason to learn. However, there is definitely a reason to learn when it is hungry. And especially when it is regularly hungry. Similar arguments hold for further paramount resources besides food, and survival itself. In the following, the chapter will deal only with the extremes of herbivores (animals that feed on plants) and carnivores (animals that feed on meat), assuming that omnivores are in relation to the ideas presented below a simple mixture of the two extremes. The chapter will also make occasional references to plants, however, a theory of biocommunication in plants of the authors is on-going work and will be reported elsewhere. Springer devoted its extensive book series Signaling and Communication in Plants to this theme, with one of them concentrating on biocommunication (Witzany and Baluska, 2012).

Paramount properties of carnivorous animals such as lions and eagles are their mobility and competitiveness. For herbivores such as cattle or sheep the need to detect habitats of plants that serve as food and the need to open up new habitats is of utmost importance. In both cases biocommunication is of great significance.

3.1 Adaptations in biocommunication

The tree types of adaptations in herbivores and carnivores where biocommunication is of paramount importance are adaptations in the food chain, adaptations to the environment (historical, selection) and interspecies competition. Below, each of them will be treated in more detail. Generally it can be said that the faster such adaptations are represented in the genome of the respective animal, the higher is its chance of survival. Reticulate evolution makes perfect sense: the relationships between species or higher taxa cannot be represented in simple forking diagrams as previously assumed (phylogenetic network as opposed to the older concept of a phylogenetic tree). Sometimes even parts of the fork fuse again (hybridization and combination of two species).

3.2 Adaptations in the food chain

Adaptations in the carnivore food chain are equivalent to adaptations between prey and predator. The authors propose that carnivores adapt their genetic code to keep pace with improvements of their prey and vice versa. Such adaptation might take place via interspecies exchange of information. The classical paper on horizontal gene transfer (which would be one way to transport hereditary information across) starts with the sentence "*Gene flow from parent to child is the basic of heredity. Gene flow between unrelated organisms, even across biological kingdoms, may be a cornerstone of evolution.*" (Amábile-Cuevas and Chicurel, 1993).

Horizontal gene transfer (HGT, also known as lateral gene transfer) denotes the movement of genetic information between species, via selectable and easily transferred genomic islands (Peter and Townsend, 2005). Heritable HGT has been shown between Eubacteria and Archaea (Nelson *et al.*, 1999), between prokaryotes and unicellular eukaryotes (Andersson, 2005; Doolittle, 1998), from bacteria to multicellular eukaryotes (Hottopp *et al.*, 2007), from eukaryotes to prokaryotes (HGT to *Escherichia coli* from a eukaryotic host, Doolittle *et al.*, 1990) and between eukaryotes (Daniels *et al.*, 1990), the latter one being even by far not as rare as previously assumed (Andersson *et al.*, 2003). HGT facilitates the acquisition of novel gene functions (such as antibiotic resistance) and can speed up evolution, either for the recipient or in case of parasitic genetic elements for themselves (Peter *et al.*, 2002).

HGT can also be detrimental if the gene has no function, if it is incompatible with existing genes, or if it is a selfishly replicating mobile element (Vogan and Higgs, 2011). Possible pathways for interspecies gene transfer are phagocytosis, introgression via interspecies hybrids (Goodman *et al.*, 1999), symbiosis (intimate contact of the host with symbionts, such as gene transfer between eukaryotic cells with plastids or mitochondria) and transfection (Peter, 2003). Such interspecies exchange of information increases the organism's knowledge about the environment. As Keeling and Palmer put it in their 2008 Nature Reviews Genetics article on HGT in eukaryotic evolution, *you are what you eat, what you live on, what lives on you and what lives in you* (Keeling and Palmer, 2008). Members of the same species were found to differ dramatically in gene content: in a study by Welch *et al.* (Welch *et al.*, 2002) three *E. coli* genomes were sequenced. Less than 40% of the genome was common to all three bacteria!

Herbivores on the other hand do not adapt their genetic code to keep pace with improvements of their predators, but adapt to new food sources: They use the existing genetic potential in their genes to change themselves — their historic gene pool (permutation by recession). See below for ideas on non-coding DNA (junk DNA) and how it might be storage for the gene pool of the ancestors of a certain individual.

Plants might do it completely different altogether. HGT was found to be especially frequent in the early evolution of land plants, when a completely new environment had to be conquered by the organisms, and fast genetic learning was of great significance (Yue *et al.*, 2012). Keeling and Palmer (Keeling and Palmer, 2008) stress the important role of HGT in adaptation to certain specialized niches and the highly variable impact of HGT in different lineages.

An interesting example for HGT is the 70-kDa heat-shock protein (HSP70) gene that is found in various organisms in three domains of life (archaebacteria, eubacteria, and eukaryotes), but not in their common ancestor (Gupta and Golding, 1993). HSP70 is one

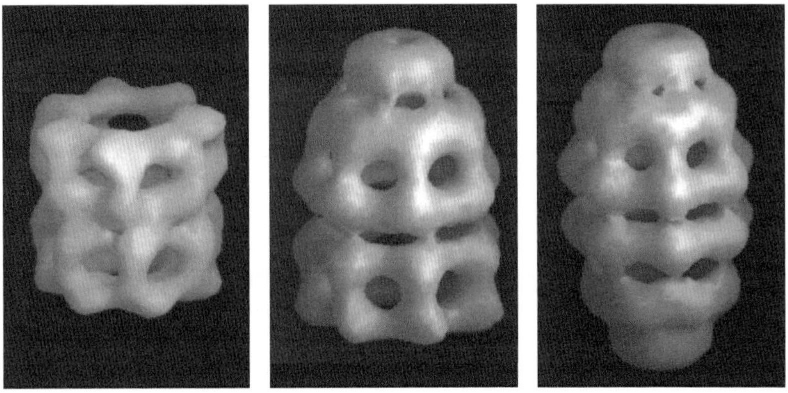

Figure 1. Chaperonins such as GroEL/GroES (HSP60) and HSP70 are proteins that help misfolded proteins to fold correctly. The gene that codes for HSP70 appears in all three domains of life, but not in the common ancestor, and is a classic example for horizontal gene transfer (Gupta and Golding, 1993). The structure in the images consists of a double ring chaperonin GroEL (left image) that can be capped with one or two chaperonins GroES (middle and right images) and that provides safe space for protein repair (Roseman *et al.*, 1996). © Cell Press. Image reproduced with permission.

of the greatest inventions in nature: it is a chaperonin protein (Fig. 1) that helps misfolded proteins to fold correctly (Hartl, 1996). Chaperonins were also the first molecules for which protein–protein interactions were imaged in real time on the single molecule level (Viani *et al.*, 2000). HSP70 is the most conserved protein present in all organisms that is known to date. What if the most substantial inventions are generally not kept within one species, but distributed to all species? Would this be a hint that all life on Earth works together and helps each other to sustain? Not on the individual level, but on the species level and beyond? After all, we are altogether here on Spaceship Earth (as Buckminster Fuller, one of the great thinkers of the 20th century, used to call it in his classic 1968 book Operating Manual for Spaceship Earth) and need each other for survival.

3.3 Adaptations to the environment

Adaptations to the environment can happen via two different ways: first, historically, by going back in the genome to genes that proved to be successful in the given environment (permutation by recession), and secondly, via innovation and selection. Intraspecies lateral gene transfer (see, e.g., Schubert *et al.*, 2009) and hereditary learning from each other (for reproduction, for the exchange of food sources) are important related aspects.

3.4 Intraspecies competition

In intraspecies competition genetic communication with the own species becomes important. The basic question is how to become attractive for potential mates.

4. Basic Design Principles in Organisms: Main Ideas

Two main ideas that are proposed in this chapter: firstly, a basic design principle in organisms, which ensures that parts of the genetic code are interchangeable in single processing units that are readable across species, even across biological kingdoms. Secondly, active (subconscious) selection of the probability for the sex of the offspring by fathers and mothers (at different times though), combined with continuous adaptations to the environment of the genetic material in sperm of the fathers.

The genetic code is uniform in most organisms: the same codons stand for the same proteins (Griffiths *et al.*, 2000). Due to this common language across species, it is possible to exchange readable blocks of genetic information, not just within species, but across all branches of the phylogenetic tree (or rather phylogenetic network, given all the cross-cuttings, short-cuts, and connections that have been found so far).

This readability and uniformity of the genetic code across known life forms (from viruses to bacteria to humans, with a few exceptions, such as in plastids and mitochondria, and in in the nuclear genome of some protozoans) can be called a universal basic design principle of organisms. The nature and message of DNA represent a universal language of life on Earth.

The interdependence and interconnectedness of all life forms on Earth shall be illustrated in one example: even pests such as socially parasitic ants turn out to be protectors when the host needs protection in face of attacks by a predator (Adams *et al.*, 2013). Such switching from parasite to protector, from being harmful to being beneficial, might happen often in the interaction of organisms, and might therefore bring sense to bilateral horizontal gene transfer between predator and prey. In the end, all living species form an ecosystem, and depend on each other. We need to sustain each other. It cannot be that herbivores eat all the plants; because otherwise no breathable oxygen would be produced any more and all the herbivores would asphyxiate and/or starve to death. There might be very elaborate higher order control systems written in our genomes, control systems that ensure the survival of life itself. Mutualism, an association between two organisms, often from different species, that benefits both partners, might be the rule, and not the exception.

4.1 The selective element of evolution and its lateral transfer: single processing units that are readable across species

The authors propose that the information exchange takes place in single processing units that are readable across species. Why this has evolved we cannot say. But it is striking that the genetic code is the same across species, with just a few notable exceptions. Only mitochondria (who have their own genetic material, which is handed down from the mother to the offspring) use a slightly varied form, as do plastids, and also some microorganisms (Kay, 2000).

4.2 The permutative element of evolution: permutation by recession or innovation

The authors propose that in the process of courtship the DNA of man and woman is compared for similar blocks, not just in the current coding genome of the future parents, but back in time for many generations (this information is partially stored in the non-coding DNA, see below). The more generations one has to go back to detect similarities, the more valuable this individual is concerning its genetic material. In a large gene pool, such permutation is not necessary, because the information is readily available from various individuals (i.e., many parallel evolution strains). However, in cases of a small gene pool, i.e., close relationships in past generations, many similar data blocks exist and permutation by recession or innovation becomes a necessity. Small genetic gene pools speed up evolution, either via differentiation into various species, who would occupy different ecological niches (cf. Darwin finches) or by a better adaptation to the environment. In a large gene pool, the common traits are confirmed. The constructive element of evolution is more dominant in large gene pools, whereas the permutative element is dominant in small gene pools.

In the proposed approach concerning the permutative element in evolution, permutation by innovation and selection is accompanied by permutation by recession. Permutation by recession indicates that if the genome provided by the parents is not good enough, the organism can access earlier genes, e.g., the ones from grandparents or other ancestors, perhaps going back in time very very long, and incorporate them in its genome (Weigel and Jurgens, 2005; Pearson, 2005; Lolle *et al.*, 2005; Ledford, 2013).

The genome can repair itself and can even reverse mutations. Examples for this are transposable elements and rare cases of children who inherit disease-causing mutations but show only mild symptoms, perhaps because some of their cells have reverted to a normal and healthier genetic code (Pearson, 2005). A transposable element (transposon, jumping gene, Nobel prize 1983 for Barbara McClintock) is non-coding DNA that has been unambiguously shown to be important in higher plant genome function and evolution (Buchers *et al.*, 2012). Such elements can change their position within the genome, sometimes creating or reversing mutations and altering the cell's genome size. Transposable elements might be involved in the permutation by recession. They can also be transferred horizontally. Schaack and co-workers (Schaack *et al.*, 2010) argue that the introduction of transposable elements by horizontal transfer in eukaryotic genomes has been a major force propelling genomic variation and biological innovation.

One example for permutation by recession is a scientific study that shows that in certain plants (Arabidopsis, cress), where genes from the parent generation were modified so that they are bad for the daughter generation, the daughter generation goes back to the respective gene of the grandparents or even great-grandparents (Pearson, 2005; Lolle *et al.*, 2005). Cress overturns textbook genetics! Mendel's laws describe the inheritage of traits linked to single genes on chromosomes in the cell nucleus. Sometimes traints do not segregate in accordance with Mendel's laws. Examples for such non-Mendelian inheritage are extranuclear inheritage via genetic material present in mitochondria and plastids, gene conversion, infectious heredity (new genetic material introduced via infectious particles such as viruses), genome imprinting, where genes are epigenetically marked before transmission, altering their levels of expression, mosaicism, where the individual has different genetic material in its cells, due to mutations — if this occurs in egg or sperm cells, the new information is hereditatry, and trinucleotide repeat disorders.

The research result by Lolle *et al.* (2005) on non-Mendelian genetics in Arabidopsis implicates that apart from the genes inherited from the parents, there is some kind of storage for the genes from previous generations. The authors propose to investigate non-coding DNA (also known as junk DNA) as a candidate for the storage space for the genomes of the ancestors. In some organisms, non-coding DNA takes up a major fraction of the genome: In the case of humans, for example, up to 98% of the DNA is non-coding.

Parts of the non-coding DNA are transcribed and result in RNA pieces that do not code for proteins. What if they are not at all junk, but in fact very important and contain important information concerning, e.g., our past, the genome of our ancestors (and thereby a backup genome for our offspring), information about environments, etc.? Important information might be stored in the DNA at various hierarchical levels, some of which not yet described by researchers. 76% of the human genome's non-coding DNA sequences are transcribed and nearly half of the genome is accessible to genetic regulatory proteins such as transcription factors. The research project ENCODE for example has revealed that some 80% of the human genome is biochemically active (see http://www.genome.gov/10005107 and references there).

Another storage option proposed in the scientific literature is a template directed process that makes use of an ancestral RNA-sequence cache (Lolle, 2005). The results presented in the Lolle *et al.*, 2005 paper elicited various rebuttal papers, and major discussions (Peng *et al.*, 2006; Mercier *et al.*, 2008; Ledford, 2013; Hopkins *et al.*, 2013). What is true we do not know yet.

4.3 The constructive element of evolution: tuneable probability of the sex of the offspring and continuous adaptations to the environment

Natural selection and permutation by recession are just two of the proposed factors, and can alone not explain the superfast speed at which evolution takes place. The authors therefore propose a constructive element in evolution, consisting of quantitative and qualitative factors (Figs. 2 and 3).

The assumption from which the authors start their train of thinking is that when a woman gives birth to a girl it is a proof of genetic advantages of the respective partners, whereas when she gives birth to a boy, the boy first has to prove in the next generation that his genetic material is good enough to produce a girl with his partner. The authors propose a four-quadrant selection matrix for the constructive element of evolution (Fig. 2), containing quantitative and qualitative elements. Quantitatively, the woman selects the partner, and the man eliminates competitors. The quantitative constructive elements of evolution depend on the environment. Qualitatively, the woman can control via changing the permeability of her egg hull if she will get a male or female child (by either letting a spermatozoon bearing the Y or the X chromosome pass and fertilize the egg, Fig. 4). On the side of the male, such

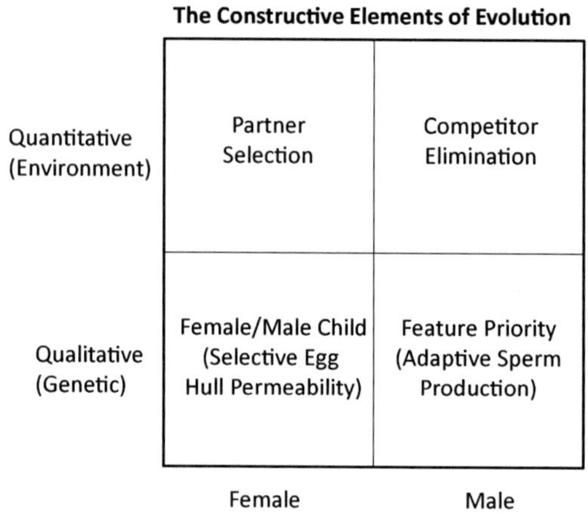

Figure 2. The constructive elements of evolution. Male and female contribute to the speed of evolution of their offspring, via actions that either depend on genetics or the environment.

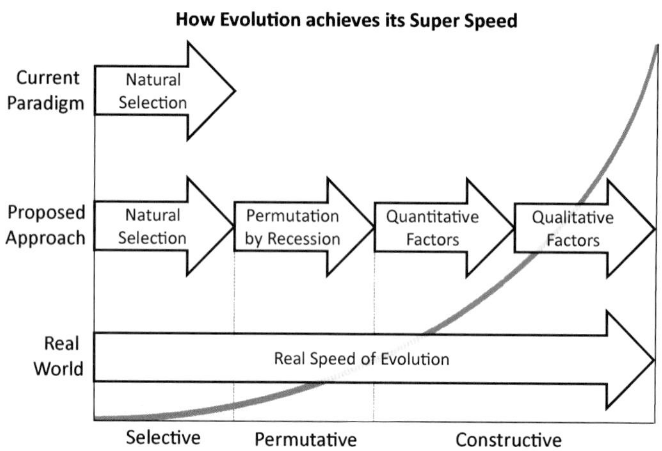

Figure 3. The main hypotheses put forward in this chapter. Evolution depends on selective, permutative, and constructive elements. Together they determine the speed of evolution, as we know it. See text.

qualitative contributions stem from feature priority coming from adaptive sperm production. Female humans are born with about 200,000 eggs in their ovaries, whereas men always produce sperm afresh.

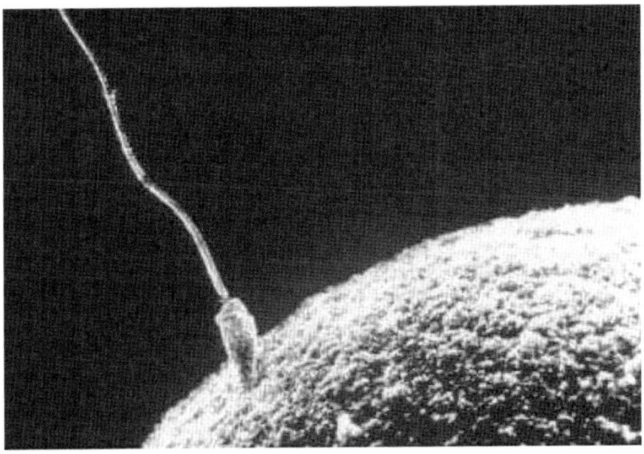

Figure 4. Sperm and egg fusing. Here, the authors put forward the hypothesis that properties of the parents, the environment and compatibility of their genes actively determine the probability whether the joint child will be male or female. Public domain image.

In most countries, for as long as records have been kept, more boys than girls have been born. The secondary sex ratio (the ratio between male and female offspring) in people for the entire world population is 101 males to 100 females (CIA, 2015).

In times of a large choice of potential male partners (which is often equivalent to times of peace) a woman has the possibility to select the most compatible man as father of her child. In case of an excellent compatibility, the probability to give birth to a female is higher (80% as opposed to 50%), whereas if they are not compatible, it would be lower (20% as opposed to 50%). Reason for that is that every healthy female has the ability to reproduce and pass on her genetic material. Males on the other hand are in competition. Martinez *et al.* (2012) showed that in the US the number of children per woman (for women aged 22–44 years) crucially depends on education: A woman with no high school diploma statistically gives birth to 2.5 children, a woman with high school diploma gives birth to 1.8 children, and a woman with some college gives birth to 1.5 kids. With a Bachelor degree or higher, every woman mothers 1.1 kids. If one looks at the respective numbers of the males fathering these children, one notices a discrepancy. Men admit the fathering of much fewer children. The respective percentage of illegitimate children becomes higher when looking at lower social classes. The probability of a male to produce legitimate offspring therefore increases with social class. With decreasing educational background the share of illegitimate children increases, which means that the genetic competition between males are higher. In times of crises such as wars, when fewer men are available, the compatibility between couples is not as likely. In cases of low genetic (and social) compatibility, there is a higher likelihood for a baby boy to be born (see treatment of the wartime sex ratio below). A lot of this selection process takes place via smell, where the women prefer the smell of men with genes different to their own, enlarging the gene portfolio of their offspring (Wedekind, 1994).

Wedekind showed in the so-called smelly T-shirt experiment that women preferred the smell of T-shirts worn by men with different genes from themselves, and thereby unconsciously select sex partners who help to put the offspring at some genetic advantage, due to the new and different genes from the partner. In times of war, when the sample of eligible men is smaller, women need to take what they get to produce offspring.

Wartime increases the sex ratio (Bernstein, 1958; Gellatly, 2009) between boys and girls: there is a statistically significant increase in the fraction of male births toward the end and immediately after both world wars. The year after World War I ended, an extra two boys were born for every 100 girls in the UK, compared to the year before the war started (Gellatly, 2009). In a subgroup of especially fertile couples after World War I, for example, from 1,178 children 55.3% were sons (Bernstein, 1958). The sex ratio remained high for about 18 months after WWI (Bernstein, 1958).

The fact that more male offspring are born after a war might corroborate the hypothesis of a tuneable probability of the sex of the offspring: at times during which not too many men with suitable genetic material for the respective women might be available, more boys are born, whose genetic material needs to go through one more iteration (generation) before a girl is born.

Various attempts have been made to determine differences in Y and X chromosome bearing spermatozoa, and using them in influencing the sex of the offspring, e.g., in *in vitro* fertilization. Where on the other hand the partner might know best if a baby boy or a baby girl would be best with the respective partner under the given circumstances. The classical phase-contrast microscopy work by Shettles on the nuclear morphology of human spermatozoa (Shettles, 1960) gave rise to the widely held idea that spermatozoa bearing the X chromosome (resulting in girls) swim faster than spermatozoa carrying the smaller Y chromosome (resulting in boys). Until recently, modern methods showed no morphological difference between human X and Y spermatozoa (Hossain *et al.*, 2001; Grant, 2006) and also no difference in swimming velocity of X and Y bearing bovine sperm (Penfold *et al.*, 1998). However, measurements of sperm surfaces with the atomic force microscope, a powerful instrument that allows for measurements on living cells down to the nanoscale, revealed that the associations of several structural measurements of the sperm cell head are promising candidates for development of a new method of sperm sexing (Carvalho *et al.*, 2013).

The authors also propose that men, under the influence of smell and other environmental factors, can influence the race of their X or Y bearing sperm by active control of the viscosity of their prostate liquid and thereby also actively contribute to the probability of the sex of the offspring. Sperm is continuously produced based on the most recent information, but for control of the sex of the offspring this system is too slow — therefore the authors propose the variable medium provided by the prostate liquid as selective medium. The prostate system seems much too complex to simply provide a low friction environment for the spermatozoa. It might be that it is a highly elaborately system controlled by lymphatic information, with environmental feedback, contributing to male reproduction and the tuneable probability of the sex of the offspring. It remains to be investigated if changes in the viscosity of prostate fluid and any resulting motility of sperm yielding male or female offspring exist. If so, another constructive element of evolution has been identified. The potential involvement of a lymphatic information system for the prostate shall be investigated by further research.

Further aspects of interest: correlations of the secondary sex ratio with the frequency of incest over many generations need to be established in further research. Via the alignment via the junk DNA, the system is well informed about the size of the gene pool and accordingly adopts the implementation velocity of new genetic material. In this respect, specific properties of offspring resulting from incest and biological bastards are also of high interest. It would also be interesting to take a closer look at the sex ratio in children who result from a rape. However, there were no data to be found on the sex ratio of children after a rape. This might be an area of no-go research, which is societally taboo. Also of interest is research on how populations with a limited gene pool such as the Inuit acquire new genetic material. Isolated Inuit groups formerly had their young girls spend the night with strangers; the resulting children enlarged the gene pool. Also here information on the sex ratio would be very informative.

5. Summary, Discussion and Outlook

Various ideas regarding biocommunication within and amongst species, even across time, have surfaced in this chapter. The proposed part of the permutative element of evolution, where single organisms can access the genomes of their ancestors and accordingly even reverse back mutations, is an example for biocommunication with the past. The authors also treated a certain selective aspect of evolution, where parts of the genetic code are transferred to other organisms, even across biological kingdoms, in single processing units — this is an example for genetic communication with other organisms, and might even be a sign of the importance of sustaining life as a whole as opposed to single individuals — the highly conserved gene for the heat shock protein HSP70, which is now — thanks to lateral gene transfer — present in all three domains of life, but not in the common ancestor, is an intriguing example for collaboration of life forms. The third type of biocommunication treated here refers to information exchange with the environment and between the partners, and is dealt with in the authors' proposal concerning a constructive element of evolution.

The major new ideas put forward by the authors are that the real speed of evolution results from three major elements: the selective element of evolution (which comprises mutation and selection, and parts of the genetic code that are interchangeable in single processing units across species, even across biological kingdoms), the permutation by recession element of evolution (where the organism can actively change bad genes back to the ones of ancestors, here the authors propose that non-coding DNA might provide the required storage space for such large amounts of data) and a constructive element of evolution, that is comprised of quantitative and qualitative elements, and is mainly concerned with which traits are passed on via male or female offspring.

What is life is still a question we cannot answer, let alone what is the meaning and goal of life (if any). Is there a motivation of evolution? Why is the genetic code (i.e., the rules by which genetic material is translated into proteins in organisms) uniform across species? Did this uniformity stay the same with time, with evolution, to keep on ensuring that genetic material can be exchanged across biological kingdoms?

Many questions remain, and a lot of research needs to be done. Thinking about biocommunication and evolution will certainly provide interesting pieces for the puzzle we need to assemble to understand the complex network of life and its elaborate ways of communication.

References

Adams RMM, Liberti J, Illum AA, Jones JH, Nash DR and Boomsma JJ. 2013. Chemically armed mercenary ants protect fungus-farming societies. *Proceedings of the National Academy of Sciences*, 110(39), 15752–15757. doi: 10.1073/pnas.1311654110.

Amábile-Cuevas CF and Chicurel ME. 1993. Horizontal gene transfer. *American Scientist*, 81(4), 332–341.

Andersson JO. 2005. Lateral gene transfer in eukaryotes. *Cellular and Molecular Life Sciences*, 62(11), 1182–1197. doi: 10.1007/s00018-005-4539-z.

Andersson JO, Sjögren AM, Davis LA, Embley TM and Roger AJ. 2003. Phylogenetic analyses of diplomonad genes reveal frequent lateral gene transfers affecting eukaryotes. *Current Biology*, 13(2), 94–104. doi:10.1016/S0960-9822(03)00003-4.

Bernstein ME. 1958. Studies in the human sex ratio. 5. A genetic explanation of the wartime increase in the secondary sex ratio. *The American Journal of Human Genetics*, 10(1), 68–70.

Bucher E, Reinders J and Mirouze M. 2012. Epigenetic control of transposon transcription and mobility in Arabidopsis. *Current Opinion in Plant Biology*, 15(5), 503–510. doi:10.1016/j.pbi.2012.08.006.

Carvalho JO, Silva LP, Sartori R and Dode MAN. 2013. Nanoscale differences in the shape and size of X and Y chromosome-bearing bovine sperm heads assessed by atomic force microscopy. *PLoS One*, 8(1), e59387. doi: 10.1371/journal.pone.0059387.

CIA, The World Fact Book, Sex Ratio World 2015. Available from: https://www.cia.gov/library/publications/the-world-factbook/fields/2018.html (Accessed on March 11 2015).

Daniels SB, Peterson KR, Strausbaugh LD, Kidwell MG and Chovnick A. 1990. Evidence for horizontal transmission of the P transposable element between *Drosophila* species. *Genetics*, 124(2), 339–355.

Doolittle RF, Feng DF, Anderson KL and Alberro MR. 1990 A naturally occurring horizontal gene transfer from a eukaryote to a prokaryote. *Journal of Molecular Evolution*, 31(5), 383–388.

Doolittle WF. 1998. You are what you eat: a gene transfer ratchet could account for bacterial genes in eukaryotic nuclear genomes. *Trends in Genetics*, 14(8), 307–311.

Fuller B. 1968. *Operating Manual for Spaceship Earth*. Simon and Schuster, New York.

Gellatly C. 2009. Trends in population sex ratios may be explained by changes in the frequencies of polymorphic alleles of a sex ratio gene. *Evolutionary Biology*, 36(2), 190–200. doi: 10.1007/s11692-008-9046-3.

Goodman SJ, Barton NH, Swanson G, Abernethy K and Pemberton JM. 1999. Introgression through rare hybridization: a genetic study of a hybrid zone between red and sika deer (genus cervus) in Argyll, Scotland. *Genetics*, 152(1), 355–371.

Grant VJ. 2006. Entrenched misinformation about X and Y sperm. *British Medical Journal*, 332(7546), 916. doi: 10.1136/bmj.332.7546.916-b.

Griffiths AJF, Miller JH, Suzuki DT, Lewontin RC and Gelbart WM. 2000. *An Introduction to Genetic Analysis*. 7th edition, Chapter 10: Universality of genetic information transfer. Available from: http://www.ncbi.nlm.nih.gov/books/NBK21915/.

Gupta RS and Golding GB. 1993. Evolution of HSP70 gene and its implications regarding relationships between archaebacteria, eubacteria, and eukaryotes. *Journal of Molecular Evolution*, 37(6), 573–582.

Hartl FU. 1996. Molecular chaperones in cellular protein folding. *Nature*, 381, 571–580. doi:10.1038/381571a0.

Hopkins MT, Khalid AM, Chang P-C, Vanderhoek KC, Lai D, Doerr MD and Lolle SJ. 2013. *De novo* genetic variation revealed in somatic sectors of single Arabidopsis plants. *F1000 Research*, 2, 5. doi: 10.12688/f1000research.2-5.v1.

Hossain AM, Barik S and Kulkarni PM. 2001. Lack of significant morphological differences between human X and Y spermatozoa and their precursor cells (spermatids) exposed to different prehybridization treatments. *Journal of Andrology*, 22(1), 119–123. doi:10.1002/j.1939-4640.2001.tb02161.x.

Hotopp D, Julie C, Clark ME, Oliveira DCSG, Foster JM, Fischer P, Muñoz Torres MC, Giebel JD, Kumar N, Ishmael N, Wang S, Ingram J, Nene RV, Shepard J, Tomkins J, Richards S, Spiro DJ, Ghedin E, Slatko BE, Tettelin H and Werren JH. 2007. Widespread lateral gene transfer from intracellular bacteria to multicellular eukaryotes. *Science*, 317(5845), 1753–1756. doi: 10.1126/science.1142490.

Kay LE. 2000. *Who Wrote the Book of Life? A History of the Genetic Code*. Stanford University Press, Stanford.

Keeling PJ and Palmer JD. 2008. Horizontal gene transfer in eukaryotic evolution. *Nature Reviews Genetics*, 9, 605–618. doi:10.1038/nrg2386.

Ledford H. 2013. Do plants 'veto' bad genes? Latest evidence fails to quell doubt about whether plants can access 'ancestral' genes outside their parents' DNA. *Nature News*, Available at: http://www.nature.com/news/do-plants-veto-bad-genes-1.12401. doi:10.1038/nature.2013.12401.

Lolle SJ, Victor JL, Young JM and Pruitt RE. 2005. Genome-wide non-mendelian inheritance of extra-genomic information in Arabidopsis. *Nature*, 434(7032), 505–509. doi: 10.1038/nature03380.

Martinez G, Daniels K and Chandra A. 2012. Fertility of men and women aged 15–44 years in the United States: national survey of family growth, 2006–2010. *National Health Statistics Report*, 51, 1–29.

Mercier R, Jolivet S, Vignard J, Durand S, Drouaud J, Pelletier G and Nogué F. 2008. Outcrossing as an explanation of the apparent unconventional genetic behavior of *Arabidopsis thaliana* hth mutants. *Genetics*, 180(4), 2295–2297. doi:10.1534/genetics.108.095208.

Nelson, KE, Clayton RA, Gill SR, Gwinn ML, Dodson RJ, Haft DH, Hickey EK, Peterson JD, Nelson WC, Ketchum KA, McDonald L, Utterback TR, Malek JA, Linher KD, Garrett MM, Stewart AM, Cotton MD, Pratt MS, Phillips CA, Richardson D, Heidelberg J, Sutton GG, Fleischmann RD, Eisen JA, White O, Salzberg SL, Smith HO, Venter JC and Fraser CM. 1999. Evidence for lateral gene transfer between Archaea and bacteria from genome sequence of *Thermotoga maritima*. *Nature*, 399(6734), 323–329. doi: 10.1038/20601.

Peter GJ and Townsend JP. 2005. Horizontal gene transfer, genome innovation and evolution. *Nature Reviews Microbiology*, 3(9), 679–687. doi:10.1038/nrmicro1204.

Peter GJ, Senejani AG, Zhaxybayeva O, Olendzenski L and Hilario E. 2002. Inteins: structure, function, and evolution. *Annual Review of Microbiology*, 56(1), 263–287. doi:10.1146/annurev.micro.56.012302.160741.

Peter GJ. 2003. Gene transfer: Gene swapping craze reaches eukaryotes. *Current Biology*, 13(2), R53–R54. doi:10.1016/S0960-9822(02)01426-4.

Pearson H. 2005. Cress overturns textbook genetics. Surprise finding shows that plants rewrite genetic code. *Nature News*. doi:10.1038/news050321-8.

Penfold LM, Holt C, Holt WV, Welch GR, Cran DG and Johnson LA. 1998. Comparative motility of X and Y chromosome–bearing bovine sperm separated on the basis of DNA content by flow sorting. *Molecular Reproduction and Development*, 50(3), 323–327. doi:10.1002.

Peng P, Chan SWL, Shah GA and Jacobsen SE. 2006. Increased outcrossing in hothead mutants. *Nature*, 443, E8–E9. doi:10.1038/nature05252.

Roseman AM, Chen S, White H, Braig K and Saibil HR. 1996. The chaperonin ATPase cycle: mechanism of allosteric switching and movements of substrate-binding domains in GroEL. *Cell*, 87(2), 241–251. doi:10.1016/S0092-8674(00)81342-2.

Schaack S, Gilbert C and Feschotte C. 2010. Promiscuous DNA: horizontal transfer of transposable elements and why it matters for eukaryotic evolution. *Trends in Ecology and Evolution*, 25(9), 537–546. doi:10.1016/j.tree.2010.06.001.

Schubert S, Darlu P, Clermont O, Wieser A, Magistro G, Hoffmann C, Weinert K, Tenaillon O, Matic I and Denamur E. 2009. Role of intraspecies recombination in the spread of pathogenicity islands within the *Escherichia coli* species. *PLoS Pathogens*, 5(1), e1000257. doi:10.1371/journal.ppat.1000257.

Shettles LB. 1960. Nuclear morphology of human spermatozoa. *Nature*, 186, 648–649. doi:10.1038/186648a0.

Viani MB, Pietrasanta LI, Thompson JB, Chand A, Gebeshuber IC, Kindt JH, Richter M, Hansma HG and Hansma PK. 2000. Probing protein-protein interactions in real time. *Nature Structural Biology*, 7(8), 644–647. doi:10.1038/77936.

Vogan AA and Higgs PG. 2011. The advantages and disadvantages of horizontal gene transfer and the emergence of the first species. *Biology Direct*, 6, 1. doi:10.1186/1745-6150-6-1.

Wedekind C. 1994. Mate choice and maternal selection for specific parasite resistances before; during and after fertilization. *Philosophical Transactions of the Royal Society of London*, 346(1317), 303–311.

Weigel D and Jurgens G. 2005. Genetics: hotheaded healer. *Nature*, 434(7032), 443. doi:10.1038/434443a.

Welch RA, Burland V, Plunkett III G, Redford P, Roesch P, Rasko D, Buckles EL, Liou SR, Boutin A, Hackett J, Stroud D, Mayhew GF, Rose DJ, Zhou S, Schwartz DC, Perna NT, Mobley HLT, Donnenberg MS and Blattner FR. 2002. Extensive mosaic structure revealed by the complete genome sequence of uropathogenic *Escherichia coli*. *Proceedings of the National Academy of Sciences of the United States of America*, 99(26), 17020–17024.

Witzany G. 2010. *Biocommunication in Soil Microorganisms (Soil Biology)*. Springer, Berlin, Heidelberg.

Witzany, G and Baluska F. 2012. *Biocommunication of Plants (Signaling and Communication in Plants)*. Springer, Berlin, Heidelberg.

Witzany G. 2012. *Biocommunication of Fungi*. Springer, Heidelberg, Dortrecht, London, New York.

Witzany G. 2014. *Biocommunication of Animals*. Springer, Germany.

Yue J, Xiangyang H, Hang S, Yongping Y and Jinling H. 2012. Widespread impact of horizontal gene transfer on plant colonization of land. *Nature Communications*, 3, 1152. doi:10.108/ncomms2148.

7. Channel Capacity and Rate Distortion in Amino Acid Networks

Boaz Tamir*,‡ and Avner Priel†,§

*Faculty of Interdisciplinary Studies,
Bar-Ilan University Ramat Gan, Israel. IYAR,
Israel Institute for Advanced Research, Rehovot, Israel
†Department of Physics
University of Alberta, Edmonton, Canada
‡canjlm@actcom.co.il
§apriel@ualberta.ca

In this paper, the channel of amino acids is analyzed in the framework of classical information theory. Given a distance measure in the space of amino acids and a distortion D, a condition on the channel's properties is calculated that is sufficient to transmit the information with a distortion less than D, for all distributions of amino acids at the channel's input. Furthermore, the rate of point mutation of the bases is related to the lowest amount of distortion that the system can tolerate over a large dynamic range of probability distributions.

1. Introduction

The application of information theory to the analysis of amino acids sequences has a long history. Early works on that subject focused on finding quantitative measures of the large sequences such as information entropy, long-range correlations, spectral analysis etc. (Roman-Roldan *et al.*, 1996; Leitao *et al.*, 2005). As the amount of genomic data increased considerably, and sequence data banks become abundant, other tools were applied to study repeating patterns and their possible biological role. The field of Bioinformatics is one outcome of these studies (Higgs and Attwood, 2005).

Another line of investigation uses the analogy between the process of translating deoxyribonucleic acid (DNA) sequences to proteins and the process of communicating information via a noisy channel. Several researchers (Gong *et al.*, 2011) and (Bouaynaya and Schonfeld, 2007) pointed out that the same three elements of a communication channel appear in this process, namely: a **source** — being the DNA sequence, a **channel**, represented by the replicating/translating mechanism, and the **target** which in our case would be the outcome protein. Intuitively these are the transmitter, the medium through which the information passes, and the receiver. The information sources are messages composed of words (codons) based on the 4 letter alphabet (bases). As such, one can define their probability distribution, and their conditional probability matrix; clearly the actual values are experimentally determined. Both electrical communication and biological channels are noisy, however, while in the first kind the designer does its best to minimize the error, the second have an inherent and essential noise that originates from the evolutionary driving force. Recent research also suggests that stochastic fluctuations, or 'noise' is also vital for the normal functioning of the cell (see for example, Eldar and Elowitz, 2010).

In this paper we study some properties of the channel of amino acids in the framework of classical information theory (Cover and Thomas, 1991). The aim is to relate the rate of point mutation (nucleotide substitution where we assume that transitions and transversions have the same rate) to the maximal distortion of genomic information the system can sustain. On one hand there is the property of channel capacity (Cover and Thomas, 1991), i.e., the highest transmission rate of information, measured in bits per unit time. On the other hand there is the signal distortion rate, which reflects the level of approximation (or deviation) in the reconstructed message at the target side. Low distortion is achieved by fine quantization of the signal, which in the language of information theory requires a high 'rate'. Therefore if we want to reduce the overall distortion of the source's signal we have to increase the channel capacity (if that signal is to be transmitted through the channel). One way to increase channel capacity is by reducing the noise, namely, reducing the point mutation. Hence, we can obtain a lower bound on the level of distortion the system can achieve for a given rate of point mutation. Such a relation between the point mutation rate and the distortion depends on the distribution of the signal, i.e., the distribution of amino acids. In the following sections we shall calculate a relation between the point mutation and the distortion that will hold for all signal distributions.

Mathematically, the channel capacity could be easily increased by simply lowering the point mutation rate, however, biological systems must maintain a certain level of noise to facilitate the evolutionary process (Steiner, 2012). In this respect biological systems differ from engineered communication systems. Hence the relation we present here should be understood as a lower bound on the overall distortion that the system can sustain, **given** a point mutation rate. This argument applies to earliest systems of life. When complexity grows up several error correction complicated mechanisms evolve.

The competition between the two requirements, capacity and distortion, is at the heart of this manuscript. The bounds we calculate for the distortion and the capacity are given in terms of a single parameter — the point mutation rate, which is the probability for a single nucleotide flip to occur. Normally, this parameter has a very low value due to various error correcting mechanisms in the cell. However, our analysis is performed under the assumption that such mechanisms are absent, which is supposed to be the case early in the primordial Earth.

In the next section we lay the foundations of the protein channel properties for the subsequent sections. Using the mutual information function and the transition matrix of the proteins, we define the capacity and the rate distortion. We then calculate the relation between the two quantities, starting from a 1-bit channel as an example, then generalize it for an n-bit channel. Finally, we evaluate the lower bound for the distortion as a function of the point mutation rate. Possible implications of the results are discussed in the last section.

2. Properties of the Protein Channel

The capacity (and distortion) rates are defined by maximizing (minimizing) the mutual information over the channel, whereas a channel is defined by its transition matrix, i.e., the statistical mapping between each pair of proteins. There are several known point accepted mutation (PAM) matrices that could serve as transition matrices (Henikoff and Henikoff, 1992). However, we will consider a codon transition matrix that is induced from the rates of point mutation between the bases A, T, C and G (for example, mutation rates for ribonucleic acid (RNA) viruses could be 10^{-3} to 10^{-5} per base per generation see Drake and Holland, 1999). If we denote the mutation probability by α, then the entries of the transition matrix of the channel would all be functions of α. Such a channel was presented and analyzed in (Gong *et al.*, 2011) and (Bouaynaya and Schonfeld, 2007) (where the authors assume $\alpha^2 \approx 0$). Hence our channel is '**induced**' from the point mutation channel.

The amino acid transition matrix is a first order approximation of the protein channel. The ideal protein channel is characterized by the transition between all possible protein sequences in the cell, however, this will demand a large and complex transition matrix. In the language of information theory (Cover and Thomas, 1991), the capacity (or the rate) of this channel is its ability to separate between different inputs. If we consider n applications of the channel, then for large n the capacity would be a function of the number of sequences of amino acids that such a channel could reliably separate. In other words the capacity is a function of the number of proteins that the channel can separate. Clearly the cell can separate two proteins even if very few amino acids are changed, however, this is due to other mechanisms. Since the channel is noisy, it could mistakenly switch between two amino acids which limits its separation power.

Let

$$A = A(\alpha) = A(\hat{x}/x)$$

be the channel transition matrix, where the random variables X and $\hat{X}$ denote the input and output of the channel respectively. Given any two random variables X and $\hat{X}$ then the mutual information $I(X, \hat{X})$ is the amount of information that one random variable contains about the other:

$$I(X, \hat{X}) = H(X) - H(X/\hat{X}).$$

Let us now write the capacity of the channel in terms of the mutual information between the input to the channel and its output:

$$C(\alpha) = \max_{p} I_A(Xp, \hat{X}), \tag{1}$$

where $I_A(X_p, \hat{X})$ denotes the mutual information of X_p and $\hat{X}$; X_p and $\hat{X}$ are related by the channel transition matrix A and the maximum of I_A is taken over all probability distributions p (Cover and Thomas, 1991). The capacity is therefore attributed to the physical (or biological) system (the medium) and not to the probability distribution at its input.

Consider now the following metric in the space of amino acids. Each amino acid can be represented as a vector in a high dimensional space where the coordinates correspond to the amino acid's physico-chemical properties like volume, bulkiness, polarity, hydrophobicity, etc. (Higgs and Attwood, 2005). The distance between two amino acid could be taken as the Euclidean distance of the corresponding vectors, however, one should make the coordinates of such a space dimensionless. Having such a metric we could write a distance matrix $d_{x,\hat{x}} = d(x, \hat{x})$, and compute the rate distortion. The rate distortion is defined by fixing the input probability distribution p and minimizing over all transition matrices. In particular, given a probability distribution p and a distortion D then the rate distortion function $R_p(D)$ is the minimal rate that is needed for the quantization (approximation of p using a discrete set of points) of p with an error which is smaller than D (Cover and Thomas, 1991; Berger, 1971). The rate distortion can be written as (Cover and Thomas, 1991):

$$R_p(D) = \max_{a(\hat{x}/x);\, E(d(\hat{x}/x)) \leq D} I_{a(\hat{x}/x)}(Xp, \hat{X}), \tag{2}$$

where

$$E(d(\hat{x}, x)) = \sum_{\hat{x}} \sum_{x} a(\hat{x}/x) p(x) d(\hat{x}, x)$$

and the minimum is taken over all transition matrices $a(\hat{x}/x)$ that satisfy the expectation condition $E(d(x, \hat{x})) \leq D$.

Given α, the above amino acid channel transmission matrix $A(\hat{x}/x) = A(\alpha)$ defines a quantization of p, that is, a discrete approximation of p (Cover and Thomas, 1991). In the following we will show (this is also intuitive) that if α is low enough then the distribution p at the input to the channel $A(\alpha)$ would be distorted by no more than the amount D, in other words

$$\sum_{\hat{x}} \sum_{x} A(\hat{x}/x) p(x) d(\hat{x}, x) \leq D.$$

Since the rate distortion is minimal over all such transition matrices $a(\hat{x}/x)$ it is clear that:

$$I_{A(\alpha)}(X_p, \hat{X}) \geq R_p(D).$$

Notice now that the channel capacity $C(\alpha)$ equals the maximal mutual information over all probability distribution p at the input while the transition matrix $A = A(\alpha)$ is fixed:

$$C(\alpha) \geq I_{A(\alpha)}(X_p, \hat{X}).$$

Using the above two inequalities we conclude that given a distribution p we can find an α such that $C(\alpha) > R_p(D)$.

The question we address in this paper is whether it is possible to obtain an α which is optimal for all probability distributions p, i.e., an α such that

$$C(\alpha) > R_p(D) \tag{3}$$

for all p. The coefficient α represents the rate of the point mutations, and it is hard to approach a zero point mutation. How far could we let the point mutation increase (decreasing the channel capacity $C(\alpha)$) and still overcome the distortion D for all p. One can also rephrase the question differently: given a point mutation a what is the lowest amount of distortion D (highest distortion rates) such that the channel capacity $C(\alpha)$ is still larger than the rate distortion $R_p(D)$ for all p.

In the context of amino acid the question can be rephrased: can we calculate the point mutation α such that the amino acid channel would pass the amino acid information with a distortion which is less than D given any probability p of amino acids. Or if we use the channel n times for large n: is there a point mutation a such that the channel would pass all proteins with a distortion no more that D, given any protein probability distribution (induced from the amino acid distribution). Again we should remark at this point that *in vivo* proteins are transmitted with a very low distortion but all the questions here are discussed from the point of view of information theory considering only the channel and no other mechanism.

In the next section we formulate the relation between the rate distortion and channel capacity. We start with a simple example for one bit symmetric channel for which we calculate the channel capacity $C(\alpha)$ and the rate distortion $R_p(D)$. Then we apply the above constraint (Eq. (3)) and obtain a relation between α and D.

Next we pose the question for a general channel. In general, it is harder to obtain the rate distortion and channel capacity. We shall use a lower bound on the rate distortion function which is a version of Shannon's lower bound (Shannon, 1959). Then we will present the above constraint in terms of an inequality between entropy functions.

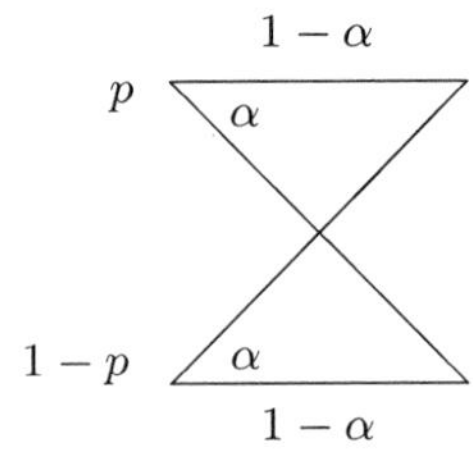

Figure 1. One bit symmetric channel.

3. The Relation Between Channel Capacity and Rate Distortion

3.1 1-bit channels

Let $A(\alpha)$ be the transition matrix of a one bit symmetric channel (see the Fig. 1). It is easy to see that $C(\alpha) = 1 - H(\alpha)$ where $H(\alpha)$ is the entropy function:

$$H(\alpha) = \alpha\log(\alpha) + (1 - \alpha)\log(1 - \alpha).$$

Let D denote a distortion. Given a probability distribution p on a one bit space $X = \{0, 1\}$, then it is also easy to show that $R_p(D) = H(p) - H(D)$ (for $D < \min\{p, 1 - p\}$) (Cover and Thomas, 1991).

For any distortion D, and any probability distribution p there exists an α such that the channel transition matrix $A(\alpha) = A(\hat{x}/x)$ satisfies:

$$E(d(\hat{x},x)) = \sum_{\hat{x}}\sum_{x} A(\hat{x}/x)p(x)d(\hat{x},x) < D.$$

Note that if $\hat{x} \neq x$ or $d(\hat{x}, x) \neq 0$ then the coefficients of the channel transmission matrix $A(\hat{x}/x) = \alpha$ could be decreased such that the above inequality is satisfied ($A(\hat{x}/x)$ is close to 1 only when $\hat{x} = x$ or $d(\hat{x}, x) = 0$).

Now the rate distortion $R_p(D)$ minimizes the rate over all transition matrices that satisfy the above expectation constraint and therefore:

$$I_A(\alpha) = (X_p, \hat{X}) \geq R_p(D).$$

On the other hand the capacity maximizes the rate over all p while fixing the transition matrix $A(\alpha)$:

$$C(\alpha) \geq I_{A(\alpha)}(X_p, \hat{X}).$$

Using the above two inequalities we conclude that given a distribution p and a distortion D we can find an α such that:

$$C(\alpha) > R_p(D).$$

We wish to calculate the coefficient a such that the above inequality is satisfied for all p. We can reformulate the above constraint using the expressions obtained for the channel capacity and rate distortion. In other words we need to calculate an α such that

$$1 - H(\alpha) = C(\alpha) > R_p(D) = H(p) - H(D)$$

for all p. A sufficient condition would be:

$$1 - H(\alpha) = C(\alpha) > \max_p\{R_p(D)\} = 1 - H(D)$$

or equivalently:

$$H(\alpha) < H(D). \tag{4}$$

This is a relation between the transition rate of the channel and the amount of distortion the system can suffer. The relation is expressed in terms of entropy functions and is independent on the distribution variable p. In simple words, given an overall distortion D, the point mutation rate α is bounded from above by Eq. (4) if we want to assure that all distributions at the input are distorted by no more than D.

We will now generalize the above result to the channel of amino acids $A(\alpha)$ in (Gong *et al.*, 2011).

3.2 *n*-bit channels

The following theorem is a generalization of the above results for a 1-bit channel.

Theorem. Consider the channel of amino acids that is induced from the point mutation α. Let X denote its input random variable,

$\hat{X}$ its output and $C(\alpha)$ its capacity. Let $d(x, \hat{x})$ be a distance measure between amino acids, and D a given distortion. Define

$$\phi(D) = \max_{p(x,\hat{x});\, E(d(x,\hat{x})) \leq D} H(p(x, \hat{x})). \tag{5}$$

To satisfy the condition $C(\alpha) > R_p(D)$ for all p it is sufficient that

$$H_A\,(X_u/\hat{X}) < \phi(D), \tag{6}$$

where $H_A\,(X_u/\hat{X})$ is the relative entropy of X_u with respect to $\hat{X}$ for a uniformly distributed variable X_u and a relative probability $p(x/\hat{x}) = A^t$, where A^t is the transpose matrix of A.

Proof: First we will show a simple version of the Shannon lower bound for rate distortion (Shannon, 1959). By the definition of the rate distortion:

$$R_p(D) = \max_{a(\hat{x}/x);\, E(d(x,\hat{x})) \leq D} I_{a(\hat{x}/x)}(Xp, \hat{X}).$$

Using the properties of the mutual information we can write:

$$\begin{aligned} I(X_p, \hat{X}) &= H(X_p) + H(\hat{X}) - H(X_p, \hat{X}), \\ &> H(X_p) - H(X_p, \hat{X}). \end{aligned}$$

If $a(\hat{x}/x)$ satisfies $E(d(x, \hat{x})) \leq D$ then by the above definition of $\phi(D)$ and the above property of the mutual information:

$$I_{a(\hat{x}/x)}\,(X_p, \hat{X}) \geq H(X_p) - \phi(D)$$

and therefore

$$R_p(D) > H(X_p) - \phi(D). \tag{7}$$

This result is very similar to Shannon's lower bound on the rate distortion (Shannon, 1959), however, our definition of $\phi(D)$ uses distributions on the higher dimensional space $X \times \hat{X}$ (we will explain the reason for that in the next section).

Extending the argument of the example above we can find an α such that:

$$C(\alpha) > R_p(D).$$

We will now apply the following constraint:

$$C(\alpha) > \max_p R_p(D).$$

Assume now that the channel capacity for such an α is achieved at the probability p^*:

$$C(\alpha) = \mathrm{H}(X_{p^*}) - H_A\,(X_{p^*}/\hat{X}).$$

Therefore we need

$$C(\alpha) = H(X_{p^*}) - H_A(X_{p^*} / \hat{X}) > \max_p R_p(D).$$

Using Eq. (7) above we have:

$$\max_p R_p(D) > \log |X| - \phi(D)$$

and therefore:

$$H(X_{p^*}) - H(X_{p^*}/\hat{X}) > \log |X| - \phi(D).$$

Since $H(X_{p^*}) - \log |X| < 0$ the constrain we wish to force would be equivalent to:

$$H_A\,(X_{p^*}/\hat{X}) < \phi(D). \tag{8}$$

Note however, that the above inequality depends on the distribution p^* which is unknown. We will therefore write a sufficient condition which is somewhat weaker but more straightforward to calculate.

Consider $H_A(\alpha)(X_u/\hat{X})$ where X_u is uniformly distributed. It depends only on the coefficients of the transition matrix of the channel. Moreover:

$$I_A(X_u, \hat{X}) \le I_A(X_{p^*}, \hat{X}),$$

since p^* was the distribution where $I_A\,(X_{p^*}, \hat{X})$ was maximal. Using the definition of mutual information we have:

$$H(X_u) - H_A(X_u/\hat{X}) \le H(X_{p^*}) - H_A\,(X_{p^*}/\hat{X})$$

or

$$H(X_u) - H(X_{p^*}) \le H_A(X_u/\hat{X}) - H_A\,(X_{p^*}/\hat{X}).$$

Since u is uniformly distributed

$$H(X_{p*}) \leq H(X_u)$$

and so

$$H_A\,(X_{p*}/\hat{X}) \leq H_A(X_u/\hat{X}).$$

Hence a sufficient (although weaker) condition to satisfy our constrain would be:

$$H_A\,(X_u/\hat{X}) \leq \phi(D). \tag{9}$$

The left-hand side of the above inequality depends only on the coefficients $A(\hat{x}/x)$ of the channel transition matrix. This completes the proof of the theorem.

The above condition defines a relation between α and D in terms of entropy functions. Given D it fixes the maximal value of a for which the channel's distortion is lower than D. Alternatively given α it fixes a lower bound on the distortion D that the system can sustain for a large family of amino acid distributions.

It is possible however, that a finer inequality exists between α and D. In fact we have used such inequality in the derivation of the above condition, however its explicit calculation is harder to obtain. We leave this for future research.

4. Evaluation of the Lower Bound for the Distortion *D*

The function $H_A(X_u/\hat{X})$ described above should be interpreted as a lower bound for $\phi(D)$. Given a point mutation rate α and a distance measure $d(x, \hat{x})$ for the amino acids (we assume 20 amino acids), then the overall distortion D cannot be lower than a certain value $D(a)$ such that $\phi(D(\alpha)) = H_A(\alpha)(X_u/\hat{X})$. In this section we will evaluate this lower bound.

It is straightforward to calculate $H_A(X_u/\hat{X})$ using the transition matrix for amino acid in (Gong, Bouaynaya and Schonfeld, 2011). Let $A(\alpha) = A(\hat{x}/x)$. Let $\beta_{\hat{x}}$ be the normalizing factor for each of the columns:

$$\beta_{\hat{x}} = \sum_x A(\hat{x}/x).$$

By a simple Bayesian argument:

$$P(x/\hat{x}) = \frac{P(x/\hat{x})P(x)}{P(\hat{x})}$$

Since X_u is distributed uniformly:

$$P(\hat{x}) = \sum_x P(\hat{x}/x)P(x) = P(x)\beta_{\hat{x}} = \frac{1}{20}\beta_{\hat{x}}.$$

Therefore,

$$P(x/\hat{x}) = \frac{P(x/\hat{x})}{\beta_{\hat{x}}} = \frac{A(\hat{x}/x)}{\beta_{\hat{x}}}.$$

Then

$$H_A(X_u/\hat{x}) = \frac{1}{\beta_{\hat{x}}}\sum_x A(\hat{x}/x)\log\left(\frac{A(\hat{x}/x)}{\beta_{\hat{x}}}\right) \tag{10}$$

and

$$H_A(X_u/\hat{X}) = \frac{1}{20}\sum_x \beta_{\hat{x}} H_A(X_u/\hat{x})$$

$$= \frac{1}{20}\sum_{\hat{x}}\sum_x A(\hat{x}/x)\log\left(\frac{A(\hat{x}/x)}{\beta_{\hat{x}}}\right),$$

where we have used the fact that X_u is uniformly distributed over the space of 20 amino acids, and therefore $p(\hat{X} = \hat{x}) = \frac{1}{20}\beta_{\hat{x}}$. It is apparent that $H_A(X_u/\hat{X})$ is a simple function of α. Figure 2 describes a graph of $H_A(X_u/\hat{X})$ as a function of α.

It is more complicated to evaluate $\phi(D)$ in general. To use the original Shannon lower bound on the rate distortion one needs a distance matrix of amino acids such that all columns are permutations of one column (Shannon, 1959).

This is the reason we use a different and more general version of the bound. To calculate $\phi(D)$ one needs to find a maximal entropy while going over all distribution on $X \times \hat{X}$. The result depends only

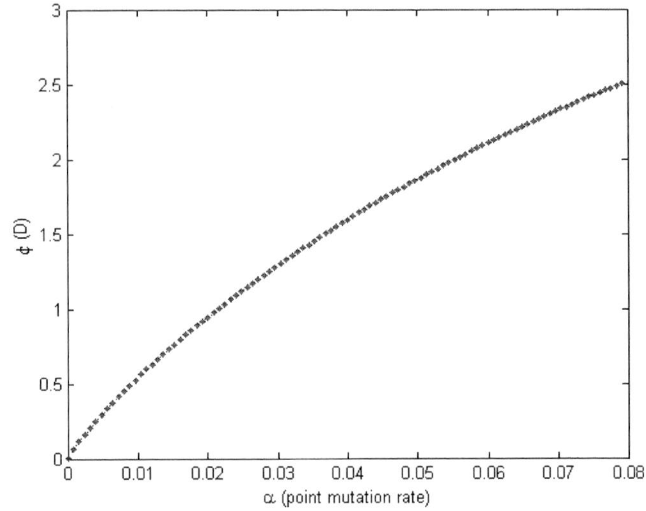

Figure 2. A lower bound on $\phi(D)$ as a function of the point mutation rate.

on the distortion rate D and not on any of the probability distributions.

Several distance measures for amino acids were discussed in (Higgs and Attwood, 2005). Each amino acid could be presented as a vector in a high dimensional space where the coordinates corresponds to the amino acid's physicochemical properties like volume, bulkiness, polarity, hydrophobicity, etc. The distance between two amino acid could be taken as the Euclidean distance of the corresponding vectors. One can write a distance matrix of amino acids.

In the notations above $D(\alpha)$ is the lowest amount of distortion for which we can still guarantee that the channel capacity is sufficient to transmit all signal distributions with a distortion less than D. To transmit signals with a lower amount of distortion we shall need a lower mutation rate α; this cannot be guaranteed for evolutionary reasons.

5. Discussion

In this paper we developed a relation between α, the rate of point mutation, and the amount of distortion D, which in our case is the average distance measure between amino acids. Given a distortion D we present a higher bound for the point mutation rate, $\alpha = \alpha(D)$, that can guarantee the properties of the channel, i.e., the transmission of all signal distributions with a distortion less than D. A higher rate of point mutation will degrade the properties of the channel. Conversely, given a point mutation rate α, we calculate a lower bound on the distortion $D = D(\alpha)$ that the system can sustain for all amino acid distributions. Using a lower amount of distortion requires a lower point mutation rate, which is hard to guarantee for biological systems.

The relations we present are not direct but are given in terms of inequalities between entropy functions. It is a generalization of a simple entropy inequality for the one bit symmetric channel. The function $\phi(D)$ resembles the Shannon lower bound for rate distortion. We leave its computation for future research.

The lower bound on $\phi(D)$ presented here guarantees the channel properties. However, such a bound could possibly be improved. A better estimate for the bound appears in the proof of the theorem above but is considerably more difficult to obtain.

Channel capacity and rate distortion properties are formulated in an n-product space and are defined by the power to differentiate between typical sequences. The correct interpretation of the multiple use of the codon channel is as a channel for proteins, which makes the notions of channel capacity and rate distortion natural.

Note that assuming the channel's capacity is sufficient to transmit one signal with the given distortion, while its capacity is too low for some other signal at the same distortion rate, it could still be a useful channel. However, the time it would take to transmit the signals will vary from one signal to another, even for the best possible coding. This could have evolutionary consequences.

The question we pose here has a general meaning in information theory. Given a signal and a channel we wish to use for the transmission of the signal; is the channel capacity large enough to transmit the signal with a very low distortion? This distortion rate depends on the distribution of the signal. Is it possible to guarantee that the channel will have sufficient capacity to support a very low distortion for a large family of signals? As discussed, the answer is that the time it takes to transmit a signal may vary if the channel's capacity is insufficient.

References

Berger T. 1971. *Rate Distortion Theory: A Mathematical Basis for Data Compression*. Prentice Hall, New Jersey.

Bouaynaya N and Schonfeld D. 2007. Protein communication system: evolution and genomic structure. *Algorithmica*, 48(4), 375–397.

Cover TM and Thomas JA. 1991. *Elements of Information Theory, Wiley-Interscience*. John Wiley & Sons, Hoboken, New Jersey.

Drake JW and Holland JJ. 1999. Mutation rates among RNA viruses. *Proceedings of the National Academy of Sciences*, 96(24), 13910–13913.

Eldar A and Elowitz MB. 2010. Functional roles for noise in genetic circuits. *Nature*, 467, 167–173.

Gong L, Bouaynaya N and Schonfeld D. 2011, Information-theoretic model of evolution over protein communication channel. *IEEE/ACM Transaction on Computational Biology and Bioinformatics*, 8(1), 143–151.

Henikoff S and Henikoff, JG. 1992. Amino acid substitution matrices from protein blocks. *Proceedings of the National Academy of Sciences USA*, 89(22), 10915–10919.

Higgs PG and Attwood TK. 2005. Bioinformatics and Molecular Evolution. Blackwell, Oxford, UK.

Leitao HCG, Pessoa LS and Stolfi J. 2005. Mutual information content of homologous DNA sequences. *Genetics and Molecular Research*, 4(3) 553–562.

Roman-Roldan R, Bernaola-Galvan P and Oliver JL. 1996. Application of information theory to DNA sequence analysis: a review. *Pattern Recognition*, 29(7), 1187–1194.

Shannon CE. 1959. Coding theorems for a discrete source with a fidelity criterion, Institute of Radio Engineers. *International Convention Record*, 7, 325–350.

Steiner CF. 2012. Environmental noise, genetic diversity and the evolution of evolvability and robustness in model gene networks. *PLoS ONE*, 7(12).

8. Communication Languages and Agents in Biological Systems

Subhash Kak

School of Electrical and Computer Engineering
Oklahoma State University, Stillwater, OK 74078, USA
subhash.kak@okstate.edu

Summary

This chapter provides a general discussion of the hierarchy of languages and metalanguages that may be associated with a biological system. At the bottom of this hierarchy are physical structures whose behavior is governed by quantum mechanics. In order for this hierarchy to be valid, we postulate a language whose primitives are quantum mechanical objects. In a complementary relationship to this physical hierarchy is a hierarchy of languages where the quantum language resides at the top. We have show how communication and cognitive agents may be conceived of as collectives of quantum particles.

1. Introduction

In biological systems there exist complex, nonlinear structures and self-organized entities which cannot be described by models based on statistical approaches. Different scales (both spatial and temporal) must be considered when analyzing interactions and activity must be viewed as directed by agents. For example, in the processing in the human brain, there exist autonomous cognitive agents. The emergence of these agents, and the capacity for self-organization that they possess, must be viewed in the context of the evolutionary history of the biological system. Their self-organization requires strong dynamical nonlinearity but this cannot be seen to occur due to macroscopic processes alone. Underlying these macroscopic processes are physical structures that have a quantum basis at the deepest level of description. The need to consider different kinds of objects, including quantum ones, in considering a general biological system from the perspectives of communication and complementarity was emphasized recently (Gautam and Kak, 2013) and a theory to represent agents and memories as quantum collectives was also advanced (Kak, 2013a). In this chapter we review these results and also ask questions about the hierarchy of languages and metalanguages within the biological system.

The biological system may be examined at different scales and there exist layers for which a classical representation is appropriate and others for which we must choose description using quantum mechanics. The quantum formalism emerged in order to deal with fundamental complementarity in Nature and it is based on superpositions of mutually exclusive attributes with collapse to one of these upon measurement and associated with non-locality, entanglement and coherent behavior. A determinant of non-locality is entanglement and in certain situations the violation of Bell Inequalities identifies a process as being quantum mechanical (Kak, 2013b). In other words, there exist macroscopic, probabilistic tests that make it possible, in principle, to determine what representation is correct for the layer.

The molecular scale represents the usual divide between the classical and the quantum. Quantum effects appear if the concentration of particles $(N/V) \geq nq$, where nq is the quantum concentration and the interparticle distance is indicated by the de Broglie wavelength $1 = h/p = h/(m_0 v)$, which for thermalized electrons in a non-metal at room temperature is about 8×10^{-9} m. The smallest molecules have size of about 10^{-10} m, but quantum effects can be exhibited at much larger distances by entangled photons and by virtual particles, which is how they have been proposed for certain biological processes. Macromolecule vibrations create quasiparticles and therefore quantum effects associated with such quasiparticles are contingent on specific macrostructures.

But the matter of the classical/quantum divide is not quite straightforward and it is not based on scale alone. While scholars upholding the orthodox Copenhagen Interpretation (CI) of quantum mechanics assume a split between the quantum and the classical worlds, scholars not subscribing to this interpretation hold other positions. In CI, the wavefunction of the object, or the system, represents the expectations associated with the physical process and when the measurement is made there is a collapse in the expectation related to it within the mind of the observer. In this interpretation, the divide between the quantum and classical is largely based on the scale at which the system is being examined and it is appropriate to consider the biological system at the smallest-scale to be quantum mechanical.

In contrast to the above view in the CI, the divide between the classical and the quantum is considered somewhat artificial in the many worlds interpretation (MWI) of quantum theory (Tegmark, 1998) which has become increasingly influential in many areas of physics, especially quantum computing. In this interpretation all systems are fundamentally quantum mechanical and if we do not see effects such as non-locality and entanglement in a macroscopic system, that is a consequence of decoherence. MWI does not view the measurement process to be cause of the collapse of the wavefunction; rather this process is associated with conditions that cause decoherence (Zurek, 2003). The consideration of a biological system as fundamentally quantum mechanical is, therefore, natural in this interpretation.

In MWI the wavefunction is the primary reality and according to it a wavefunction may be assigned not only to biological and other physical systems, but to the universe itself. If CI is an inside-out view of the universe where the reality is constructed out of the perceptions of the experimenter, MWI is an outside-in view in which the mathematical function of the universe, and its subfunctions, represent the primary reality. When considering macroscopic systems, the two interpretations are very different in their assessment of the applicability of quantum theory. But in biological systems considered at the smallest scales, both the interpretations favor the application of quantum theory.

In recent years, experiments and theory both have come out on the side of quantum mechanics playing a direct or indirect role in photosynthesis (Collini *et al.*, 2010) olfaction (Turin, 1996), vision (Polli *et al.*, 2010), long-range electron transfer (Gray and Winkler, 2005), and bird navigation (Ritz *et al.*, 2004). In some of these areas the evidence is not concerning coherence but rather on quantization and discrete energy levels (Lambert *et al.*, 2013). In photosynthesis, the light-harvesting chlorosome of green-sulphur bacteria collects and then transfers energy to the reaction center through the Fenna–Matthews–Olson (FMO) complex with nearly 100% efficiency even though the intermediate electronic excitations are very short-lived (~1 ns). Although alternative explanations for this coherence have been advanced (Briggs and Eisfeld, 2011; Miller, 2012), this is considered the most successful example of quantum effects in biology (Ishizaki *et al.*, 2010). In many situations, probabilistic tests to separate quantum from classical behavior are quite complicated due to the effects of decoherence on the quantum process.

Several proposals describe brain functioning as a classical/quantum hybrid system. Fröhlich argued (Fröhlich, 1968) that electric and elastic forces within the dense arrangement of dipolar molecules of the biological cells will interact leading to vibrations at characteristic frequencies that couple electrical displacements to physical deformations. These vibrations may be viewed as collective behavior of phonons that extends correlations across macroscopic distances within the organism. Ricciardi and Umezawa proposed (Ricciardi and Umezawa, 1967) a mechanism of memory storage and retrieval in terms of virtual bosons associated with the physiological structures of the brain in which long-term memory is related to the ground state and short-term memory to the metastable excited states. Jibu, Yasue and Pribram further developed these ideas and considered implications for consciousness (Jibu *et al.*, 1996).

The dissipative quantum field model (Freeman and Vitiello, 2006) associates memory with the non-zero energy states coupled to the infinitely many minimal energy states of the system. This model is different from the model (Hameroff and Penrose, 2003) that takes microtubules to be the place where quantum coherence is maintained. It has been recently proposed that extended Calcium/Calmodulin-dependent protein kinase II (CaMKII) kinases can potentially encode synaptic Ca^{2-} information via phosphorylation as ordered arrays of binary bits and candidate sites for CaMKII phosphorylation-encoded molecular memory include microtubules (Craddock *et al.*, 2012).

In our own previous consideration of this problem (Kak, 1995, 1996, 2000), the structure of the biological system itself is taken to be in a state of superposition in the sense that it is in one of many metastable states based on internal and external conditions. In particular, in brain dynamics state transitions occur when the cortex switches abruptly from one metastable state to another. These metastable states are like the component states of a superposition but they form a discrete spectrum and not a continuous set of values. Brain dynamics can be seen either from the perspective of these discrete states or, complementarily, from the perspective of attractive basins associated with nonlinear dynamics.

The actual brain state cannot be identified excepting in its linguistic projection in the mind, which may be non-verbal or verbal. When the state is quizzed by the conscious mind, it can do no better than speak of a transition state that may be one spatial pattern from which it goes to another as frames in a cinema film. The existence of the many metastable states makes it possible for the system to adapt to the environment. The consequence of this is that the cognitive system is quantum at a deeper level but it is coupled to the conscious system which is classical: the quantum system is defined by collectives of quasiparticles; the classical system comprises of the neural networks of the brain. This dual system of the cognitive system can thus describe individual and social behavior. We also propose that the linguistic mind creates local explanations for phenomena, and since quantum processes are non-local, this is a consequence of a principle of veiled non-locality (Kak, 2014).

Recently, quantum probability has been used in modeling cognition and decision (e.g., Busemeyer and Bruza, 2012) sidestepping the question of its physical basis. This line is an offshoot of the proposal made early by the pioneers of the CI that the unconscious mind is quantum mechanical and different from classical mind of the internal dialog within the individual. As stated by Niels Bohr: "The quantum postulate implies that any observation of atomic phenomena will involve an interaction with the agency of observation not to be neglected. Accordingly, an independent reality in the ordinary physical sense can neither be ascribed to the phenomena nor to the agencies of observation. After all, the concept of observation is in so far arbitrary as it depends upon which objects are included in the system to be observed. Ultimately, every observation can, of course, be reduced to our sense perceptions. The circumstance, however, that in interpreting observations use has always to be made of theoretical notions entails that for every particular case it is a question of convenience at which point the concept of observation involving the quantum postulate with its inherent *irrationality* is brought in" (Bohr, 1928).

In the quantum cognition field (e.g., Busemeyer and Bruza, 2012; Khrennikov, 2010; Haven and Khrennikov, 2013) it has been argued that human judgments do not always follow classical logic and cognition has order and interference effects. If it is accepted that there exist non-classical aspects to human cognition and decision, there is then need to discuss how these non-classical agents and memories are structured. In the global workspace theory, it is consciousness that provides access to the cognitive agent to its memory sources.

If subsystems within biological systems are quantum mechanical, what is the nature of communication between these subsystems? In analogy with classical systems, one would expect that the communication will have syntactic, pragmatic, and semantic aspects. We also stress that biocommunication includes associative (classical) and reorganizational (adaptive) elements in addition to those that are quantum (Kak, 1996). Here the associative language is the ordinary language of communication for which the medium is chemical in terms of a variety of neurotransmitters and electrical in the manner in which signals are transmitted; the reorganizational language is the one in which the subsystems reorganize and in doing so change their modes of communication; and the quantum language is the one that is characterized by a quantum process (see e.g., Ball, 2011).

2. Representations and Languages

From an operational point of view, we need to speak not only signs and codes and their representation but also interpretation within the organism, which requires an appropriate metalanguage. Proper linguistic expressions are different from noise and this has important implications in that it makes it possible for the organism to recognize self from non-self (Sanabria *et al.*, 2008). Furthermore, unlike a formal language and quite like natural languages, the biocommunication language is adaptive. The biological organism cannot be seen in isolation. It shares an identity with others in a group, it uses signs in a coherent manner, it belongs to a habitat where it shares space with other organisms, its signs have meanings that can be interpreted differently based on context, and its semiotic rules evolve based on the nature of interactions with other organisms. The nature of intra-level interaction, together with the governing grammar, has implications for the capacity of the organism to operate within its ecological environment.

One can see limitations of representations from a variety of perspectives, including that of information and computability (Kak, 2012). No representation can be looked at in isolation because interactions occur both within a hierarchy and across hierarchy. For example, top-down influences shape cortical and thalamic processing of sensory information in highly interconnected ways. Cortical structures work as adaptive processors subject to attention and

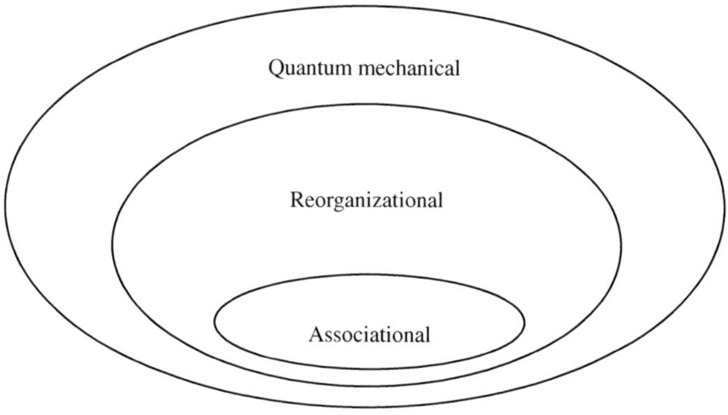

Figure 1. The language hierarchy with adaptive subsets.

expectation related to the perceptual task, and brain states are a consequence of complex interactions between multiple structures and the modulation of intrinsic circuits by feedback connections.

Representations are not static but context-sensitive, which context is provided by the appropriate top-down signals and expectations. The cumulative repertoire of behaviors at all the levels has infinite variety. The communications between different nodes within the system must therefore have all the elements of the hierarchy of Fig. 1, where it is remembered that for biological systems it is the subsets of the elements under certain internal and external conditions that define the boundaries. We call the highest level quantum (Kak, 1996) to stress that it comes with identification of wholeness and awareness of self as distinct from non-self. The three languages interface with each other adding to the complexity of the interaction between the communicating elements. The languages are also in a complementary relationship with their respective metalanguages. Note that this hierarchy is different from the Chomsky hierarchy of formal languages although the two hierarchies may very well have points of convergence. For example, recursively enumerable languages (Type-0 in the Chomsky hierarchy) could very well correspond to what we call quantum mechanical type languages since they are associated with unrestricted grammars, but this is a matter that requires additional research.

The elements of this communication are in directed relationships amongst its components. A cortical state engenders expectations about the subsequent state, and the expectation affects the way the incoming sensory information is processed. This broadens the functional diversity of each area and makes the way information is exchanged across areas more important than the contents of any specific area. The exchange of information across areas increases the capacity of the structures. It has been "proposed that this interaction is manifest in the way that feedback connections address subsets of intrinsic cortical connections, and the functional properties of a neuron depend on which subset is gated at any given time. The mechanism of perceptual learning may involve the setting up of this addressing, such that both the encoding and recall of learned information involves the appropriate selection of the inputs that convey information about the stimulus being discriminated" Gilbert and Sigman, 2007). In genetic information, the situation is similarly complex (Witzany and Baluska, 2012).

Now consider the complementarity between the neural and linguistic processing in the brain. The processing in neural networks, which are defined in a hierarchical sense with complex interaction amongst structures, is associated with what one might call *neuronal concepts*. On the other hand, there is the system of logic and language at work within systems that informs cognitive agents in their higher-order processing. The human mind creates linguistic models of reality whose elements are interconnected grammatically and logically. Paralleling the structure of the brain, language is organized hierarchically. The functional elements of the linguistic system are *logical or linguistic concepts*. The neuronal concepts are non-linguistic as contrasted from the logical concepts of the linguistic objects (Fig. 2).

The neuronal and the linguistic systems are interconnected. The linguistic system is a product of the brain, but, in turn, the linguistic dialog also influences the functioning of the brain. The relationship between neuronal and logical concepts is an important problem. Neuroscience provides clues about the lowest levels of

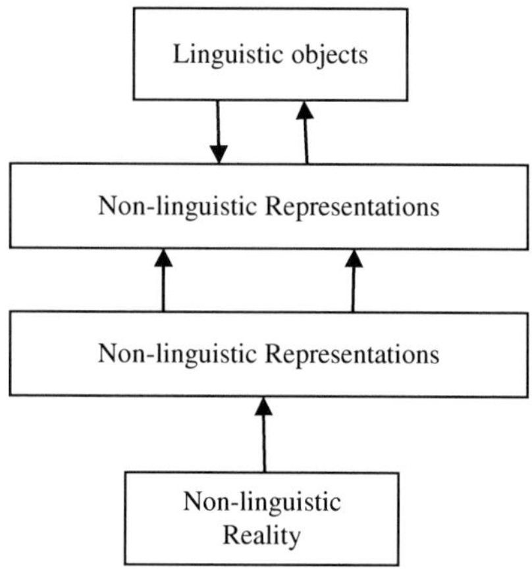

Figure 2. Language and reality.

the neuronal hierarchy; about the higher levels we do not possess knowledge that is independent of language. Since language is a behavior associated with the brain, a close relationship should exist between the highest levels of the neuronal hierarchy and the lowest stages of linguistic concepts. The logical structure of an object must have a neuronal correlate. The initial skeleton of the concepts that a child comes to have is a consequence of interaction with caregivers that is essentially before the use of speech. But this skeleton is filled out as the child learns additional concepts that are language-dependent. Thus the problem of the mutual relations of neuronal and logical concepts has an interesting developmental side since the more complex concepts in the child emerge under the influence of language.

The complementarity of the neuronal and the linguistic concepts has significant implications for structure and performance. The nature of language and the subject's proficiency of it play a role in the understanding of reality. Indeed, this should have an influence even in the manner in which the brain reorganizes itself based on experience and activity.

Since linguistic objects become a part of social reality that play a role not very different from material objects, it becomes necessary to create other linguistic objects that describe this reality of which linguistic objects are components necessitating a metasystem transition (Turchin, 1977). Such transitions can occur at many different hierarchical levels thus leading to a variety of nested metalanguages.

An example of metasystem transition is the one from unicellular to multicellular organisms. A different transition is the emergence of eusociality or symbolic thought (Costa and Fitzgerald, 2005). A metasystem is formed by the integration of a number of initially independent components and the collective of components becomes a new, goal-directed individual, capable of acting in a coordinated way. This metasystem is more complex, more intelligent, and more flexible in its actions than the initial component systems. One can also speak of a metasystem transition in the development of a language.

3. Quantum Mechanics in Memory

Now we consider how quantum mechanical and linguistic structures may be correlated. Specifically, we consider linkages between neuronal structures and the specific linguistic concept of memory. Several years ago, the Bose–Einstein quantum probability distribution was proposed as model of human memory and shown to describe experimental results related to Piaget's developmental stages (Pascual-Leone, 1970). This approach was taken further and it was shown how such memories might be structurally conceived (Kak, 2013a).

The starting point of this approach was to conceptualize Piaget's cognitive-developmental variable as a processor with capacity M, which was the maximum number of chunks of information or schemes that can be controlled or integrated in a single act. M was supposed to grow with age and therefore it became a quantitative measure of each developmental stage.

Particles and quasiparticles cannot be agents or memories in themselves because they are too numerous and also indistinguishable. In physics the property of most interest concerning particles or quasiparticles is energy, but going from physics to chemistry and biology, molecular structure and shape play an increasingly significant role in processes. When considering cognitive agents, informational attributes of these collectives can be expected to play a part in their identity. We assume that bosonic quasiparticles or fermions are assembled in different arrangements to become cognitive agents. The three-dimensional structures of the brain will come with corresponding quantum space that will define agents and memories.

Agents and memories should be set apart by number, structure, and informational content. In classical computers, they are both represented by binary sequences and they are differentiated by context. Agents, unlike memories, are linked to sensors and actuators. As sequences shorn of context, they ought to be very similar. An agent (or a memory) must be invariant to certain types of transformation and it should be resistant to noise so long as the noise is within a certain limited range.

Quantum objects and quasiparticles associated with macrostructures, in collectives representing different cognitive agents and memories, have long range correlations. The property of resistance to noise implies that there is a minimum *separation* between patterns representing them. This distance must be defined in an abstract space that is different from the three-dimensional geometry associated with the particles. To study such agents one needs to go beyond statistics related to energy distributions as the properties of collectives should be invariant with certain translations of energy.

4. Statistical Considerations

Classical and quantum objects may be differentiated based on their statistics. Indeed the very search for a quantum theory began as statistics associated with black body radiation were different from that of classical thermodynamics. The Planck radiation formula is an example of the distribution of energy according to Bose–Einstein statistics.

Classical objects are distinguishable whereas quantum objects are not. For N classical particles distributed over M single-particle distinct states (which could be energy states), the number of possible arrangements is $n = M^N$, and this is the basis of the Maxwell–Boltzmann distribution. Quantum statistics must be considered in cognitive processes if cognitive agents are collectives of quasiparticles.

As there are two different classes of quantum particles, bosons and fermions, we have two different quantum statistics. Bosons are governed by the Bose–Einstein statistics and fermions by the Fermi–Dirac statistics. For N bosons associated with M single-particle states, the number of arrangements is

$$n_{\text{bosons}} = \frac{(N+M-1)!}{N!(M-1)!}.$$

For fermions, the number of arrangements is reduced further due to the Pauli Exclusion Principle according to which no two such particles can be in the same state, and we obtain:

$$n_{\text{fermions}} = \frac{M!}{N!(M-N)!}.$$

The probability of observing any particular state is determined by the number of copies of that state divided by the total number of arrangements. The number of permissible arrangements goes down as we move from classical to bosonic to fermionic states. In statistical distributions we are interested in the probability of arrangements corresponding to specific energy values since these either represent temperature or some other measurable characteristic of the ensemble.

Example 1. Given two particles, the classical state can be one of the following four: 00, 01, 10, and 11. The bosonic states will be $|0\rangle|0\rangle$, $|1\rangle|1\rangle$ and $\frac{1}{\sqrt{2}}(|0\rangle|1\rangle + |1\rangle|0\rangle)$; and the fermionic state will be

$\frac{1}{\sqrt{2}}(|0\rangle|1\rangle - |1\rangle|0\rangle)$. The probability of finding different outputs for the three situations is summarized as shown in Table 1.

Table 1. Probability of different outputs for two particles.

	Classical	Bosons	Fermions
Both 0	0.25	0.33	0
Both 1	0.25	0.33	0
One 0 and one 1	0.50	0.33	1

Example 2. Consider $N = 3$ and $M = 4$. This may be viewed as 3 particles in four boxes (distinct, quantized, energy levels) that corresponds to $4^3 = 64$ classical arrangements. Let us label the four boxes 0, 1, 2 and 3 (with energy levels equal to the index) and each of the three balls can be shown at a different location. The 64 arrangements will then be the sequences 000, 001, 012, ..., 222, ..., 333. Here the arrangement 000 means that all the three particles are in the energy state 0. Each of these sequences has the probability 1/64 of being observed. The energy in the system will vary from a minimum of 0 (corresponding to the arrangement 000) to a maximum of 9 (arrangement 333). For bosons, the number of arrangements with the same number of particles and boxes is 20, and for fermions it is 4.

Let the energy of a particle in a box be determined by the label of the box (0 has energy of 0; 1 has energy of 1; 2 has energy of 2; 3 has energy 3), then the distribution is through a spectrum of values ranging from 0 to 6 of Table 2.

The probability that the state will be one of these arrangements has gone up to 1/20 for bosons and 1/4 for fermions from the 1/64 for classical particles.

The distribution of Table 2 is shown in Fig. 3 which highlights the fact that the number of arrangements decreases as we go from classical to quantum distributions.

The specific quantum probability distribution is obtained by choosing a certain energy value associated with the system in

Table 2. Distribution of energy in three different distributions.

Energy level	Number of arrangements in Maxwell–Boltzmann distribution	Number of arrangements in Bose–Einstein	Number of arrangements in Fermi–Dirac
0	1	1	0
1	3	1	0
2	6	2	0
3	10	3	1
4	12	3	1
5	12	3	1
6	10	3	1
7	6	2	0
8	3	1	0
9	1	1	0

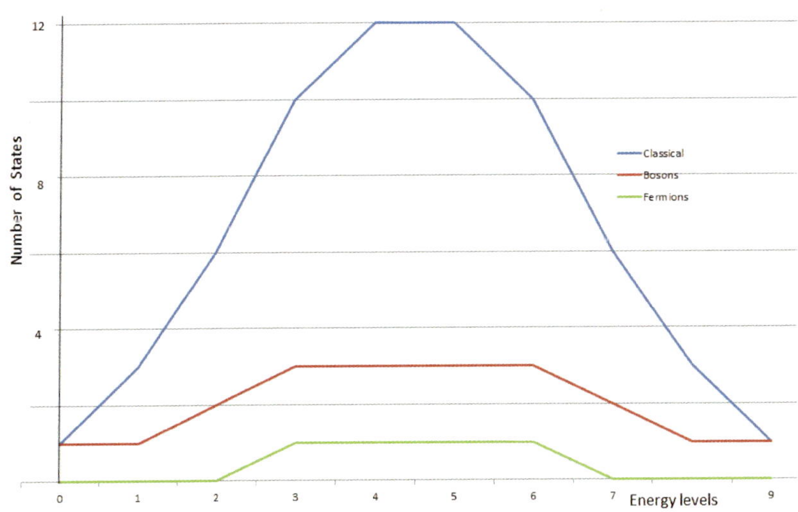

Figure 3. Number of states for different types of particles. Top curve: Classical, middle curve: Bosons, bottom curve: Fermions.

equilibrium and then seeing how many of the arrangements correspond to each of the energy levels.

In principle, the statistics can reveal if quantum modeling of a cognitive phenomenon is correct. It is significant that Pascual-Leone found the Bose–Einstein statistics to be the correct model for memory in the Piaget's developmental model on the compound-stimuli visual information tasks (Pascual-Leone, 1970). This work assumed that the tasks solved at about the same age by normal children involve formulas of equal maximum complexity, which was denoted by $m = a + k$, where a stands for the processing space required, and k is the number of independent cognitions required by the task. The value k varied with each Piagetian stage: i.e., $a + 2$, $a + 3$, $a + 4$, $a + 5$,.... The value m was a measure of the computing space M, which was taken to depend on the subject's representation of the task instructions, the testing situation, and the information needed to generate the correct logical response.

Cognitive tasks require that the representation be done in one space and binding be done in another higher dimensional space and the requirements for this seem particularly matched to quantum representation models. Furthermore, quantum dissipation models are associated with fractal behavior that has been observed in brain states (Vitiello, 2009).

Example 3. Consider 9 bosons that are associated with 6 energy states (ranging from values 0 through 5) for a total energy of 8 units. The arrangements associated with this are shown in Table 3. Each indexed value is a unique representation of the energy partition for the total value of 8, where the level 0 serves to account for the particles that do not contribute to the energy total. The 18 cases of Table 3 represent the different partitions of number 8.

The averages for the different columns are according to the Bose-Einstein distribution which is shown in Fig. 4.

Example 4. Consider 6 fermions of spin-½ with 7 energy states (ranging from values 0 through 6) for a total energy of 10 units. The arrangements associated with this are shown in Table 4. The 8 cases

Table 3. Arrangements for 6 energy levels and 9 bosons for total energy = 8 (Columns are energy levels 0 through 5 and rows are arrangements 1–18).

Index ↓	0	1	2	3	4	5
1	5	3	0	0	0	1
2	6	1	1	0	0	1
3	7	0	0	1	0	1
4	4	4	0	0	1	0
5	5	2	1	0	1	0
6	6	0	2	0	1	0
7	6	1	0	1	1	0
8	7	0	0	0	2	0
9	3	5	0	1	0	0
10	4	3	1	1	0	0
11	4	0	1	2	0	0
12	5	1	2	1	0	0
13	5	2	0	2	0	0
14	2	6	1	0	0	0
15	3	4	2	0	0	0
16	4	2	3	0	0	0
17	5	0	4	0	0	0
18	1	8	0	0	0	0
Averages →	4.66	2.33	1	0.5	0.33	0.17

of Table 4 represent the different unique partitions of number 10. Since each particle can be either in spin-up or spin-down, each energy state can have at most two particles.

Low energy states are more probable with the Bose–Einstein statistics and less probable with the Fermi–Dirac statistics as compared to the Maxwell–Boltzmann statistics. At very low energies, bosons can condense into the lowest energy states.

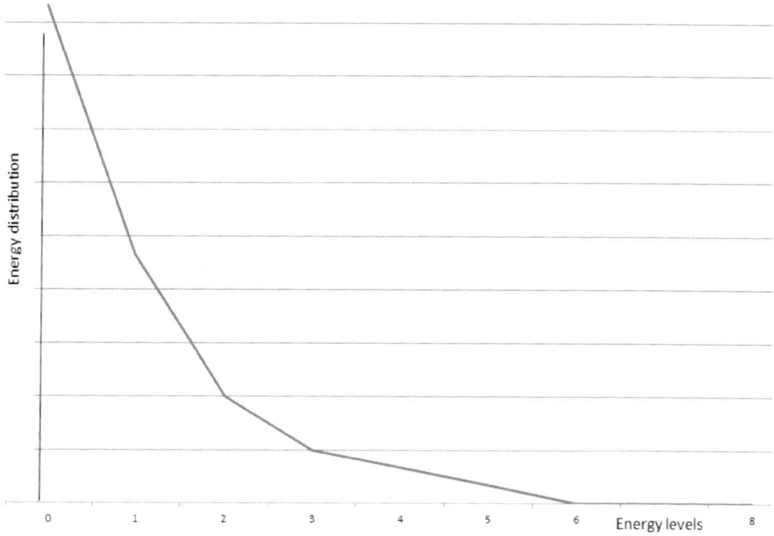

Figure 4. Bose–Einstein distribution for example 3.

Table 4. Arrangements for 7 energy levels and 6 fermions for total energy = 10 (Columns are energy levels 0 through 6 and rows are arrangements 1–8).

Index ↓	0	1	2	3	4	5	6
1	2	2	1	0	0	0	1
2	2	1	2	0	0	1	0
3	2	2	0	1	0	1	0
4	2	1	1	1	1	0	0
5	1	2	2	0	1	0	0
6	2	0	2	2	0	0	0
7	1	2	1	2	0	0	0
8	2	2	0	0	4	0	0
Averages	1.75	1.50	1.12	0.75	0.50	0.25	0.12

5. Quantum States as Memories and Agents

Classical objects are uniquely defined in terms of their many attributes. On the other hand, since quantum objects are indistinguishable, they must be defined in terms of arrangements (or patterns) associated with quantum states. A memory or cognitive agent as a collective of quantum particles must have a unique structure. In biological systems the structure is likely to be three-dimensional, but here, for simplicity, we consider structure for one-dimensional sequences.

As example, consider three particles in four energy levels. Let the agents (or memories) be defined in terms of whether all the particles are same (Agent 1); there are two of one kind and one of another (Agent 2); and they are all different (Agent 3). If each of the arrangements occurs with equal probability, the frequencies of the three agents would be as given below by the number of elements of each set:

Agent 1 : 4 cases — 000, 111, 222, 333
Agent 2 : 12 cases — 001, 002, 003, 011, 022, 033, 112, 113, 122, 133, 223, 233
Agent 3 : 4 cases — 012, 023, 023, 123

Each of these maps several energy values to a single agent.

Energy alone cannot be a marker of memory or agent. To see this, consider now a quantum system that is associated with five energy levels, which we label as Level 0 through Level 5, where the energy of a particle in Level *k* is taken to be *k*. The quantum objects will belong to one of these five levels and the arrangements would correspond to the energy associated with the system. There can be many instances of different arrangements each with the same total energy. These arrangements from an informational content appear to be very different for there is no discernible pattern associated with them.

We propose than an algorithmic approach to information content in which the length of the program required to generate the patterns is a measure of the information (Li and Vitanyi, 1997). This is equivalent to a structural view of the problem, and its special merit is that such structure can mimic the object of information in form.

The number of quantum collectives in a system with n energy levels may be computed in a manner analogous to sphere packing

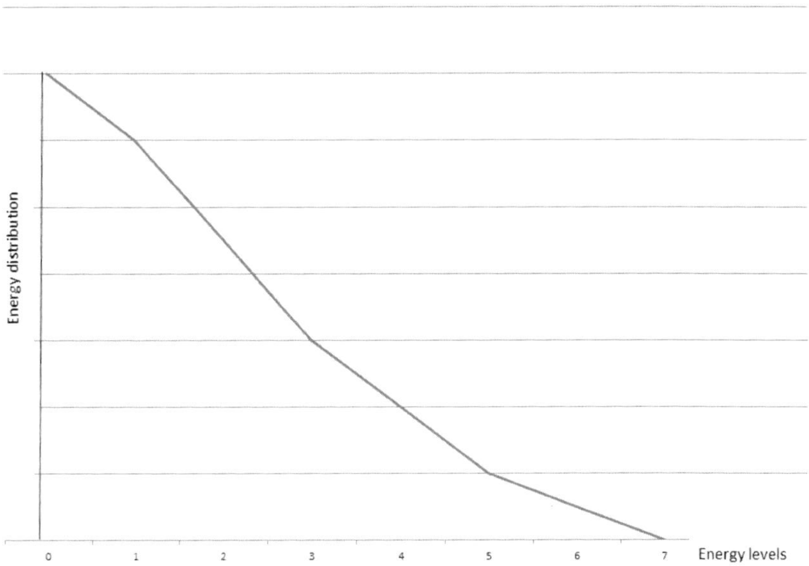

Figure 5. Fermi–Dirac distribution for example 4.

arguments. In linear codes, the number of elements of the collective, M, is given by the following relation (van Lint, 1975):

$$M \leq \frac{q^n}{\sum_{i=0}^{s} \binom{n}{i} (q-1)^i},$$

where q is the number of energy states (or quantization levels), n is the size of the collective, and s is a measure of the noise resistance that the collectives possess. In the example of Fig. 5, $q = 5$ and $n = 14$. Therefore, for small noise resistance the number of collectives can be very many. The number of agents (that may well be equivalent to the number of chunks of memory) can grow in number as the neuronal structures grow and mature.

In number of chunks of short-term memory, the numbers four to six has been given (Cowan, 2000), although that of chimpanzees seems to be much higher (Inoue and Matsuzawa, 2007; Matsuzawa, 2013). It is interesting to speculate if the processing in the workspace corresponding to short-term memory has a quantum basis and whether the number of its chunks can help determine its physical structural correlates.

6. Conclusions

This chapter has provided a general discussion of the hierarchy of languages and metalanguages that must be associated with a biological system. At the bottom of this hierarchy are physical structures whose behavior is governed by quantum mechanics. In order for this hierarchy to be valid, we must postulate a language whose primitives are quantum mechanical objects. In a complementary relationship to this physical hierarchy is a hierarchy of languages where the quantum language resides at the top.

We have shown how communication (and cognitive) agents may be conceived of as collectives of quantum particles. We associate agents and memories with patterns that belong to unique classes of quantum collectives. The structural approach to the definition of memory has the good property that it can mimic features of the two- or three-dimensional original it seeks to represent. It is plausible that the fundamental character of agents and memories is different and that agents are fermion collectives whereas memories are boson collectives. If agents reside in quantum physical structures, that supports the view that they are fermion collectives.

It should be possible to devise experiments to test the theory advanced in this chapter. The evidence for order and interference effects in probability associated with cognition and decision supports the broad idea discussed here, but additional tests concerning quantum statistics must be devised. These tests will reveal the nature of the particles in the underlying physical structures and they will raise new questions on the physical basis of the quantum collectives.

References

Ball P. 2011. The dawn of quantum biology. *Nature*, 474, 272–274.

Bohr N. 1928. The quantum postulate and the recent development of atomic theory. *Nature*, 121, 580–590.

Briggs JS and Eisfeld A. 2011. Equivalence of quantum and classical coherence in electronic energy transfer. *Physical Review E*, 83, 051911–051914.

Busemeyer, JR and Bruza PD. 2012. *Quantum Models of Cognition and Decision*. Cambridge University Press, Cambridge.

Costa JT and Fitzgerald TD. 2005. Social terminology revisited: where are we ten years later? *Annales Zoologici Fennici*, 42, 559–564.

Collini E, Wong CY, Wilk KE, Curmi PMG, Brumer P and Scholes GD. 2010. Coherently wired light-harvesting in photosynthetic marine algae at ambient temperature. *Nature*, 463, 644–648.

Cowan N. 2000. The magic number 4 in short term memory: a reconsideration of mental storage capacity. *Behavioral and Brain Sciences*, 24, 87–125.

Craddock TJA, Hameroff SR and Tuszynski JA. 2012. Cytoskeletal signaling: is molecular memory encoded in microtubule lattices by CaMKII phosphorylation. *PloS Computational Biology*, 8(3), e1002421.

Freeman W and Vitiello G. 2006. Nonlinear brain dynamics as macroscopic manifestation of underlying many-body dynamics. *Physics of Life Reviews*, 3, 93–118.

Fröhlich H. 1968. Long range coherence and energy storage in biological systems. *International Journal of Quantum Chemistry*, 2, 641–649.

Gautam A and Kak S. 2013. Symbols, meaning, and origins of mind. *Biosemiotics*, 6, 301–309.

Gilbert CD and Sigman M. 2007. Brain states: top-down influences in sensory processing. *Neuron*, 54, 677–696.

Gray HB and Winkler JR. 2005. Long-range electron transfer. Proceedings of the National Academy of Sciences USA 102, 3534–3539.

Hameroff S and Penrose R. 2003. Conscious events as orchestrated spacetime selections. *Neuroquantology*, 1, 10–35.

Haven E and Khrennikov AY. 2013. *Quantum Social Science*. Cambridge University Press, Cambridge.

Inoue S and Matsuzawa T. 2007. Working memory of numerals in chimpanzees. *Current Biology*, 17, R1004–1005.

Ishizaki A, Calhoun TR, Schlau-Cohen GS and Fleming GR. 2010. Quantum coherence and its interplay with protein environments in photosynthetic electronic energy transfer. *Physical Chemistry Chemical Physics*, 12, 7319–7337.

Jibu M, Pribram KH and Yasue K. 1996. From conscious experience to memory storage and retrieval: the role of quantum brain dynamics

and boson condensation of evanescent photons. *International Journal of Modern Physics B*, 10, 1735–1754.
Kak S. 1995. Quantum neural computing. In *Advances in Imaging and Electron Physics*, Hawkes P (ed.), Vol. 94, 259–313. Academic Press, USA.
Kak S. 1996. The three languages of the brain: quantum, reorganizational, and associative. In *Learning as Self-Organization*, Pribram K and King J (eds.), 185–219. Lawrence Erlbaum Associates, Mahwah, NJ.
Kak S. 2000. Active agents, intelligence, and quantum computing. *Information Sciences*, 128, 1–17.
Kak S. 2012. Hidden order and the origin of complex structures. In *Origin(s) of Design in Nature*, Swan L, Gordon R and Seckbach J (eds.), 643–652. Springer, Dordrecht.
Kak S. 2013a. Cognitive agents as quantum collectives. *NeuroQuantology*, 11, 391–398.
Kak S. 2013b. Probability constraints and the classical/quantum divide. *NeuroQuantology*, 11, 600–606.
Kak S. 2014. From the no-signaling theorem to veiled non-locality. *NeuroQuantology*, 12, 12–20.
Khrennikov AY. 2010. *Ubiquitous Quantum Structure: From Psychology to Finance*. Springer, Berlin.
Lambert N, Chen YN, Cheng YN, Li CM, Chen GY and Nori F. 2013. Quantum biology. *Nature Physics*, 9, 10–18.
Li M and Vitanyi P. 1997. *An Introduction to Kolmogorov Complexity and its Applications*. Springer-Verlag, New York.
Matsuzawa T. 2013. Evolution of the brain and social behavior in chimpanzees. *Current Opinion in Neurobiology*, 23, 443–449.
Miller WH. 2012. Perspective: Quantum or classical coherence? *The Journal of Chemical Physics*, 136, 210901.
Pascual-Leone J. 1970. A mathematical model for the transition rule in Piaget's developmental stages. *Acta Psychologica*, 32, 301–345.
Polli D, Altoe P, Weingart O, Spillane KM, Manzoni C, Brida D, Tomasello G, Orlandi G, Kukura P, Mathies RA, Garavelli M and Cerullo G. 2010. Conical intersection dynamics of the primary photoisomerization event in vision. *Nature*, 467, 440–443.
Ricciardi LM and Umezawa H. 1967. Brain physics and many-body problems. *Kibernetik*, 4, 44–48.
Ritz T, Thalau P, Phillips JB, Wiltschko R and Wiltschko W. 2004. Resonance effects indicate a radical pair mechanism for avian magnetic compass. *Nature*, 429, 177–180.

Sanabria N, Goring D, Nürnberger T and Dubery I. 2008. Self/nonself perception and recognition mechanisms in plants: a comparison of self-incompatibility and innate immunity. *New Phytologist*, 178(3), 503–514.

Tegmark M. 1998. The interpretation of quantum mechanics: many worlds or many words? *Fortschritte Physics*, 46, 855–862.

Turchin V. 1977. *The Phenomenon of Science: A Cybernetic Approach to Human Evolution*. Columbia University Press, New York.

Turin L. 1996. A spectroscopic mechanism for primary olfactory reception. *Chemical Senses*, 21, 773–791.

Van Lint, JH. 1975. A survey of perfect codes. *Rocky Mountain Journal of Mathematics*, 5(2), 199–224.

Vitiello G. 2009. Coherent states, fractals and brain waves. *New Mathematics and Natural Computation*, 5, 245–264.

Witzany G and Baluska F. 2012. Life's Code Script doesn't code itself. EMBO Reports.

Zurek W. 2003. Decoherence, einselection, and the quantum origins of the classical. *Review of Modern Physics*, 75, 715–775.

Part II

Experimental Approaches

9. Chemical Communication

Ally R. Harari

Department of Entomology, ARO,
Volcani Center, Bet Dagan, Israel
aharari@volcani.agri.gov.il

R. Sharon

Northern Research & Development,
MIGAL Research Institute, Kiryat Shemona,
Ohalo College, Katsrin, Israel
rakefetsh@gmail.com

1. Introduction

1.1 What is chemical communication

Chemicals were present before "individuals" in the primordial soup, and ever since have a role in information transfer in all living creatures (Stowers and Marton, 2005). Thus, chemical communication is probably the most ancient method of information transfer between individuals. Communication that is based on the ability to sense chemicals, both as attractant or deterrent, is shared by all organisms from plants and bacteria to mammals and is therefore probably the oldest sense (Wilson, 1970). Semiochemicals-mediated communication (infochemicals) can be found throughout evolution, in aquatic systems of both marine and freshwater leaving species, in soil dwelling organism and land occupying taxa. Whereas some infochemicals can be detected through a physical contact of both the signaler and the receiver, other chemicals need to be highly soluble in water or air borne to reach other individuals located downstream or downwind. In lakes and ponds or in still air, the odor may disperse relatively slowly and it is often actively pushed by stationary individuals using ventilating or fanning organs as gills, wings, or legs, to reach a nearby conspecific (Weissburg 2000; Kamio *et al.*, 2008).

Chemical communication is superior over other communication routes as vision and sound in several aspects that include: (1) the endless possible options for conveying various simple as well as complex messages, (2) exploitation of the message is costly and involves the development of specific receptors which are adapted to receive and process the information, (3) the temporary nature of the volatile substances ensures updated information, (4) the information can continue to be transferred in the absence of the signaler, and (5) the information is broadcasted to any potential receiver in the active range. Chemical information (infochemicals) can be transferred intentionally (signals) or unintentionally (cues). The diversity, volatility, and stability increase the possibilities of complex massages (Wyatt, 2003).

1.2 Chemical structure

Signals and complex infochemicals have evolved from basal chemical molecules. The olfactory systems in most animals contain nonspecific receptors that allow them to detect a broad range of chemicals. Infochemicals have probably evolved from chemical compounds that originally had other uses, such as host plant odors or waste products. These chemical blends may have led to the increase in survival or reproductive success of the releaser, the receiver, or both individuals of the same species (Wyatt, 2003). The chemicals in use are wildly divers and are largely dependent on the molecule carrying media (aquatic or terrestrial habitats), the function of the information transferred (a cue or a signal, sex pheromone or either of the allelomones, see Fig. 1) and the availability of the infochemical's building materials.

1.3 Infochemicals

Information may be transferred between cells within an organism (hormones, neurotransmitters) or between organisms (semiochemicals) of different species (allelomones), or of the same species (pheromones) (Dicke and Sabelis, 1988; Wyatt, 2003) (Fig. 1).

1.4 Allelomones

The information transferred by allelomones may benefit the emitter (signaler) but not the receiver (allomones), the receiver but not

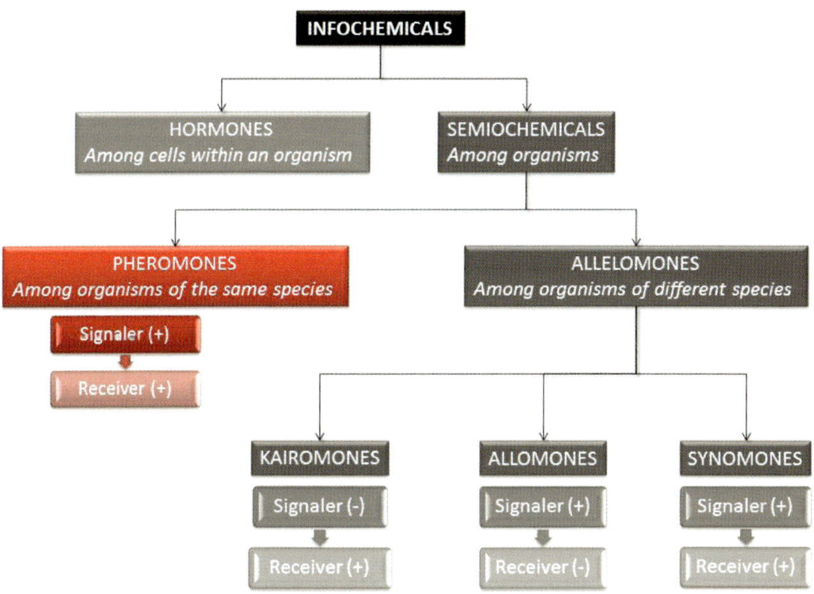

Figure 1. Infochemicals' scheme of cost and benefit. The signaler releases the semiochemical either as a signal or a cue. The receiver accepts the semiochemical and responds. The released and received semiochemicals may benefit (+) or harm (–) the signaler and/or the receiver.

the emitter (kairomones), or both the emitter and the receiver (synomones).

1.4.1 *Allomones*

Volatile chemicals that benefit the emitter but harm the receiver. A typical use of these semiochemicals is to attract prey. Predatory plants, for example, lure their prey, usually insects that feed on decaying animals, by mimicking the smell of rotten meat (Gallie and Chang, 1997). Similarly, plants often use allomones to deceive potential pollinators or prey. The dead horse arum plant, *Helicodiceros muscivorus*, attracts flies by emitting volatile components that very much resemble the smell of a dead horse. This volatile attracts the female blowfly (Calliphoridae) to oviposit on the "dead animal" — an oviposition substance for the female fly. When it leaves the flower, the insect carries the pollen to another plant and fertilizes the female flowers (Stensmyr *et al.*, 2002).

Orchids are known to use volatiles to lure their pollinators by means of sexual deception. For example, the flowers of the spider orchid, *Ophrys sphegodes*, mimic the pheromone blend of its female pollinator bee, *Andrena nigroaenea* (Schiestl *et al.*, 2000). Only males are attracted to the mimetic flower. Upon landing on the flower, the pollenium (a coherent body of pollen grains) attaches to the male bee and is transferred to a new flower that the male bee, again, mistakes for a female bee. Pollination in *Ophrys* is highly specific, with each orchid species pollinated by one or a few closely related bees (Paulus and Gack, 1990).

The bolas spider, *Mastophora hutchinsoni*, uses a similar trick; by mimicking female moth pheromones, the spider attracts its prey, the male moths *Lacinipolia renigera* and *Tetanolita mynesalis*. A mature female spider produces a hanging vertical thread with a terminal sticky globule that emits the pheromones. By swinging the thread through the air the spider female captures the mate-seeking male moths (Haynes *et al.*, 2002).

1.4.2 *Kairomones*

Volatile materials that benefit the receiver at the expense of the emitter. Information that is transferred by one species aimed at eliciting a certain expected response (i.e., a signal) can be used, or exploited (i.e., a cue), by a different species (Carroll *et al.*, 1998; Weissburg *et al.*, 2002; Bruce *et al.*, 2005; Smallegange *et al.*, 2005). Sex pheromones, for example, are aimed at attracting a mate, but are frequently used by eavesdropping predators as kairomones to locate their prey. As an example, the bug *Elatophilus hebraicus* (Heteroptera) uses the sex pheromone of its prey, the pine bast scale *Matsucoccus josephi* (Homoptera), as a kairomone to locate the scale insect on the pine tree. Interestingly, the scale's sex pheromone is also utilized by the predator to aggregate near the scale as a mating point (Dunkelblum *et al.*, 1996).

The male large cabbage white butterfly, *Pieris brassicae*, is attracted to the female-produced sex pheromone. After copulation, the male transfers a "marking" pheromone to the female that informs other males of the mating status of the female (deterring pheromone), and thus promotes female monogamy. The tiny egg parasitoid, *Trichogramma brassicae*, detects the "anti-aphrodisiac" pheromone and uses it as a kairomone: *T. brassicae* rides on the mated female butterfly to the oviposition site, and then parasitizes the freshly oviposited clutch of more than 20 eggs (Fatouros *et al.*, 2005; Huigens *et al.*, 2010).

Bark beetles attack weakened trees using the tree's volatile kairomone. The bark beetles *Dendroctonus brevicomis* (Pearson, 1931) and *Ips paraconfusus* (Elkinton *et al.*, 1981) are attracted to volatile compounds produced by weakened ponderosa pines. When excavating a nuptial chamber in the tree, male *Ips paraconfusus* beetles release an aggregation pheromone (Wood and Bushing, 1963; Pitman *et al.*, 1965; Wood *et al.*, 1966) aimed at recruiting other *I. paraconfusus* beetles in order to weaken the tree's defense and subjugate it for the benefit of beetle colonization (Wood and Vite, 1961; Sallé *et al.*, 2005). However, the aggregation pheromone is also used by the parasitoid wasp *Tomicobia tibialis* to locate and parasitize the adult beetles that are aggregated on the ponderosa pine (Bedard, 1965; Rice, 1969; Brown *et al.*, 1970; Borden, 1974).

1.4.3 *Synomones*

Benefit both the emitter and the receiver; thus, typical synomones are those that are released by a prey to attract the predators of its predator. Several ecosystems involve a prey (plant), its predator (herbivore), and the herbivore's predator. In such tritrophic interactions, the herbivore is attracted by a kairomone emitted by the host plant. The damaged plant releases a pheromone to warn nearby conspecific plants of the presence of herbivores; the informed plants, then, release a synomone, a long-range volatile substance that informs the potential herbivore's predator of the location of its prey (Turlings *et al.*, 1995; Turlings and Fritzsche, 1999; Hilker and Meiners, 2006). This is the story of the lima bean, among others. Lima bean leaves attract the herbivore spider mite *Tetranychus urticae* (with a kairomone); the damaged lima bean leaves emit a blend of volatiles that induce the production and release of a similar blend by undamaged conspecific lima beans (thus a pheromone); this volatile blend attracts the predatory mite, *Phytoseiulus persimilis*, to attack the herbivorous spider mite (thus a synomone) (Choh *et al.*, 2004).

Other examples consist of warning signals that are used to inform a predator that the releaser is aware of its presence and that contact with the emitter bears a potential cost (e.g., poison) to the intruder (Stowe *et al.*, 1995).

1.4.4 *Pheromones*

Communication tools used by many taxa to convey information among individuals of the same species (Karlson and Luscher, 1959; Wyatt, 2003). Accordingly, pheromones are species-specific and can elicit a response in another individual even when emitted in small quantities (Wyatt, 2003). Pheromones are major behavior modifiers; information carried by pheromones has multiple ways to affect a species from egg to adult. Pheromones may provide information about the physical conditions of the releaser, similar to information conveyed though visual or acoustic signals (Insects: Moore *et al.*, 1997; Worden *et al.*, 2000; Rantala *et al.*, 2003; Jaffe *et al.*, 2007; Harari *et al.*, 2011; Lizards: Olsson *et al.*, 2003; López *et al.*, 2006; Mammals: Zala *et al.*, 2004). The unique properties of species-specific pheromones allow for diverse information delivered to the conspecific receivers. The insect class will illustrate the endless routes in which information is transferred using pheromones.

1.5 Pheromones as behavior modifiers in insects

Insect pheromones are often comprised of a blend of several components. The component ratio is crucial for conveying a particular message. For example, females of the sheep moths *Hemileuca eglanterina* and *Hemileuca nuttalli* in geographically distinct (allopatric) populations produce the same active compound in their sex pheromone blend. However, when they share the same geographical area (sympatric populations), *H. eglanterina* females do not produce one of the components, and *H. eglanterina* males actively avoid females which retain this component in their pheromone blend, thereby minimizing mating mistakes and hybrid offspring (McElfresh and Millar, 2001).

1.5.1 *Sex pheromone*

It is generally agreed that sex pheromones provide information about the signaler species and gender (Svensson, 1996; Johansson and Jones, 2007). Recently however, accumulating studies confer that sex pheromones are also deliver information concerning the releaser phenotypic conditions, thereby the chosen mate may maximize the receiver reproductive success (see review by Harari and Steinitz, 2013). Typically, in most sexual organisms, males compete with other males for mating with females: Males, in most organisms, invest less than females in production and upbringing of offspring and invest more in competition over the limited and choosier females (Trivers, 1972). Accordingly, males assume costly displays and attract females for mating, exposing themselves to "eavesdropping" predators and parasites. In contrast, in most moth species, sex pheromone is released by females. The pheromone is highly species specific, released in minute amounts (nanograms) and fast to evaporate, thus selecting against specialist predators and for high quality searching males (Harari and Steinitz, 2013). Male produce sex pheromone occurs in the minority of insect species, and most notably in beetles and flies. In both taxa male produce pheromones are associated with local food availability, such that arriving females gain mate, food and ovipositing sites at the same place and time (Landolt, 1997). Male sex pheromones are often produced in large amounts relative to that of female sex pheromones and are synergized by the host plant volatiles (Landolt, 1997). Often, male produced pheromones attract both males and females to form aggregations in the vicinity of the host plants. This pheromone is termed "aggregation pheromone".

1.5.2 *Aggregation pheromone*

Aggregation pheromones attract conspecific males and females in contrast to sex pheromones that attract the opposite sex only. The aggregation pheromones' release point may serve as a rendezvous site and promote matings. Often the aggregated individuals benefit from other adaptive components of group living, as reduce predation and overcoming host resistance. For example, nymphs of the green stink bug, *Nazara viridula*, aggregate in response to a density regulated pheromone that attracts conspecifics at low to intermediate concentrations but acts as a repellent when concentration is high. The aggregated bugs benefit form synchrony secretion of a repellent bitter substance that deters predators. Natural enemies are significantly less efficient when exposed to aggregated bugs than when preying on single individuals (Lockwood and Story, 1985). The response to aggregation pheromones is often dose dependence, low and intermediate doses are highly attractive whereas high dose of the same pheromone deter conspecifics. Aposematic insects may also benefit from increasing number of individuals in aggregations, having their signal of unpalatable more conspicuous for potential predators (Riipi *et al.*, 2001).

Ovipositing *Drosphila melanogster* into a fruit often escorted with the addition of aggregation pheromone that calls to other females to oviposit at the same fruit. Group feeding of larvae reduces fungal growth in the fruit and increase survival rate of larvae. In addition, ovipositing females introduce yeast to the fruit that also enhance larval growth and survival (Wertheim *et al.*, 2002).

Many bark beetle species use aggregation pheromone to locate a mate in the vicinity of food and ovipositing sites. When approaching a healthy tree, the recruitment of large number of individuals to a synchronized attack, facilitates the subjugation of the defence mechanism of the attacked tree, an impossible task for one individual or a small group. However, the benefit of mass attack in overcoming the host defences reaches an optimum over which additional arriving individuals compete for the limited resources needed for larval development (Schlyter and Birgersson, 1999).

Aggregated individuals are less prone to desiccation (Benoit *et al.*, 2007) and better protected from predators (Wertheim *et al.*, 2005). The common bed bug, *Cimex lectularius* L. (Hemiptera: Cimicidae) is a worldwide ectoparasite, blood feeding pest of human (Panagiotakopulu and Buckland, 1999). After foraging for blood meal all individuals, immature and adults retreat back to their hidden shelter marked by a complex aggregation pheromone of 10 essential components, released primarily by males (Siljander *et al.*, 2008). Furthermore, in the absence of sex pheromone in this species, aggregation may facilitate mate finding (Wertheim *et al.*, 2005) when females are engorged with blood that is essential for egg laying.

1.5.3 *Trail marking pheromone*

Group leaving activities may involve a coordinating response of numerous individuals in the absence of control center. Trail pheromone deposits by forager ants often recruits other nest-mates to assist in carrying a large item or exploiting a patch of valuable food (Beckers *et al.*, 1993). The type of food, whether is long lasting resource or a short time episode, dictates the nature of the trail pheromone; volatile of low molecular weight will fast evaporate, whereas heavier molecules will last for longer time. Army ants colonies consist of up to 20 million individuals are practically blind, thus communication among individuals is restricted to chemical and tactile stimuli. Swarms of ten to hundreds of thousands raiding foragers are formed daily in a search for food that is brought back to the nest. This massive movement is coordinated by a single pheromone. A number of scout ants leave the nest, those that have found a food patch return home, depositing a trail pheromone. The pheromone volume characterizes the food size and quality and according to the information provided recruits mass followers that further enhance the volume of the signal reinforcing its reliability (Deneubourg *et al.*, 1989).

Trail pheromone is also used for nest relocation in *Myrmecina nipponica* ants. First individual ants search, independently, for a suitable place for a new nest. When a proper site is located a trail pheromone recruits other conspecifics to assess the property. When returning to the origin nest, each deposits more of the trail pheromone that elicits more visits of ants in the new location. When a certain quorum is achieved a consensus decision is made and the nest is relocated to the new site by transferring the brood and the queen (Cronin, 2012).

1.5.4 *Territorial marking pheromone*

The desert ant, *Cataglyphis niger,* does not rely on trail pheromone to recruit nest-mates for food collection. Alternatively they exploit ephemeral and unpredicted food sources such as insect carcasses. The highly valuable stored food in the otherwise harsh semi-desert environment calls for intruders attempt to monopolize the resources. The owner ants mark their territory with a pheromone, originated in the cloacal gland. In the presence of the pheromone, the resource holders, having the "owner advantage", win in combats with conspecific intruders regardless of size asymmetry (Wenseleers *et al.*, 2002).

The European beewolf, *Philanthus triangulum,* male wasps apply pheromone on small plants in the perimeter and inside their small territories and defend it against competing males. Females approach the territory and mating occurs within the territory although no resources are provided to the females. Apparently females choose the males based on information carried through the volatile pheromone as male age (Kaltenpoth, 2006), kinship and geographical distance in order to avoid inbreeding (Herzner *et al.*, 2006; Kaltenpoth *et al.*, 2007).

1.5.5 *Host marking pheromone*

In certain occasions in an insect's life time a species or individual recognition has an adaptation. For example, when an egg is oviposited into a small host a fierce intraspecific competition for food may evolve, resulting with the death of both larvae. Often, ovipositing females apply host marking pheromone to deter self-re-ovipositing and other ovipositing females (Roitberg and Mangel, 1988). This is the case of the European berry moth, *Lobesia botrana*, when ovipositing one egg on a grape berry (Gabel and Thiéry, 1992), and of the parasitoid wasp *Anaphes victus*, that oviposits one egg in an egg of the weevil *Listronotus oregonensis*. A conspecific female learns to associate the marking pheromone with an "occupied" egg. Interestingly, this associative learning is faster when the two females are close kin (van Baaren and Boivin, 1998).

1.5.6 *Individual odor recognition*

Social insects may recognize their species, nest-mates, cast, age, and gender based on hydrocarbon carrying pheromones. The released pheromone can be genetic and/or environmentally induced by food, nest activities or other nest-mates. Acknowledging a nest-mate or kin can be learned through multiple encounters or matching own's odor to that of others ("armpit" effect). The latter is more common for individuals that are not reared among kin but may benefit from inbreeding avoidance if kin are encountered when sexually mature. In honeybee hives, forager bees leave the hive daily and are back with food. Nest-mate recognition is essential to avoid invasion of parasites and predators. Bees, guarding at the hive entrance, recognize intruders through their different pheromone and prevent their entry (Smith and Breed, 1995). There is evidence that honeybees in hives of multiply mated queens can distinguish between full sisters, half-sisters, no-sister nest-mates and foreigners based on their own "armpit" (genetic) pheromone and the complex, exogenic, environmentally based pheromone (Getz *et al.*, 1986). Parasite species that manage to enter the hive change the chemical profile of their own cuticle to match that of their host (Lenoir *et al.*, 2001).

1.5.7 *Alarm pheromones*

Alarm signals are aimed to alert conspecific of potential danger. The pheromone is usually produced by one individual and affects the behavior of several conspecifics. Alarm pheromones are wide spread in insects, second only to sex pheromones. Alarm pheromone is often released by offspring-tending bug-mothers in the presence of a predator. In clonal aphids alarm signals are well developed. In response to pheromone which can be released by an attacked larvae or the mother all individuals stop feeding, run away from the predator or jump from the leaf. Moreover, frequent exposure to alarm pheromone induces females in a group to produce a higher proportion of winged morphs in their offspring increasing their potential to fly away from the risk-prone site (Kunert *et al.*, 2005).

1.6 Pheromone activities through evolution

The use of pheromones is not unique for insects and can be detected throughout the evolutionary phylogenetic tree of the animal kingdom (Fig. 2).

Following are some examples demonstrating the use of pheromones as infochemicals in different taxa along the evolution tree:

1.6.1 *Bacteria — Autoinduced quorum sensing pheromone*

Bacteria, single cell prokaryots, use pheromones to manipulate other bacterial populations in the community and to alter their

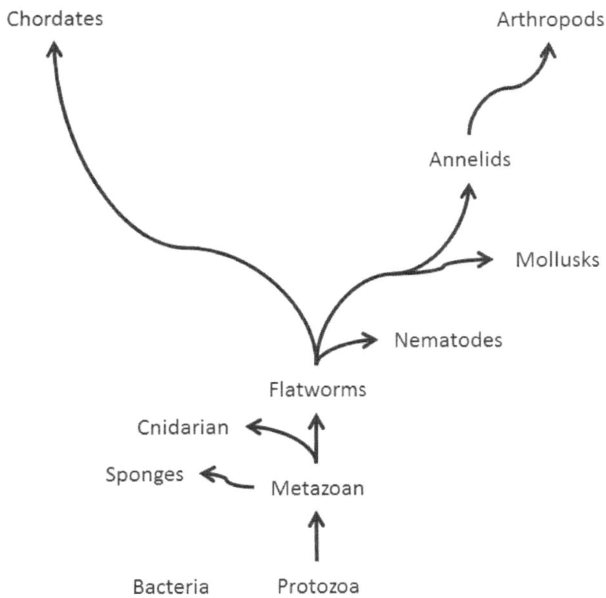

Figure 2. Evolutionary tree of the animal kingdom (Taken form Ruppert E.E. and Barmes R.D. 1994. Invertebrate Zoology, 6th Ed. Saunders College Publishing).

host's traits (Tashiro *et al.*, 2013). Some bacteria secrete small, species-specific, signaling molecules called autoinducers. These molecules enable the individual organism to monitor the density of its population by measuring the concentration of an autoinducer which correlates with population density. The mechanism of sensing and secreting autoinducers is called quorum sensing (Miller and Bassler, 2001; Sbarbati and Osculati, 2006). In pathogenic bacteria, quorum sensing can help minimize host immune responses by delaying the production of virulence factors until there are sufficient bacteria to overwhelm the host's defense mechanisms (Kaufmann *et al.*, 2005).

1.6.2 *Protista — aggregation pheromone in amoeba*

The amoeba (Entamoeba invadens; Phylum Rhizopoda; subphylum Sarcodina), single cell eukaryote protozoans, use aggregation pheromone to recruited neighboring 'midwife' cells to mechanically intervene and help daughter cells that were stuck in the process of cell division (Biron *et al.*, 2001). The postmitotic cell divides into two daughter cells through a process termed cytokinesis (Nagasaki and Uyeda, 2008). In amoebas, most of the asexual division is spontaneous and cells reproduce by themselves but, as has been shown for *Entamoeba invadens* (Biron *et al.*, 2001), a third of the cells become stuck in the process: the last phase of cleavage furrow formation may stop before separation is complete. In this situation, a 'midwife' is called to help complete the separation process. More specifically, the two daughter cells are attached by a tubular tether that needs to be mechanically severed to complete the division process (Biron *et al.*, 2001). About one-third of the dividing cells cannot sever the leash. The environment of these dividing cells is enriched with phosphatidylinositol trisphosphate which calls in other cells (midwives) to help. Thus the 'midwife' cells are chemotactically recruited to intervene mechanically. Perception of the signal is connected to the "b" subunit of the trimeric "G" protein (Gb), as evidenced by the fact that cells lacking the Gb subunit do not respond to the call for help. The 'midwife' then proceeds to rupture the connection by bouncing into the connecting tether; once it has been severed, all three amoebas move apart.

A similar phenomenon has been observed in dividing cells of the slime mold *Dictyostelium discoideum* (Nagasaki and Uyeda, 2008). Insall *et al.* (2001) and Rohlfs *et al.* (2007) reported that interphase cells of *D. discoideum* also often move toward the connecting region of a neighboring dividing cell, although in this organism most of the mitotic cells complete division without help from 'midwife' cells. As *Entamoeba* and *Dictyostelium* are phylogenetically distinct, these reports suggest that 'midwifing' is a widespread feature in the cytokinesis of animal and amoebic cells.

1.6.3 *METAZOA*

1.6.3.1 *Cnidaria — alarm pheromone in sea anemone*

Organisms in the phylum Cnidaria are the first to form cells that are organized into true tissues. *Anthopleura elegantissima* is a sea anemone (Cnidaria: anthozoa) that is about 2 cm in diameter and feeds on intertidal rocks. When a sea anemone, *Anthopleura elegantissima*, is wounded by a predator, it releases an alarm pheromone that warns neighboring anemones of the nearby danger. Upon receiving the signal the anemone shakes its tentacles and fast pulled them into its mouth cavity and close the opening (Agosta, 1994). When the alarm signal is no longer present, the anemone opens its mouth and resumes its normal activity in about two hours.

A wounded anemone at the upper end of a stream was found to affect downstream anemones (Howe and Sheikh, 1975). The anemones in the group nearest the wounded animal contracted first, and then each group downstream contracted as the flowing water carrying the pheromone reached it. This reaction provided a sensitive bioassay for the pheromone and with its help, the investigators were able to fractionate an extract of homogenized sea anemone and isolate colorless needle like crystals which were named anthopleurine. This is an exceptional pheromone in that it is an ionic compound. The cation is the active signal, and it triggers an alarm response in seawater.

1.6.4 *Nematodes — larval development pheromone*

Worms are the next branch in the evolutionary tree and are presented here by the nematode species *Caenorhabditis elegans* which uses pheromones to mediate larval development under harsh environmental conditions (Butcher *et al.*, 2007; Bone and Shorey, 1978). When growth conditions are good for larval stages precede the sexually matured adult. In harsh conditions (i.e., high population density, low food availability, high temperature), however, a dauer larva is produced. These larvae have a thickened cuticle, they do not feed and survive on fat stores. Dauers of *C. elegans* can survive for several months and precede maturation to adults only when conditions improve (Bone and Shorey, 1978; Butcher *et al.*, 2007).

The fatty acid-like pheromone enhances dauer larval formation and inhibits progression to the adult stage. The signal is probably a combination of pheromone concentration and availability of food, monitored by the chemosensory organs of the stage 2 and dauer larvae. The integration of these opposing signals determines the course of larval development (Bone and Shorey, 1978). Pheromone-mediated control of dauer larval formation is not unique to *C. elegans*: at least two other dauer-forming species (of unknown genus) produce a different pheromone (Ku, 1982).

1.6.5 *Mollusca — aggregation pheromone in gastropod*

Mollusks can be found in terrestrial, marine and freshwater habitats, although most of them are marine. Pheromones are associated with the control of mollusk behavior. One example is a family of waterborne pheromonal attractant peptides (attractions) in the gastropod *Aplysia*. *Aplysia* is a marine hermaphrodite mollusk that does not fertilize its own eggs (Cummins *et al.*, 2006). During reproductive time a pheromone (attractin) is released to attract the otherwise solitary *Aplysia* and aggregate into 'brothels' to form egg-laying and mating aggregations. The attractions are the first discovered waterborne peptide or protein pheromone family in invertebrates and vertebrates that are not species-specific (Cummins *et al.*, 2006).

1.6.6 *Arthropoda — sex pheromone in crustaceans*

In the arthropods, crustaceans, pheromones are often excreted in the urine and detected downstream by conspecifics, or the molecules are located on the sender's body surface (contact pheromones) and are detected by conspecifics only upon contact with the releaser.

In *Telmessus cheiragonus*, the females release a sex pheromone that elicit mate searching, following by precopulatory guarding, in which the male grasps and carries the female until she molts. Mating occurs right after molting and before the carapace hardens. After copulation, the male guards the female for a while and then leaves, but not before depositing plugs into the female gonopores, presumably to prevent her from remating (Kamio *et al.*, 2002). In the swimming crab, *Arenaeus cribrarius*, the male carries the female in the precopulatory position for 30 days, until she finally molts. After copulation, which may last 17 hours, the male continues to carry his soft-shelled mate for an additional 30 days until her carapace hardens, guarding her from male competitors (Pinheiro and Fransozo, 1999).

1.6.7 *Vertebrates — central nervous system involvement in pheromone detection*

In vertebrates (phylum chordata, subphylum vertebrata) the main olfactory system is devoted chiefly to the detection of volatile substances, while the accessory (vomeronasal) olfactory system is involved in the detection of pheromones (Sbarbati and Osculati, 2006; Witt and Hummel, 2006).

Reaction to chemical cues in vertebrates involves the central nervous system, in particular the amygdala and hippocampus (Takahashi *et al.*, 2005). Typically, semiochemicals emitted by vertebrates are based on steroids, which are modified triterpenes (Singer, 1991; Sbarbati and Osculati, 2006).

1.6.8 *Fish — pheromone-mediated sex change*

Some fish have sex changes. Pheromone-mediated sex changes is common in fish. Many fish species, including wrasse, parrotfish and groupers, change their sex during their lifetime, in accordance with higher potential reproductive success they may achieve as a male or female. In the cleaner wrasse, a group of fish typically has one big male and many smaller females. The larger male, which monopolizes all of the females, has a larger reproductive success than each individual female in the group. The male releases a pheromone advertising his quality and suppressing the development of other males. Smaller males are chased away from the group and do not mate. Thus, small fish will maximize their reproductive success as females by mating with the larger male. When the male dies, the pheromone titer in the water rapidly declines and the largest female quickly changes its sex; in a matter of hours she becomes a male that suppresses the development of other fish into males (Roughgarden, 2004).

1.6.9 *Amphybia — sex recognition and signaling phromones*

In amphibian urodeles (tritons, newts and salamanders), sex recognition and courtship are regulated through pheromones (Kikuyama *et al.*, 2002). During courtship, in salamanders, which is performed on land, the female follows the male's tail, resting her chin on his tail base. In the red-legged salamander, *Plethodon shermani*, males utilize proteinaceous pheromones from the mental (submandibular) gland (Rollmann *et al.*, 1999) that affects female behavior and reduces the duration of courtship. The two major proteins, PRF and plethodontid modulating factor (PMF), exist as multiple isoforms within the gland of an individual male (Kikuyama *et al.*, 2002). A remarkable diversity of isoform ratios was found among males of a single population, with probable functional effects on reproductive behavior in this family of salamanders. The male turns back and slaps his chin on the female's nares (Chouinard *et al.*, 2013) and then deposits a spermatophore containing his sperm on the ground. The female picks it up through her cloaca for internal fertilization. The pheromone situated in a gland underneath the male's chin enhances the female's receptivity and decreases the time spent on courtship.

1.6.10 *Mammals — reproduction and social behavior pheromones*

The most studied pheromones in mammals other than human are in the house mice (*Mus musculus*/*Mus domesticus*). The pheromones in the urine of *M. musculus* were the first mammalian signals to be studied. These pheromones mediate diverse behavioral responses, such as sex recognition, suppression or synchronization of estrus or delayed puberty among females (Vandenbergh *et al.*, 1974, 1975;

Thompson *et al.*, 2004; Kimoto *et al.*, 2005), and advertisement of territory and aggression responses between males (Hurst and Beynon, 2004).

A most interesting pheromone was found in male's mice urine, a protein pheromone named Darcin, after Darcy, the attractive hero in Jane Austen's novel "Pride and Prejudice" from 1813. Females' contact with the volatile pheromone elicits a learned attraction, such that exposed females will repeatedly attract to the scent of this specific male, the signaler of the pheromone, but not to other conspecific males (Roberts *et al.*, 2010). Darcin is the first species-specific pheromone that stimulates learning and repeatedly attracts females to individual males. This pheromone is the closest resemblance, so far, to "pheromone of love".

2. Conclusions

This chapter illustrates the universal use of infochemicals to transfer information whether as signals or cues between and within the various biological taxa, allowing for multiplicity, complexity and simplicity of the communicated information.

Semiochemicals are used in many taxa across all life forms as the main mode of information transfer. Infochemicals are wildly divers and are largely dependent on the molecules carrying media (whether the habitat is aquatic or terrestrial), the function of the information transferred (a cue or a signal, a pheromone, or either of the allelomones) and the available building materials. In water, soluble components should be used as small steroid-based and large polar molecules, whereas, in air fast evaporating carbon molecules with low molecular weight should be implemented. Synomones released by plants to recruit the predators of their herbivores share a common molecular structure, while sex pheromone should be highly species-specific to avoid mating mistakes resulting in reduced fitness of hybrid offspring. The expected life time of the information also influences the chosen molecules; alarm pheromones should be fast evaporated but fast dissolved when danger has disappeared, whereas mammalian territorial scent marking and ant trails should persist for a longer time (Wyatt, 2003). Infochemicals are products of coevolution between the signaler and the receiver, sometimes for the benefit of them both (Synomones and pheromones), but in other times for the benefit of the signaler alone (allomones) or that of the receiver (kairomones). In all cases however, coevolution is a continuous process leading to a subtle but constant changes in the infochemical characteristic under pressure of selection.

Further reading: To further understand the evolution and function of pheromones as infochemicals we suggest the excellent book of T.D Wyatt (2003) Pheromone and animal behavior, communication by small and taste. For infochemicals in aquatic systems we highly recommend on Bronmark and Hansson edited book (2012) *Chemical Ecology in Aquatic Systems*. For further reading on chemical communication we suggest d'Ettorre and Hughes edited book (2008) *Sociobiology of communication*.

Acknowledgment

We thank Matthew Symonds, Deakin University, Australia, for his comments on earlier draft of the chapter.

References

Agosta WC. 1994. Using chemicals to communicate. *Journal of Chemical Education*, 3, 242–246.

Beckers R, Deneubourg JL and Goss S. 1993. Modulation of trail laying in the antLasius niger (Hymenoptera: formicidae) and its role in the collective selection of a food source. *Journal of Insect Behavior*, 6(6), 751–759.

Bedard WD. 1965. The biology of *Tomicobia tibialis* parasitizing *Ips confusus* in California. *Contribution from Boyce Thompson Institute*, 23, 77–81.

Benoit JB, Del Grosso NA, Yoder JA and Denlinger DL. 2007. Resistance to dehydration between bouts of blood feeding in the bed bug, Cimex lectularius, is enhanced by water conservation, aggregation, and quiescence. *The American Journal of Tropical Medicine and Hygiene*, 76, 987–993.

Biron D, Libros P, Sagi D, Mirelman D, Moses E. 2001. Asexual reproduction 'Midwives' assist dividing amoebae. *Nature*, 410, 430–430.

Bone LW, Shorey HH. 1978. Nematode sex pheromones. *Chemical Ecology*, 4, 595–612.

Borden JH. 1974. Aggregation pheromones in the Scolytidae. Pheromones. North-Holland, Amsterdam. 135–160.

Brönmark C and Hansson LA. 2012. *Chemical Ecology in Aquatic Systems*. Oxford University Press, Oxford, New York.

Brown WL Jr., Eisner T and Whittaker RH. 1970. Allomones and kairomones: trans-specific chemical messengers. *Bioscience*, 20, 21–22.

Bruce TJ, Wadhams LJ and Woodcock CM. 2005. Insect host location: a volatile situation. *Trends in Plant Science*, 10, 269–274.

Butcher RA, Fujita M, Schroeder FC and Clardy J. 2007. Small-molecule pheromones that control dauer development in Caenorhabditis elegans. *Nature Chemical Biology*, 3, 420–422.

Carroll JF, Mills GD. Jr. and Schmidtmann ET. 1998. Patterns of activity in host-seeking adult *Ixodes scapularis* (*Acari*: *Ixodidae*) and host-produced kairomones. *Journal of Medical Entomology*, 35, 11–15.

Choh Y, Shimoda T, Ozawa R, Dicke M and Takabayashi J. 2004. Exposure of lima bean leaves to volatiles from herbivore-induced conspecific plants results in emission of carnivore attractants: active or passive process? *Journal of Chemical Ecology*, 30, 1305–1317.

Cronin AL. 2012. Consensus decision making in the ant Myrmecina nipponica: house-hunters combine pheromone trails with quorum responses. *Animal Behavior*, 84(5), 1243–1251.

Cummins SF, Nichols AE, Schein CH and Nagle GT. 2006. Newly identified water-borne protein pheromones interact with attractin to stimulate mate attraction in *Aplysia*. *Peptides*, 27, 597–606.

Deneubourg JL, Goss S, Franks N and Pasteels JM. 1989. The blind leading the blind: modeling chemically mediated army ant raid patterns. *Journal of insect behavior*, 2(5), 719–725.

d'Ettorre P and Huges DP. 2008. *Sociobiology of Communication, an Interdisciplinary Perspective*. Oxford University Press, Oxford, New York.

Dicke M and Sabelis MW. 1988. Infochemical terminology: based on cost-benefit analysis rather than origin of compounds? *Functional Ecology*, 2, 131–139.

Dunkelblum E, Mendel Z, Gries G, Gries R, Zegelman L, Hassner A and Mori K. 1996. Antennal response and field attraction of the predator *Elatophilus hebraicus* (Hemiptera: Anthocoridae) to sex pheromones and analogues of three *Matsucoccus* spp. (Homoptera: matsucoccidae). *Bioorganic and Medical Chemistry*, 4, 489–494.

Elkinton JS, Wood DL and Browne LE. 1981. Feeding and boring behavior of the bark beetle, *Ips paraconfusus*, in extracts of ponderosa pine phloem. *Journal of Chemical Ecology*, 7, 209–220.

Fatouros NE, Huigens ME, van Loon JJ, Dicke M and Hilker M. 2005. Chemical communication: butterfly anti-aphrodisiac lures parasitic wasps. *Nature*, 433, 704–704.

Gabel B and Thiéry D. 1992. Biological evidence of an oviposition-deterring pheromone in Lobesia botrana, Den. et Schiff. (Lepidoptera, Tortricidae). *Journal of Chemical Ecology*, 18(3), 353–358.

Gallie DR and Chang SC. 1997. Signal transduction in the carnivorous plant *Sarracenia purpurea*: regulation of secretory hydrolase expression during development and in response to resources. *Plant Physiology*, 115, 1461–1471.

Getz WM, Brfickner D and Smith KB. 1986. Conditioning honeybees to discriminate between heritable odors from full and half-sisters. *Journal of Comparative Physiology A*, 159, 251–256.

Harari AR and Steinitz H. 2013. The evolution of female sex pheromones. *Current Zoology*, 59(4), 569–578.

Harari AR and Steinitz H. 2013. The evolution of female sex pheromones. *Current Zoology*, 59, 567–576.

Harari AR, Zahavi T and Thiery D. 2011. Fitness cost of pheromone production in signaling female moths. *Evolution*, 65, 1572–1582.

Haynes KF, Gemeno C, Yeargan KV, Millar JG and Johnson KM. 2002. Aggressive chemical mimicry of moth pheromones by a bolas spider: how does this specialist predator attract more than one species of prey? *Chemoecology*, 12, 99–105.

Herzner G, Schmitt T, Heckel F, Schreier P and Strohm E. 2006. Brothers smell similar: Variation in the sex pheromone of male European beewolves *Philanthus triangulum* F. (Hymenoptera: crabronidae) and its implications for inbreeding avoidance. *Biological Journal of the Linnean Society*, 89, 433–442.

Hilker M and Meiners T. 2006. Early herbivore alert: insect eggs induce plant defense. *Journal of Chemical Ecology*, 32, 1379–1397.

Howe NR and Sheikh YM. 1975. Anthopleurine, a sea anemone alarm pheromone. *Science*, 189, 386–388.

Huigens ME, Woelke JB, Pashalidou FG, Bukovinszky T, Smid HM and Fatouros NE. 2010. Chemical espionage on species-specific butterfly anti-aphrodisiacs by hitchhiking *Trichogramma* wasps. *Behavioral Ecology*, 21, 470–478.

Hurst JL and Beynon RJ. 2004. Scent wars: the chemobiology of competitive signalling in mice. *Bioessays*, 26, 1288–1298.

Insall R, Muller-Taubenberger A, Machesky L, Kohler J, Simmeth E, Atkinson SJ, Weber I and Gerisch G. 2001. Dynamics of the Dictyostelium Arp2/3 complex in endocytosis, cytokinesis, and chemotaxis. *Cell Motil Cytoskeleton*, 50, 115–128.

Jaffe K, Mir´as B and Cabrera A. 2007. Mate selection in the moth *Neoleucinodes elegantalis*: evidence for a supernormal chemical stimulus in sexual attraction. *Animal Behavior*, 73, 727–734.

Johansson BG, Jones TM. 2007. The role of chemical communication in mate choice. Biol. Rev. Camb. Philos. Soc, 82, 265–289.

Kaltenpoth M and Strohm E. 2006. The scent of senescence: age-dependent changes in the composition of the marking pheromone of the male European beewolf, *Philanthus triangulum*. *Journal of Insect Science*, 6, 20.

Kaltenpoth M, Kroiss J and Strohm E. 2007. The odor of origin: kinship and geographical distance are reflected in the marking pheromone of male beewolves (*Philanthus triangulum* F., hymenoptera, crabronidae). *BMC Ecology*, 7(1), 11.

Kamio M, Matsunaga S and Fusetani N. 2002. Copulation pheromone in the crab Telmessus cheiragonus (Brachyura: Decapoda). *Marine Ecology Progress Series*, 234, 183–190.

Kamio M, Reidenbach MA and Derby CD. 2008. To paddle or not: context dependent courtship display by male blue crabs, *Callinectes sapidus*. *Journal of Experimental Biology*, 211, 1243–1248.

Karlson P and Luscher M. 1959. Pheromones: A new term for a class of biologically active substances. *Nature*, 183, 55–56.

Kaufmann GF, Sartorio R, Lee RSH, Rogers CJ, Meijler MM, Moss JA, Clapham B, Brogan AP, Dickerson TJ and Janda KD. 2005. Revisiting quorum sensing: discovery of additional chemical and biological functions for 3-oxo-N-acylhomoserine lactones. *Proceedings of the National Academy of Science* USA, 102, 309–314.

Kikuyama S, Yamamoto K, Iwata T and Toyoda F. 2002. Peptide and protein pheromones in amphibians. *Comparative Biochemistry and Physiology*. Part B, 132, 69–74.

Kimoto H, Haga S, Sato K and Kazushige T. 2005. Sex-specific peptides from exocrine glands stimulate mouse vomeronasal sensory neurons. *Nature*, 437, 6898-6901.

Ku DD. 1982. A pheromone influences larval development in the nematode caenorhabditis elegans. *Science*, 218, 578–580.

Kunert G, Otto S, Röse US, Gershenzon J and Weisser WW. 2005. Alarm pheromone mediates production of winged dispersal morphs in aphids. *Ecology Letters*, 8(6), 596–603.

Landolt PJ and Phillips TW. 1997. Host plant influences on sex pheromone behavior of phytophagous insects. *Annual Review of Entomology*, 42, 371–391.

Lenoir A, D'Ettorre P, Errard C and Hefetz A. 2001. Chemical ecology and social parasitism in ants. *Annual Review of Entomology*, 46(57346), 57. López P, Amo L and Martin J. 2006. Reliable signaling by chemical cues of male traits and health state in male lizards, *Lacerta monticola*. *Journal of Chemical Ecology*, 32, 473–488.

Lockwood JA and Story RN. 1985. Bifunctional pheromone in the first instar of the southern green stink bug, *Nezara viridula* (L.) (Hemiptera: pentatomidae): its characterization and interaction with other stimuli. *Annuals of the Entomological Society of America*, 78, 474–479.

McElfresh JS and Millar JG. 2001. Geographic variation in the pheromone system of the saturniid moth *Hemileuca eglanterina*. *Ecology*, 82, 3505–3518.

Moore PJ, Reagan-Wallin, NL, Haynes KF and Moore AJ. 1997. Odour confers status on cockroaches. *Nature*, 389, 25.

Nagasaki A and Uyeda TQP. 2008. Chemotaxis-mediated scission contributes to efficient cytokinesis in dictyostelium. *Cell Motility and Cytoskeleton*, 65, 896–903.

Olsson M, Madsen T, Nordby J, Wapstra E, Ujvari B and Wittsell H. 2003. Major histocompatibility complex and mate choice in sand lizards. *Proceedings of the Royal Society of London* B, 270(2), S254–S256.

Panagiotakopulu E and Buckland PC. 1999. Cimex lectularius L., the common bed bug from Pharaonic Egypt. *Antiquity*, 73, 908–911.

Paulus HF and Gack C. 1990. Pollinators as prepollinating isolation factors: evolution and speciation in *Ophrys* (Orchidaceae). *Israel Journal of Botany*, 39, 43–79.

Pearson HL. 1931. Theory in explanation of the selection of certain trees by the western pine beetle. *Journal of Forestry*, 29, 696–699.

Pinheiro MAA and Fransozo A. 1999. Reproductive behavior of the swimming crab *Arenaeus cribrarius* (Lamarck, 1818) (Crustacea, brachyura, portunidae) in captivity. *Bulletin of Marine Science*, 64, 243–253.

Pitman GB, Kliefoth RA and Vite IP. 1965. Studies of the pheromone of *Ips confusus* (leConte). II. Further observations on the site of production. Contributions Boyce Thompson Institute, 23, 13–17.

Rantala MJ, Kortet R, Kotiaho JS, Vainilla A and Suhonen J. 2003. Condition dependence of pheromones and immune function in the grain beetle *Tenebrio molitor. Functional Ecology*, 17, 534–540.

Rice RE. 1969. Response of some predators and parasites of *Ips confusus* (LeC.) to olfactory attractants. *Contributions from* Boyce Thompson Institute, 24, 189–194.

Riipi M, Alatalo RV, Lindstrom L and Mappes J. 2001. Multiple benefits of gregariousness cover detectability costs in aposematic aggregations, *Nature*, 413, 512–514.

Rohlfs M, Arasada R, Batsios P, Janzen J and Schleicher M. 2007. The Ste20-like kinase SvkA of dictyostelium discoideum is essential for late stages of cytokinesis. *Journal of Cell Science*, 120, 4345–4354.

Roitberg BD and Mangel M. 1988. On the evolutionary ecology of marking pheromones. *Evolutionary ecology*, 2(4), 289–315.

Rollmann SM, Houck LD and Feldhoff RC. 1999. Proteinaceous pheromone affecting female receptivity in a terrestrial salamander. *Science*, 285, 1907–1909.

Roughgarden J. 2004. *Evolution's Rainbow: Diversity, Gender, and Sexuality in Nature and People.* University of California Press, Berkeley, CA.

Ruppert EE and Barmes RD. 1994. Invertebrate Zoology. 6th Edition. Saunders College Publishing, USA.

Sallé A, Monclus R, Yart A, Garcia J, Romary P and Lieutier F. 2005. Fungal flora associated with *Ips typographus*: frequency, virulence, and ability to stimulate the host defence reaction in relation to insect population levels. *Canadian Journal of Forest Research*, 35, 365–373.

Sbarbati A and Osculati F. 2006. Allelochemical communication in vertebrates: kairomones, allomones and synomones. *Cells Tissues Organs*, 183, 206–219.

Schiestl FP, Ayasse M, Paulus HF, Löfstedt C, Hansson BS, Ibarra F and Francke W. 2000. Sex pheromone mimicry in the early spider orchid (*Ophrys sphegodes*): patterns of hydrocarbons as the key mechanism for pollination by sexual deception. *Journal of Comparative Physiology*. A, 186, 567–574.

Schlyter F and Birgersson G. 1999. Forest Beetles. *In Pheromones in Non-Lepidopteran Insects Associated with Agricultural Plans*. Hardie J and Minks AK (eds.), CAB International, Oxford, United Kingdom.

Siljander E, Gries R, Khaskin G and Gries G. 2008. Identification of the airborne aggregation pheromone of the common bed bug, Cimex lectularius. *Journal of Chemical Ecology*, 34(6), 708–718.

Singer AG. 1991. A chemistry of mammalian pheromones. *The Journal of Steroid Biochemistry and Molecular Biology*, 39, 627–632.

Smallegange RC, Qiu YT, van Loon JJ and Takken W. 2005. Synergism between ammonia, lactic acid and carboxylic acids as kairomones in the host-seeking behavior of the malaria mosquito Anopheles gambiae sensu stricto (Diptera: Culicidae). Chemical Senses, 30, 145–152.

Smith BH and Breed MD. 1995. The chemical basis for nestmate recognition and mate discrimination in social insects. In *Chemical Ecology of Insects* 2, 287–317. Springer, US.

Stensmyr MC, Urru I, Collu I, Celander M, Hansson BS and Angioy AM. 2002. Pollination: rotting smell of dead-horse arum florets. *Nature*, 420, 625–626.

Stowe MK, Turlings TC, Loughrin JH, Lewis WJ and Tumlinson JH. 1995. The chemistry of eavesdropping, alarm, and deceit. *Proceeding of the Natlonal Academy of Sciences*, USA, 92, 23–28.

Stowers L and Marton TF. 2005. What is a pheromone? Mammalian pheromone reconsidered. *Neuron*, 46, 699–702.

Svensson M. 1996. Sexual selection in moths: the role of chemical communication. *Biological Reviews*, 71(1), 113–135.

Takahashi LK, Nakashima BR, Hong H and Watanabe K. 2005. The smell of danger: a behavioral and neural analysis of predator odor-induced fear. *Neuroscience and Biobehavioral Reviews*, 29, 1157–1167.

Tashiro Y, Yawata Y, Toyofuku M, Uchiyama H, and Nomura N. 2013. Interspecies interaction between *Pseudomonas aeruginosa* and other microorganisms. *Microbes and Environment*, 28, 13–24.

Thompson RN, Robertson BK, Napier A and Wekesa K. 2004. Sex specific response to urinary chemicals by the mouse vormeronasal organ. *Chemical Senses*, 29, 749–754.

Trivers RL. 1972. Parental investment and sexual selection, 136–179. *In Sexual Selection and the Descent of Man*, Campbell B (ed.), 1871–1971. Heinemann, London.

Turlings TC and Fritzsche ME. 1999. Attraction of parasitic wasps by caterpillar-damaged plants. Novartis Foundation Symposia, 223, 21–32.

Turlings TC, Loughrin JH, McCall PJ, Rose US, Lewis WJ and Tumlinson JH. 1995. How caterpillar-damaged plants protect themselves by attracting parasitic wasps. *Proceedings of the National Academy of Sciences*. USA, 92, 4169–4174.

van Baaren J and Boivin G. 1998. Learning affects host discrimination behavior in a parasitoid wasp. *Behavioral Ecology and Sociobiology*, 42(1), 9–16.

Vandenbergh JG. 1994. Pheromone and mammalian reproduction. In *Physiology of Reproduction*, Knobil E and Neill JD (eds.). Vol. 2, 343–362. Raven Press, New York.

Vandenbergh JG, Whitsett JM and Lombardi JR. 1975. Partial isolation of a pheromone accelerating puberty in female mice. *Journal of Reproduction and Fertility*, 43, 515–523.

Weissburg MJ. 2000. The fluid dynamical context of chemosensory behavior. *The Biological Bulletin*, 198, 188–200.

Weissburg MJ, Ferner MC, Pisut DP and Smee DL. 2002. Ecological consequences of chemically mediated prey perception. *Journal of Chemical Ecology*, 28, 1953–1970.

Wenseleers T, Billen J and Hefetz A. 2002. Territorial marking in the desert ant Cataglyphis niger: does it pay to play bourgeois? *Journal of Insect Behavior*, 15(1), 85–93.

Wertheim B, Marchais J, Vet LEM and Dicke M. 2002. Allee effect in larval resource exploitation in drosophila: an interaction among density of adults, larvae, and micro-organisms. *Ecological Entomology*, 27, 6.

Wertheim B, van Baalen EJA, Dicke M and Vet LE. 2005. Pheromone-mediated aggregation in nonsocial arthropods: an evolutionary ecological perspective. *Annual Review of Entomology*, 50, 321–346.

Wilson EO. 1970. *The Insect Societies*. Harvard University Press, Cambridge, MA.

Witt M and Hummel T. 2006. Vomeronasal versus olfactory epithelium: is there a cellular basis for human vomeronasal perception? International Review of Cytology, 248, 209–259.

Wood DL, Browne LE, Silverstein RM and Rodin JO. 1966. Sex pheromones of bark beetles. I. Mass production, bio-assay, source, and isolation of the sex pheromones of *Ips confusus* (LeC.). *Journal of Insect Physiology*, 12, 523–536.

Wood DL and Bushing RW. 1963. The olfactory response of *Ips confusus* (LeC.) to the secondary attraction in the laboratory. Canadian Entomologist, 95, 1066–1078.

Wood DL and Vite JP. 1961. Studies on the host selection behavior of *Ips confusus* (LeConte) attacking *Pinus ponderosa*. Contributions from Boyce Thompson Institute, 21, 79–96.

Worden BD, Parker PG and Papas PW. 2000. Parasites reduce attractiveness and reproductive success in male grain beetles. *Animal Behavior,* 59, 543–550.

Wyatt TD. 2003. *Pheromones and Animal Behavior: Communication by Taste and Smell.* Cambridge University Press, Cambridge, UK.

Zala SM, Potts WK and Penn DJ. 2004. Scent-marking displays provide honest signals of health and infection. *Behavioral Ecology,* 15, 338–344.

10. *Paenibacillus vortex* — A Bacterial Guide to the Wisdom of the Crowd

Alin Finkelshtein[*,¶], Alexandra Sirota-Madi[*,†], Dalit Roth[*], Colin J. Ingham[‡,||,¶], and Eshel Ben Jacob[*,§]

[*]*Sackler School of Medicine, Physics and Astronomy, Tel Aviv University, Tel Aviv-Yafo, Israel*

[†]*Department of Molecular Genetics, Weizmann Institute of Science, Rehovot 76100, Israel*

[‡]*MicroDish BV, Utrecht, NL*

[§]*Center for Theoretical Biological Physics, Rice University, Houston, TX, USA*

[¶]*alinf@post.tau.ac.il*

[||]*colinutrecht@gmail.com*

Complex organizations of bacteria emerge through the communication-based interplay between individuals. Each bacterium is, by itself, a biotic autonomous system with its own internal cellular informatics capabilities (storage and processing of information triggering an appropriate response). The cell plasticity allows the bacteria to select its response to biochemical messages it receives, including self-alteration (exposure to antibiotics causes enrichment of one subpopulation in *Paenibacillus vortex*) and the broadcasting of messages to influence on other bacteria. New features can collectively emerge during self-organization from the intracellular level to the whole colony. *P. vortex*, is a highly social, motile, pattern forming bacteria. Its information processing abilities are supported by exceptionally high number of communication related genes (two-component system (TCS), transcription factors (TFs), transport and defense related genes) suggesting a high "Bacterial Social Intelligence Quotient (IQ) Score". *P. vortex* contains two subpopulations with different locations and functions, that contribute to creation of complex pattern colonies. The coexistence of two subpopulations within *P. vortex* colonies allows flexibility and adjustment to heterogeneous environments including other microorganisms such as fungi. An unusual intraspecies temporary mutualism facilitates the spreading of both species. While *P. vortex* transporting fungal conidia with his swarming masses, the fungi bridges air gaps with its mycelia and allows bacteria to traverse these and colonize new niches.

1. Introduction

Under natural growth conditions, bacteria can utilize intricate communication capabilities (e.g., quorum sensing, chemotactic signaling and plasmid exchange) to cooperatively form (self-organize) complex colonies with elevated adaptability — the colonial pattern is collectively engineered according to the environmental conditions.

Quorum sensing (Bassler, 2002; Kolenbrander *et al.*, 2002; Miller *et al.*, 2002; Mok *et al.*, 2003; Xavier and Bassler, 2003) is a well-studied method of communication-based cooperation using bacterial sensing faculties. Many bacterial species use quorum sensing molecules to switch the expression of a variety of genes once the species density exceeds a critical threshold (e.g., virulence genes, and competence versus sporulation decision circuits). In *Vibrio fischeri* production of a membrane-permeable homoserine lactone (diffusible signaling molecule) is sensed and this regulates genes expressing the proteins required for luminescence (Nealson and Hastings, 1979). Bacterial communication-based also contributes to colony morphogenesis, which requires coordinated gene expression, regulated cell differentiation and division of tasks (Ben-Jacob and Shapira, 2005; Ben-Jacob, 2003; Ben-Jacob *et al.*, 1988, 2000; Shapiro, 1995, 1998; Shapiro and Dworkin, 1997; Franco *et al.*, 2002).

Many bacteria have intricate intra-cellular signaling mechanisms involving signal transduction networks (Ptashne and Gann, 2002) and genetic language (Searls, 2002). These mechanisms are part of the intra-cellular functional complexity which is used to generate intrinsic meaning for contextual interpretations of the chemical messages and for formulating appropriate complex responses. One such example would be transitions between polar and lateral flagella in different extracellular matrices (Ben-Jacob *et al.*, 2004; Ben-Jacob, 2003; Monahan *et al.*, 2014; McCarter, 2004). Biochemical messages are also used in bacterial cell–cell talk to exchange information across colonies and species. For example, biosurfuctants produced by one bacteria inhibit quorum sensing of its competitors. Another example is a quorum sensing molecules produces by *Pseudomonas aeruginosa* can be recognized by mammalian bitter taste receptors (Kolenbrander *et al.*, 2002; Ibacache-Quiroga *et al.*, 2013; Viswanathan, 2013).

1.1 Bacterial linguistics

Linguistic metaphors have begun to penetrate the literature a few decades ago in the description of various biological processes, from "protein linguistics" and "language of genes" (Searls, 2002; Pevznera *et al.*, 1989; Gelfand *et al.*, 1992) to "bacterial chemical language" (Bassler 2002; Kolenbrander *et al.*, 2002; Miller *et al.*, 2002; Mok *et al.*, 2003; Xavier and Bassler, 2003; Wirth *et al.*, 1996; Dunny and Winans, 1999). The current usage of "language" with respect to intra- and inter-bacteria communication is mainly in the sense that one would use in, for example, "computer language" or "language of algebra". Namely, it refers to structural aspects of communication, corresponding to the structural (lexical and syntactic) linguistic motifs (Green, 1997, 1999; Hurford *et al.*, 1998; Gauker, 2003; Gardner, 1993; Allwood, 1981; Hauser, 1996; Warnow, 1997). Higher linguistic levels — assigning contextual meaning to words and sentences (semantic) and conducting meaningful dialogue (pragmatic) — are typically associated with humans, higher primates or social insects (Bonabeau and Theraulaz, 1999; Camazine, 2003; Humphrey, 2002; Queller and Strassmann, 2002; Sternberg, 2000), but is an unusual idea in thinking about bacterial communication. This view is now being reconsidered (Ben-Jacob and Shapira, 2005; Ben-Jacob *et al.*, 2004; Stephens, 1998; Strassmann, 2000; Crespi, 2001; Velicer, *et al.*, 2000; Mire *et al.*, 2004, Velicer, 2003).

1.2 Communication-based cognition

This ability to form cooperative collectives serves an important survival strategy when bacteria are exposed to harsh and changing conditions. New functional features that support the foundation of cognition appear at every level of colonial self-organization — from the internal cellular gel to the growth of the colony as a whole — thus facilitating a high level of functional complexity. To form such multicellular super-organisms, the respective units (the individual bacteria) assume newly co-generated traits and faculties that are triggered by expression of genetic information of the individuals. The bacteria glean information from both the environment and neighboring bacteria. Information in both cases is gathered as signaling molecules that induce the bacteria to express specific genome-encoded functions that act to produce the multicellular structures and creating colonial patterns (Toguchi *et al.*, 2000; Shank and Kolter, 2011).

Bacteria use their intra-cellular flexibility, involving signal transduction networks and genomic plasticity, to collectively create the colony and maintain its integrity by sharing interpretations of chemical cues and exchanging chemical messages. The ensuing dialogues appears to be meaning-based communication (Ben-Jacob and Shapira, 2005; Ben-Jacob, 2003; Ben-Jacob, *et al.*, 2004;

Green, 1997, 1999; Hauser, 1996; Hurford *et al.*, 1998; Gauker, 2003; Gardner, 1993; Allwood, 1981; Warnow, 1997), which allows the colony to alters of structures and influence decision-making. For example, human pathogen *Pseudomonas aeruginosa* when treated with a β-lactam antibiotic transforms from rod-shaped cells to viable cell wall-defective spherical cells as a strategy to survive antibiotic exposure (Monahan *et al.*, 2014). Another example is swarming motility of *Pseudomonas putida* that becomes possible only at low temperatures, that induce cell elongation and surfactin secretion (Matilla *et al.*, 2007). These features represent social intelligence and fundamental (primitive) elements of cognition (Crespi, 2001; Stephens, 1998; Strassmann, 2000; Velicer, 2003; Velicer *et al.*, 2000, Bonabeau and Theraulaz, 1999; Camazine, 2003; Humphrey, 2002; Queller and Strassmann, 2002; Sternberg, 2000).

1.3 The Paenibacillus genus

The genus *Paenibacillus*, including *P. vortex*, was originally considered a part of the genus *Bacillus* but was later reclassified as a separate group in 1993 (Ash *et al.*, 1993). These facultative anaerobic, spore-forming bacteria are found in a variety of heterogeneous environments, such as soil, the rhizosphere, insect larvae and clinical samples (Lal and Tabacchioni, 2009; McSpadden Gardener, 2004; Montes *et al.*, 2004; Ouyang *et al.*, 2008). To face the challenges posed by these environments, *Paenibacillus* spp. (Fig. 1a) produce a wealth of enzymes and proteases as well as a great variety of antimicrobial substances that affect a wide range of microorganisms (Konishi and Maruhashi, 2003; Watanapokasin *et al.*, 2009; Raza *et al.*, 2008). The possession of these advanced defensive and offensive strategies render the *Paenibacillus* spp. bacteria a rich source of useful genes for agricultural, medical and industrial applications. The most familiar representatives of the genus include *P. larvae* which is known to cause American foulbrood in honeybees (Alippi *et al.*, 2002). *P. alvei* is a secondary invader during European foulbrood of honeybees (Djukic *et al.*, 2012) and

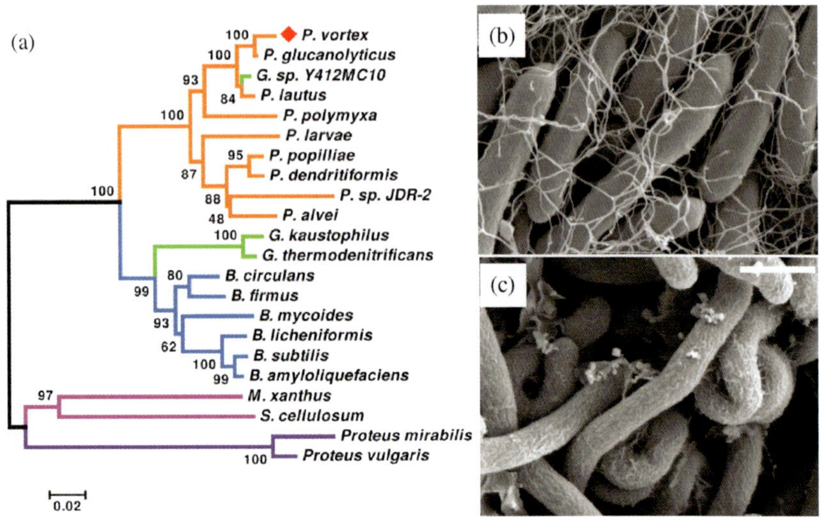

Figure 1. (a) Phylogenetic tree based on 16 S rRNA. The abbreviated genera P — *Paenibacillus*, B — Bacillus, G — Geobacillus, M — Myxococcus, S — Sorangium. Bootstrap values are shown next to the branches (Sirota-Madi *et al.*, 2010). (b) SEM of actively swarming *P. vortex* cells cultured on 1.5% (w/v) agar MH plates, showing entangled flagella. Scale bar — 1 μm (Ingham and Ben-Jacob, 2008). (c) SEM of cells cultured with Congo Red 400 μg/ml with frequent convoluted/bent cells and absence of flagella. Scale bar — 1 μm (Ingham and Ben-Jacob, 2008).

P. polymyxa is capable of fixing nitrogen and therefore is used in agriculture and horticulture (Nelson *et al.*, 1976), the *Paenibacillus* spp. JDR-2 which is known to be a rich source of chemical agents for biotechnology applications (StJohn *et al.*, 2006). The most fascinating pattern-forming strains are *P. vortex* (Ben-Jacob, 2003) and *P. dendritiformis* (Ben-Jacob, 2003; Ben-Jacob *et al.*, 1994). Under laboratory growth conditions these bacteria can develop colonies that behave much like a multicellular organism, with cell differentiation and task distribution (Aguilar *et al.*, 2007; Ben-Jacob *et al.*, 2004).

2. The Social Bacteria *P. Vortex*

P. vortex is a bacterial species discovered in the early 90s (Ben-Jacob *et al.*, 1995; Sirota-Madi *et al.*, 2010). It is a social microorganism that forms colonies with remarkably complex and dynamic architectures (Ben-Jacob, 2003; Ben-Jacob *et al.*, 1995, 1998; Ben-Jacob and Levine, 2006). *P. vortex* possesses advanced social motility employing cell–cell attractive and repulsive chemotactic signaling. An attraction-repulsive signaling mechanism regulated by Slf and subtilisin has been identified in a closely related *P. dendritiformis* (Béer *et al.*, 2010). When grown on agar surfaces, the collective motility is reflected by the formation of foraging swarms (Ingham and Ben-Jacob, 2008) that act as arms sent out in search for food. These swarms have an aversion to crossing each other's trail and collectively change direction when food is sensed. The swarms can even split and reunite when detecting scattered patches of nutrients (Ingham and Ben-Jacob, 2008). When plated on hard agar surfaces (≥2% agar) *P. vortex* generates swarming patterns with leading groups of cells containing tens to thousands of bacteria, named vortices, that cooperatively whirl around a common center. The vortex expands in size and moves outward as a unit, leaving behind a trail — the vortex branch (Fig. 3(a)). The dynamics of the vortices is quite complicated and includes attraction, repulsion, merging, and splitting (Ingham and Ben-Jacob, 2008; Ben-Jacob and Levine, 2006; Ben-Jacob *et al.*, 2000; Shklarsh *et al.*, 2012). The vortices serve as building blocks of colonies with modular organization.

Individual cells of *P. vortex* are distinctly curved average 18.6° deviation from a straight line, with the maximum bend of 101° (Ingham and Ben-Jacob, 2008). The size of the cells in the population also varies significantly 0.5–145 μm long with a mean of 9.15 μm and median of 6 μm (unpublished data based on Image Stream). Cell length can change dramatically in different environments and conditions, for example, it increases on agar at 42°C (>33% are above 16 μm long) and decreases in salt base (M9) liquid medium (>60% are below 3 μm). Swarming cells present multiple peritrichous flagella (2 to 8 flagella per μm of cell length) with a strong tendency to be entwined with those of the near neighbors. Swarming performed by vortices initiates 5–8 h post inoculation from liquid culture but potentially within tens of minutes when inoculated from already swarming cultures. The vortices are comprised of cell rafts that are held together with flagella creating single rotating mass to drive cell motility (Fig. 1(b)), that can move outwards at >1 cm/h under optimal conditions on agar. In agreement with previous studies (McCarter, 2004; McCarter, 2006) liquid growth led to the loss of both flagella and motility, presumably due to expression of polar instead of peritrichous flagella (Ingham and Ben-Jacob, 2008).

Addition of 400 μg/ml Congo Red (inhibitor of flagella production (Arnold and Shimkets, 1988)) completely inhibited swarming (>72 h), but still permitted growth. Examination of cells grown with Congo Red by, scanning electron microscope (SEM) and anti-flagellin antibodies staining showed an absence of flagella observed in motile cells (Fig. 1(c)) indicating flagella are necessary for *P. vortex* motility (Ingham and Ben-Jacob, 2008).

2.1 Realizing social Intelligence of Bacteria

Social intelligence is an individual's capacity to perceive and understand the environment — from local surroundings to what is happening in the world — and to respond to that understanding in a personally and socially effective manner. Bacteria, communicate with each other, with other colonies and different species. They glean relevant latent information from the complex environment, interpret the information in an existential "meaningful" way, develop common knowledge, and in some situations the physiological state of a bacteria (and therefore its experiences) influence the development of its progeny. We know that bacteria record their past experiences and that ancestor bacteria can influence progeny in ways other than by stable mutations. There are a number of mechanisms including the carry over signaling compounds, damage or stress. Whilst it is necessary to be very careful about the interpretation there clearly is long term (as bacteria generation times are measured) plasticity in that past events can influence current phenotypes. Whilst clearly not a neuronal process and not learning in the sense of more complicated organisms it is also important to note that bacteria have at least some of the characteristics related to "learning" (Ben-Jacob *et al.*, 2004, 2006;

Ben-Jacob, 2003, 2009; Ben-Jacob and Levine, 2006; Veening *et al.*, 2008). That is why bacterial social intelligence can be estimated by their ability to communicate, sense their environment and adapt accordingly.

It was previously suggested that bacteria successful in heterogeneous and competitive environments often contain extensive signal transduction and regulatory networks (Dwyer, *et al.*, 2008a; Galperin, 2005a; Galperin and Gomelsky, 2005a). These observations, and the fact that bacteria enriched with specific set of "tools" (such as TCS, TF, Transport, and defense related genes) has a greater potential to survive in highly fluctuating environments, led to the notion that the combined score reflecting the enrichment of these "tools", can be used as a measure for the "Bacteria IQ" (Sirota-Madi *et al.*, 2010).

The *P. vortex* genome sequence was analyzed by comparative analysis with a dataset of 500 complete bacterial genomes. This comparison was done with regard to four groups of genes which are related to complex bacterial lifestyle and adaptability to fluctuating environments: TCS, TFs, defense mechanisms and transport systems (Fig. 2). The analysis also revealed genes required for competition over resources (e.g., iron, amino acids, and sugars), for offensive compounds production (antibiotics and lytic enzymes) and for defense (resistance to pollutants, antibiotics and other toxins). These genes can support traits needed for thriving in the heterogeneous and highly competitive environments (Sirota-Madi *et al.*, 2010).

"Bacterial IQ" was visualized by plotting the distribution of the combined score (non-normalized per genome size), calculated for 500 bacterial genomes. Representatives of *Paenibacillus* spp., pathogenic, predator and model organism bacteria were mapped and highlighted based on their combined score value (the *x*-axis) (Fig. 2(a)). The results suggest that *P. vortex* (with three other *Paenibacillus* bacteria strains) has the highest Social-IQ score among all 500 sequenced bacteria, over 3 standard deviations higher than average, indicating a capacity for exceptionally brilliant social skills (Fig. 2(b)) (Sirota-Madi *et al.*, 2010).

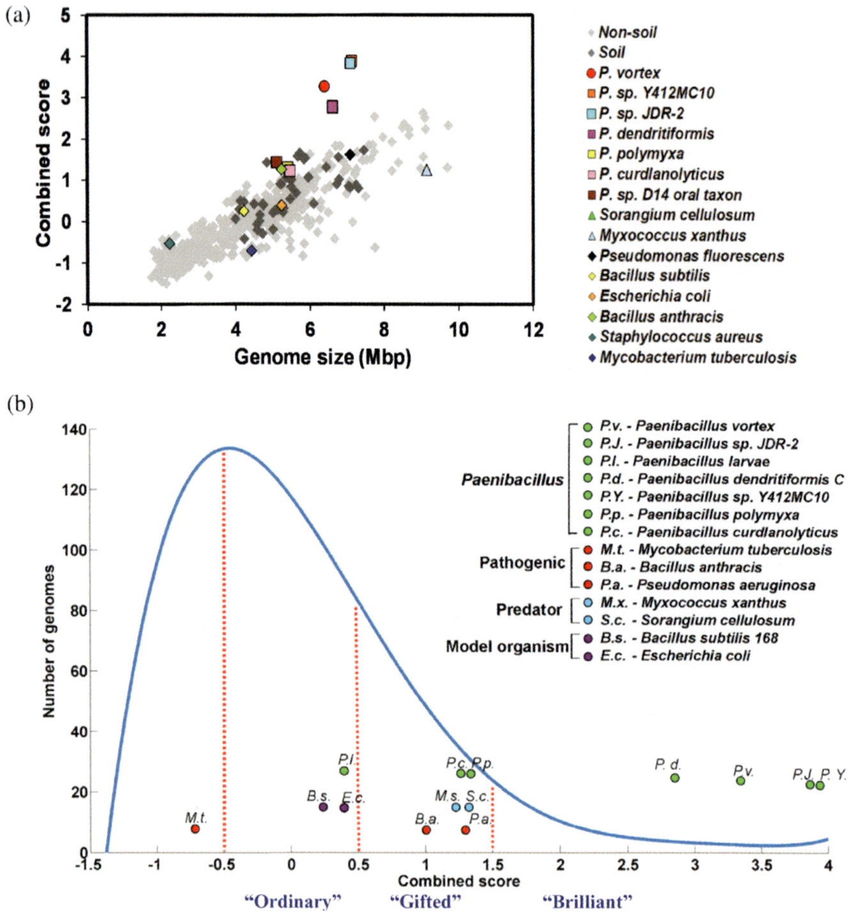

Figure 2. (a) Non-normalized combined score as function of genome size. The combined score is calculated as an average of the standard deviation of TCS, TF, transport, and defense genes for the dataset of 500 bacterial genomes. (b) Bacterial IQ score. Plot of the non-normalized combined score distribution is presented. Example of several species is presented on the x-axis. *Paenibacillus* marked green; Pathogenic bacteria were marked red; Predator bacteria were marked blue; Model organism bacteria were marked purple (Sirota-Madi *et al.*, 2010).

3. Building a Complex Colony — Task Distribution

P. vortex is one of the most fascinating pattern forming bacteria, generating complex colony pattern on hard agar with leading

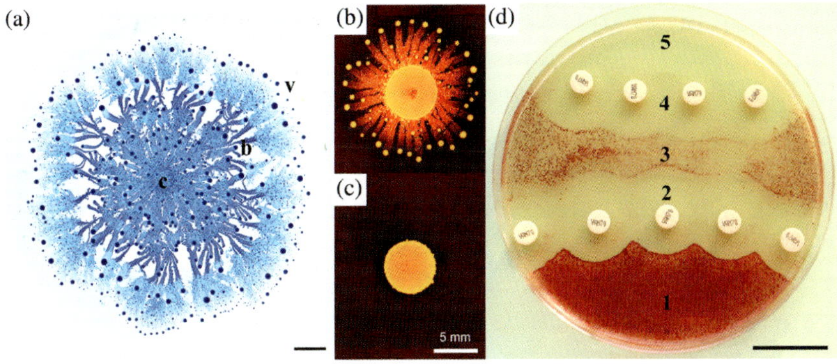

Figure 3. (a) *P. vortex*'s colony, grown on peptone agar (20 g/l peptone, 2.25% w/v agar) for 96 h at 30°C. v — vortices; b — branches; c — center. (b) Demonstrating replica plating from the vortices (v), grown on peptone for 48 h at 30°C. The colony is characterized by accelerated expansion rate (0.098 mm h^{-1}) in comparison to mixed colony expansion rate (0.0625 mm h^{-1}). The colors were inverted to emphasize higher cell densities using the brighter shades of yellow. (c) Replica plating from the center (c), grown on peptone for 48 h at 30°C. The colony has shown no expansion as opposed to isolations from vortices. Scale bar — 5 mm (Roth *et al.*, 2013). (d) Experiment showing the ability of pioneering swarming masses of *P. vortex* to cross regions of an agar plate where growth is inhibited by antibiotics to reach more favorable niches. RMHA plate (14 cm diameter; 1.5% wt./vol agar) inoculated with *P. vortex* within the center of zone 1. Picture was taken after 24 h incubation at 37°C using a dye (INT) to stain metabolically active bacteria in red. Zones 2 and 4 are two zones of inhibitory concentrations of vancomycin (VAN) created by diffusion of this antibiotic from a series of NeoSensi Tabs. The bacteria rapidly swarm and occupy zone 1, but most were limited by VAN from growing and swarming into zone 2. However, a limited number of swarming colonies could swarm across zone 2 to colonize zone 3. The bacteria swarming from zone 3 to 4 were not resistant, since most of them were affected by VAN in zone 4. This finding suggests that stable VAN resistant mutants were not the explanation for the ability of *P. vortex* to cross zone 2. As with zone 2, limited numbers could cross to zone 5. By 36 h, zones 1, 3, and 5 were heavily colonized by *P. vortex*, whereas only small numbers of bacteria could be found in zones 2 and 4. Scale bar — 3 cm (Ingham *et al.*, 2011).

groups, named vortices and a common center (Fig. 3(a)) (Ben-Jacob, 1997, 2003; Ben-Jacob *et al.*, 1998). We have recently discovered that one of *P. vortex* tricks used to create such extraordinary colonies is to maintain two co-existing phenotypic subpopulations, referred as *"builders"* and *"explorers"*. Under normal growth conditions, the *builders* constitute the majority of the population; they are recognized by a relatively high growth rate in liquid and reduced swarming capabilities on solid medium (Fig. 3(c)). The *explorers,* on the other hand, have a reduced growth rate in liquid and are hyperflagellated with elevated swarming capabilities on hard surfaces (Fig. 3(b)). The explorers are mainly localized to the vortices while builders comprise the central parts of the colony. The population is thus regulated by the fast reproduction of builders and the expansion rate of explorers (Roth *et al.*, 2013). Similar phenomena were described in *B. subtilis and P. aeruginosa,* several subpopulations were identified within a swarming colony (tendril tips versus swarm center). Similar with explorers, *B. subtilus* swarmers were found at the tendril tips and these show a two-fold increase in flagella and elevated *hug* (encoding flagellin) expression. However the swarmers were smaller than non-swarmers and expressed low levels of extracellular polysaccharide and surfactin genes (Tremblay and Deziel, 2010) . The *P. aeruginosa* tip subpopulation shows up-regulation of genes involved in energy metabolism (Hamze *et al.*, 2011; Tremblay and Deziel, 2010). In all cases main purpose of the tip subpopulation appears to rapidly spread on uncolonized surfaces while swarm center population are in a state allowing a permanent settlement of the colonized area (Hamze *et al.*, 2011; Roth *et al.*, 2013; Tremblay and Deziel, 2010).

3.1 Identification of subpopulations

The enrichment process of each subpopulation from the mixed culture (wild type *P. vortex* containing all subpopulations) is based on different characterizations of each subpopulation. Builders, due to their enhanced growth rate can be enriched by serial dilutions of the mixed culture at the early log phase (~4 h). After 4 such dilutions enriched builders are obtained. The explorers are the minority (~30%) of the mixed culture during liquid growth, however, they are enriched during exposure to kanamycin reaching

100% in 18 h. This enrichment process becomes possible due to explorers high resistance to kanamycin (Roth *et al.*, 2013).

Both morphotypes coexist in the culture but their ratio changes depending on culture age and environmental conditions. Each morphotype is capable to convert to the other and has a different stability. While builders are stable (without explorers) only for 8 h, explorers remain stable as long as they are kept in liquid culture (days–weeks). Both subpopulations convert to the mixed culture composition; *builders* 12 h post inoculation and explorers in a few hours on solid medium (Roth *et al.*, 2013).

3.2 Hyperactive nature of the explorers

An additional characteristic of explorers is hyper flagellation; that has been shown by TEM imaging, flagellin enhanced expression levels and high concentrations of flagellin protein. Treatment with 400 μg/ml Congo red (an inhibitor of motility, including flagella production), inhibited explorers spreading on solid surface. Interestingly, cellular Adenosine triphosphate (ATP) levels of explorers were found to be reduced up to 45% in comparison to the mixed culture. It is possible that decrease in cellular ATP causes reduction in metabolism and thus reproduction rate and uptake of antibiotics from the environment, explains the increased resistance to kanamycin and decreased reproduction rates of the explorers. This conclusion is supported by the fact that explorers resistance to antibiotics is not unique to kanamycin but applies to a wide range of structurally unrelated antibiotics (Roth *et al.*, 2013). Similar phenomena, of cell dormancy with low reproduction rates and elevated antibiotic resistance, were observed in bacteria with reduced cellular ATP e.g., Small Colony Variants (SCV) and persisters (Melter and Radojevič, 2010; Proctor *et al.*, 1998; Lewis, 2010). However, unlike the dormant cells (SCV and persisters) explorers show hyper flagellation and high motility actively communicating with their environment (Roth *et al.*, 2013).

The ability of both morphotypes to interconvert allows the colony to send explorers to seek for new and favorable niches for colonization. It was previously shown that *P. vortex* is capable of collectively crossing regions containing a large variety of antibiotics by using "pioneering, swarming masses". After reaching an antibiotic-free area the pioneering bacteria recreate the colony (Fig. 3(d)) (Ingham *et al.*, 2011). Based on the characteristics of explorers we assume they are those — "pioneering, swarming masses" that are capable to overcome environmental barriers such as antibiotics (being both resistant and highly motile) and once they reach a favorable condition to recreate a colony with both subpopulations by converting to mixed culture composition (Roth *et al.*, 2013).

4. Swarming Intelligence

Another intricate aspect of the *P. vortex* swarming intelligence is an ability to transport fungal conidia (fungal spores) with its swarming masses (vortices) and in return benefit from the fungal ability to bridge air gaps. This is an interesting temporary mutualism between bacteria and fungi (Ingham *et al.*, 2011).

4.1 *A. fumigatus* dispersal by *P. vortex*

A. fumigatus is a saprophytic fungus with mycelial mode of growth. The mycelia produce specialized fruiting bodies (conidiophores) containing conidia (non-motile asexual fungal spores) that are passively dispersed by airborne. Unlike highly motile *P. vortex*, *A. fumigatus* is immobile. Direct transport of *A. fumigatus* conidia by swarming *P. vortex* was observed when the microorganisms were coinoculated in the center of an reduced strength Mueller–Hinton agar (RMHA) plate. Individual and aggregated conidia over 25 μm across could be observed in motion within masses of *P. vortex* several centimeters away from the origin of swarming within 1–4 h of inoculation. Swarming *P. vortex* carried viable conidia, which were successfully recovered by replating (Ingham *et al.*, 2011).

Transport was only seen in multilayered masses of cells. Imaging by SEM showed that the conidia were incorporated within the rafting mass of moving and interconnected bacteria, with the flagella of *P. vortex* both linking adjacent bacteria and entrapping the conidia. Each conidium was captured by 6 to 30 flagella, typically derived from 2 to 9 bacteria (Fig. 4(a)). *P. vortex* flagella on one side and proteins on the surface of the fungal conidia on the other, are involved in capturing specificity. It was shown by subsequent transport inhibition following either SDS/proteinase K treatment of the conidia or addition of purified flagella from *P. vortex*. The transport of conidia from other fungi was relatively inefficient compared with *A. fumigatus* conidia. Similarly other species of swarming bacteria were unable to transport *A. fumigatus* conidia, implying there is a cargo-transporter specificity (Ingham *et al.*, 2011).

Conidial transport rates were similar to the movement rates of *P. vortex* masses alone (10.8 mm h^{-1}). After loaded, conidia were transported over long distances, reaching the edge of 14 cm diameter RMHA plates within 7 h. While purified *A. fumigatus* conidia inoculated onto RMHA without bacteria, formed a slowly expanding colony, typically reached the edge of 14 cm diameter plate in 6 to 12 days (Ingham *et al.*, 2011).

4.2 Benefits for *A. fumigatus*

To see if *P. vortex* can rescue fungal conidia from unfavorable growth conditions, an antifungal agent Voriconazole (VOR) (effective against many strains of *A. fumigatus* (Verweij *et al.*, 2007)) was used to create regions (a niche) where conidial outgrowth and mycelial extension were inhibited. When *P. vortex* bacteria and *A. fumigatus* conidia were coinoculated fungal growth was inhibited inside the VOR containing region (2–3 cm radius of the inoculation point). However, conidia were transported by swarming *P. vortex* to regions outside the zone of inhibition (Fig. 4(b)). This transport allowed successful conidial germination and outgrowth. In this case fungal dispersal was dependent on the presence of swarming *P. vortex* (Ingham *et al.*, 2011).

Fungal sporulation is inhibited by *P. vortex* that can actually distinguish between ungerminated and germinating conidia. Thus

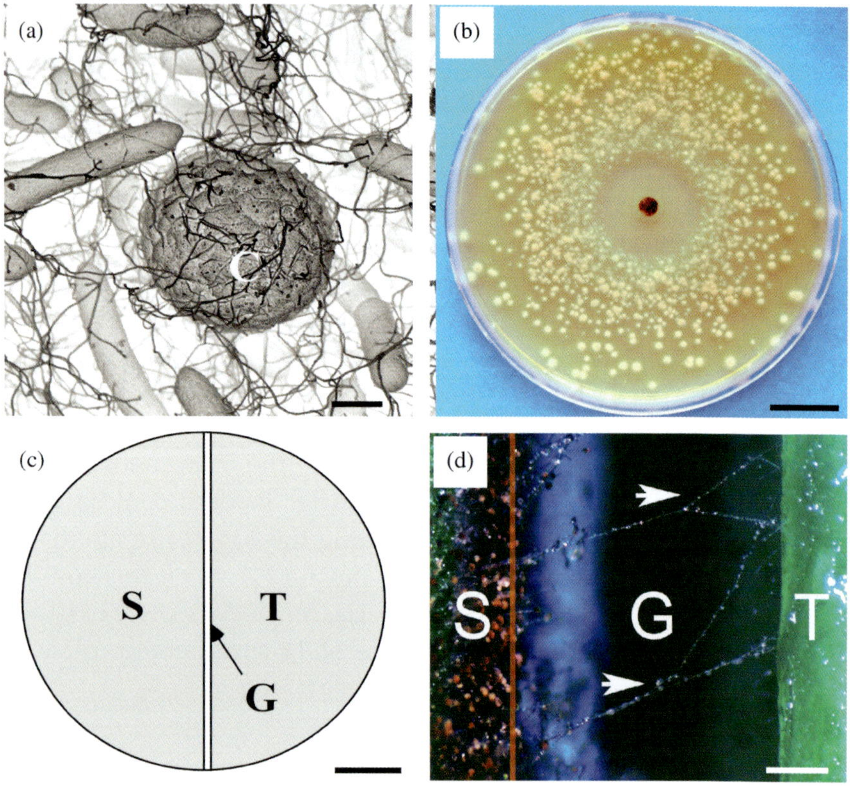

Figure 4. (a) Imaging of transport by scanning EM. The gray scale was inverted to emphasize flagella. Conidium (C) of *A. fumigatus* during transport by a *P. vortex* swarm, imaged 3 h after inoculation at a location 24 mm from the inoculation point. Scale bar — 1 μm. (b) *P. vortex* was coinoculated into the central point of a 14 cm diameter RMHA plate with conidia and 30 μg VOR. This amount of VOR was sufficient to inhibit fungal growth within a 27 mm radius from the central point. Despite this fact, *P. vortex* was able to transport conidia, which were then able to germinate and form colonies, outside the zone of inhibition by VOR. In this image (taken 48 h after inoculation), the fungi are visible as white colonies in a background of near-confluent *P. vortex*. The ring-like patterning around the central point is composed of *P. vortex* cells. Scale bar — 15.5 mm. (c) Testing the ability of *P. vortex* to cross air gaps with the assistance of *A. fumigatus*. Experimental setup: A 90 mm diameter agar plate was filled with RMHA. A sterile razor blade was used to remove the RMHA from the right-hand one-half of the plate, leaving the inoculation region S (Start). A second RMHA block T (Target) was positioned to create an air gap (G) 0.8 mm wide. *P. vortex* with *A. fumigatus* conidia (or as a control, the bacteria alone) were inoculated in the

fungal microcolonies above 10–20 μm in diameter were not transported by *P. vortex*. Interestingly, *P. vortex* was able to transport conidia into locations in which bacterial growth was inhibited by the presence of antibiotics. In this situation, fungal outgrowth from the *P. vortex* colonies was rapid, and *A. fumigatus* quickly came to dominate the niche. This findings suggests that circumstances may exist in the soil where *A. fumigatus* conidia can be transported by *P. vortex* from niches where the fungus cannot thrive to more favorable locations and even reach niches where bacteria are inhibited (Ingham *et al.*, 2011).

4.3 Benefits for *P. vortex*

Fungal mycelia are known to grow across air-filled regions between soil particles. Such air gaps are a significant barrier to many species of bacteria (Warmink and van Elsas, 2009; Nazir *et al.*, 2010). Indeed, *P. vortex* (alone) was unable to cross a 0.5–0.8 mm wide air gap in an RMHA plate. No bacteria were detected on the other side. When the swarming bacteria were loaded with *A. fumigatus* conidia, *P. vortex* crossed the air gap in half of the cases. A successful crossing of the mycelia by bacteria was facilitated by germination of transported conidia near the gap with a small number of mycelia bridging the gap (Figs. 4(c) and 4(d)). Hexidium Iodide stained bacteria were detected on the mycelia bridges. These data suggest that there is a degree of mutualism between *P. vortex* and *A. fumigatus*. Each one can assist the dispersal of the other in the right situation (Ingham *et al.*, 2011). Both *Paenibacillus* and *Aspergillus* are widely found in nature, including the soil, and therefore are expected to have common habitats. Its not clear if more specific and

Figure 4. (*Continued*) inoculation region (S). This setup was incubated in a humidity chamber at 37°C for 5 d. Scale bar — 17 mm C. (d) Mycelial growth across a 0.8 mm air gap (G) from the starting region (S) to the target region (T). The arrows are pointing at the fungal micellia bridges over the gap. The left-hand edge of the air gap is obscured by fungal growth and is indicated by a vertical orange line. Scale bar — 0.3 mm (Ingham *et al.*, 2011).

closer interactions exist in the soil, since currently this is the best transport interaction known between *Paenibacillus* and fungus.

5. Discussion

Bacteria are the most prolific organisms on Earth. Some are pathogens, but many more are indispensible to our survival. In order to understand bacterial interactions with the environment and their neighbors we have chosen a highly social, motile and complex colony forming *P. vortex* which lives in heterogeneous environments. Comparative genomics and comparative network biology emerged as key tools in understanding of how bacteria respond cooperatively to challenging complex environments. In particular, it was suggested that bacteria successful in heterogeneous and competitive environments often contain extensive signal transduction and regulatory networks (Dwyer *et al.*, 2008; Galperin, 2005). These observations, and the fact that signal transduction networks afford intracellular information processing (Hellingwerf, 2005), led us to calculate the "Bacteria IQ" for 500 fully sequenced bacteria (Galperin and Gomelsky, 2005; Galperin, 2005). Four "brilliant" species were revealed, all belong to the *Paenibacillus* family and interestingly all are motile. The fact that motile bacteria are "smarter" than others make sense, since motile bacteria face rapid environmental changes and thus has to respond to those changes much faster.

Colony composition, containing at least two subpopulations with different locations and task distribution, further supports *P. vortex* complexity and high social IQ. Functional differentiation in swarming bacteria has also been shown in *B. subtilis* (Hamze *et al.*, 2011; Shank and Kolter, 2011), *Pseudomonas aeruginosa* (Tremblay and Deziel, 2010), *Salmonella enteric, Proteus* (Toguchi *et al.*, 2000; Dienes, 1970, 1942) and others (Weibull and Gyllang, 1965). A *P. vortex* colony contains two genetically identical subpopulations; builders in the central parts of the colony, responsible for house-keeping and proliferation and explorers at the edges of the colony with highly motile and resistance traits. The ability of each subpopulation to revert into the wild type colony composition within hours, allows flexibility in heterogeneous environments. The highly motile and resistant explorers, can bypass areas with unfavorable conditions and recreate the colony then reach a new niche (Roth *et al.*, 2013; Ingham and Ben-Jacob, 2008). In the soil, bacteria often cope with many fluctuations in the environment. This may be particularly true for swarming ('traveling') bacteria, such as passage from watery to hard surfaces. The coexistence of *explorers* and *builders* in a planktonic population may be essential for a rapid translocation and colonization in changing environment (Roth *et al.*, 2013).

Additional characteristic of heterogeneous environment such as the rhizosphere, is a close proximity and coexistence of many species side by side in combination with gradients of nutrients. Thus interspecies interactions such as competition over resources, mutualism and even cannibalism, are not rare. However, *P. vortex* was found to interact with *A. fumigatus* in a very uncommon way; to transport fungal conidia. This provides *A. fumigatus* 20 to 40 times faster dispersal (locally, without having to grow and generate new airborne spores), conidia rescue from niches of unfavorable growth conditions and transport of conidia into locations of limited bacterial growth with later dominance in the niche by the fungi. *P. vortex*, on the other hand, can use the ability of the fungi to bridge the air gaps by mycelia to overpass obstacles and colonize new niches. This way each one can assist the dispersal of the other in the right situation (Ingham *et al.*, 2011).

Today we are facing another major health problem worldwide: bacteria are becoming multidrug resistant and we can't invent new drugs fast enough to beat them. To change this threat to our health, we must realize they have social intelligence. Only if we accept how smart they are can we find ways to destroy the pathogenic bacteria and at the same time find new ways to better exploit the capabilities of friendly bacteria for our benefit. Deeper understanding of bacteria's social intelligence, will allow us to develop novel strategies to fight with pathogens for example, by tampering with their communication or by turning toxic materials they produce against them (Sirota-Madi *et al.*, 2010).

References

Aguilar C, Vlamakis H, Losick R and Kolter R. 2007. Thinking about bacillus subtilis as a multicellular organism. *Current Opinion Microbiology*, 10(6), 638–643.

Alippi AM, López AC and Aguilar OM. 2002. Differentiation of Paenibacillus larvae subsp. larvae, the Cause of American Foulbrood of Honeybees, by Using PCR and Restriction Fragment Analysis of Genes Encoding 16S rRNA. *Applied and Environmental Microbiology*, 68(7), 3655–3660. doi: 10.1128/aem.68.7.3655-3660.2002.

Allwood J. 1981. On the Distinctions between semantics and pragmatics. In *Crossing the Boundaries in Linguistics*, Wolfgang K and Willem L (eds.), 177–189. Springer, Netherlands.

Arnold JW and Shimkets LJ. 1988. Inhibition of cell-cell interactions in Myxococcus xanthus by congo red. *Journal of Bacteriology*, 170(12), 5765–5770.

Ash C, Priest FG and Collins MD. 1993. Molecular identification of rRNA group 3 bacilli (Ash, Farrow, Wallbanks and Collins) using a PCR probe test. Proposal for the creation of a new genus Paenibacillus. *Antonie Van Leeuwenhoek*, 64(3–4), 253–260.

Bassler BL. 2002. Small talk: cell-to-cell communication in bacteria. *Cell*, 109(4), 421–424.

Béer A, Ariel G, Kalisman O, Helman Y, Sirota-Madi A, Zhang HP, Florin EL, Payne SM, Ben-Jacob E and Swinney HL. 2010. Lethal protein produced in response to competition between sibling bacterial colonies. *Proceedings of the National Academy of Sciences*, 107(14), 6258–6263. doi: 10.1073/pnas.1001062107.

Ben-Jacob E, Shochet O, Tenenbaum A and Avidan O. 1995. Evolution of complexity during growth of bacterial colonies. *In NATO Advanced Research Workshop; Santa Fe, USA*, Cladis PE and Palffy-Muhorey P (eds.), 619–633. Addison-Wesley Publishing Company, UK.

Ben-Jacob E. 1997. From snowflake formation to growth of bacterial colonies II: cooperative formation of complex colonial patterns. *Contemporary Physics*, 38(3), 205–241.

Ben-Jacob E and Shapira Y. 2005. *Meaning-based natural intelligence versus information-based artificial intelligence. In The Cradlse of Creativity, Shaarey Mishpat*, Ben-Nun C (ed.). Jerusalem, Israel.

Ben-Jacob E. 2003. Bacterial self-organization: co-enhancement of complexification and adaptability in a dynamic environment. *Philosophical*

Transactions of the Royal Society of London. Series A: Mathematical, Physical and Engineering Sciences, 361(1807), 1283–1312. doi: 10.1098/rsta.2003.1199.

Ben-Jacob E. 2009. Learning from bacteria about natural information processing. *Annals of the New York Academy of Sciences*, 1178(1), 78–90. doi: 10.1111/j.1749-6632.2009.05022.x.

Ben-Jacob E, Becker I, Shapira Y and Levine H. 2004. Bacterial linguistic communication and social intelligence. *Trends in Microbiology*, 12(8), 366–372. doi: http://dx.doi.org/10.1016/j.tim.2004.06.006.

Ben-Jacob E, Cohen I, Golding I, Gutnick DL, Tcherpakov M, Helbing D and Ron IG. 2000. Bacterial cooperative organization under antibiotic stress. *Physica A: Statistical Mechanics and its Applications*, 282(1–2), 247–282.

Ben-Jacob E, Cohen I and Gutnick DL. 1998. Cooperative organization of bacterial colonies: from Genotype to Morphotype. *Annual Review of Microbiology*, 52(1), 779–806. doi: doi:10.1146/annurev.micro.52.1.779.

Ben-Jacob E, Cohen I and Levine H. 2000. Cooperative self-organization of microorganisms. *Advances in Physics*, 49(4), 395–554. doi: 10.1080/000187300405228.

Ben-Jacob E and Levine H. 2006. Self-engineering capabilities of bacteria. *Journal of The Royal Society Interface*, 3(6), 197–214. doi: 10.1098/rsif.2005.0089.

Ben-Jacob E, Schochet O, Tenenbaum A, Cohen I, Czirok A and Vicsek T. 1994. Generic modeling of cooperative growth patterns in bacterial colonies. *Nature*, 368(6466), 46–49.

Ben-Jacob E, Shapira Y and Tauber AI. 2006. Seeking the foundations of cognition in bacteria: from Schrödinger's negative entropy to latent information. *Physica A: Statistical Mechanics and its Applications*, 359(0), 495–524. doi: http://dx.doi.org/10.1016/j.physa.2005.05.096.

Bonabeau E and Theraulaz MD. 1999. *Swarm Intelligence: From Natural to Artificial Systems*, Oxford University Press, Oxford, UK.

Camazine S. 2003. *Self-Organization in Biological Systems* Princeton, Princeton University Press, USA.

Crespi BJ. 2001. The evolution of social behavior in microorganisms. *Trends in Ecology and Evolution*, 16(4), 178–183. doi: http://dx.doi.org/10.1016/S0169-5347(01)02115-2.

Dienes L. 1942. The Significance of the Large Bodies and the Development of L Type of Colonies in Bacterial Cultures. *Journal of Bacteriology*, 44(1), 37–73.

Dienes L. 1970. Permanent Alterations of the L-Forms of Proteus and Salmonella Under Various Conditions. *Journal of Bacteriology*, 104(3), 1369–1377.

Djukic M, Becker D, Poehlein A, Voget S and Daniel R. 2012. Genome Sequence of Paenibacillus alvei DSM 29, a Secondary Invader during European Foulbrood Outbreaks. *Journal of Bacteriology*, 194(22), 6365. doi: 10.1128/jb.01698-12.

Dunny GM and Winans SC. 1999. *Cell–Cell Signaling in Bacteria*. ASM Press, Herndon, USA.

Dwyer DJ, Kohanski MA and Collins JJ. 2008. Networking opportunities for bacteria. *Cell*, 135(7), 1153–1156. doi: S0092-8674(08)01577-8 [pii] 10.1016/j.cell.2008.12.016.

Franco CD, Beccari E, Santini T, Pisaneschi G and Tecce G. 2002. Colony shape as a genetic trait in the pattern-forming Bacillus mycoides. *BMC Microbiology*, 2(1), 33.

Galperin MY. 2005. A census of membrane-bound and intracellular signal transduction proteins in bacteria: bacterial IQ, extroverts and introverts. *BMC Microbiology*, 5, 35. doi: 1471-2180-5-35 [pii] 10.1186/1471-2180-5-35.

Galperin MY and Gomelsky M. 2005a. Bacterial signal transduction modules: from genomics to biology. *ASM News*, 71(7), 326–333.

Gardner H. 1993. *Multiple Intelligences: The Theory in Practice*. Basic Books, New York, USA.

Gauker C. 2003. *Words Without Meaning*. The MIT Press, Cambridge, USA.

Gelfand MS, Kozhukhin CG and Pevzner PA. 1992. Extendable words in nucleotide sequences. *Computer Applications in the Biosciences*, 8(2), 129–135.

Green MS. 1999. Illocutions, implicata, and what a conversation requires. *Pragmatics & Cognition*, 7(1), 65–91.

Green MS. 1997. On the autonomy of linguistic meaning. *Mind*, 106(422), 217–244. doi: 10.1093/mind/106.422.217.

Hamze K, Autret S, Hinc K, Laalami S, Julkowska D, Briandet R, Renault M, Absalon C, Holland IB, Putzer H and Séror SJ. 2011. Single-cell analysis in situ in a Bacillus subtilis swarming community identifies distinct spatially separated subpopulations differentially expressing hag (flagellin), including specialized swarmers. *Microbiology*, 157(9), 2456–2469. doi: 10.1099/mic.0.047159-0.

Hauser MD. 1996. *The Evolution of Communication*. The MIT Press, Cambridge, USA.

Hellingwerf KJ. 2005. Bacterial observations: A rudimentary form of intelligence? *Trends in Microbiology*, 13(4), 152–158.

Humphrey N. 2002. *The Inner Eye Social Intelligence in Evolution*. Oxford University Press, Oxford, UK.

Hurford JR, Studdert-Kennedy M and Knight C. 1998. *Approaches to the Evolution of Language: Social and Cognitive Bases*. Cambridge University Press, Cambridge, UK.

Ibacache-Quiroga C, Ojeda J, Espinoza-Vergara G, Olivero P, Cuellar M and Dinamarca MA. 2013. The hydrocarbon-degrading marine bacterium Cobetia sp. strain MM1IDA2H-1 produces a biosurfactant that interferes with quorum sensing of fish pathogens by signal hijacking. *Microbial Biotechnology*, 6(4), 394–405.

Ingham C and Ben-Jacob E. 2008. Swarming and complex pattern formation in Paenibacillus vortex studied by imaging and tracking cells. *BMC Microbiology*, 8(1), 36.

Ingham CJ, Kalisman O, Finkelshtein A and Ben-Jacob E. 2011. Mutually facilitated dispersal between the nonmotile fungus Aspergillus fumigatus and the swarming bacterium Paenibacillus vortex. *Proceedings of the National Academy of Sciences*, 108(49), 19731–19736. doi: 10.1073/pnas.1102097108.

Kolenbrander PE, Andersen RN, Blehert DS, Egland PG, Foster JS and Palmer RJ. 2002. Communication among Oral Bacteria. *Microbiology and Molecular Biology Reviews*, 66(3), 486–505. doi: 10.1128/mmbr.66.3.486-505.2002.

Konishi J and Maruhashi K. 2003. 2-(2′-Hydroxyphenyl)benzene sulfinate desulfinase from the thermophilic desulfurizing bacterium Paenibacillus sp. strain A11-2: purification and characterization. *Applied Microbiology and Biotechnology*, 62(4), 356–361. doi: 10.1007/s00253-003-1331-6.

Lal S and Tabacchioni S. 2009. Ecology and biotechnological potential of Paenibacillus polymyxa: a minireview. *Indian Journal of Microbiology*, 49(1), 2–10.

Lewis K. 2010. Persister Cells. *Annual Review of Microbiology*, 64(1), 357–372. doi: 10.1146/annurev.micro.112408.134306.

Matilla MA, Ramos JL, Duque E, de Dios Alché J, Espinosa-Urgel M and Ramos-González MI. 2007. Temperature and pyoverdine-mediated iron acquisition control surface motility of Pseudomonas putida. *Environmental Microbiology*, 9(7), 1842–1850. doi: 10.1111/j.1462-2920.2007.01286.x.

McCarter LL. 2004. Dual flagellar systems enable motility under different circumstances. *Journal of Molecular Microbiology and Biotechnology*, 7(1–2), 18–29.

McCarter LL. 2006. Regulation of flagella. *Current Opinion in Microbiology*, 9(2), 180–186. doi: http://dx.doi.org/10.1016/j.mib.2006.02.001.

McSpadden Gerdener BB. 2004. Ecology of Bacillus and Paenibacillus spp. in Agricultural Systems. *Phytopathology*, 94(11), 1252–1258. doi: 10.1094/phyto.2004.94.11.1252.

Melter O and Radojevič B. 2010. Small colony variants of Staphylococcus aureus — review. *Folia Microbiologica*, 55(6), 548–558. doi: 10.1007/12223-010-0089-3.

Miller MB, Skorupski K, Lenz DH, Taylor RK and Bassler BL. 2002. Parallel quorum sensing systems converge to regulate virulence in vibrio cholerae. *Cell*, 110(3), 303–314. doi: http://dx.doi.org/10.1016/S0092-8674(02)00829-2.

Mire CE, Tourjee JA, O'Brien WF, Ramanujachary KV and Hecht GB. 2004. Lead precipitation by vibrio harveyi: evidence for novel quorum-sensing interactions. *Applied Environmental Microbiology*, 70(2), 855–864.

Mok KC, Wingreen NS and Bassler BL. 2003. Vibrio harveyi quorum sensing: a coincidence detector for two autoinducers controls gene expression. *EMBO Journal*, 22(4), 870–881.

Monahan LG, Turnbull L, Osvath SR, Birch D, Charles IG and Whitchurch CB. 2014. Rapid conversion of pseudomonas aeruginosa to a spherical cell morphotype facilitates tolerance to carbapenems and penicillins but increases susceptibility to antimicrobial peptides. *Antimicrobial Agents and Chemotherapy*, 58(4), 1956–1962. doi: 10.1128/aac.01901-13.

Montes MJ, Mercadé E, Bozal N and Guinea J. 2004. Paenibacillus antarcticus sp. nov., a novel psychrotolerant organism from the Antarctic environment. *International Journal of Systematic and Evolutionary Microbiology*, 54(5), 1521–1526. doi: 10.1099/ijs.0.63078-0.

Nazir R, Warmink JA, Boersma H and Van Elsas JD. 2010. Mechanisms that promote bacterial fitness in fungal-affected soil microhabitats. *FEMS Microbiology Ecology*, 71(2), 169–185. doi: 10.1111/j.1574-6941.2009.00807.x.

Nealson KH and Hastings JW. 1979. Bacterial bioluminescence: Its control and ecological significance. *Microbiological Reviews*, 43(4), 496–518.

Nelson AD, Barber LE, Tjepkema J, Russell SA, Powelson R, Evans HJ and Seidler RJ. 1976. Nitrogen fixation associated with grasses in Oregon. *Canadian Journal of Microbiology*, 22(4), 523–530. doi: 10.1139/m76-078.

Ouyang J, Pei Z, Lutwick L, Dalal S, Yang L, Cassai N, Sandhu K, Hanna B, Wieczorek RL, Bluth M and Pincus MR. 2008. Paenibacillus thiaminolyticus: a new cause of human infection, inducing bacteremia in a patient on hemodialysis. *Annals of Clinical & Laboratory Science*, 38(4), 393–400.

Pevzner PA, Borodovskyb MY and Mironov AA. 1989. Linguistics of nucleotide sequences II: stationary words in genetic texts and the zonal structure of DNA. *Journal of Biomolecular Structure and Dynamics*, 6(5), 1027–1038.

Pevznera PA, Borodovskyb MY and Mironova AA. 1989. Linguistics of nucleotide sequences I: the significance of deviations from mean statistical characteristics and prediction of the frequencies of occurrence of words. *Journal of Biomolecular Structure and Dynamics*, 6(5), 1013–1026.

Proctor RA, Kahl B, Von Eiff C, Vaudaux PE, Lew DP and Peters G. 1998. Staphylococcal small colony variants have novel mechanisms for antibiotic resistance. *Clinical Infectious Diseases*, 27(1), S68–S74. doi: 10.1086/514906.

Ptashne M and Gann A. 2002. Genes and signals. In *Cold Spring Harbor*. Cold Spring Harbor Press, USA.

Queller DC and Strassmann JE. 2002. The many selves of social insects. *Science*, 296(5566), 311–313. doi: 10.1126/science.1070671.

Raza W, Yang W and Shen QR. 2008. Paenibacillus polymyxa: antibiotics, hydrolytic enymes and hazard assessment. *Journal of Plant Pathology*, 90(3), 419–430.

Roth D, Finkelshtein A, Ingham C, Helman Y, Sirota-Madi A, Brodsky L and Ben-Jacob E. 2013. Identification and characterization of a highly motile and antibiotic refractory subpopulation involved in the expansion of swarming colonies of Paenibacillus vortex. *Environmental Microbiology*, 15(9), 2532–2544. doi: 10.1111/1462-2920.12160.

Searls DB. 2002. The language of genes. *Nature*, 420(6912), 211–217.

Shank EA and Kolter R. 2011. Extracellular signaling and multicellularity in Bacillus subtilis. *Current Opinion in Microbiology*, 14(6), 741–747.

Shapiro JA. 1995. The significances of bacterial colony patterns. *BioEssays*, 17(7), 597–607. doi: 10.1002/bies.950170706.

Shapiro JA. 1998. Thinking about bacterial populations as multicellular organisms *Annual Review of Microbiology*, 52(1), 81–104. doi: 10.1146/annurev.micro.52.1.81.

Shapiro JA and Dworkin M. 1997. *Bacteria as Multicellular Organisms*. University Press, Oxford, USA.

Shklarsh A, Finkelshtein A, Ariel G, Kalisman O, Ingham C and Ben-Jacob E. 2012. Collective navigation of cargo-carrying swarms. *Interface Focus*, 2(6), 786–798. doi: 10.1098/rsfs.2012.0029.

Sirota-Madi A, Olender T, Helman Y, Ingham C, Brainis I, Roth D, Hagi E, Brodsky L, Leshkowitz D, Galatenko V, Nikolaev V, Mugasimangalam R, Bransburg-Zabary S, Gutnick D, Lancet D and Ben-Jacob E. 2010. Genome sequence of the pattern forming Paenibacillus vortex bacterium reveals potential for thriving in complex environments. *BMC Genomics*, 11(1), 710.

Stephens C. 1998. Bacterial sporulation: A question of commitment? *Current Biology*, 8(2), R45–R48.

Sternberg RJ. 2000. *Handbook of Intelligence*. Cambridge University Press, Cambridge, UK.

StJohn FJ, Rice JD and Preston JF. 2006. Paenibacillus sp. strain JDR-2 and XynA1: a novel system for methylglucuronoxylan utilization. *Applied and Environmental Microbiology*, 72(2), 1496–1506. doi: 10.1128/aem.72.2.1496-1506.2006.

Strassmann JE. 2000. Evolution: bacterial cheaters. *Nature*, 404(6778), 555–556.

Toguchi A, Siano M, Burkart M and Harshey RM. 2000. Genetics of swarming motility in salmonella enterica Serovar Typhimurium: critical Role for Lipopolysaccharide. *Journal of Bacteriology*, 182(22), 6308–6321. doi: 10.1128/jb.182.22.6308-6321.2000.

Tremblay J and Deziel E. 2010. Gene expression in Pseudomonas aeruginosa swarming motility. *BMC Genomics*, 11(1), 587.

Veening J-W, Stewart EJ, Berngruber TW, Taddei F, Kuipers OP and Hamoen LW. 2008. Bet-hedging and epigenetic inheritance in bacterial cell development. *Proceedings of the National Academy of Sciences USA*, 105(11), 4393–4398.

Velicer GJ. 2003. Social strife in the microbial world. *Trends in Microbiology*, 11(7), 330–337.

Velicer GJ, Kroos L and Lenski RE. 2000. Developmental cheating in the social bacterium Myxococcus xanthus. *Nature*, 404(6778), 598–601.

Verweij PE, Mellado E and Melchers WJG. 2007. Multiple-Triazole–Resistant Aspergillosis. *New England Journal of Medicine*, 356(14), 1481–1483. doi:10.1056/NEJMc061720.

Viswanathan VK. 2013. Sensing bacteria, without bitterness? *Gut Microbes*, 4(2), 91–93.

Warmink JA and van Elsas JD. 2009. Migratory response of soil bacteria to lyophyllum sp. strain karsten in soil microcosms. *Applied and Environmental Microbiology*, 75(9), 2820–2830. doi: 10.1128/aem.02110-08.

Warnow T. 1997. Mathematical approaches to comparative linguistics. *Proceedings of the National Academy of Sciences*, 94(13), 6585–6590.

Watanapokasin RY, Boonyakamol A, Sukseree S, Krajarng A, Sophonnithiprasert T, Kanso S and Imai T. 2009. Hydrogen production and anaerobic decolorization of wastewater containing Reactive Blue 4 by a bacterial consortium of Salmonella subterranea and Paenibacillus polymyxa. *Biodegradation*, 20(3), 411–418. doi: 10.1007/s10532-008-9232-0.

Weibull C and Gyllang H. 1965. Metabolic properties of some L forms derived from gram-positive and gram-negative bacteria. *Journal of Bacteriology*, 89(6), 1443–1447.

Wirth R, Muscholl A and Wanner G. 1996. The role of pheromones in bacterial interactions. *Trends in Microbiology*, 4(3), 96–103. doi: http://dx.doi.org/10.1016/0966-842X(96)81525-3.

Xavier KB and Bassler BL. 2003. LuxS quorum sensing: more than just a numbers game. *Current Opinion in Microbiology*, 6(2), 191–197. doi: http://dx.doi.org/10.1016/S1369-5274(03)00028-6.

11. The Crosstalk Between Plants and Their Arbuscular Mycorrhizal Symbionts: A Mycocentric View

Cristiana Sbrana[*,‡], Alessandra Turrini[†,§], and Manuela Giovannetti[†,||]

[*] *Institute of Agricultural Biology and Biotechnology CNR, University of Pisa, Via del Borghetto 80, 56124 Pisa, Italy*

[†] *Department of Agriculture, Food and Environment University of Pisa, Via del Borghetto 80, 56124 Pisa, Italy*

[‡] *sbrana@ibba.cnr.it*

[§] *alessandra.turrini@unipi.it*

[||] *manuela.giovannetti@unipi.it*

Arbuscular mycorrhizal fungi (AMF) are beneficial microorganisms fundamental for soil fertility and plant nutrition, which establish mutualistic symbioses (mycorrhiza) with the roots of 80% of land plants, including many agricultural crops. AMF obtain carbon compounds from plant cells, which, as chemoheterotrophs, are unable to synthesize, and provide their host plants with soil mineral nutrients, mainly phosphorus and nitrogen, absorbed and translocated to the root cells by the extensive extraradical mycelium spreading from roots to the surrounding soil. In addition, mycorrhizal plants are more tolerant to biotic and abiotic stresses, and show higher production of secondary metabolites with antioxidant properties. During symbiosis establishment a complex dialogue occurs between the two partners: AMF recognize compounds secreted by host roots, such as strigolactones, which elicit an extensive hyphal branching in the vicinity of the roots, before the formation of appressoria on the root surface. On the other hand, AMF release symbiotic signals, a mixture of sulphated and non-sulphated simple lipochitooligosaccharides (LCOs), triggering mycorrhizal formation in different plant families. Here, we will focus on (i) fungal release of symbiotic signals during the asymbiotic phase of AMF life cycle and fungal recognition responses to host-derived signals during the presymbiotic phase, (ii) appressorium formation on the root surface; (iii) the symbiont/host crosstalk during cell-to-cell interactions leading to the formation of a functional symbiosis, (iv) fungal behavior in the presence of non-host plants.

1. Introduction

Arbuscular mycorrhizal fungi (AMF, phylum Glomeromycota) are beneficial soil microbes which establish symbiotic associations with the roots of most land plants and play a fundamental role in plant growth, soil fertility, and productivity (Smith and Read, 2008). AMF provide many essential ecosystem services, such as an efficient use of fertilizer and soil nutrients, protection against biotic and abiotic stresses, increased N-fixation in legume crops and improved soil aggregation (Gianinazzi *et al.*, 2010), reducing the need of chemical fertilizers and pesticides and leading to less environmental damages. In addition, AMF increase the synthesis of diverse beneficial plant secondary metabolites, thus contributing to the production of safe and high quality food (Giovannetti *et al.*, 2013). They enhance plant nutrient uptake by means of extensive mycelial networks which spread from colonized roots into the nearby soil and efficiently absorb and translocate mineral nutrients, mainly phosphorus (P), nitrogen (N), copper (Cu), and zinc (Zn) (Smith and Read, 2008). In exchange, they obtain plant carbon compounds, which they are not able to synthesize, as chemoheterotrophic organisms.

AMF are obligate biotrophs, depending on the establishment of a functional symbiosis with host plants for the completion of their life cycle, that involves key developmental switches highly regulated by different edaphic and environmental conditions. After spore germination, the asymbiotic mycelium undergoes a differential morphogenesis and branching pattern in the presence of host roots, which preludes the formation of specialized structures, the appressoria, able to adhere to and penetrate the roots by means of intraradical hyphae. Such hyphae grow intercellularly before developing intracellular austoria-like branchings, called arbuscules, which represent the key structure of the symbiosis where nutrient exchanges between the host plant and the fungal symbiont occur. After arbuscule formation the carbon compounds obtained by the host plant allow AMF to develop an extensive extraradical hyphal network and to complete their life cycle by producing new spores (see Giovannetti *et al.*, 2010).

A complex chemical dialogue occurs between the two partners of the symbiosis: fungal symbionts secrete diffusible signals stimulating mycorrhizal establishment and host plants release root exudates, acting as molecular signals activating early fungal recognition responses, represented by chemotropic hyphal growth and differential hyphal morphogenesis and branching in the vicinity of roots (Akiyama *et al.*, 2005; Sbrana and Giovannetti, 2005). Hyphal branching is followed by the formation of appressoria, which are the subsequent additional signs of fungal host recognition leading to root colonization and to symbiosis formation. Thus, the establishment of a functional AMF symbiosis can be considered the product of multifold signals of different nature, acting at different steps of the interaction between fungal symbionts and host plants.

In this chapter, we will review the different steps leading to the formation of a functional mycorrhizal symbiosis focusing on the AMF symbionts. We will examine (i) AMF release of symbiotic signals and fungal recognition responses to host-derived signals allowing the establishment of the symbiosis, (ii) appressorium formation on the root surface, (iii) the symbiont/host crosstalk during cell-to-cell interactions leading to the formation of a functional symbiosis, (iv) fungal behavior in the presence of non-host plants, hindering mycorrhizal establishment.

2. AMF Release of Symbiotic Signals and Fungal Recognition Responses to Host-Derived Signals

The first step of AMF life cycle is represented by spore germination, which is not regulated by host-derived signals. This inconsistent behavior for obligately biotrophic organisms like AMF, which are considered evolutionary successful living fossils, having survived 460,000,000 to possibly 980,000,000 years (Redecker *et al.*, 2000; Blair, 2009), is balanced by diverse survival strategies (Sbrana *et al.*, 2011). The germination of spores is triggered by different edaphic and environmental factors, such as pH, temperature, moisture, mineral and organic nutrients, host/non-host plants and microorganisms (see Giovannetti *et al.*, 2010). After germination, germling hyphae follow a linear growth with a strong apical dominance, and give rise to a coenocytic mycelial network, containing many nuclei and organelles. In several genera — *Acaulospora*, *Claroideoglomus*, *Glomus* and *Rhizophagus* — when coming into contact, hyphae are able to fuse and form anastomosis at rates ranging from 9% to 91% (Giovannetti *et al.*, 1999; de Novais *et al.*, 2013). Species belonging to *Gigaspora* and *Scutellospora* are often unable to fuse, and the rare anastomoses observed mostly occurred within the same hypha (de la Providencia *et al.*, 2005). Anastomoses are fundamental for the establishment of a highly interconnected mycelium where protoplasm flow allows the maintenance of the physiological and genetic continuity within individual germlings.

Many important biochemical changes occur in the asymbiotic phase, including the synthesis of proteins, lipids, sterols, polyamines, mRNA, adenosine triphosphate (ATP) and enzymes of the central metabolic pathways, necessary for germling growth. Concurrently, different genes are expressed in germlings showing homology to those involved in primary metabolism, transport processes and polarized growth (see Giovannetti *et al.*, 2010). Nevertheless, AMF germlings are not capable of extensive hyphal development and, in the absence of the host, their growth is arrested within three weeks, when protoplasm is retracted from hyphal tips and reallocated in the viable germ tube close to the mother spore, to maintain the limited energy resources (Logi *et al.*, 1998; Bago *et al.*, 1998). During spore germination and germling growth arrest, protoplasm flow is always associated with cytoskeletal microtubules, microfilaments and proteins, functional to nuclear and cellular organelles migration (Åström *et al.*, 1994; Logi *et al.*, 1998). No information is available on the molecular steps leading to protoplasm retraction and growth arrest in the absence of the host plant, although it has been hypothesized that the *Funneliformis mosseae* gene *GmGIN* 1, expressed in asymbiotic mycelium, may be involved in the signaling cascade controlling such phenomenon (Requena *et al.*, 2002).

Many authors demonstrated that during asymbiotic growth, in the absence of host plants, AMF germlings constitutively release compounds (the so-called "Myc factors") which, similarly to Nod factors, are able to activate arbuscular mycorrhizal (AM)-related signaling pathways. The release of diffusible symbiotic signals by different AMF species induced calcium spiking in root epidermal cells of *Medicago truncatula* and *Daucus carota*, the accumulation of starch in *Lotus japonicus* roots and the expression of symbiotic genes, such as early nodulins, involved in both nodulation and mycorrhization, and of a membrane-bound steroid-binding protein (MtMSBP1) (Kosuta *et al.*, 2003; Kuhn *et al.*, 2010; Chabaud *et al.*, 2011). Recent findings provided a better understanding of the molecular mechanisms leading to mycorrhizal establishment, revealing that *Rhizophagus intraradices* germinating spores are able to secrete a mixture of sulphated and non-sulphated simple LCOs, with structural similarities with rhizobial Nod factors, which stimulated mycorrhizal symbiosis formation in plant species of diverse families (*Fabaceae, Asteraceae,* and *Umbelliferae*) (Maillet *et al.*, 2011). The plant protein RAM1 has been suggested to be involved in LCO-induced responses through the formation of heterodimers with the factor NSP2 (Gobbato *et al.*, 2012), while a lysin motif receptor-like kinase LYR3, showing high-affinity binding to LCOs but not to chito-oligosaccharides released by plant pathogenic fungi, was proposed as a putative LCO-binding protein (Fliegmann *et al.*, 2013). Other authors showed that short-chain chitin oligomers, simple compounds possibly released by AMF hyphae, are able to mimic Ca^{2+} spiking elicited by germinated spores exudates through the involvement of both DMI1 and DMI2, indicating a role of such compounds as fungal symbiotic signals (Genre *et al.*, 2013). Interestingly, their levels in spore exudates are enhanced by exposure to synthetic strigolactones, suggesting that, in natural environment, plant-derived signals stimulation of germlings may boost subsequent host root responses.

The presymbiotic phase of AMF life cycle starts with the perception of signals released by host roots which are able to elicit a differential hyphal morphogenesis in the fungal symbiont, represented by the development of a very densely branched network whose density is 4–20-fold that of the mycelium growing apart from the roots (Giovannetti *et al.*, 1993 a; Buée *et al.*, 2000; Nagahashi and Douds, 2000).

Such fan-like or arbuscule-like structures may be functional to locate suitable infection sites on the root surface, and occur exclusively in the proximity of host roots, either on membranes or on agar.

Host-specific chemotropic growth was detected in aerial hyphae of the AM fungus *G. gigantea*, which were able to contact maize roots over a distance of up to 11 mm, and in mycelium of *F. mosseae*, able to perceive host signals and reorient hyphal growth toward host roots through interposed membranes up to 1 mm distance (Koske, 1982; Sbrana and Giovannetti, 2005).

The chemical identification of "branching factors" exuded from host roots has been widely studied. The use of dialysis membranes allowed the detection of their maximum molecular weight, 500 Da (Giovannetti *et al.*, 1996), while an *in vitro* bioassay was fundamental for the determination of some chemical characteristics and of their distribution in the plant kingdom (Buée *et al.*, 2000). The first identified branching factor was 5-Deoxystrigol, detected in *L. japonicus* root exudates (Akiyama *et al.*, 2005), then many host plant exudates were found to contain compounds belonging to strigolactones able to stimulate hyphal branching and to change fungal physiology and mitochondrial activity (Besserer *et al.*, 2006). An interaction between strigolactones and auxin in mycorrhizal development has been proposed, since IAA positively regulates the expression of strigolactone biosynthesis genes and strigolactone exudation (Foo, 2013). Interestingly, the branched strigolactone-induced fungal hyphae seem necessary for calcium oscillation in plant root cells (Kosuta *et al.*, 2008) and for inducing the expression of MtMSBP1, a *M. truncatula* gene critical for mycorrhizal establishment (Kuhn *et al.*, 2010).

Mutants or transgenic plants with reduced expression of the strigolactone key biosynthetic enzymes and *Petunia hybrida* defective in pdr1 ABC transporter, functioning as strigolactones exporter, showed decreased AMF colonization (Koltai *et al.*, 2010; Kretzschmar *et al.*, 2012). The same study showed that neither plant defence responses or altered colonization processes were elicited, suggesting that the lower mycorrhizal susceptibility was the result of a limited fungal commitment to symbiosis. On the other hand, *Lycopersicon esculentum* (*pmi*) mutants unable to be colonized by AMF germlings released low levels of the branching inducers strigolactones (David-Schwartz *et al.*, 2003; Koltai *et al.*, 2010), indicating that pre-contact recognition events represented by hyphal branching are needed for fungal physiological switch from the asymbiotic to the presymbiotic stage. Such an hypothesis is supported by studies performed on pre-branching activation of fungal genes encoding for putative pyruvate carboxylase and mitochondrial adenosine diphosphate (ADP)/ATP translocase and of oxygen consumption, in *G. rosea* and *R. intraradices* treated with host root exudates (Tamasloukht *et al.*, 2003).

Plant-exuded compounds belonging to 2-hydroxy fatty acids (2-OH-FA) stimulate a different kind of hyphal ramification, characterized by low branching indices and occurring at a larger distance from roots with respect to strigolactones-induced branching (Nagahashi and Douds, 2011), suggesting that the activity of root exudates on AMF is the result of a synergy of signaling compounds.

3. Appressorium Formation on the Root Surface

The first physical contact between AMF symbionts and host plants occurs on the root epidermis, where the hyphae form appressoria, which are inflated, multinucleate structures representing the key signs of fungal recognition of potential host plants (Staples and Macko, 1980). They vary in number, shape, and size and are produced in different numbers at different times by the different AMF species. For example, *F. mosseae* was able to produce appressoria on *Ocimum basilicum* roots as early as 36 hours after the beginning of the interaction, and their number was 22 per plant after 72 hours (Giovannetti and Citernesi, 1993) (Fig. 1). The event of hyphal adhesion to root cells has not been adequately investigated: it is not known whether adhesion is mediated by compounds occurring in root mucilages, or by the interaction between root and hyphal surfaces. Studies on factors affecting appressorium formation reported that even in the presence of host root exudates, thigmotropic stimuli, such as nylon, silk cellulose, polyamide or glass threads were not able to elicit differentiation of infection structures (Giovannetti *et al.*, 1993 b), while they were detected on isolated cell walls of carrot roots (Nagahashi and Douds, 1997). Likewise, scanty information is available on the nature of signals eliciting the formation of infection structures: it has been hypothesized that specific

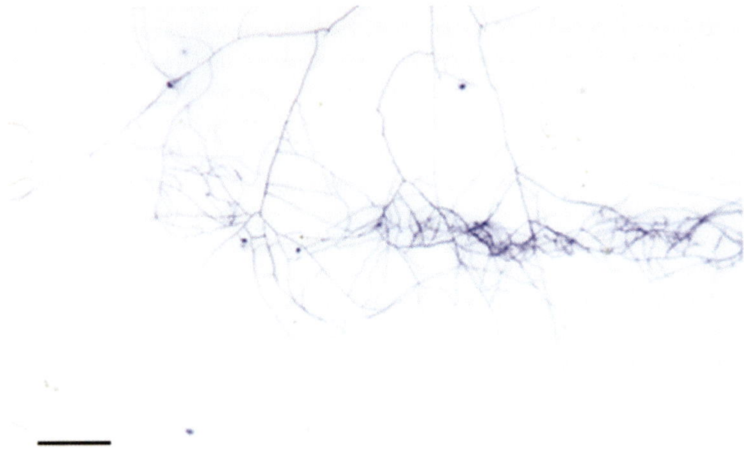

Figure 1. *Funneliformis mosseae* hyphae showing enhanced branching along the axis of a *Ocimum basilicum* root growing underneath a membrane filter. Scale bar = 200 μm.

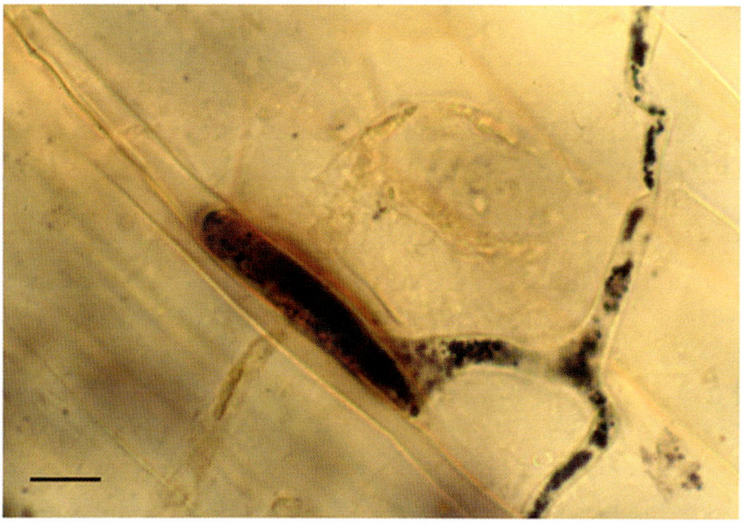

Figure 2. Appressorium of *Funneliformis mosseae* formed in epidermal grooves on the surface of a *Pisum sativum* P2 mutant root, showing no hyphal penetration of root tissues and red staining of phenolic depositions in cell walls. Scale bar = 10 μm.

topographical signals may be involved, since hyphae often grow along epidermal grooves and appressoria are differentiated on clinal and anticlinal wall junctions between adjacent epidermal cells (Giovannetti *et al.*, 1993 b), as observed in other biotrophic fungi. The development of appressoria is considered the key event discriminating host from non-host plants. The detection of host mutants eliciting the formation of appressoria, but hindering root colonization, allowed the dissection of such recognition event and stimulated studies on mutant genotypes of host plants showing the disruption of the colonization process at further steps of the interaction (Duc *et al.*, 1989; Bradbury *et al.*, 1991). In the mycorrhiza-resistant (*myc-*) P2 pea mutant (*locus a*), identified amongst *nod-* mutants, the AM fungus *F. mosseae* formed on the root epidermis many viable appressoria producing short infection hyphae which rapidly aborted, while eliciting wall thickenings and paramural deposition of phenolics and glucans in adjacent root cells, indicating the activation of defence responses (Gollotte *et al.*, 1993) (Fig. 2).

Several other mutations were identified by using two model legumes, *M. truncatula* and *L. japonicus*, where fungal penetration was hindered and AMF growth was blocked on epidermal cells (Parniske, 2008). Different plant mutants allowed the cloning of a set of plant genes, belonging to the so-called common symbiosis (SYM) pathway that perceive both Nod (from Rhizobia) and Myc (from AMF) factors and encode for proteins essential for Ca^{2+} spiking production and decodification, the starting point for activating downstream responses that lead to the symbiosis establishment (Haage and Parniske, 2013). Recently, the *M. truncatula ram1* and *ram2* (*required* for *arbuscular mycorrhization*) mutants were shown to be defective in the production of a GRAS-type transcriptional regulator and of a glycerol-3-phosphate acyl-transferase (GPAT), respectively (Gobbato *et al.*, 2012). RAM1 protein, whose absence produces a defect in appressorium formation, interacts with GRAS-TR to regulate the expression of RAM2/GPAT, promoting the biosynthesis of cutin monomers, which can act as signals for the fungal symbionts, in analogy with foliar fungal pathogens (Liu *et al.*, 2011). In addition, such proteins can also play a fundamental role in arbuscule formation, since arbuscules are often abortive in *ram2* mutants (Wang *et al.*, 2012). Interestingly, RAM1/GRAS-TR is lacking in some *Brassicales* (*Arabidopsis thaliana, Arabidopsis lyrata,* and *Brassica rapa*), which are non-host plants, suggesting an important function of this protein in the establishment of AM symbiosis.

Whilst signals involving Ca^{2+} spiking and gene activation pathways have been well documented in plants, signaling systems that promote the beginning of symbiotic interactions have been only hypothesized to occur in AMF (Breuninger and Requena, 2004; Liu *et al.*, 2013). Actually, genes related to Ca^{2+} signaling, coding for a putative *p*-type Ca^{2+}-ATPase, a calmodulin, a leucine zipper EF-hand protein, and a Ca^{2+}-induced Ras inactivator (CAPRI) are transcribed in *F. mosseae* during appressorium development (Breuninger and Requena, 2004), whilst an endoplasmic reticulum *p*-type ATPase and a Vcx1-like vacuolar Ca^{2+} ion transporter is activated in intraradical hyphae and arbuscules of *R. intraradices* (Liu *et al.*, 2013): a conserved signaling cascade regulating changes in

Ca^{2+} concentration and triggering downstream pathways seems a rule for symbiotic partner interaction, but no information is available on molecules involved in Ca^{2+} spiking activation in AMF during appressorium development. Interestingly, Seddas *et al.* (2009) reported that the inactivation of SYMRK/DMI2 and CCaMK/DMI3, which are involved in plant Ca^{2+} spiking signal at the stage of appressorium development, was associated with inhibition/inactivation of some fungal genes, such as the fungal proteins U1 and U2 of unknown function, the protein RHO (putatively implicated in cellular signaling), a stearoyl-CoA desaturase, a peptidyl-prolyl isomerase and 26S proteasome regulatory subunit involved in lipid metabolism and protein turnover. Such findings suggest that some stimulatory signals and/or inhibitory factors in the fungus are regulated by the host plant (Seddas *et al.*, 2009).

In host plants, 4–5 hours after the formation of appressoria a subcellular structure precedes the intracellular colonization by the fungal symbiont, the so-called prepenetration apparatus (PPA), which contains cytoskeletal microtubules and microfilaments forming a transcellular tunnel together with dense endoplasmic reticulum (Genre *et al.*, 2005). The formation of PPA requiring the common Sym genes *dmi2* and *dmi3*, is correlated to the activation of Ca^{2+} spiking decodification protein CCaMK and is concomitant with the activation of a expansin-like gene, involved in plant cell wall modification (Siciliano *et al.*, 2007). Unfortunately, nothing is known on the signals triggering the production of such a structure. During the cellular rearrangements following the Ca^{2+} spiking signal, Vapyrin, a novel protein that mediate protein–protein interactions may play an important structural role in plant cell rearrangement, as it is induced transiently in the epidermis concurrent with appressorium development, and is detected in the cortex during arbuscule formation (Pumplin *et al.*, 2010). Indeed, such protein was not detected in the non-host plant *Arabidopsis*. In Vapyrin RNA interference (RNAi) mutant roots, the fungal symbionts form appressoria and attempt cell penetration, suggesting that the early signaling in the crosstalk is intact, while the mutation affects the plant cellular machinery that sustains fungal colonization, possibly those linked to membrane invagination (Pumplin *et al.*, 2010). A role in membrane re-organization was hypothesized recently also for MtMSBP1 activated when branched hyphae, originating from a runner hypha, reach the root surface (Khun *et al.*, 2010). MtMSBP1 RNAi roots showed thick and septated appressoria, decreased numbers of arbuscules and distorted arbuscule morphology, providing evidences for a critical role of this gene during mycorrhizal symbiosis establishment. By the fungal side, the activation of genes involved in structural remodeling are supposed to be switched on as well. Hence, recently, Tisserant *et al.*, (2012) detected in the intraradical mycelium a high activity of enzymes involved in sterol and steroid metabolism that may be involved in membrane invagination and in the penetration of cortical cells by fungal symbionts.

A recent genome-wide analysis, based on the use of laser microdissection and *M. truncatula* GeneChip, provided novel information on plant and fungal genes specifically activated in samples of root cells with appressoria and in other four types of cells, containing and non-containing the fungal symbiont during arbuscular mycorrhizal symbiosis establishment (Hogekamp and Küster, 2013). In cells with appressoria most fungal sequences encoded proteins with unknown functions, the remaining sequences codifying housekeeping genes and genes for amino acid transport and metabolism: unfortunately, the authors were unable to discriminate between genes expressed in the extraradical mycelium and those expressed in the appressoria. Similar data were found by the first large-scale report of fungal symbiosis-related genes, most coding for unknown proteins, the remainder coding for proteins associated with ion transport, lipid/steroid metabolism and DNA replication (Tisserant *et al.*, 2012). Among the unknown proteins, the authors found a significant set of small secreted proteins (SSP) which were found only in the intraradical mycelium. The most expressed orphan SSP was step3_c3163, highly similar to sequences found in several biotrophic basidiomycetes. The effector protein SP7, the first SSP discovered in an AMF (*R. intraradices*), interacts with the ethylene-responsive factor ERF19, suppressing host defences. SP7 transcript was increased in the pre-symbiotic mycelium and in intraradical hyphae, and its constitutive expression *in planta* accelerated mycorrhizal colonization (Kloppholz *et al.*, 2011). Given the significance of proteins like SP7 in mycorrhizal establishment and the huge amount of SSP found by Tisserant *et al.* (2012), their identity and role in the symbiont/host crosstalk should be further investigated.

4. The Symbiont/Host Crosstalk During Cell-to-Cell Interactions

After the differentiation of appressoria, AMF colonize host roots forming intracellular and intercellular hyphae, vesicles and arbuscules. The activation of symbiont/host crosstalk pathways at the stage of arbuscule development is very complex, since the two partners must regulate a differential set of cell processes including nutrient exchange, membrane biogenesis and metabolism, cell wall re-organization and hormone balance. Molecules such as phosphate and lysophosphatidylcholine have been supposed to be signals involved at this stage of AMF cycle (Balzergue *et al.*, 2011; Drissner *et al.*, 2007). Actually, phosphate (P) import into plant cell by the *M. truncatula* P transporter MtPt4 was demonstrated to be essential for a functional mycorrhizal symbiosis, since MtPt4 mutants showed a premature arbuscule senescence and fungal growth arrest (Javot *et al.*, 2007). Besides, lysophosphatidylcholine, a bioactive molecule in plants and mammals, was shown to be able to activate two potato phosphate transporters specifically induced in mycorrhizal roots (Drissner *et al.*, 2007). Some authors have shown that P transporters are present contemporaneously both in plants and fungi in arbusculated cells (Balestrini *et al.*, 2007), but despite the importance of nutrient exchanges in the symbiosis, where partners offering the best rate of exchange are rewarded, little is known on the mechanisms that regulate such processes. In *F. mosseae* and *R. intraradices* other nutrient (ammonium, sulphate, potassium) transporters have been found in arbusculated cells and, at least for ammonium, a probable regulatory role in arbuscule functioning was suggested (Javot *et al.*, 2011). Recently, a monosaccharide transporter (MST2) was detected in *R. irregularis* intercellular hyphae and arbuscules: when the expression of this transporter was reduced, arbuscules were malformed and also MtPt4 expression decreased, supporting the hypothesis that the exchange of carbon for phosphate is tightly linked and required both fungal and plant control (Helber *et al.*, 2011).

In plant cells, arbuscule development requires the presence of two AM-specific subtilases, SbtM1 and SbtM3, since their suppression by RNAi caused a decrease in intraradical hyphae and arbuscule colonization (Takeda *et al.*, 2009). On the other hand, in fungal cells a set of enzymes essential for the maintenance of fungal cell wall plasticity, including chitin synthases, chitin deacetylases, chitinases, and proteins involved in fungal membranes development, and the acetyl-CoA carboxylase were upregulated (Tisserant *et al.*, 2012).

5. Fungal Behavior in the Presence of Non-Host Plants

The inability of plant species belonging to *Chenopodioideae*, *Brassicaceae*, *Caryophyllaceae*, *Polygonaceae* and *Urticaceae*, including many agricultural weeds, and to the genus *Lupinus* to establish symbiotic relationships with AMF remains poorly understood, although different hypotheses have been posed. Some studies reported that non-host plants are not able to elicit mycelial branching, and that the differentiation of functional appressoria on root surfaces is hindered (Avio *et al.*, 1990; Giovannetti *et al.*, 1993 b, 1994), suggesting the absence of factors needed for host recognition and symbiosis development. Moreover, inhibitory compounds reducing fungal growth have been detected in non-hosts and in tomato mutant root exudates (Nagahashi and Douds, 2000; David-Schwartz *et al.*, 2003).

Appressoria-like structures, often producing abortive entry points, or hyphal swellings were detected in viable non-host root cells challenged with different AMF, suggesting a constitutive ability of AMF hyphae to recognize root surfaces (Giovannetti *et al.*, 1994; Wagg *et al.*, 2011).

Results obtained from analyses of non-host roots both from the field or AMF-inoculated in microcosm suggest that the outcome of fungal interactions with non-hosts may be modulated by fungal identity and experimental systems. In a microcosm experiment, three *F. mosseae* isolates were not able to establish symbioses with the non-host weed species *Cerastium arvense, Chenopodium album* and *Stellaria media*. Nevertheless, their hyphae showed apical swellings and/or appressoria-like formation on roots (Ramasamy, 2011). The mycorrhizal status of *C. album* is controversial, since field roots or experimentally inoculated plants showed either AMF colonization, with or without arbuscules, or no colonization (Wang *et al.*, 2006; Ramasamy, 2011). Field samples of different species belonging to the non-host genus *Thlaspi*, showing colonized roots with vesicles and arbuscules, produced many AMF sequences clustering with those of the worldwide distributed species *R. intraradices*, though when inoculated with *R. intraradices* isolates in the greenhouse, *Thlaspi* roots did not show appressoria formation or colonization (Regvar *et al.*, 2003).

Hosts of ectomycorrhizas, like *Abies alba, Pinus nigra, Pinus strobus, Pinus banksiana, Pinus contorta* and *Picea glauca* × *Picea engelmannii*, showed limited root colonization with intracellular coils, swellings, vesicles or spores, without any arbuscule formation (Giovannetti *et al.*, 1994; Wagg *et al.*, 2008). Similarly, *Arbutus unedo* and *Vaccinium myrtillus*, hosts of arbutoid and ericoid mycorrhizas, respectively, were not able to establish AMF symbiosis, although a larger number of hyphal swellings was detected on their roots, compared with those found on roots of non-host plants (Giovannetti *et al.*, 1994).

Interestingly, non-host plants grown in the presence of a mycorrhizal companion plant (or nurse plant) often showed variable extent of root colonization, although lacking arbuscules (Ocampo *et al.*, 1980; Veiga *et al.*, 2013), suggesting that symbiotic mycelium does not need host-derived activation to colonize roots. A recent study reported that *A. thaliana* was unable to respond to AMF exudates and that its roots were colonized by *R. irregularis* hyphae and vesicles only in the presence of a mycorrhizal host plant (Veiga *et al.*, 2013).

Strigolactones have been detected at lower levels in root exudates of non-host plants compared to host plants, and in strigolactone-deficient mutants of pea and tomato, showing highly reduced root colonization (Koltai *et al.*, 2010). Interestingly, the synthetic strigolactone GR24 increased AMF colonization in the strigolactone-deficient *rms1* mutant of pea to a similar level as in the wild type plants, but it did not induce colonization by *R. intraradices* in the non-host plant species *A. thaliana, Spinacia oleracea, Lupinus polyphillus* and *Fagopyrum esculentum*, suggesting that strigolactones are not involved in non-host immunity (Illana *et al.*, 2011).

6. Conclusions

The establishment of a functional AM symbiosis is the outcome of multifold signals of different nature, acting at different steps of the interaction between fungal symbionts and host plants. The complex chemical dialogue occurring between the two partners in the early phase of symbiosis establishment has been widely investigated and allowed the dissection of the molecular events leading to mutual recognition. During the asymbiotic growth phase, in the absence of host plants, AMF germlings constitutively release a mixture of sulphated and non-sulphated simple LCOs, able to stimulate mycorrhizal symbiosis formation in plant species of diverse families. During the presymbiotic phase of their life cycle, AMF perceive chemical signals released by host roots, which elicit fungal chemotropism and a differential morphogenesis represented by the development of a very densely branched hyphal network. The further step leading to the establishment of the symbiosis, appressorium development on the root epidermis, the key event discriminating host from non-host plants, has been poorly investigated, as scanty information is available on the nature of signals eliciting the formation of infection structures. The detection of mutants of host species where fungal penetration was hindered and AMF growth was blocked on epidermal cells allowed the cloning of plant genes belonging to the SYM pathway that perceive both Nod (from Rhizobia) and Myc (from AMF) factors. A recent genome-wide analysis provided novel information on plant and fungal genes activated in samples of root cells with appressoria during symbiosis establishment: unfortunately, it was not possible to discriminate between genes expressed in the extraradical mycelium and those expressed in the appressoria. Moreover, a significant set of SSP was detected only in the intraradical mycelium: further studies should investigate their role in mycorrhizal establishment and in the symbiont/host crosstalk. Challenges remain concerning non-host plants, which could be utilized as a tool for the identification of signals triggering appressorium formation and for the dissection of early fungal developmental switches necessary for the establishment of a functional symbiosis.

References

Akiyama K, Matsuzaki K and Hayashi H. 2005. Plant sesquiterpenes induce hyphal branching in arbuscular mycorrhizal fungi. *Nature*, 435, 824–827. doi: 10.1038/nature03608.

Åström H, Giovannetti M and Raudaskoski M. 1994. Cytoskeletal components in the arbuscular mycorrhizal fungus *Glomus mosseae*. *Molecular Plant-Microbes Interactions*, 7, 309–312.

Avio L, Sbrana C and Giovannetti M. 1990. The response of different species of *Lupinus* to VAM endophytes. *Symbiosis*, 9, 321–323.

Bago B, Azcón-Aguilar C, Goulet A and Piché Y. 1998. Branched absorbing structures (BAS): a feature of the extraradical mycelium of symbiotic arbuscular mycorrhizal fungi. *New Phytologist*, 139, 375–388.

Balestrini R, Gómez Ariza J, Lanfranco L and Bonfante P. 2007. Laser microdissection reveals that transcripts for five plant and one fungal phosphate transporter genes are contemporaneously present in arbusculated cells. *Molecular Plant-Microbe Interactions*, 20, 1055–1062. doi: 10.1094/MPMI-20-9-1055.

Balzergue C, Puech-Pagès V, Bécard G and Rochange SF. 2011. The regulation of arbuscular mycorrhizal symbiosis by phosphate in pea involves early and systemic signalling events. *Journal of Experimental Botany*, 62, 1049–1060. doi: 10.1093/jxb/erq335.

Besserer A, Puech-Pagès V, Kiefer P, Gomez-Roldan V, Jauneau A, Roy S, Portais J-C, Roux C, Bécard G and Séjalon-Delmas N. 2006. Strigolactones stimulate arbuscular mycorrhizal fungi by activating mitochondria. *PLoS Biology*, 4, 1239–1247. doi: 10.1371/journal.pbio.0040226.

Blair, JE. 2009. Fungi. In *The Timetree of Life,* Hedges SB and Kumar S, 215–219. Oxford University Press, Oxford.

Bradbury SM, Peterson RL and Bowley SR. 1991. Interactions between three alfalfa nodulation genotypes and two *Glomus* species. *New Phytologist,* 119, 115–120.

Breuninger M and Requena N. 2004. Recognition events in AM symbiosis: analysis of fungal gene expression at the early appressorium stage. *Fungal Genetics and Biology,* 41, 794–804. doi: 10.1016/j.fgb.2004.04.002.

Buée M, Rossignol M, Jauneau A, Ranjeva R and Bécard G. 2000. The pre-symbiotic growth of arbuscular mycorrhizal fungi is induced by a branching factor partially purified from plant root exudates. *Molecular Plant-Microbe Interactions,* 13(6), 693–698.

Chabaud M, Genre A, Sieberer BJ, Faccio A, Fournier J, Novero M, Barker DG and Bonfante P. 2011. Arbuscular mycorrhizal hyphopodia and germinated spore exudates trigger Ca^{2+} spiking in the legume and nonlegume root epidermis. *New Phytologist,* 189, 347–355 doi: 10.1111/j.1469-8137.2010.03464.x.

David-Schwartz R, Gadkar VJ, Wininger S, Bendov R, Galili GAD, Levy AA and Kapulnik Y. 2003. Isolation of a premycorrhizal infection (*pmi2*) mutant of tomato, resistant to arbuscular mycorrhizal fungal colonization. *Molecular Plant-Microbe Interactions,* 16(5), 382–388.

de la Providencia IE, De Souza FA, Fernández F, Séjalon-Delmas N and Declerck S. 2005. Arbuscular mycorrhizal fungi reveal distinct patterns of anastomosis formation and hyphal healing mechanisms between different phylogenic groups. *New Phytologist,* 165(1), 261–271. doi: 10.1111/j.1469-8137.2004.01236.x.

de Novais CB, Sbrana C, Orivaldo JS Jr., Siqueira JO and Giovannetti M. 2013. Vegetative compatibility and anastomosis formation within and among individual germlings of tropical isolates of arbuscular mycorrhizal fungi (Glomeromycota). *Mycorrhiza,* 23, 325–331.

Drissner D, Kunze G, Callewaert N, Gehrig P, Tamasloukht M'B, Boller T, Felix G, Amrhein N and Bucher M. 2007. Lyso-phosphatidylcholine is a signal in the arbuscular mycorrhizal symbiosis. *Science,* 3118, 265–268. doi: 10.1126/science.1146487.

Duc G, Trouvelot A, Gianinazzi-Pearson V and Gianinazzi S. 1989. First reports of non-mycorrhizal plant mutants (*Myc-*) obtained in pea (*Pisum sativum*) and fababean (*Vicia faba*). *Plant Science,* 60, 215–222.

Fliegmann J, Canova S, Lachaud C, Uhlenbroich S, Gasciolli V, Pichereaux C, Rossignol M, Rosenberg C, Cumener M, Pitorre D, Lefebvre B, Gough C, Samain E, Fort S, Driguez H, Vauzeilles B, Beau J-M, Nurisso A,

Imberty A, Cullimore J and Bono J-J. 2013. Lipo-chitooligosaccharidic symbiotic signals are recognized by LysM receptor-like kinase LYR3 in the legume *Medicago truncatula*. *ACS Chemical Biology*, 8(9), 1900–1906. dx.doi.org/10.1021/cb400369u.

Foo E. 2013. Auxin influences strigolactones in pea mycorrhizal symbiosis. *Journal of Plant Physiology*, 170, 523–528.

Genre A, Chabaud M, Balzergue C, Puech-Pagès V, Novero M, Rey T, Fournier J, Rochange S, Bécard G, Bonfante P and Barker DG. 2013. Short-chain chitin oligomers from arbuscular mycorrhizal fungi trigger nuclear Ca^{2+} spiking in *Medicago truncatula* roots and their production is enhanced by strigolactone. *New Phytologist*, 198, 179–189. doi: 10.1111/nph.12146.

Genre A, Chabaud M, Timmers T, Bonfante P and Barkerb DG. 2005. Arbuscular mycorrhizal fungi elicit a novel intracellular apparatus in *Medicago truncatula* root epidermal cells before infection. *The Plant Cell*, 17, 3489–3499.

Gianinazzi S, Gollotte A, Binet MN, Tuinen Dv, Redecker D and Wipf D. 2010. Agroecology: the key role of arbuscular mycorrhizas in ecosystem services. *Mycorrhiza*, 20, 519–530.

Giovannetti M and Citernesi AS. 1993. Time-course of appressorium formation on host plants by arbuscular mycorrhizal fungi. *Mycological Research*, 98, 1140–1142.

Giovannetti M, Avio L and Sbrana C. 2010. Fungal spore germination and pre-symbiotic mycelial growth. Physiological and genetic aspects. In *Arbuscular Mycorrhizas: Physiology and Function*, Koltai H and Kapulnik Y (eds.), 3–32. Springer Science+Business Media, Dordrecht.

Giovannetti M, Avio L and Sbrana C. 2013. Improvement of nutraceutical value of food by plant symbionts. In *Natural Products*, Ramawat KG and Mérillon J-M (eds.), 2641–2662. Springer-Verlag, Berlin Heidelberg.

Giovannetti M, Avio L, Sbrana C and Citernesi AS. 1993b. Factors affecting appressorium development in the vesicular-arbuscular mycorrhizal fungus *Glomus mosseae* (Nicol. & Gerd.) Gerd. & Trappe. *New Phytologist*, 123, 114–122.

Giovannetti M, Azzolini D and Citernesi AS. 1999. Anastomosis formation and nuclear and protoplasmic exchange in arbuscular mycorrhizal fungi. *Applied and Environmental Microbiology*, 65(12), 5571–5575.

Giovannetti M, Sbrana C and Logi C. 1994. Early processes involved in host recognition by arbuscular mycorrhizal fungi. *New Phytologist*, 127, 703–709.

Giovannetti M, Sbrana C, Avio L, Citernesi AS and Logi C. 1993a. Differential hyphal morphogenesis in arbuscular mycorrhizal fungi during pre-infection stages. *New Phytologist,* 125, 587–594.

Giovannetti M, Sbrana C, Citernesi AS and Avio L. 1996. Analysis of factors involved in fungal recognition responses to host-derived signals by arbuscular mycorrhizal fungi. *New Phytologist,* 133(1), 65–71.

Gobbato E, Marsh JF, Vernié T, Wang E, Maillet F, Kim J, Miller JB, Sun J, Bano SA, Ratet P, Mysore KS, Dénarié J, Schultze M and Oldroyd GED. 2012. A GRAS-type transcription factor with a specific function in mycorrhizal signaling. *Current Biology,* 22, 2236–2241.

Gollotte A, Gianinazzi-Pearson V, Giovannetti M, Sbrana C, Avio L and Gianinazzi S. 1993. Cellular localization and cytochemical probing of resistance reactions to arbuscular mycorrhizal fungi in a 'locus a' *myc*– mutant of *Pisum sativum* L. *Planta,* 191, 112–122.

Haage K and Parniske M. 2013. Molecular mechanisms governing arbuscular mycorrhiza development and function. In *Molecular Microbial Ecology of the Rhizosphere,* de Brujin FJ (ed.), Vol. 1, 459–465, John Wiley and Sons, USA.

Helber N, Wippel K, Sauer N, Schaarschmidt S, Hause B and Requena N. 2011. A versatile monosaccharide transporter that operates in the arbuscular mycorrhizal fungus *Glomus* sp is crucial for the symbiotic relationship with plants. *The Plant Cell,* 23, 3812–3823. doi:10.1105/tpc.111.089813.

Hogekamp C and Küster H. 2013. A roadmap of cell-type specific gene expression during sequential stages of the arbuscular mycorrhiza symbiosis. *BMC Genomics,* 14, 306. doi:10.1186/1471-2164-14-306.

Illana A, García-Garrido JM, Sampedro I, Ocampo JA and Vierheilig H. 2011. Strigolactones seem not to be involved in the nonsusceptibilty of arbuscular mycorrhizal (AM) nonhost plants to AM fungi. *Botany,* 85, 285–288. doi:10.1139/B11-014.

Javot H, Penmetsa RV, Breuillin F, Bhattarai KK, Noar RD, Gomez SK, Zhang Q, Cook DR and Harrison MJ. 2011. *Medicago truncatula mtpt4* mutants reveal a role for nitrogen in the regulation of arbuscule degeneration in arbuscular mycorrhizal symbiosis. *The Plant Journal,* 68, 954–965. doi: 10.1111/j.1365-313X.2011.04746.x.

Javot H, Penmetsa RV, Terzaghi N, Cook DR and Harrison MJ. 2007. A *Medicago truncatula* phosphate transporter indispensable for the arbuscular mycorrhizal symbiosis. *Proceedings of the National Academy of Sciences of the United States of America*, 104, 1720–1725. doi:10.1073/pnas.0608136104.

Kloppholz S, Kuhn H and Requena N. 2011. A secreted fungal effector of *Glomus intraradices* promotes symbiotic biotrophy. *Current Biology*, 21, 1204–1209. doi:10.1016/j.cub.2011.06.044.

Koltai H, LekKala SP, Bhattacharya C, Mayzlish-Gati E, Resnick N, Wininger S, Dor E, Yoneyama K, Yoneyama K, Hershenhorn J, Joel DM and Kapulnik Y. 2010. A tomato strigolactone-impaired mutant displays aberrant shoot morphology and plant interactions. *Journal of Experimental Botany*, 61(6), 1739–1749. doi:10.1093/jxb/erq041.

Koske RE. 1982. Evidence for a volatile attractant from plant roots affecting germ tubes of a VA mycorrhizal fungus. *Transactions of the British Mycological Society*, 79, 305–310.

Kosuta S, Chabaud M, Lougnon G, Gough C, Dénarié J, Barker DG and Bécard G. 2003. A diffusible factor from arbuscular mycorrhizal fungi induces symbiosis-specific MtENOD11 expression in roots of *Medicago truncatula*. *Plant Physiology*, 131, 952–962. doi:10.1104/pp.011882.

Kosuta S, Hazledine S, Sun J, Miwa H, Morris RJ, Downie JA and Oldroyd GED. 2008. Differential and chaotic calcium signatures in the symbiosis signaling pathway of legumes. *Proceeding of the National Academy of Science USA*, 105, 9823–9828.

Kretzschmar T, Kohlen W, Sasse J, Borghi L, Schlegel M, Bachelier JB, Reinhardt D, Bours R, Bouwmeester HJ and Martinoia E. 2012. A petunia ABC protein controls strigolactonedependent symbiotic signaling and branching. *Nature*, 438, 346. doi:10.1038/nature 10873.

Kuhn H, Küster H and Requena N. 2010. Membrane steroid-binding protein 1 induced by a diffusible fungal signal is critical for mycorrhization in *Medicago truncatula*. *New Phytologist*, 185, 716–733. doi:10.1111/j.1469-8137.2009.03116.x.

Liu W, Zhou X, Li G, Li L, Kong L, Wang C, Zhang H and Xu J-R. 2011. Multiple plant surface signals are sensed by different mechanisms in the rice blast fungus for appressorium formation. *PLoS Pathogens*, 7(1), e1001261. doi:10.1371/journal.ppat.1001261.

Liu Y, Gianinazzi-Pearson V, Arnould C, Wipf D, Zhao B and Tuinen Dv. 2013. Fungal genes related to calcium homeostasis and signaling are upregulated in symbiotic arbuscular mycorrhiza interactions. *Fungal Biology*, 117(1), 22–31. doi:10.1016/j.funbio.2012.11.002.

Logi C, Sbrana C and Giovannetti M. 1998. Cellular events involved in survival of individual arbuscular mycorrhizal symbionts growing in the absence of the host. *Applied and Environmental Microbiology,* 64(9), 3473–3479.

Maillet F, Poinsot V, André O, Puech-Pagès V, Haouy A, Gueunier M, Cromer L, Giraudet D, Formey D, Niebel A, Martinez EA, Driguez H, Bécard G and Dénarie J. 2011. Fungal lipochitooligosaccharide symbiotic signals in arbuscular mycorrhiza. *Nature,* 469, 58–63. doi:10.1038/nature09622.

Nagahashi G and Douds DD. Jr. 1997. Appressorium formation by AM fungi on isolated cell walls of carrot roots. *New Phytologist,* 136, 299–304. doi:10.1046/j.1469-8137.1997.00739.x.

Nagahashi G and Douds DD. Jr. 2000. Partial separation of root exudate components and their effects upon the growth of germinated spores of AM fungi. *Mycological Research,* 104(12), 1453–1464. doi:10.1017/S0953756200002860.

Nagahashi G and David DD. Jr. 2011. The effects of hydroxy fatty acids on the hyphal branching of germinated spores of AM fungi. *Fungal Biology,* 115, 351–358. doi:10.1016/j.funbio.2011.01.006.

Ocampo JA, Martin J and Hayman DS. 1980. Influence of plant interactions on vesicular-arbuscular mycorrhizal infections. I. host and non-host plants grown together. *New Phytologist,* 84, 27–35.

Parniske M. 2008. Arbuscular mycorrhiza: the mother of plant root endosymbiosis. *Nature Reviews Microbiology,* 6, 763–775. doi:10.1038/nrmicro1987.

Pumplin N, Mondo SJ, Topp S, Starker CG, Gantt JS and Harrison MJ. 2010. *Medicago truncatula* Vapyrin is a novel protein required for arbuscular mycorrhizal symbiosis. *The Plant Journal,* 61, 482–494. doi:10.1111/j.1365-313X.2009.04072.x.

Ramasamy CK. 2011. Arbuscular mycorrhizal fungi in low external input agroecosystems. PhD diss., Scuola Superiore Sant'Anna, Pisa.

Redecker D, Kodner RB and Graham LE. 2000. Glomalean fungi from the Ordovician. *Science,* 289, 1920–1921. doi:10.1126/science.289.5486.1920.

Regvar M, Vogel K, Irgel N, Wraber T, Hildebrandt U, Wilde P and Bothe H. 2003. Colonization of pennycresses (*Thlaspi* spp.) of the Brassicaceae by arbuscular mycorrhizal fungi. *Journal of Plant Physiology,* 160, 615–626. doi:10.1078/0176-1617-00988.

Requena N, Mann P, Hampp R and Franken P. 2002. Early developmentally regulated genes in the arbuscular mycorrhizal fungus *Glomus mosseae*: identification of GmGIN1, a novel gene with homology to the C-terminus of metazoan hedgehog proteins. *Plant and Soil,* 244, 129–139.

Sbrana C and Giovannetti M. 2005. Chemotropism in the arbuscular mycorrhizal fungus *Glomus mosseae*. *Mycorrhiza,* 15(7), 539–545. doi: 10.1007/s00572-005-0362-5.

Sbrana C, Fortuna P and Giovannetti M. 2011. Plugging into the network: belowground connections between germlings and extraradical mycelium of arbuscular mycorrhizal fungi. *Mycologia,* 103(2), 307–316. doi: 10.3852/10-125.

Seddas PMA, Arias CM, Arnould A, Tuinen Dv, Godfroy O, Benhassou HA, Gouzy J, Morandi D, Dessaint F and Gianinazzi-Pearson V. 2009. Symbiosis-related plant genes modulate molecular responses in an arbuscular mycorrhizal fungus during early root interactions. *Molecular Plant-Microbe Interactions,* 22(3), 341–351. doi: 10.1094/MPMI-22-3-0341.

Siciliano V, Genre A, Balestrini R, deWit PJGM, Bonfante P. 2007. Prepenetration apparatus formation during AM infection is associated with a specific transcriptome response in epidermal cells. *Plant Signaling and Behavior,* 2, 533–535. doi: 10.1104/pp.107.097980.

Smith SE and Read DJ. 2008. *Mycorrhizal Symbiosis*. 3rd edition. Academic Press, London.

Staples RC and Macko V. 1980. Formation of infection structures as a recognition response in fungi. *Experimental Mycology,* 4, 2–16.

Takeda N, Sato S, Asamizu E, Tabata S and Parniske M. 2009. Apoplastic plant subtilases support arbuscular mycorrhiza development in *Lotus japonicus*. *The Plant Journal,* 58, 766–777. doi: 10.1111/j.1365-313X.2009.03824.x.

Tamasloukht, M'B, Séjalon-Delmas N, Kluever A, Jauneau A, Roux C, Bécard G and Franken P. 2003. Root factors induce mitochondrial-related gene expression and fungal respiration during the developmental switch from asymbiosis to presymbiosis in the arbuscular mycorrhizal fungus *Gigaspora rosea*. *Plant Physiology,* 131, 1468–1478. doi: 10.1104/pp.012898.

Tisserant E, Kohler AC, Dozolme-Seddas P, Balestrini R, Benabdellah K, Colard A, Croll D, Da Silva CM, Gomez SK, Koul R, Ferrol N, Fiorilli V, Formey D, Franken PH, Helber N, Hijri M, Lanfranco L, Lindquist EA, Liu Y, Malbreil M, Morin E, Poulain J, Shapiro HJ, Tuinen Dv, Waschke A, Azcón-Aguilar C, Bécard G, Bonfante P, Harrison MJ, Küster H,

Lammers PJ, Paszkowski U, Requena N, Rensing SA, Roux CP, Sanders IR, Shachar-Hill Y, Tuskan GA, Young JPW, Gianinazzi-Pearson V and Martin FGP. 2012. The transcriptome of the arbuscular mycorrhizal fungus *Glomus intraradices* (DAOM 197198) reveals functional tradeoffs in an obligate symbiont. *New Phytologist*, 193, 755–769. doi: 10.1111/j.1469-8137.2011.03948.x.

Veiga R, Faccio A, Genre A, Pieterse CMJ, Bonfante P and van der Heijden MGA. 2013. Arbuscular mycorrhizal fungi reduce growth and infect roots of the non-host plant *Arabidopsis thaliana*. *Plant, Cell and Environment*, 36, 1926–1937.

Wagg C, Antunes PM and Peterson LR. 2011. Arbuscular mycorrhizal fungal phylogeny-related interactions with a non-host. *Symbiosis*, 53, 41–46. doi: 10.1007/s13199-011-0107-5.

Wagg C, Pautler M, Massicotte HB and Peterson LR. 2008. The co-occurrence of ectomycorrhizal, arbuscular mycorrhizal, and dark septate fungi in seedlings of four members of the *Pinaceae*. *Mycorrhiza*, 18, 103–110. doi: 10.1007/s00572-007-0157-y.

Wang B and Qiu Y-L. 2006. Phylogenetic distribution and evolution of mycorrhizas in land plants. *Mycorrhiza*, 16, 299–363.

Wang E, Schornack S, Marsh JF, Gobbato E, Schwessinger B, Eastmond P, Schultze M, Kamoun S and Oldroyd GED. 2012. A common signaling process that promote mycorrhizal and oomycete colonization of plants. *Current Biology*, 22, 2242–2246.

12. Attraction of Preferred Prey by Carnivorous Plants

Douglas W. Darnowski

Department of Biology
Indiana University Southeast, 4201 Grant Line Road
New Albany, Indiana 47150, USA
ddarnows@ius.edu

1. Introduction

Carnivorous plants thrive in nutrient poor habitats around the world, and in order to go about their business of obtaining nitrogen and phosphorous from their animal prey, they have to falsely communicate with that prey. That means that the plants must lie to their prey to fool the prey into their traps, which range from slippery pits to sticky leaves to suction machines. How they do so, using enticing odors, delicious rewards, and bright markings, will be the subject of this chapter. Also, carnivorous plants can choose among a range of potential prey to obtain the nitrogen and phosphorous which they so desperately need in their nutrient deprived habitats. The question has arisen as to just what carnivorous plants are looking to trap with their deceptive communication, which implies a kind of communication between predator plant and its prey. This chapter proposes a new mathematical index, prey preference index, to ascertain just which potential prey are sought by these carnivorous plants and shows a simple example of its use.

2. How This is Biocommunication

Carnivorous plants must, by their very nature, do quite a bit of communicating with other organisms. That is because carnivorous plants create structures which are designed to trap prey organisms. The communication from these plants to their potential prey consists of the structure of these traps, as well as molecules which they may produce to help entice prey into their traps.

All of this implies not only communication, but also deceptive communication — carnivorous plants are liars! The deception is required so that the prey will come near enough to be trapped. Carnivorous plants expend large amounts of energy and precious nutrients to build structures to trap and digest prey, and the return from simply waiting for prey to blunder into the traps at random will not suffice to support the plants.

This is particularly important for carnivorous plants since almost all of them grow in low-nutrient habitats, usually ones which are brightly lit and very wet or at least seasonally inundated. Carnivorous plants' advantage in obtaining nutrients, especially N and P, from prey allows them to succeed in these habitats, and when placed in habitats with more abundant nutrients, most carnivorous plants are quickly outcompeted by non-carnivores and perish.

These habitats favored by carnivorous plants also often involve ancient, heavily-weathered soils or oligotrophic waters, and so nutrients like N and P are at very low abundance. The plants, in these conditions, need a strong return on their investment in the structures involved in carnivory in order to survive and complete their lifecycle.

These same carnivorous plants must communicate non-deceptively with their pollinators, since consuming one's pollinators provides a short-term nutritional benefit but is ultimately self-defeating as no seed can be produced and the species will disappear. At the same time, the deceptive communication leading to trapping must not be made with these pollinators, so that they do not fall prey to the plant's hunger for nutrients.

It is important to briefly review what traits are required for a plant to be carnivorous. In order to be a carnivorous plant, a plant must attract, trap, digest, and absorb prey, or, at least, traditionally, this was the view. However, in the last few years, based on the work of Adamec (2011), this definition has been modified. The reason for this is that it can be difficult, even in the case of the most obviously carnivorous plant such as the Venus Flytrap (*Dionaea muscipula*), to prove definitely a particular attractive mechanism, even if, as in that case just noted, new details such as scent production, have been shown. Usually, what is found is a trap with many prey, or the evidence of many prey, such as empty exoskeletons, having been caught and digested.

The traps of carnivorous plants can take a range of forms, though certain general categories can be identified. Passive traps do not exert any force or show any movement in trapping, rather waiting for prey to blunder into the trap. These passive traps include sticky leaves to which prey adhere and pitfall traps filled with water and digestive enzymes and acid. Also, there are the coiled traps of the Corkscrew plants (*Genlisea*) which contain hairs pointed in such a fashion that small prey which enter can only move unidirectionally, toward a bulb-shaped chamber in which they are digested (Joel *et al.*, 1985).

Some traps of carnivorous plants are active, expending energy in the trapping of prey. These include the terrestrial Venus Flytrap and its aquatic sister genus, the Waterwheel plant (*Aldrovanda*), which have modified leaves that snap shut on prey. Bladderworts have suction traps which maintain a negative internal pressure, which pulls in prey when the trap door is opened by prey jostling the trap's trigger hairs. Some sundews also have partly active traps, with the passive sticky hairs being assisted by motile hairs which bend over trapped prey and help to secure them to the leaf.

This difficulty in proving an attractive mechanism is not true in all cases, as in the sun pitchers (*Heliamphora* spp.), which exude nectar from a spoon-shaped part of their modified trapping leaves (McPherson, 2007), the nectar being an easily proven attractive mechanism. Also, many non-carnivorous plants, such as wild tomatoes, may trap many prey, implying an attractive mechanism. These mechanisms could, in either case, include visual clues, such as reflected light or odors. However, it still remains that the attractive part of a mechanism for trapping, whether in carnivorous or in non-carnivorous plants, is most difficult to prove.

Also, it should be noted that carnivorous plants are generally restricted to the Flowering Plants, though a few probable examples have been found outside the angiosperms (e.g., Barthlott *et al.*, 2000).

By the newer definition, carnivorous plants are plants which trap and digest their prey, with a significant ecological benefit being derived from this (Adamec, 2011). This benefit can be measured as increased production of seed or of other organs intended for propagation. These latter could include the gemmae of Australia's pygmy sundews (Drosera spp.), tiny winter produced plants derived from modified leaves and mimicking seeds. This would also include certain butterworts (Pinguicula spp.) which produce plantlets from leaf tips. The benefit of carnivory could also be seen in increased size of the trapping plants, or of some other measurable feature of that plant.

Digestion must be performed by the plant itself for it to be fully carnivorous, and a range of enzymes have been found to operate in the various kinds of carnivorous plants. These enzymes include proteases, nucleases, lipases, and various carbohydrate-degrading enzymes. Acid is also usually involved (Juniper *et al.*, 1989). Some carnivorous plants wait to produce these enzymes until they are sure that prey have been caught well enough to prevent escape. This occurs in the Sundew Family, as when sundews and Venus Flytraps do not express and secrete material to effect digestion until chemical and physical signs of prey are assured. Venus flytrap and its close cousin the Waterwheel plant (*Aldrovanda vesiculosa*) also have a tightening of their traps which occurs only after they have been physiologically assured that prey has been trapped. In contrast, other carnivorous plants produce the enzymes ahead of time and simply exocytose them once prey are trapped.

Once digestion occurs, nutrients are absorbed, but this is in many ways the least critical step for the proof of full carnivory since so many plants readily absorb material placed on their surfaces, especially leaves. This can be seen in the use of herbicides, which work well due to the generally rapid uptake which occurs across plants' surfaces, and in the use of foliar feeds of fertilizer in agriculture. These foliar feeds are taken up very rapidly and are especially useful for highly deficient plants which have an acute need for minerals.

After absorption, nutrients must be transported within plants, but, again, this is an easy thing to have happen, given how readily plants do this with whatever they absorb. The very popular herbicide Roundup™ acts by being absorbed by leaves, being transported readily to the roots of the plant, and there killing the roots, and thus the whole plant.

Thus, the critical steps in proving carnivory in a given plant are demonstration of trapping, digestion, and benefit, with uptake and transport being the typically easier steps to demonstrate.

Some plants colloquially referred to as being carnivorous are more precisely either sub-/hemicarnivorous or protocarnivorous. These also have to communicate deceptively with their prey, but it is worthwhile briefly discussing these groups. It should also be noted that some scientists prefer other terms, such as "pseudocarnivorous," and that a wide range of plants exist, from those which may be evolving towards the carnivorous syndrome to those possessing it now to those which may have lost some or all of the syndrome for various reasons.

Protocarnivorous plants are so classified because they have some of the features of carnivory but do not yet fully use these features for obtaining a significant benefit. The wild tomatoes mentioned above would be a good example of this — they may evolve in future to extract nutrients from the insects which they trap. Most probably, protocarnivorous plants trap insects as a means of stopping these animals from damaging the plants by their herbivorous activities. It is thought that many carnivorous plants, especially those with sticky leaves, may have started as non-carnivorous plants which used sticky leaves to stop herbivorous insects and then evolved the other features of full carnivory.

Given the demonstration of carnivorous behavior in plants not previously known as such over the last decade or so, plants in genera such as *Triphyophyllum* (Dionocophyllaceae) or *Phillcoxia* (Plantaginaceae), some, perhaps many, plants now termed protocarnivorous may actually stalk prey and benefit from them, making them truly carnivorous. More work remains to be done on genera like *Stylidium* (Stylidiaceae), the triggerplants, which may be more than just protocarnivores (Darnowski, 2002; Darnowski *et al.*, 2006). These latter plants have both an extremely fast pollination mechanism and sticky hairs on the backs of flower parts and inflorescence stems which trap much smaller insects than those which effect pollination.

3. How Do They Communicate

Carnivorous plants communicate with their prey in several ways, making use of the prey's senses. These senses which are duped by the plants include visual and olfactory senses as well as taste. There is, of course, variation among organisms in terms of what they can detect with a given sense, and carnivorous plants use signals according to the abilities of their intended prey. For example, the ability of insects to see in the ultraviolet (UV) range makes UV markings, similar to the nectar guides of flowers, useful in attracting them, while these markings would be useless with vertebrates, which do not have the ability to see UV patterns. It should also be briefly noted that odors are probably the least well examined aspect of carnivorous plants' traps. There are papers which show that there are some chemical signals detected by prey (e.g., Darnowski and Fritz, 2010), but this aspect of prey attraction needs much more coverage.

When they use visible patterns in attracting prey, carnivorous plants are simply following what flowering plants have done since their origins, making use of visual cues to communicate with animals. This is probably first important in the history of flowering plants in the attraction of pollinators, and today we can see things like nectar guides in many flowers. Nectar guides are markings made of flavonoid pigments with UV elements invisible to the human eye. These UV markings point pollinators to nectar rewards and to the part of the flower where they pick up pollen to transfer to another flower of the same species.

A particularly interesting example of a complex visual guide comes from the Northern Australian sundew species *Drosera hartmeyerorum*, a member of the much more broadly distributed *Drosera indica* complex of species (Hartmeyer and Hartmeyer, 2001, 2010; Schlauer, 2001). *D. hartmeyerorum* was first noticed because, in addition to the red, mucilage-secreting trapping emergences, often incorrectly and colloquially referred to as "hairs," which are found generally in its genus, it also has stalked, yellow, mulberry-shaped appendages near the base of its leaves. These appendages reflect light, and it has been proposed that their purpose is to help attract insects. The discoverers of this species have photographs which clearly show how these structures stand out very clearly and would present an interesting item for prey to investigate. Once prey reached these yellow structures, or on the way to them, they would likely be caught in the sticky hairs on the leaves and then consumed by the plants.

Another interesting example, really a pair of intriguing visual attractions by carnivorous plants can be seen in the two pitcher-plants from the Sarraceniaceae, the family of pitcherplants which is found in North and South America. This family makes pitchers by having their leaves curl back around so that the two leaf edges meet and fuse, and a small part of the leaf blade also hangs over the opening as a semi-open lid. The pitcher fills with water, and then acid and digestive enzymes are used to break down prey which enter these pitfall traps and cannot exit due to the smooth, waxy leaves and downward pointing hairs on the inside top of the pitchers.

One of the pitcherplants in the current example is *Darlingtonia californica* from northern California and southern Oregon, and the other is *Sarracenia psittacina* from the southeastern part of the United States. Both have rounded lids to their pitchers with clearer, white areas compared to the red and green tissue making up much of the lids. These clearer areas transmit much more light than the surrounding colored material, and thus, to an insect, it appears that these pieces of bright, but solid, tissue are actually openings in the top of the pitcher with sun shining through them. Insects then fly towards the "openings," strike the lid, and fall into the digestive fluid below. This is quite a dramatic example of false communication intended to help trap prey.

A number of carnivorous plants, notably some of the world's pitcher plants, use a nectar reward, like that offered by many flowers, to draw prey to themselves. This includes species like those of *Heliamphora* and *Nepenthes* which secrete nectar from either the top of the pitcher or the lid which is suspended above the pitcher opening.

4. Symbiosis

There is also the question of symbiosis in several examples from carnivorous plants. Briefly, one should mention that some arthopods are adapted to live in the traps of carnivorous plants either for the protected environment found there, with protective walls and shade, or to share in the plant's catch of prey. In the former case there are examples like some mosquitos, whose larvae have adaptations to live in *Sarracenia* pitchers without being digested (Juniper *et al.*, 1989; McPherson, 2007), and in the latter case there are the spiders found living in pitchers of *Nepenthes* which take some of insects trapped by these pitcher plants.

Roridula is a sticky-leaved plant from South Africa, and the genus contains two species with superficial morphological differences but which operate in a similar way. These species have glandular structures, like very large and more complex versions of epidermal hairs, which secrete sticky material, similar to what sundews do to catch prey. However, this sticky material is resinous, and therefore hydrophobic. This means that the secretions of *Roridula* cannot allow enzymatic reactions, preventing *Roridula* itself from performing digestion of prey. Instead, *Roridula* plants have symbionts, called *Roridula* bugs, which live at their bases. These insects have adaptations to avoid being caught in the secreted resin, and, when they detect vibrations in the plant from trapped prey, they scurry up the plant. When they reach the trapped prey, they consume and digest material from the prey, defecating a particularly rich mixture of nutrients in their feces. Thus, the plant traps, the insect digests, and both benefit (McPherson, 2009). There must be not only the false communication from the plant to the prey but also some form of communication, probably chemical, between *Roridula* and its symbionts.

Another interesting case of symbiosis, or at least one suggested as a symbiosis, comes with the aquatic bladderworts, genus *Utricularia* (Richards, 2001). In some species, there is a large range of algae present in the traps, and various hypotheses have been advanced over time to explain their presence. These include algae present due to accidental uptake when traps are triggered by other organisms while algae are merely nearby. It has been suggested that there may be some symbiotic relationship in which the algae might sit in the traps as if in aquaria, being protected but also nourishing the plants, and there is good evidence of this (Sirova *et al.*, 2009), but no specific communication has been demonstrated in this case. However, one could imagine some communication being important for such a relationship, since other cases of internal symbiosis usually involve signaling between host and guest.

This can be seen in the case of nodulation in legumes. These members of the Bean Family signal back and forth with soil bacteria of the genus *Rhizobium* and its close cousins. The bacteria approach the roots hairs of the legume, which forms a hollow tube from its root hair many cells layers into the cortex of the root. The bacteria take up residence extracellularly, but inside the root, in the cortex, protected and fed by the root. In return, especially due to the legume protecting the bacteria from high concentrations of oxygen, the bacteria fix gaseous, atmospheric nitrogen into biologically useful forms. The plant cannot do this on its own, so it derives great benefit from this fertilization. This symbiosis is probably the main reason for the phenomenal success of the Bean Family in terms of the number of species and individual plants found all around the world.

Thus, any similar symbiotic relationship between bladderworts and the algal residents of their traps might also be very beneficial.

One possible argument against this in the case of at least some bladderworts might be the rarity of the very species first suggested to host such a symbiosis — *utricularia purpurea*, one of the loveliest of North America's bladderworts, with bright purple flowers held above the water. It is very sensitive to water conditions, disappearing quickly when water passes from oligotrophic to even merely mesotrophic, and thus is not widespread, in sharp contrast to the legumes, with the benefit that they derive from their symbiosis. *U. purpurea* was never the most widely distributed species, in contrast to others like *U. gibba* and *utricularia macrorhiza*, even before the decline in water quality in its habitats in eastern North America which has been so rapid in the last century or so, showing that any symbiotic benefit did not give *U. purpurea* the strong competitive advantage that the legumes received from their rhizobial symbionts. It is quite possible that this is also true for other species of *Utricularia* noted above.

It is also important to note that some carnivorous plants actually get some of their nutrients from non-carnivorous communication with organisms that have replaced, or, in an evolutionary sense, are in the process of replacing their prey. An excellent example of this is the bird toilet pitcher plant from the tropical genus *Nepenthes* (Nepenthaceae). This species, *Nepenthes lowii,* has a pitcher with a very small lid and a wide opening to what, in other species of its genus, is the trap. The shape of the pitcher of this particular plant turns out to form an ideal perch for tree shrews and also, probably, birds. These tree shrews will pause during and after eating from nearby trees and bushes, defecating into the pitcher (Phillipps and Lamb, 1996; Clarke *et al.*, 2009). Excess N and P excreted by these shrews is the major source of these nutrients for this plant, and so its shape, which is immediately recognizable as being similar to a Western toilet, communicates to the birds that a rest is available and provides information on how they should position themselves. This in turn directs the birds' feces into the pitcher and the nutritional benefit to *N. lowii*.

This is not the only species of *Nepenthes* evolving in this direction (Juniper *et al.*, 1989), so perhaps the evolution from protocarnivorous plant to post-carnivorous plant, to coin a term, could be found in other families and genera of carnivorous plants. This forms an interesting avenue for further research, perhaps starting with the other families which produce pitchers, the Sarraceniaceae and Cephalotaceae.

5. Prep Preference Index

The author of this work suggests a new mathematical index to help determine whether carnivorous plants are preferentially targeting prey and which prey they are targeting. This is, in part, in response to the questioning of whether some of the most obviously carnivorous plants, and those traditionally recognized as such, are truly carnivorous. Also, this is in response to questioning of whether the more obvious prey of a particular carnivorous plant like the Corkscrew plants (Genlisea spp.; Lentibulariaceae) are their true prey (Barthlott *et al.*, 1998; Plachno *et al.*, 2008). The latter paper has some especially important flaws, such as the lack of supporting work with field-grown plants or even herbarium specimens.

For a given carnivorous plant, one can imagine a range of possible prey based on the size of the trap and the kind of attractants and trapping mechanism. Some possible prey can be excluded from consideration because they are too large, too strong, or too fast to be caught by a particular plant.

However, there are still many possible prey, but not all possible/conceivable prey may be what is/are mainly being caught by that carnivorous plant. For example, an erect-leaved sundew is likely to catch smaller, more delicate prey which cannot escape from the mucus which the sundew produces. These sundews lack the motile hairs and semi-active trapping mechanism described above for ground-hugging rosette sundews. This is because more robust insects like beetles, which are able to break free of mucilage, are usually found crawling on the ground, while more delicate prey like some smaller flies, gnats, and the like, fly through the air at the level of the sticky leaves of an erect sundew.

The prey preference index (PPI) can help to indicate what a plant is preferentially trapping. Calculation begins with the counting of the number of potential prey from some group of potential prey in a set volume of the air, water, or soil around a carnivorous plant's traps. Since the carnivores discussed so far are, for the most part, terrestrial or floating aquatics, it should be noted that all or some species in some carnivorous genera have traps immersed in soil, often saturated soil. These include the genera *Utricularia* and *Genlisea* of the Lentibulariaceae as well as the genus *Philcoxia*, the carnivorous nature of which was only recently described (Fritsch *et al.*, 2007; Pereira *et al.*, 2012).

This count of potential prey in air, water, or soil could be a count of some particular species or of a defined group of organisms likely to be caught by a particular species of carnivorous plant, such as members of the Diptera of less than a given body weight. Once the count is complete, the relative proportions of each of these organisms is determined for each group counted — for example, 10% butterflies, 25% flies, etc. It would be helpful to have regular redetermination of available prey or, better, of the PPI, since the prey spectrum could, and probably would, change throughout a growing season. Traps with longer lives, say those of *Sarracenia* rather than those of *Utricularia*, would be exposed to different prey at different times. This is also important for the examination of trapped insects, discussed more fully below, since longer-lasting traps would contain the results of perhaps a season's worth of trapping, but a single determination of PPI would only compare that trapped population with a short snapshot of the overall prey spectrum. Studies using PPI really should look at traps over a wide range of times and places, for the same species.

Then, the same procedure is followed for the actual trapped insects, counting them group by group and determining the percentage of trapped organisms coming from each group of potential prey. Finally, the percentage trapped is divided by the percentage in the air or water, leading to the unitless PPI. If this number is 1 for a particular group, no preference is being shown for that group by the carnivorous plant under consideration since equal portions are found in what is available and what is trapped. If PPI is greater than 1, then

PPI is calculated as follows:

$$\text{PPI for prey species, X} = \frac{\text{\% of trapped, biotics which are X}}{\text{\% of biotics in water sample which are X}}$$

where

if PPI<1, X is selected against
if PPI=1, X is neither selected against nor preferred
if PPI>1, X is preferred

Figure 1. How PPI is calculated.

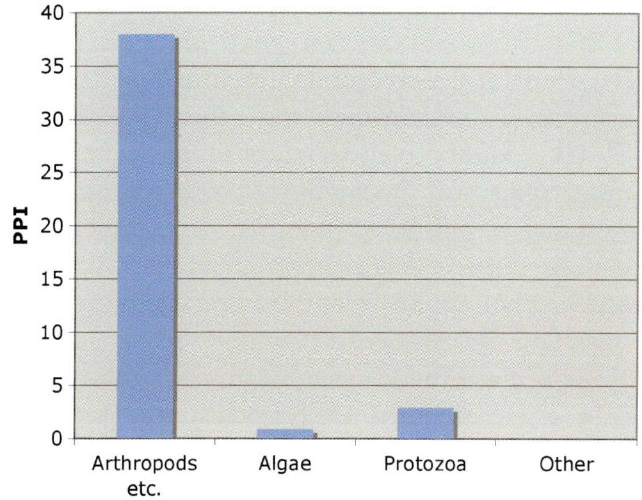

Figure 2. PPI calculated for several categories of prey items in a sample of traps of the bladderwort *U. macrorhiza* growing in a natural lake in Indianapolis, Indiana, USA, in shallow water. 100 ml of water from around the plants was taken and sampled 20 times, 10 μl of water per time, for potential prey, after vigorous mixing. 20 traps were examined for trapped organisms.

that prey group is preferred, with greater PPI indicating stronger preference since that group is being trapped in disproportionate numbers. Conversely, PPI less than 1 shows a distaste for that prey, and the smaller PPI, the stronger the dislike since the plant's traps are somehow discriminating against what seems to be a likely prey organism (see Fig. 1 for a summary of this calculation).

Figure 2 shows a sample calculation for one site taken in central Indiana, USA, involving plants of *U. macrorhiza* growing in shallow water at the side of a mesotrophic lake. The "Other" category is a grab-bag of other prey, principally rotifers.

It is important in using PPI to be careful in selecting the distance from the plant at which to stop sampling for making the count of available potential prey. This is a common sense decision based on the behavior of the potential prey, such as how high they fly compared to the height of carnivorous vegetation. In the same way, the behavior of aquatic potential prey can be observed and the normal distance which they travel, including the distance from the plants that they travel and the average distance traveled. This will also involve a look at the kind of habitat in which a particular species of carnivorous plant grows and where the PPI concept is applied. To take aquatic bladderworts as an example, some are found floating freely near the top of the water column, as is true for *U. macrorhiza* and *U. foliosa,* while others only prefer to grow tangled among other species, as *U. olivacea* is only found lost among larger species. Other aquatic bladderworts often grow in shallow water on mudflats, as can be true for *U. resupinata,* and yet others have floating photosynthetic tissue combined with non-photosynthetic shoots where all traps are found, submerged in the loose substrate, as is the case with *U. intermedia.* All of those different scenarios could involve very different prey traveling very different distances. Thus, one must be careful to observe potential prey and the preferred habitat, and specific details of that, when using PPI with a particular species.

PPI also has the potential to be skewed in favor of smaller prey, since a single large prey item might provide far more nutrients than many smaller ones, though if the smaller prey were trapped in high numbers, they might seem, when looking with the help of PPI, to be preferred when the plant might actually derive much more benefit from the single large prey and item and have traps designed to catch that larger, lower abundance item. It is to correct for just such issues that PPI is made as the ratio of two percentages. At least to some large degree, PPI ought to self-correct for such potential bias for smaller, more abundant prey since they are likely to form a larger percentage of what is counted for both numerator and denominator in the PPI calculation, and, similarly, larger prey are likely to be a smaller percentage, but in both numerator and denominator.

Even if PPI does not tend to skew in favor of smaller, less valuable prey, comparing PPI with the nutrient status of the plants — that is, what internal concentrations of P, N, etc., exist — to see what kind of benefit plants accrue. For example, a plant with a strong PPI but not showing a benefit nutritionally might just be trapping insects as a defensive mechanism. For such experiments, the use of heavy isotopes of C and N in prey would be quite helpful in confirming the origin of C and N found in plants to which PPI was being applied.

There are other ways in which this could be corrected. Just as studies to look at prey travel near carnivorous plants are needed to be sure that sampling for potential prey occurs in the right places which make sense for a given combination of plant and potential prey, controlled studies could be done in the lab to look at the proportional benefit of a potential prey based on size. For example, a group of sundews could be cultivated, some fed nothing as a control, other given delicate flies to show the effects of lighter prey, still others provided with beetles as examples of larger, more nutrient-rich prey. Non-living model "prey" of various sizes could be used — perhaps agar blocks containing powdered milk and salts. The benefit in terms of mass, flower production, seed set, etc. could then be observed to help determine the potential for size-based bias in PPI for that kind of sundew.

In addition, PPI could also be used to help identify carnivores in the first place. If possible prey organisms are in some habitat, they might occasionally be found in narrow spaces such as constricted leaf bases. Calculating PPI for these kinds of spaces as well as similarly sized natural spaces, such as cracks in rocks, could help point out whether these are truly prey being trapped by plants. If PPI is greater than one for the plant spaces but not for the non-biological spaces, this would be one indication of deliberate trapping.

One could even make non-natural mimics of such plant spaces, place them in the landscape in similar positions to the plants in question, and then calculate the PPI for the non-natural mimics. This could help indicate whether mere physical structure is key to trapping, if PPI was greater than 1 for the non-natural mimics, or whether some biological aspect of the trap was also critical, with PPI for the non-natural mimics being 1 or less than 1, or at least much less than PPI for the natural, biological structures.

In short, PPI has a lot of potential to help in identifying which prey carnivorous plants communicate with deceptively, and just how strong the preference and deception are.

Carnivorous plants are excellent liars!

References

Adamec L. 2011. Ecophysiological look at plant carnivory: why are plants carnivorous? In *All Flesh is Grass*, Seckbach J and Dubinski Z, 455–489. Springer, New York.

Barthlott W, Fischer E, Frahm JP and Seine R. 2000. First experimental evidence for zoophagy in the hepatic. *Colura. Plant Biology*, 2, 93–97.

Barthlott W, Porembski S, Fischer E and Gemmel B. 1998. First protozoa-trapping plant found. *Nature*, 392(6675), 447. doi: 10.1038/33037.

Clarke CM, Bauer U, Lee CC, Tuen AA, Rembold K and Moran JA. 2009. Tree shrew lavatories: a novel nitrogen sequestration strategy in a tropical pitcher plant. *Biology Letters*, 5(5), 632–635.

Darnowski DW. 2002. *Triggerplants*. Rosenberg Press, Dural, Australia.

Darnowski DW, Carroll DM, Plachno B, Klebanoff E and Cinnamon E. 2006. Protocarnivory in triggerplants is timed to flowering. *Plant Biology*, 8, 805–812.

Darnowski DW and Fritz S. 2010. Prey preference in genlisea small crustaceans, not protozoa. *Carnivorous Plant Newsletter*, 39(4), 114–116.

Fritsch PW, Almeda F, Martins AB, Cruz BC and Estes D. 2007. Rediscovery and phylogenetic placement of *philcoxia minensis* (plantaginaceae), with a test of carnivory. *Proceedings of the California Academy of Sciences*, 4th Series, 58(21), 447–467.

Hartmeyer I and Hartmeyer SRH. 2010. Snap-tentacles and runway lights: summary of comparative examination of *Drosera* tentacles. *Carnivorous Plant Newsletter*, 39(4), 101–113.

Hartmeyer S and Hartmeyer I. 2001. Observations on a new *drosera* species in the Ord River Region (Australia). *Carnivorous Plant Newsletter*, 30(4), 107–110.

Joel DM, Juniper BE and Dafni A. 1985. Ultraviolet patterns in the traps of carnivorous plants. *New Phytologist*, 101(4), 585–593.

Juniper BE, Robins RJ and Joel D. 1989. *The Carnivorous Plants*. Academic Press, London.

McPherson S. 2007. *Pitcher Plants of the Americas*. McDonald & Blacksburg, Woodward Publishing Company, USA.

McPherson S. 2009. *Glistening Carnivores — The Sticky-Leaved Insect-Eating Plants*. Redfern Natural History Productions, Poole, England.

Pereira CG, Almenara DP, Winter CE, Fritsch PW, Lambers H and Oliveira RS. 2012. Underground leaves of *philcoxia* trap and digest nematodes. *Proceedings of the National Academy of Sciences USA*, 109(4), 1154–1158. doi: 10.1073/pnas.1114199109.

Phillipps A and Lamb A. 1996. *Pitcher-Plants of Borneo*. Natural History Publications (Borneo), Kota Kinabalu.

Plachno BJ, Kiszkurno K, Swiatek MP and Douglas DW. 2008. Prey attraction in carnivorous *Genlisea* (Lentibulariaceae). *Acta Biologica Cracoviensia*, 50(2), 87–94.

Richards JH. 2001. Bladder function in *Utricularia purpurea* (Lentibulariaceae): is carnivory important? *American Journal of Botany*, 88, 170–176.

Schlauer J. 2001. *Drosera hartmeyerorum* spec. nov. (droseraceae), a new sundew in sect. *Arachnopus* from Northern Australia. *Carnivorous Plant Newsletter*, 30(4), 104–106.

Sirová D, Borovec J, Černá B, Rejmánková E, Adamec L and Vrba J. 2009. Microbial community development in the traps of aquatic *utricularia* species. *Aquatic Botany*, 90, 129–136. doi: 10.1093/jxb/erp286.

13. Animal Communication: Competition for Acoustic Space in Birds and Fish

Hans Slabbekoorn

Institute of Biology, Leiden University, The Netherlands
h.w.slabbekoorn@biology.leidenuniv.nl

Many animals communicate with acoustic signals as sound is an excellent medium to convey information in natural habitats in air and in water. However, ambient noise from abiotic, biotic, and anthropogenic sound sources in the environment can interfere with the signaling function of bird and fish sounds. Consequently, competition for acoustic space has affected the historical evolution of vocal variety and is still a current habitat-dependent selection pressure. Studies in cities, along highways, and in noisy waters provide fundamental insights, but also raise awareness of noise pollution and make us discover the intrinsic beauty and ecological value of natural soundscapes in air and water.

1. Evolution of Vocal Variety in the Animal Kingdom

Biodiversity in the animal kingdom is well reflected in acoustic diversity. Animals use a wide variety of sounds for communication over short and long ranges. All species that generate sounds for communication have species-specific signals and also vary acoustically within species. Repertoires of songs and calls may concern just a few vocalization types but can also be elaborate and complex. Acoustic signals can serve animals well in many circumstances and, for example, provide obvious benefits to individuals that are looking for partners in the night, competing over territory boundaries in dense vegetation, or aggregating for spawning underwater. This is the case because perception of acoustic details is independent of light conditions, penetration of leaf layers is relatively easy for low-frequency sounds, and water is an excellent medium for sound transmission. Furthermore, receivers of these acoustic messages can be engaged in other activities and do not need to pay directional attention for auditory detection. Sound will not spread in all directions equally, but still can reach an audience of receivers at any angle from the sound-generating animal. For these reasons, evolution has often led to the emergence and elaboration of acoustic displays and natural selection has resulted in many sound solutions to communicative challenges.

The evolution of all the variety in acoustic signals is driven and delimited by the ability to produce and hear sounds within the context of the natural environment (Wiley and Richards, 1982; Endler, 1992; Slabbekoorn, 2004a). Evolutionary changes are typically gradual and determined by phylogenetic constraints on morphological, physiological, and neurobiological changes that affect species-specific sound production. Signal evolution beyond what can be heard by conspecifics is also very unlikely as this would obviously prevent communication. Furthermore, although acoustic signals may often penetrate natural habitats easier than signals of other modalities, the environment can still have a strong impact on their audibility. Sound propagation may vary across habitats as sound will attenuate faster through dense vegetation than through air. Trees, twigs and leaves will also filter out high frequencies disproportionally and generate reverberations through reflection of echoes. Furthermore, more open habitat often has more wind and air turbulence, which will select against relatively long duration of sound elements. Consequently, sound propagation can drive the evolutionary design of acoustic signals towards the use of low and slow notes in relatively dense habitat and high and short repetitive notes in relatively open habitat (Wiley and Richards, 1982; Slabbekoorn, *et al.*, 2002).

Similar to sound propagation issues in air, frequency-dependent filtering effects can also be expected underwater. Low frequencies attenuate, for example, less quickly in general, but more quickly for shallow water conditions (Forrest *et al.*, 1993). Also underwater vegetation, such as seagrass meadows, can have significant effects on frequency-dependent propagation and modify environmental selection pressures on the use of sound underwater (Wilson *et al.*, 2013). In both air and water there are, however, also many other dominant sound sources that can interfere with the perception of acoustic signals. Natural habitat is certainly not quiet and potentially limiting the use of sound for communication in air and in water. Consequently, competition for acoustic space between animal signals and abiotic sounds, and among signals from different animals that share the same habitat, concerns an important selection pressure driving evolutionary change in signal variety. Furthermore, human activities are often very noisy and have in many places induced a dramatic ecological change with potentially detrimental consequences for the efficiency of evolved signal variety.

In this chapter I will address the role of the environment in the evolution of acoustic signals and specifically focus on the dominant selection pressure of ambient noise. I will first briefly address the nature and function of acoustic variety in birds and fish (see Fig. 1) to stress the importance of hearing all the details of communicative sounds in these two groups of animals. I selected these two species groups because birds have been studied already for a long time and there is a recent rise in attention for fish acoustics. Individual birds and fish have the same needs for communication, although their potential for production and perception may vary. There may also be different issues with sound propagation and signal interference in their respective habitats. Treating them side-by-side in terms of signaling problems and solutions provides

Figure 1. Singing European blackbird (*T. merula*) in the city of Leiden, the Netherlands on the left (photo by Herman Berkhoudt) with a sonographic representation of his advertisement song underneath. Blackbird songs typically start with relatively low-frequency motif parts followed by more broad-band and high-frequency twitters. Variation in the relative contribution of both parts to the duration of the whole song varies with context and plays a role in motivational signaling (Ripmeester *et al.*, 2007, 2010). African Lake Victoria cichlid (*P. nyererei*) from the captive breeding stock at Leiden University on the right (photo by Kees Hofker) with a sonographic representation of his grunt-like pulse trains underneath. These cichlids produce single bursts, pairs or triplets of sounds during the lateral quiver display of male courtship to a nearby female in about a third of all cases (Verzijden *et al.*, 2010a; Estramil *et al.*, 2013).

insights into general concepts and current gaps in our knowledge. I will address issues with natural ambient and anthropogenic noise in air and in water. Natural ambient noise entails all abiotic and biotic sound sources in the natural environment, while anthropogenic noise refers to all artificial sounds added by sound-generating human activities.

1.1 Avian acoustic advertisement

Birds are likely best known for their use of sounds in typical singing behavior and often elaborate call repertoires. Songs are mostly used by birds for territorial advertisement and can keep competitors out of defended areas and attract mates to nesting locations. These functions have been experimentally confirmed in the '70s and '80s. Playback of songs in the absence of the physical presence of territory owners keeps competitors out of the territory for longer than playback of control sounds (Krebs *et al.*, 1978). Playback of songs also makes a dummy male a stronger attractant to female birds than a silent dummy (Eriksson and Wallin, 1986). The difference between songs and calls is not always clear-cut but best defined by signal function and also a little bit by acoustic structure. Although there are exceptions, bird calls do typically not function as long-distance advertisement signals and are usually less complex and less stereotypically repeated compared to songs. Calls can function in close contact agonistic interactions, bust also as warning, begging or all-clear signals (Marler, 2004). In addition, songbirds and hummingbirds learn their songs, which means they have to hear conspecific song during ontogeny and imitate the acoustic details, which is generally not the case for calls (Baptista and Schuchmann, 1990; Marler, 1997; Hultsch and Todt, 2004). However, again the situation is not clear-cut, as we tend to say that parrots call and they are also considered vocal learners (e.g., Hile *et al.*, 2000). Furthermore, at the moment we believe that the vast majority of singing species from all other avian taxa have not evolved the capacity of vocal learning and develop their songs and calls without social impact of auditory experience (e.g., Kroodsma and Konishi, 1991).

Most birds hear within a very similar range of frequencies compared to humans (we hear from 20 Hz up to 20 kHz) and typically have the best range for hearing from 1 up to 5 kHz (Dooling, 1992, Fig. 2). The often relatively small songbirds are relatively more sensitive at higher frequencies and the often larger non-songbird species are relatively more sensitive at lower frequencies, while nocturnal predators such as many owl species hear better than all

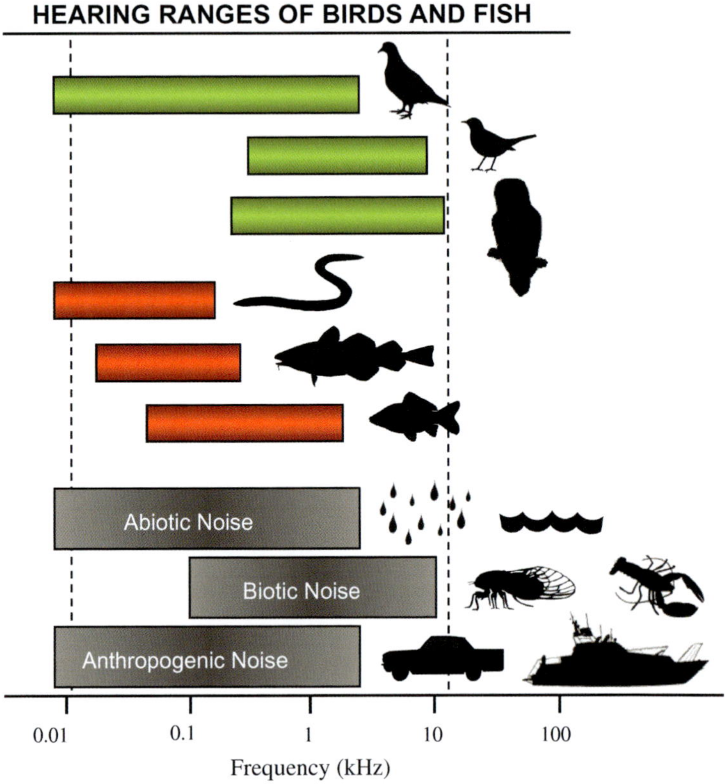

Figure 2. An overview of spectral hearing ranges of birds and fish with the potential for spectral overlap with abiotic, biotic and anthropogenic noise. Some of the typical variety among birds is depicted by three example species: Pigeon (*C. livia*), American robin (*Turdus migratorius*) and barn owl (*Tyto alba*). Some of the typical variety for fish is reflected by: European eel (*A. aguilla*), Atlantic cod (*Gadus morhua*) and Goldfish (*Carassius auratus*). The vertical dashed lines depict the hearing range for humans in air. Many abiotic and anthropogenic noise sources have most energy in relatively low frequencies and typically overlap partly with the spectral range of birds and completely with the spectral range of fish. Biotic noise sources are highly variable and cover higher frequencies. They have potential for overlapping with the whole frequency range of birds and they are partly overlapping or well above the spectral range of fish.

other birds in absolute sense across all frequencies. There are some species with exceptional hearing abilities, as some species such as pigeons (*Columba livia*) have been reported to be sensitive to atmospheric infrasound (Yodlowski *et al.*, 1977), below the frequency range audible to humans. No birds are known to be sensitive to ultrasound, above the frequency range audible to humans. Some species like the oilbird (*Steatornis caripensis*) use echolocation like bats, but only based on echoes from calls that are also audible to us (Konishi and Knudsen, 1979). There is also the mysterious case of the blue-throated hummingbird from Mexico (*Lampornis clemenciae*) generating clear ultrasonic song components, which are not harmonics of song parts in the audible range. The mystery is that hearing tests indicate that birds of this species cannot hear these high frequencies themselves (Pytte *et al.*, 2004). However, there are no other studies like this and birds typically sing within the species-specific hearing ranges.

Birds generate sounds through air-driven vibration of membranes in the syrinx, which can be found at the junction of the trachea and both bronchi (Suthers, 2004). They can generate communicative sounds that vary dramatically in amplitude and spectral and temporal patterns (Fig. 1). Some species sing a single and very simple song, while others have an elaborate repertoire of complex syllables or combine various song components with almost infinite variety. As said above, acoustic variety is related to species-specific signals, but also within species there is variation related to individual singer or context, which yields a large number of potential acoustic messages that could be relevant to other birds. Songbirds learn their songs from nearby and not from far-away conspecifics and therefore often sing different dialects in different neighborhoods (Podos and Warren, 2007). Males usually sing more often than females in temperate climate, although it was recently found out that female song is actually globally widespread and that singing by both sexes is likely the ancestral condition of all songbirds (Odom *et al.*, 2014). When both sexes do sing, males often sound differently from females, while the capacity of singing high-performance songs with respect to absolute frequency, bandwidth, trill rate, repertoire size, acoustic consistency or a combination of these may vary with individual quality (ten Cate *et al.*, 2002; Podos *et al.*, 2009; Cardoso, 2011). Birds that are highly motivated to escalate a confrontation into a fight may sing at higher rates or with longer songs and with shorter intervals. Furthermore, each individual bird has its own vocal signature. Many playback experiments have confirmed that all such acoustically encoded information is also detected and meaningful to birds that are within earshot.

Playback studies make use of recordings that are typically broadcasted within or at the boundaries of a bird's territory (McGregor, 1992; ten Cate *et al.*, 2002). They allow a test of the impact of the acoustic presence of a conspecific without any potentially confounding effects of physical presence. Well-designed studies make use of a replicated set of songs that are played back in a replicated set of territories. Each sample bird can be exposed in sequence to two different song stimuli, in alternating order, and a paired test can reveal whether the particular acoustic variation is detected and meaningful in the test conditions of that moment in the field. In this way, birds have been shown to discriminate among geographic variants (Dingle *et al.*, 2010; Ruegg *et al.*, 2012) and acoustic variation related to motivation or relative quality of competitor males (Ripmeester *et al.*, 2007; Linhart *et al.*, 2012). Also neighbors are recognized acoustically and playback of their familiar songs typically triggers lower response levels than unfamiliar songs of strangers (Godard, 1991). Also females have been shown to listen to acoustic details (Riebel, 2009) as for example great reed warbler females were shown to prefer males singing with a large syllable repertoire over males singing with small repertoires (Hasselquist *et al.*, 1996). In another study, great tit females (*Parus major*), while sitting within the four walls of a wooden nest box, were able to discriminate among almost identical song types sung by either their mate or his neighbor (Blumenrath *et al.*, 2007). All these studies confirm that acoustic details of avian signals that are carriers of information are detected and meaningful to birds in the natural context in the field. This also means that interference by ambient noise may prevent auditory detection or adequate recognition of relevant details.

1.2 The auditory world of fish

Fish also make and hear acoustic signals underwater although this is less well known by the general public (Myrberg *et al.*, 1965;

Tavolga, 1971; Ladich, 2014). Many fish species generate sounds and some even do so in air when taken out of the water by fishermen. Sounds of individual fish are often associated with agonistic interactions and courtship (Myrberg, 1997; McKibben and Bass, 1998), while chorusing fish can be heard in the context of spawning aggregations (McCauley and Cato, 2000; Aalbers, 2008) and some schooling species exhibit synchronized release of sounds in response to immediate predation risk (Wilson *et al.*, 2004). Fish are able to make sounds by rhythmic contraction of specially evolved soniferous muscles surrounding the swim bladder, by rasping with pharyngeal jaws, or by beating with pectoral fins (Kasumyan, 2008). Some species even use combinations of sound production mechanisms for highly complex acoustic signals, while others combine sound with electric signals or elaborate visual displays. Although people often refer to singing, calling, or vocalizing fish, this terminology may not be appropriate given the variety and nature of sound production mechanisms. There are no records of acoustic imitation or convergence of fish sounds like in bird vocalizations and it is unlikely that social learning plays any role in the development of adult sound characteristics. The context associated with sound production is known for quite a few fish species, but in terms of understanding of signal function, fish are 'limping' far behind birds. This is obviously due to the medium that is less familiar to human observers that are used to auditory experience in air and due to the practical problems of making underwater sounds audible in air and getting an adequate visual at the same time of the associated sound producer.

Most fish hear a much more restricted frequency range than humans or birds in air (Fig. 2). They are sensitive to the particle motion component of sound, but also to the sound pressure component if they have a pressure-to-motion transducer in the form of a swim bladder (Popper and Fay, 2011; Ladich and Schulz-Mirbach, 2013). Having a swim bladder, with or without other specially evolved structures for better hearing, improves auditory sensitivity and typically also extends the spectral hearing range. Depending on the presence of a swim bladder fish may for example detect sounds from 20 Hz up to 400 Hz or from 20 Hz up to 2 or 3 kHz. However, there is a continuum of hearing sensitivity related to a variety of evolved hearing adaptations, from relative insensitive and narrow frequency range to relatively sensitive and wide frequency range. Generally speaking, more freshwater fish than marine fish are found on the sensitive and wide spectral range side of the continuum and pelagic fish are also more likely to be on the more sensitive and wide spectral range side than benthic or demersal fish that live closer to the bottom of the sea (Ladich and Fay, 2013). We typically still have very limited insight, however, in why and how. It is for example still difficult to measure and determine the relative roles of both underwater sound components (sound pressure and particle motion) in auditory sensitivity of fish (Radford *et al.*, 2012). Fish also have yet another way of sensing nearby sources of very low frequencies. Responsible for this sensitivity are hair cells embedded in the skin, on their head and alongside their body, in the so-called lateral line system (Bleckmann and Zelick, 2009). This detection system may contribute to behavioral responsiveness to environmental infrasound, inaudible to the human ear, as for example, revealed in eel (*Aguilla aguilla*) (Sand *et al.*, 2000), while auditory detection of ultrasound, also inaudible to us, has been reported for some exceptional fish species, such as the shad (*Alosa alosa*). This species can apparently use this sensitivity to avoid echolocating dolphin predators (Mann *et al.*, 1997). Fish use sound for orientation and predator–prey interactions, but many species also listen for the communicative sounds generated by conspecifics.

Fish may be more restricted than birds, but can still generate acoustic signals of considerable variety in the temporal and spectral domain (Amorim, 2006; Ladich, 2008). Most sounds generated by fish can be described as grunts, clicks, or hums, and concern relatively broadband and low-frequency sounds that vary in temporal patterns and spectral energy distribution (Fig. 1). All of the over 800 species of fish that are now known to generate sounds have their own species-specific sounds. Closely related species can be acoustically distinct, for example, through temporal deviation in the number of pulses per sound (Verzijden *et al.*, 2010a). Different populations of the same species may exhibit geographic variation (Parmentier *et al.*, 2005). Many fish species also have small repertoires and context-associated within-species variation, just like birds. In croaking gouramis (*Trichopsis vittata*), for example, both males and females generate broadband croaking sounds during territorial interactions, while females also emit shorter, lower-intensity purring sounds, which initiate and coordinate spawning behavior (Ladich, 2007). There are also many reports on spectral correlations with size as large fish typically produce lower-frequency sounds than small fish (e.g., de Jong *et al.*, 2007). This moderate list of correlations between characteristics of sounds and the maker of sounds should make clear that variety in fish sounds stands for many potential messages for individuals of the same species that are within earshot, just like in birds. However, still relatively few playback experiments have confirmed that the acoustically encoded information is indeed detected and meaningful to fish.

Underwater playbacks to fish have shown that conspecific sounds elicit various behavioral responses. Playback studies typically report approach responses or other more subtle changes in behavior, which do not occur during the control playback of artificial or heterospecific sounds. Estramil *et al.* (2013) showed, for example, that cichlids (*Pundamilia nyererei*) in fish tanks responded to playback of recordings of short sequences of cichlid sound bursts, but not to white noise bursts in the very same temporal pattern. These findings are in line with outdoor findings on bicolor damselfish (*Pomacentrus partitus*) on coral reefs, which not only discriminated between fish sounds and artificial noise bursts, but also showed that fish are able to locate sound sources (Myrberg *et al.*, 1986). Similarly, cichlid females were shown to select the aquarium side of a male cichlid associated with playback of sound when given a choice between two males, matched in coloration and body size (Verzijden *et al.*, 2010a). The report on bicolor damselfish in outdoor conditions also suggested that size-related sound characteristics may affect such female responsiveness (Myrberg *et al.*, 1986). In round gobies (*Neogobius melanostomus*) males with a nest make grunt and drum sounds. Playbacks revealed that neither of these sounds elicit responses from other males with a nest, but grunts are attractive to females and drums elicit responses from males without a nest (Isabella-Valenzi and Higgs, 2013). Although still very few experimental tests exist, and mainly for the relatively close range function of fish sounds, there is also growing experimental support for fish larvae and juveniles responding phonotactically to environmental sounds over larger distances (Simpson *et al.*, 2005; Montgomery *et al.*, 2006). This means that fish may also be bothered by interference of auditory detection and recognition of conspecific signals or environmental cues.

1.3 Environmental noise as selection pressure

Animal signals may be loud and clear at the sound-generating animal and may be transmitted over sufficient distance, but still not be audible by receivers due to interference by other sounds (Brumm and Slabbekoorn, 2005). Any other sound in the environment can affect audibility and is typically referred to as background noise. The signal level relative to the noise level (signal-to-noise ratio) at potential receivers is critical for whether or not signals can come across in sufficient detail (Fig. 3). In general, overlap in time and frequency between the target signal and any other sound in the background can render the signal content partially or completely inaccessible ('masked'). The extent of masking determines therefore the threshold distance for detection, recognition, and comfortable hearing. As the associated signal efficiency is directly related to survival and reproductive success of individuals, consistent noise patterns may drive signal evolution towards an acoustic design that copes best with prevailing noise conditions. In the following part, I will address natural abiotic and biotic sound sources of environmental noise as well as anthropogenic noise caused by human activities. I will highlight correlative and experimental evidence of current competition for acoustic space and review cases suggestive of noise-related signal evolution both in air for birds and in water for fish.

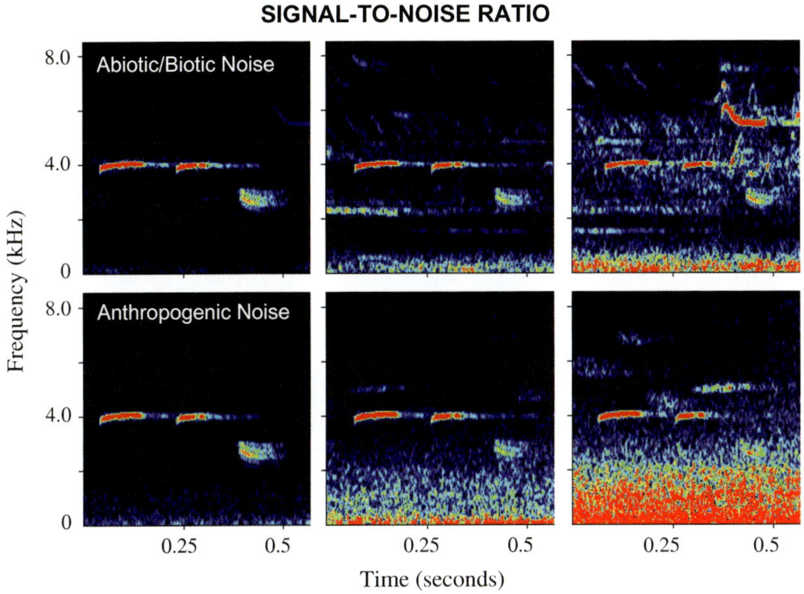

Figure 3. Sonographic representation of variation in signal-to-noise ratios for great tit song for fluctuating natural and artificial noise levels. The very same song recording of the same signal level is mixed with a background of chorusing birds in a Dutch forest (top) or a background of urban traffic noise (bottom). Both noise maskers are artificially elevated in amplitude, from left to right: from low, moderate, to high noise level. Abiotic noise in the form of wind is visible in the low-frequency band between 0–1 kHz in the sonogram on the far right in the top row. The presence of biotic noise in the form of various other singing and calling bird species is best visible in the same sonogram in a wide range between 1–8 kHz. Traffic noise forms a broad band between 0–2 kHz and fades out from 2–4 kHz best visible in the sonogram on the far right in the bottom row.

1.4 Abiotic sound sources

Major sources of ambient noise include weather conditions of wind and rain. Air turbulence will cause signal degradation and especially wind directed from potential receivers towards the singing bird will cause excess attenuation. However, wind and rain do not only affect sound transmission, but also have an audible impact on habitat-related substrate and vegetation. The spectrum of wind- and rain-generated sound is typically biased to low frequencies

(Fig. 3), although rustling tree leaves can cause considerable sound levels over a broad range of frequencies all around perching receivers. Consequently, most birds typically exhibit low singing activity under windy conditions, but some species of very open and often windy habitat may show other responses. For example, king penguins (*Aptenodytes patagonicus)* change their calling behavior during wind-related elevation of noise levels (Lengagne *et al.*, 1999). Penguins on sub-Antarctic islands in the Indian Ocean increase their calling rate and increase the number of syllables per call under windy conditions. They achieve perceptual advantages by this change in calling behavior through increased probability of detection during the challenging signal-to-noise ratios. Few species, such as European blackbirds (*Turdus merula*) and wood pigeons (*Colomba palumbus*), seem not to be bothered by rain while singing, if it is not windy at the same time. However, most species stop singing during heavy rain as the distance over which their acoustic signals would be audible is heavily constrained by the typically high sound levels of rain drops on canopy or tree branches around potential receivers. A study on relatively low-frequency hooting by tawny owls (*Strix aluco*) reported a decline of signal range from over 600 m under dry weather conditions to less than 100 meters under rainy conditions (Lengagne and Slater, 2002).

Also water turbulence at the seaside or in fast-flowing rivers can radiate sounds into the air and cause a significant rise of environmental noise levels. Such abiotic sounds can have broad spectral ranges, but are typically loud in the low frequencies. Many shorebirds have adaptively loud, piercing calls at relatively high frequencies. An interesting example of shorebird signal divergence potentially driven by surf-generated noise is found in the willet (*Catoptrophorus semipalmatus*) of North America (Douglas and Conner, 1999). Birds of a coastal breeding population of the eastern willet generate calls of higher frequency and shorter duration compared to inland breeding birds of the western willet. The higher frequency may improve signal efficiency due to masking reduction, while more, shorter calls may provide a better chance to be heard within the temporal gaps of fluctuating noise from breaking waves. Continuously high levels of ambient noise, with a bias towards relatively low frequencies, were also suggested to be responsible for the convergent evolution of several bird species living around Himalayan torrents (Martens and Geduldig, 1990). The species in this community tended to sing simple song notes with few spectral modulations at relatively high frequencies, which indeed stand out best acoustically in this high-altitude habitat with high levels of low-frequency noise.

The examples of evolution of avian signals possibly driven by abiotic sound sources are corroborated by a few examples of noise-related signal adjustments at a more short-term scale of ontogenetic development or immediate modification (Patricelli and Blickley, 2006). For example, blue-throated hummingbirds, breeding close to forest streams, sing louder during noisy conditions, both in response to natural stream noise and to experimentally elevated noise levels (Pytte *et al.*, 2003). Chaffinches (*Fringilla coelebs*) singing close to waterfalls and torrents also adjust their songs to noisy conditions by singing longer bouts of the same song type before switching to a new type and by singing with higher minimum song frequencies compared to birds singing further away from river noise (Brumm and Slater, 2006). These noise-related song modifications may improve the probability to convey signals under noisy conditions by raising the signal-to-noise ratio, increasing serial redundancy, and decreasing the extent of masking through spectral overlap avoidance.

Prominent abiotic sound sources underwater also include wind and rain, like in air. The wind-generated waves and impact accumulation of droplets cause a sheet source of sound at the water surface that can be loud in rough weather. The surf at beaches or rocky shores will cause a line source of sound, although low-frequency sounds do not travel very far through the waterfront shallows (Forrest *et al.*, 1993). In addition, there are relatively low and loud sound sources such as seismic activity and the occasional lightning events. Lakes and seas may be relatively quiet on calm days (Wysocki *et al.*, 2007), but flowing rivers and fast-flowing streams can be very noisy due to air bubbles, water turbulence and tumbling rocks (Prosperetti, 1988; Tonolla *et al.*, 2010). Both fish and abiotic sound spectra are typically heavily biased towards the lower frequencies and the restricted hearing range of most fish species seem to provide little potential for masking avoidance (Fig. 2).

Masking is likely to have similar effect sizes in both the sound pressure and particle motion component of sound, but may vary considerably dependent on species-specific hearing curves. This was revealed in a study on two cichlid species with differently sized and shaped swim bladders (Ladich and Schulz-Mirbach, 2013). The lionhead cichlid (*Steatocranus tinanti*) from western Africa was the less sensitive species with only a small swim bladder and hearing predominantly via particle motion up to 1.0 kHz. The orange chromide (*Etroplus maculates*) of southern India was the more sensitive species with a large swim bladder and extensions up to the inner ear. Hearing in this species, via particle motion directly and via sound pressure converted into particle motion, extends over a wider frequency range up to 3 kHz. An artificially generated white noise masker caused only moderate levels of masking for the less sensitive species. The more sensitive species experienced already moderate masking at low noise exposure up to 1 kHz and severe masking up to 3 kHz at the highest noise exposure. These results are in line with the hypothesis that the evolution of better hearing is not driven primarily by acoustic communication, but instead by relative noise levels in the environment. The less sensitive lionhead cichlid is a species of relatively noisy riverine habitat, while the more sensitive orange chromis lives in more quiet lagoons and estuaries. Only the less noisy environment may allow exploitation of subtle sounds and may, therefore, select for higher auditory sensitivity.

Few studies have addressed noise-dependent evolution of fish sounds. Lugli and Fine (2003) suggested that shallow streams are generally noisy, but also have spectral ranges in which it is relatively quiet. They found out that the sounds of Italian freshwater gobies (*Padogobius martensii*) fitted nicely into this sound window of low acoustic competition. Also the acoustic divergence between two darter species has been linked to differences in their respective habitats (Speares *et al.*, 2011). The fringe darter (*Etheostoma crossopterum*) lives in relatively quiet habitat and produces drum sounds that are relatively low, but broadband, and of harmonic structure. The fantail darter (*Etheostoma flabellare*) lives in much more noisy habitat and produces drum sounds that have the energy much more concentrated within a relatively narrow frequency band around 300 Hz, which matches with a relatively quiet range in the ambient spectrum. These findings clearly warrant more studies on environmentally determined acoustic differentiation and further investigation of ecological matches between sounds, sensitivity and apparent acoustic niches underwater.

1.5 Biotic sound sources

Other animals also making sounds in the same environment can be very prominent competitors. Sound-generating animals can be of the same species, competing for acoustic space and various resources. They can also be other bird species or mammals, frogs, or insects. These other species are not necessarily ecologically meaningful to each other, although there are examples in which one species exploits the sounds of another. Birds are reported, for example, to respond to hetero-specific alarm calls, which may be beneficial as multiple species can be targets of the same predator (Templeton and Greene, 2007). Sounds of other species may also be exploited to find suitable breeding or foraging grounds when the other species overlap in habitat use (Slabbekoorn and Bouton, 2008). Advertisement sounds aimed at conspecific receivers can also be exploited by predators to find their prey (Schmidt and Belinsky, 2013). However, more often sounds of other species may be neutral and just interfere with detection and recognition of signals for communication. Jamming of signals can be extreme in high-quality habitat, such as found in tropical rainforests with incredibly rich vocal communities (Slabbekoorn and Smith, 2002; Dingle *et al.*, 2008), but also during seasonal peaks in temperate habitats with vocal rush hours during the dawn chorus (e.g., La, 2012; Belinsky *et al.*, 2012).

In general, competition for acoustic space with biotic sound sources occurs at higher frequencies and is more dynamic than for abiotic sound sources (Brumm and Slabbekoorn, 2005). There is typically more variability in space and time related to the composition of local animal communities and birds are known to adjust their timing of vocal activity related to prominent competitors. Neighboring bird species can shift periods of singing in such a way that times of peak activity are not overlapping. Song overlap with vocal competitors of the same species is, therefore, likely responsible for some of the most severe cases of masking due to the synchrony of diurnal cycles and complete spectral overlap. However, birds can also adjust timing at short time scales in response to other and their own species. Eurasian wrens (*Troglodytes troglodytes*), for example, have an excellent timing ability and consistently wait for the end of a song from their neighbor to start their own song (Yang *et al.*, 2014). This behavior prevents masking of their songs for the singing competitor and third-party eavesdroppers, but also allows the wrens to listen themselves to the acoustic details of their vocal opponent. In acoustically highly complex environments like tropical rainforests, there is not a clear pattern of temporal avoidance related to the spectral overlap among species pairs. However, comparative analyses indicate that, also in the tropics, at least some bird species avoid singing at times at which there are many vocal competitors active in the same frequency range (Planqué and Slabbekoorn, 2008).

Different habitats have different vegetation types and different climatological conditions and, therefore, often harbor highly specific animal communities that yield highly specific sound profiles (Slabbekoorn, 2004b; Bormpoudakis *et al.*, 2013). Especially insect and frog species, but also some bird species, can be loud and active for long periods of time. They may block particular frequency bands for large parts of the day, approaching the continuity of some abiotic sound sources and force other species to use a different frequency channel (Slabbekoorn and Smith, 2002; Kirschel *et al.*, 2009). A particularly clear example of noise-determined frequency use in birds is found along the altitudinal gradient of the Andean slopes in South America (Dingle *et al.*, 2008; Caro *et al.*, 2013).

The forest habitat, as sampled in Ecuador and Colombia, changes stereotypically with elevation, as trees become smaller and temperatures get lower from low to high altitude. This geographically consistent altitudinal gradient is negatively correlated with the number of actively present vocal insect and frog species. Consequently, there is a lot of sound in the upper frequencies for singing birds at low elevations, while it becomes increasingly more quiet at high elevations. Different sub-species of the grey-breasted wood wren (*Henichorina leucophrys*) displace each other altitudinally and there is a clear match between the variation in song frequency use among subspecies and the relative amplitude of the ambient noise in their respective altitudinal ranges. At low elevations, wood wren songs are restricted to the low-frequency range avoiding competition from high-frequency ambient noise. In contrast, at high elevations, wood wren songs cover a much broader frequency range with overall little acoustic competition from ambient noise.

Also underwater, individuals of different species may generate sounds roughly at the same time and location and compete for acoustic space. This is especially true for the tropical coral reef systems, which are also referred to as the 'underwater rainforests'. Coral reefs are indeed like rainforests with respect to the high levels of biodiversity that are associated with ecological complexity and acoustic variety. Some species, like parrotfishes (Scaridae) and sea urchins (Echinometridae), are acoustically present through scraping activities on hard coral reef substrate, while snapping shrimps (Alpheidae) can be dominant sound producers when foraging (Au, 2002). The snapping sounds are generated by the collapse of cavitating bubbles that are ejected at high speed by their snapping claw and result in sonic stunning of prey (Versluis *et al.*, 2000). Shrimps are often so numerous and loud that they serve as acoustic beacon for pelagic stages of many fish, crustacean and coral species that are in search for substrate to settle (Simpson *et al.*, 2005; Montgomery *et al.*, 2006). A study of long-term underwater recordings at the Great Barrier Reef of Australia revealed that also fish sounds can be prominent in the underwater soundscape

(McCauley and Cato, 2000). Four distinct fish sounds were reported in the lower frequency range with dominant frequencies around 300 Hz up to 900 Hz, varying in duration and number of pulses. Acoustic activities often peaked at night and sometimes ensonified the reef collectively. Large groups of one fish species generated sounds synchronously in choruses, while individual fish of another species generated isolated sounds repeatedly for long periods of time. Dominating sound levels from snapping shrimps or chorusing fish will affect signal efficiency for other species in the community. However, as far as I know, there are no studies yet on masking impact from biotic sound sources or studies on overlap avoidance strategies in fish like addressed above for birds.

Underwater habitats also vary in type and quality, just like forest habitats in air, and the associated variation in animal communities is reflected in habitat-specific sound profiles (Kennedy *et al.*, 2010; McWilliam *et al.*, 2013). Fish density and diversity typically have the strongest correlations with sounds below 1 kHz, while coral and benthic diversity yields more sounds above 1 kHz. Coral reefs differ clearly from adjacent habitat types such as sandy beaches or seagrass beds, but there is also significant variation among reef sites. Staaterman *et al.* (2013) found differences in the frequency distribution and temporal variability between two coral reef habitats that were far apart, one in the Eastern Pacific (Panama) and one in the Caribbean (Florida Keys). Radford *et al.* (2010) also reported spectral divergence among reef sites that were more close to each other, but that were ecologically distinct. Although fish may be more restricted than birds to spectrally adjust at evolutionary or ecological time scales, the study on darters and abiotic sound sources (Speares *et al.* 2011) suggests that we can also expect habitat-dependent acoustic divergence in fish sounds related to variation in biotic sound sources.

1.6 Anthropogenic sound sources

The levels of ambient noise in animal habitat, originally only from abiotic and biotic sound sources, have been elevated artificially in the last decades. This is due to the addition of anthropogenic noise: sounds generated by human activities. Transport of passengers and cargo by car, truck, train or plane, in addition to many different kinds of motorized industrial and recreational activities, generates sounds that reach large parts of our planet (Slabbekoorn and Ripmeester, 2008; Barber *et al.*, 2009). Although environmental sounds are by no means new to animals, the nature and level of anthropogenic noise, with its evolutionarily recent and expanding distribution, can be detrimental in various ways. Birds can leave habitat that is suitable for breeding or migratory stopover as a direct response to noisy conditions (Bayne *et al.*, 2008; Francis *et al.*, 2009; McClure *et al.*, 2013) or males that still sing at a noisy site may be unable to attract a mate (Habib *et al.*, 2007). Also breeding pairs of noisy territories may experience detrimental consequences in having lower clutch sizes and lower quality of fledglings (Halfwerk *et al.*, 2011). These negative effects from artificial elevation of ambient noise levels suggest that also natural abiotic and biotic sound sources must have affected species distributions and signal evolution in this way in historical times, and likely still have this effect today.

Several lines of research on anthropogenic noise have revealed the close link between ambient noise conditions and avian signal efficiency (Slabbekoorn, 2013; Nemeth *et al.*, 2013). There are observational studies on lower breeding densities in more noisy forest plots for bird species that sing at relatively low frequencies and not for species singing high (Goodwin and Shriver, 2011). This suggests that masking by traffic noise, with most energy in the low-frequency range (Fig. 3), plays a role in the noise-dependent quality reduction of breeding habitat. There are also experimental studies showing that masking can indeed cause frequency-dependent signal efficiency. Female great tits exposed to anthropogenic noise on the nest showed a significant decline in response to relatively low-frequency song types, while their response to relatively high-frequency song types remained unaffected (Halfwerk *et al.*, 2011). These results have been corroborated by indoor studies on canaries (*Serinus canaria*), a well-established model species in which females express their extent of sexual excitement through copulation solicitation displays in response to song playback. Experimental exposure to anthropogenic noise, either alternating or overlapping with the song playback, revealed a clear impact through masking. Also here the relatively low-frequency songs elicited significantly fewer displays due to the masking noise, while the sexual responsiveness to relatively high-frequency songs remained unaffected again (Huet des Aunay *et al.*, 2013).

Singing birds have many ways in which they can respond to fluctuations in ambient noise levels, which is illustrated particularly well in studies on anthropogenic noise. Singing birds have been observed to move away upon exposure to anthropogenic noise (McLaughlin and Kunc, 2013), which may improve their own hearing conditions, but not necessarily the audibility of their song by others in the target area. Singing birds have also been reported to move closer to birds that are target receivers and experience noisy conditions. Halfwerk *et al.* (2012) showed that female great tits in nest boxes made their singing mates come closer during experimental noise exposure, by which signal-to-noise ratios in the nest box were restored to pre-exposure levels. Birds can also sing louder during more noisy conditions, a taxonomically widespread phenomenon known as the Lombard effect (Brumm and Zollinger, 2011). This is likely to occur in response to artificial as well as natural noise fluctuations. In a study on urban nightingales of Berlin, Brumm (2004) was able to show that birds of noisy territories sang louder on weekdays than in the weekend.

There are also several reports on urban birds starting to sing earlier than forest birds. Such a dawn chorus shift into the more quiet nocturnal conditions leads to masking avoidance in time. A recent study on blackbirds in and around Leipzig, Germany, showed that song onset of the dawn chorus is generally determined by natural light intensity, related to day length, cloud cover and ambient temperatures. However, birds along the noisy ring road woke up more than five hours earlier than birds in the more quiet urban forest. The shift to earlier song on-set was significantly related with street light intensities as well as traffic noise levels

(Nordt and Klenke, 2013). In an earlier study, European robins (*Erithacus Rubecula*) in urban territories of Sheffield, England, were also found to be more active at night at locations that were relatively noisy during day-time (Fuller *et al.*, 2007). The results of these correlative studies are matched by an experimental study on urban birds of Seville, Spain, which showed that two typical urban species, the spotless starling (*Sturnus unicolor*) and the house sparrow (*Passer domesticus*) woke up earlier when they were exposed earlier to local and familiar anthropogenic noise (Arroyo-Solís *et al.*, 2013).

Another widespread noise-dependent adjustment is singing higher in frequency, which was first reported for great tits and subsequently replicated for many urban species all over the world (reviewed in Slabbekoorn, 2013). Urban bird populations of the great tit in Europe sing higher on average than nearby forest populations (Slabbekoorn and den Boer-Visser, 2006), while also individual birds of noisy territories sing higher than nearby conspecifics in quiet territories (Slabbekoorn and Peet, 2003). Experimental exposure studies in the field have revealed that individual birds can even adjust almost immediately (Gross *et al.*, 2010; Verzijden *et al.*, 2010b) and great tits do so by switching to another song type more quickly when they happen to sing a low-frequency song type and continue to sing the same when they happen to sing a high-frequency song type (Halfwerk and Slabbekoorn, 2009). Different species can have different singing styles and also show a variety in mechanisms for noise-dependent frequency shifts (Bermudez-Cuamatzin *et al.*, 2010; Slabbekoorn, 2013).

Although singing at higher frequencies may improve signal detection and communication range through masking avoidance of the low-frequency traffic noise, it is by no means certain that the modification is always beneficial to the singer. A playback test in northern cardinals (*Cardinalis cardinalis*) recently revealed that songs with upward spectral shifts as found under noisy conditions are less effective in triggering a male territorial response under quiet conditions compared to songs of average frequencies (Luther and Magnotti, 2014). It has also been postulated that singing low may be important for male birds to attract females and that there is a trade-off under noisy conditions between being heard and being attractive (Slabbekoorn and Ripmeester, 2008). Great tit males sing indeed at their lowest during the period of egg fertilization and female great tits of males that sing low frequencies show higher fidelity than females of males that sing high frequencies (Halfwerk *et al.*, 2011). These field data are corroborated again in the laboratory, as also canary females become sexually less responsive in quiet conditions to male songs that are shifted upward in frequency. Consequently, although high-frequency songs outcompete low-frequency songs in terms of masking avoidance, they do not seem to outcompete low-frequency songs in quiet or noisy conditions (Huet des Aunay *et al.*, 2013; Luther and Magnotti, 2014). An advantage may still be that it may be mechanistically easier to sing louder at higher frequencies, which can positively affect both the transmission distance of a signal (Nemeth *et al.*, 2013) and its attractiveness (Ritschard *et al.*, 2010).

Human activities, in, on, or close to water, also cause considerable elevation of natural ambient sound levels underwater. Activities such as container shipping, pile driving, seismic surveys, sonar exercises, dredging, passenger transport, and motorized recreational activities have expanded considerably over the past decades (Popper *et al.*, 2004; Ladich, 2008). Impulse sounds of high power may cause immediate physical harm that may lead to increased mortality or hearing damage (Hawkins and Popper, 2012). Although the impact of moderate noise levels concerns more subtle behavioral and physiological effects compared to loud impulse sounds, more moderate noise levels are much more widespread in time and space and reach a much higher number of species and higher number of individual animals (Slabbekoorn *et al.*, 2010). Like for birds, there is some insight into effects of anthropogenic noise on fish distributions from fisheries data. For example, catching rates have been reported to decline during and well after a noisy seismic survey. Furthermore, anthropogenic noise may not only deter but also interrupt important activities such as courting, spawning, and foraging, or disturb fish during migration.

Although there are far fewer studies than in birds, several recent studies have addressed masking issues and noise impact on signaling fish. As mentioned before, fish sounds are typically broadband and relatively low in frequency, which also matches with the typical broadband nature and energy bias to low frequencies of anthropogenic noise (Fig. 2). Vasconcelos *et al.* (2007) reported considerable masking effects of ferry boat noise on the audibility of advertisement sounds of the Lusitanian toadfish (*Halobatrachus didactylus*) in Portugal, based on hearing sensitivity data and sound pressure assessments in the field of both signals and noise. Codarin *et al.* (2009) also measured ambient and boat noise and the impact on hearing sensitivity in three Mediterranean fish species. They reported that communication distances are restricted dramatically during passages of a cabin-cruiser, some times decreasing the distance over which the fish sounds are audible to conspecifics from about 500 m to just 1 m.

Another study showed that noisy conditions due to boat traffic can affect the outcome of interactions probably through interference of acoustic communication (Sebastianutto *et al.*, 2010). Round goby (*Gobius cruentatus*) males typically defend an area for food and a nest to receive egg laying females. Such resident males usually have an advantage over newcomers and are able to win the majority of encounters with intruders. However, an indoor study on captive fish revealed that residents become more often submissive under noisy than under quiet conditions. As a consequence, resident males did not achieve their usual advantage and intruders were successful in winning an equal amount of encounters. The kind of impact of anthropogenic noise, like in this example in gobies, is likely to be widespread underwater and awaiting further investigations. It has also become clear that noise impact extends to detrimental effects on soundscape orientation based on environmental cues. Holles *et al.* (2013) were recently able to show directional responses of settlement-stage larvae of the coral reef cardinal fish (*Apogon doryssa*) to playback of reef sounds. However, there was a significant reduction in this phonotactic response when reef noise was played back together with boat noise. This interference by boat noise may indicate that acoustically guided dispersal can be disrupted under natural conditions. Such a disruption can cause an extension of the pelagic phase of fish larvae, which will increase predation risk and decrease the chance of settlement on appropriate substrate.

There is still only a single case reported in which fish appear to adjust their signaling behavior in response to anthropogenic noise, which is a huge contrast to the bird literature. Brown meagres (*Sciaena umbra*), of the fish family with the appropriate name of 'drums', were exposed to boat passages to study natural activity patterns of sound production in coastal waters of the Mediterranean in Italy (Picciulin *et al.*, 2012). Although the number of active individuals could not be scored in this nocturnal study, repeated exposure to noisy boat passages resulted in an increased mean pulse rate. This behavior could positively effect detection probabilities and compensate for sounds that are masked by boat noise. Fish may also be able to generate louder sounds in response to elevated noise levels, but their moderate flexibility in terms of sound production and their limited range of hearing sensitivity may make spectral adjustments that would yield at least some masking avoidance less likely. Obviously, many more studies are needed to accumulate more findings, test whether expectations can be confirmed, and possibly to reveal alternative signaling strategies in the acoustic world of fish (Radford *et al.*, 2014).

2. Conclusions

It should be clear from this review that acoustic signals are an excellent way to communicate in air and in water as reflected by their prevalence in birds and fish. However, noise interference can be a considerable challenge and can have a serious impact on signal efficiency. The fitness consequences of signal efficiency through the effect on mate attraction and territorial defence explain why ambient noise also affects the evolution of acoustic signals. Interestingly, not many studies have investigated evolution or current signal efficiency and counter strategies with respect to abiotic sound sources like wind, rain, and water turbulence. There are a considerable number of studies that have addressed current interactions among biotic sound sources, but mainly for birds and not for fish. Also the impact on signal evolution by animal community related noise profiles has been addressed in birds but not yet in fish. It seems, therefore, that experimental exposure studies using abiotic and biotic sound sources are likely to yield interesting new fundamental insights for birds but especially for fish.

However, the most striking pattern in the literature is the recent surge in studies on current signal efficiency and the environmental shaping of bird signals in the context of anthropogenic noise. The artificial rise in ambient noise levels due to human activities has triggered interest and concern, which obviously stimulates studies with high potential for applied value. There are also many studies on the impact of anthropogenic noise that are mainly curiosity driven addressing primarily fundamental questions that also apply to natural sound sources. Such studies on the competition for acoustic space are still needed, especially for signaling fish, to get a better understanding of both natural and artificial settings. Finally, more insight into the impact of anthropogenic noise will not only provide insight into animal vulnerability but also has the potential to raise awareness about the intrinsic beauty and ecological value of natural soundscapes in air and water.

References

Aalbers SA. 2008. Seasonal, diel, and lunar spawning periodicities and associated sound production of white seabass (*Atractoscion nobilis*). *Fishery Bulletin*, 106, 143–151.

Amorim MCP. 2006. Diversity of sound production in fish. In *Communication in Fishes*, Ladich F, Collin SP, Moller P and Kapoor BG (eds.), Vol 1, 71–104, Enfield: Science Publishers.

Arroyo-Solís A, Castillo JM, Figueroa E, Lopez-Sanchez JL and Slabbekoorn H. 2013. Experimental evidence for an impact of anthropogenic noise on dawn chorus timing in urban birds. *Journal of Avian Bioliogy*, 44, 288–296.

Au WWL. 2002. Snapping shrimps: the bane of bioacoustics monitoring of coral reefs. *Journal of the Acoustical Society of America*, 112, 2202.

Baptista LF and Schuchmann KL. 1990. Song learning in the Anna hummingbird (*Calypte anna*). *Ethology*, 84, 15–26.

Barber JR, Crooks KR and Fristrup KM. 2009. The costs of chronic noise exposure for terrestrial organisms. *Trends in Ecology and Evolution*, 25, 180–189.

Bayne EM, Habib L and Boutin S. 2008. Impacts of chronic anthropogenic noise from energy-sector activity on abundance of songbirds in the boreal forest. *Conservation Biology*, 22, 1186–1193.

Belinsky KL, Hogle JL and Schmidt KA. 2012. Veeries experience more varied acoustic competition at dawn than at dusk. *Wilson Journal of Ornithology*, 124, 265–269.

Bermudez-Cuamatzin E, Ríos-Chelén AA, Gil D and Garcia CM. 2010. Experimental evidence for real-time song frequency shift in response to urban noise in a passerine bird. *Biology Letters*, 7, 36–38.

Bormpoudakis D, Sueur J and Pantis JD. 2013. Spatial heterogeneity of ambient sound at the habitat type level: ecological implications and applications. *Landscape Ecology*, 28, 495–506.

Bleckmann H and Zelick R. 2009. Lateral line system of fish. *Integrative Zoology*, 4, 13–25.

Blumenrath SH, Dabelsteen T and Pedersen SB. 2007. Vocal neighbor-mate discrimination in female great tits despite high song similarity. *Animal Behavior*, 73, 789–796.

Brumm H. 2004. The impact of environmental noise on song amplitude in a territorial bird. *Journal of Animal Ecology*, 73, 434–440.

Brumm H and Slabbekoorn H. 2005. Acoustic communication in noise. *Advances in the Study of Behavior*, 35, 151–209.

Brumm H and Slater PJB. 2006. Ambient noise, motor fatigue, and serial redundancy in chaffinch song. *Behavioral Ecology and Sociobiology*, 60, 475–481.

Brumm H and Zollinger SA. 2011. The evolution of the Lombard effect: 100 years of psychoacoustic research. *Behavior*, 148, 11–13.

Cardoso GC. 2011. Paradoxical calls: the opposite signaling role of sound frequency across bird species. *Behavioral Ecology*, 23, 237–241.

Caro LM, Caycedo-Rosales PC, Bowie RCK, Slabbekoorn H and Cadena CD. 2013. Ecological speciation along an elevational gradient in a tropical passerine bird? *Journal of Evolutionary Biology*, 26, 357–374.

Codarin A, Wysocki LE, Ladich F and Picciulin M. 2009. Effects of ambient and boat noise on hearing and communication in three fish species living in a marine protected area (Miramare, Italy). *Marine Pollution Bulletin*, 58, 1880–1887.

de Jong K, Bouton N and Slabbekoorn H. 2007. Azorean rock-pool blennies produce size-dependent calls in a courtship context. *Animal Behavior*, 74, 1285–1292.

Dingle C, Halfwerk W and Slabbekoorn H. 2008. Habitat-dependent song divergence at subspecies level in the grey-breasted wood-wren. *Journal of Evolutionary Biology*, 21, 1079–1089.

Dingle C, Poelstra J, Halfwerk W, Brinkhuizen D and Slabbekoorn H. 2010. Asymmetric response patterns to subspecies-specific song differences in allopatry and parapatry in the gray-breasted wood-wren. *Evolution*, 64, 3537–3548.

Dooling RJ. 1992. Hearing in birds. In *The Evolutionary Biology of Hearing*, Webster DB, Fay RR and Popper AN (eds.), 545–559. Springer-Verlag, New York.

Douglas III HD and Conner WE. 1999. Is there a sound reception window in coastal environments? Evidence from shorebird communication systems. *Naturwissenschaften*, 86, 228–230.

Endler J. 1992. Signals, signal conditions, and the direction of evolution. *American Nataralist*, 139, S125–S153.

Eriksson D and Wallin L. 1986. Male bird song attracts females — a field experiment. *Behavioral Ecology and Sociobiology*, 19, 197–299.

Estramil E, Bouton N, Verzijden MN, Hofker K, Riebel K and Slabbekoorn H. 2013. Cichlids respond to conspecific sounds but females exhibit no phonotaxis without the presence of live males. *Ecology of Freshwater Fish*. doi: 10.1111/eff.12081.

Forrest T, Miller G and Zagar J. 1993. Sound propagation in shallow water: implications for acoustic communication by aquatic animals. *Bioacoustics*, 4, 259–270.

Francis CD, Ortega CP and Cruz A. 2009. Noise pollution changes avian communities and species interactions. *Current Biology*, 19, 1415–1419.

Fuller RA, Warren PH and Gaston KJ. 2007. Daytime noise predicts nocturnal singing in urban robins. *Biology Letters*, 3, 368–370.

Godard R. 1991. Long-term memory of individual neighbors in a migratory songbird. *Nature*, 350, 228–229.

Goodwin SE and Shriver WG. 2011. Effects of traffic noise on occupancy patterns of forest birds. *Conservation Biology*, 25, 406–411.

Gross K, Pasinelli G and Kunc HP. 2010. Behavioral plasticity allows short-term adjustment to a novel environment. *American Naturalist*, 176, 456–464.

Habib L, Bayne EM and Boutin S. 2007. Chronic industrial noise affects pairing success and age structure of ovenbirds *Seiurus aurocapilla*. *Journal of Applied Ecology*, 44, 176–184.

Halfwerk W, Bot S, Buikx J, van der Velde M, Komdeur J, ten Cate C and Slabbekoorn H. 2011. Low songs lose potency in urban noise conditions. *Proceedings of the National Academy of Sciences of the United States of America*, 108, 14549–14554.

Halfwerk W, Bot S and Slabbekoorn H. 2012. Male great tit song perch selection in response to noise-dependent female feedback. *Functional Ecology*, 26, 1339–1347.

Halfwerk W and Slabbekoorn H. 2009. A behavioral mechanism explaining noise-dependent frequency use in urban birdsong. *Animal Behavior*, 78, 1301–1307.

Hasselquist D, Bench S and von Schantz T. 1996. Correlation between male song repertoire, extra-pair paternity and offspring survival in the great reed warbler. *Nature*, 381, 229–232.

Hawkins AD and Popper AN. 2012. (Normandeau Associates, Inc.). *Effects of noise on fish, fisheries, and invertebrates in the US Atlantic and Arctic from energy industry sound-generating activities.* A literature synthesis for the US Dept. of the Interior, Bureau of Ocean Energy Management. Contract # M11PC00031. 153.

Hile AG, Plummer TK and Striedter GF. 2000. Male vocal imitation produces call convergence during pair bonding in budgerigars *Melopsittacus undulates*. *Animal Behavior*, 59, 1209–1218.

Holles S, Simpson SD, Radford AN, Berten L and Lecchini D. 2013. Boat noise disrupts orientation behavior in a coral reef fish. *Marine Ecology Progress Series*, 485, 295–300.

Huet des Aunay G, Slabbekoorn H, Nagle L, Passas Nicolas P and Draganoiu TI. 2013. Urban noise undermines female sexual preferences for low-frequency songs in domestic canaries. *Animal Behavior*, 87, 67–75.

Hultsch H and Todt D. 2004. Learning to sing. In *Nature's Music — The Science of Birdsong*, Marler P and Slabbekorn H (eds.), 80–107. Elsevier Publishers, Academic Press, Amsterdam & Boston.

Isabella-Valenzi L and Higgs DM. 2013. Sex- and state-dependent attraction of round gobies, *Neogobius melanostomus*, to conspecific calls. *Behavior*, 150, 1509–1530.

Kasumyan AO. 2008. Sound and sound production in fishes. *Journal of Ichthyology*, 11, 981–1030.

Kennedy EV, Guzman HM, Holderied MW, Mair JM and Simpson SD. 2010. Reef-generated noise provides reliable information about habitats and communities: evidence from a Panamanian case study. *Journal of Experimental Marine Biology and Ecology*, 395, 85–92.

Kirschel ANG, Blumstein DT, Cohen RE, Buermann W, Smith TB and Slabbekoorn H. 2009. Birdsong tuned to the environment: green hylia song varies with elevation, tree cover, and noise. *Behavioral Ecology*, 20, 1089–1095.

Konishi M and Knudsen EI. 1979. Oilbird — Hearing and echolocation. *Science*, 204, 425–427.

Krebs J, Ashcroft R and Webber M. 1978. Song repertoires and territory defense in great tit. *Nature*, 271, 539–542.

Kroodsma DE and Konishi M. 1991. A subosine bird (eastern phoebe, *Sayornis phoebe*) develops normal song without auditory feedback. *Animal Behavior*, 42, 477–487.

La VT. 2012. Diurnal and nocturnal birds vocalize at night: a review. *The Condor*, 114, 245–257.

Ladich F. 2007. Females whisper briefly during sex: context- and sex-specific differences in sounds made by croaking gouramis. *Animal Behavior*, 73, 379–387.

Ladich F. 2008. Sound communication in fishes and the influence of ambient and anthropogenic noise. *Bioacoustics*, 17, 35–37.

Ladich F. 2014. Fish bioacoustics. *Current Opinions in Neurobiology*, 28, 121–127.

Ladich F and Fay RR. 2013. Auditory evoked potential audiometry in fish. *Reviews in Fish Biology and Fisheries*, 23, 317–364.

Ladich F and Schulz-Mirbach T. 2013. Hearing in cichlid fish under noise conditions. *PLoS ONE*, 8, e57588.

Lengagne T, Aubin T, Lauga J and Jouventin P. 1999. How do king penguins *Aptenodytes patagonicus* apply the mathematical theory of information to communicate in windy conditions? *Proceedings of the Royal Society London B*, 266, 1623–1628.

Lengagne T and Slater PJB. 2002. The effects of rain on acoustic communication: tawny owls have good reason for calling less in wet weather. *Proceedings of the Royal Society London B*, 269, 2121–2125.

Linhart P, Slabbekoorn H and Fuchs R. 2012. The communicative significance of song frequency and song length in territorial chiffchaffs. *Behavioral Ecology*, 23, 1338–1347. doi: 10.1093/beheco/ars127.

Lugli M. and Fine ML. 2003. Acoustic communication in two freshwater gobies: ambient noise and short-range propagation in shallow streams. *Journal of the Acoustical Society of America*, 114, 512–521.

Luther D and Magnotti J. 2014. Can animals detect differences in vocalizations adjusted for anthropogenic noise? *Animal Behavior*, 92, 111–116.

Mann DA, Lu ZM and Popper AN. 1997. A clupeid fish can detect ultrasound. *Nature*, 389, 341.

Marler P. 1997. Three models of song learning: evidence from behavior. *Journal of Neurobiology*, 33, 501–516.

Marler P. 2004. Bird calls: A cornucopia for communication. In *Nature's Music The Science of Birdsong*, Marler P and Slabbekoorn H (eds.), 132–177. Academic Press/Elsevier, San Diego.

Martens J and Geduldig G. 1990. Acoustic adaptations of birds living close to Himalayan torrents. *Proceedings of the International 100 DO-G Meeting. Bonn: Current Topics in Avian Biology*, 123–131.

McCauley RD and Cato DH. 2000. Patterns of fish calling in a nearshore environment in the Great Barrier Reef. *Philosophical Transactions of the Royal Society B*, 355, 1289–1293.

McClure CJW, Ware HE, Carlisle J, Kaltenecker G and Barber JR. 2013. An experimental investigation into the effects of traffic noise on distributions of birds: avoiding the phantom road. *Proceedings of the Royal Society B*, 280. doi: 10.1098/rspb.2013.2290.

McGregor PK. 1992. *Quantifying responses to playback: one, many, or composite multivariate measures?* In *Playback and Studies of Animal Communication*, McGregor PK (ed.), 79–96. Plenum Press, New York.

McKibben JR and Bass AH. 1998. Behavioral assessment of acoustic parameters relevant to signal recognition and preference in a vocal fish. *Journal of Acoustical Society of America*, 104, 3520–3533.

McLaughlin KE and Kunc HP. 2013. Experimentally increased noise levels change spatial and singing behavior. *Biology Letters*, 9. doi: 10.1098/rsbl.2012.0771.

McWilliam JN and Hawkins AD. 2013. A comparison of inshore marine soundscapes. *Journal of Experimental Marine Biology and Ecology*, 446, 166–176.

Montgomery JC, Jeffs A, Simpson SD, Meekan MG and Tindle C. 2006. Sound as an orientation cue for the pelagic larvae of reef fishes and decapod crustaceans. *Advances in Marine Biology*, 51, 143–196.

Myrberg AA. 1997. Sound production by a coral reef fish (*Pomacentrus partitus*): evidence for a vocal territorial "keep-out" signal. *Bulletin of Marine Sciences*, 60, 1017–1025.

Myrberg AA, Kramer E and Heinecke P. 1965. Sound production by cichlid fishes. *Science*, 149, 555–558.

Myrberg AA Jr., Mohler M and Catala JD. 1986. Sound production by males of a coral reef fish (*Pomacentrus partitus*): its significance to females. *Animal Behavior*, 34, 913–923.

Nemeth E, Pieretti N, Zollinger SA, Geberzahn N, Partecke J, Miranda AC and Brumm H. 2013. Bird song and anthropogenic noise: vocal constraints may explain why birds sing higher-frequency songs in cities. *Proceedings of the Royal Society B*, 280. doi: 10.1098/rspb.2012.2798.

Nordt A and Klenke R. 2013. Sleepless in town — Drivers of the temporal shift in dawn song in urban European blackbirds. *PLoS ONE*, 8, e71476.

Odom KJ, Hall ML, Riebel K, Omland KE and Langmore NE. 2014. Female song is widespread and ancestral in songbirds. *Nature Communications*, 5, 3379. doi: 10.1038/ncomms4379.

Parmentier E, Lagardère JP, Vandewalle P and Fine ML. 2005. Geographical variation in sound production in the anemonefish *Amphiprion akallopisos*. *Proceedings of the Royal Society London B*, 272, 1697–1703.

Patricelli GL and Blickley JL. 2006. Avian communication in urban noise: causes and consequences of vocal adjustment. *Auk*, 123, 639–649.

Picciulin M, Sebastianutto L, Codarin A, Calcagno G and Ferrero EA. 2012. Brown meagre vocalization rate increases during repetitive boat noise exposures: a possible case of vocal compensation. *Journal of the Acoustical Society of America*, 132, 3118–3124.

Planqué R and Slabbekoorn H. 2008. Spectral overlap in songs and temporal avoidance in a Peruvian bird assemblage. *Ethology*, 114, 262–271.

Podos J, Lahti DC and Moseley DL. 2009. Vocal performance and sensorimotor learning in songbirds. *Advances in the Study of Behavior*, 40, 159–195.

Podos J and Warren PS. 2007. The evolution of geographic variation in birdsong. *Advances in the Study of Behavior*, 37, 403–458.

Popper AN and Fay RR. 2011. Rethinking sound detection by fishes. *Hearing Research*, 273, 25–36.

Popper AN, Fewtrell J, Smith ME and McCauley RD. 2004. Anthropogenic sound: effects on the behavior and physiology of fishes. *Marine Technology Society Journal*, 37, 35–40.

Prosperetti A. 1988. Bubble-related ambient noise in the ocean. *Journal of the Acoustical Society of America*, 84, 1042–1054.

Pytte CL, Rush KM and Ficken MS. 2003. Regulation of vocal amplitude by the blue-throated hummingbird, *Lampornis clemenciae. Animal Behavior*, 66, 703–710.

Pytte CL, Ficken MS and Moiseff A. 2004. Ultrasonic singing by the blue-throated hummingbird: a comparison between production and perception. *Journal of Comparative Physiology A*, 190, 665–673.

Radford AN, Kerridge E and Simpson SD. 2014. Acoustic communication in a noisy world: can fish compete with anthropogenic noise? *Behavioral Ecology*, doi: 10.1093/beheco/aru029.

Radford CA, Montgomery JC, Caiger P and Higgs DM. 2012. Pressure and particle motion detection thresholds in fish: a re-examination of salient auditory cues in teleosts. *Journal of Experimental Biology*, 215, 3429–3435.

Radford CA, Stanley JA, Tindle CT, Montomery JC and Jeffs AG. 2010. Localised coastal habitats have distinct underwater sound signatures. *Marine Ecological Progress Series*, 401, 21–29.

Riebel K. 2009. Song and female choice in zebra finches: a review. *Advances in the Study of Animal Behavior*, 40, 197–238.

Ritschard M, Riebel K and Brumm H. 2010. Female zebra finches prefer high-amplitude song. *Animal Behavior*, 79, 877–833.

Ripmeester EAP, de Vries AM and Slabbekoorn H. 2007. Do blackbirds signal motivation to fight with their song? *Ethology*, 113, 1021–1028.

Ripmeester EAP, Kok J, van Rijssel J and Slabbekoorn H. 2010. Habitat-related birdsong divergence: a multi-level study on the influence of territory density and ambient noise. *Behavioral Ecology and Sociobiology*, 64, 409–418.

Ruegg K, Anderson E and Slabbekoorn H. 2012. Differences in timing of migration and response to sexual signalling drive asymmetric hybridization across a migratory divide. *Journal of Evolutionary Biology*, 25, 1741–1750.

Sand O, Enger PS, Karlsen HE, Knudsen F and Kvernstuen T. 2000. Avoidance responses to infrasound in downstream migrating European silver eels. *Anguilla anguilla. Environmental Biology of Fishes*, 57, 327–336.

Schmidt KA and Belinsky KL. 2013. Voices in the dark: Predation risk by owls influences dusk singing in a diurnal passerine. *Behavioral Ecology and Sociobiology*, 67, 1837–1843.

Sebastianutto L, Picciulin M, Costantini M and Ferrero EA. 2011. How boat noise affects an ecologically crucial behavior: the case of territoriality in *Gobius cruentatus* (Gobiidae). *Environmental Biology of Fishes*, 92, 207–215.

Simpson SD, Meekan MG, Montgomery JC, McCauley RD and Jeffs A. 2005. Homeward sound. *Science*, 308, 221–221.

Slabbekoorn H. 2004a. Singing in the wild: the ecology of birdsong. In *Nature's Music The Science of Birdsong*, Marler P and Slabbekoorn H (eds.), 178–205. Academic Press/Elsevier, San Diego.

Slabbekoorn H. 2004b. Habitat-dependent ambient noise: consistent spectral profiles in two African forest types. *Journal of the Acoustical Society of America*, 116, 3727–3733.

Slabbekoorn H. 2013. Songs of the city: noise-dependent spectral plasticity in the acoustic phenotype of urban birds. *Animal Behavior*, 85, 1089–1099.

Slabbekoorn H and Bouton N. 2008. Soundscape orientation: a new field in need of sound investigation. *Animal Behavior*, 76, e5–e8.

Slabbekoorn H, Bouton N, van Opzeeland I, Coers A, ten Cate C and Popper AN. 2010. A noisy spring: the impact of globally rising underwater sound levels on fish. *Trends in Ecology and Evolution*, 25, 419–427.

Slabbekoorn H and den Boer-Visser A. 2006. Cities change the songs of birds. *Current Biology*, 16, 2326–2331.

Slabbekoorn H, Ellers J and Smith TB. 2002. Bird song and sound transmission: the benefits of reverberations. *Condor*, 104, 564–573.

Slabbekoorn H and Peet M. 2003. Birds sing at a higher pitch in urban noise. *Nature*, 424, 267.

Slabbekoorn H and Ripmeester EAP. 2008. Birdsong and anthropogenic noise: Implications and applications for conservation. *Molecular Ecology*, 17, 72–83.

Slabbekoorn H and Smith TB. 2002. Habitat-dependent song divergence in the little greenbul: an analysis of environmental selection pressures on acoustic signals. *Evolution*, 56, 1849–1858.

Speares P, Holt D and Johnston C. 2011. The relationship between ambient noise and dominant frequency of vocalizations in two species of darters (Percidae: etheostoma). *Environmental Biology of Fishes*, 90, 103–110.

Staaterman E, Rice AN, Mann DA and Paris CB. 2013. Soundscapes from a tropical eastern Pacific reef and a Caribbean sea reef. *Coral Reefs*, 32, 553–557.

Suthers RA. 2004. How birds sing and why it matters. In *Nature's Music The Science of Birdsong*, Marler P and Slabbekoorn H (eds.), 272–295. Academic Press/Elsevier, San Diego.

Tavolga WN. 1971. Sound production and detection. In *Fish Physiology*, Hoar WS and Randall DJ (eds.), 135–205. Academic Press, New York, USA.

Templeton CN and Greene E. 2007. Nuthatches eavesdrop on variations in heterospecific chickadee mobbing alarm calls. *Proceedings of the National Academy of Sciences USA*, 104, 5479–5482.

Ten Cate C, Slabbekoorn H and Ballintijn MR. 2002. Birdsong and male–male competition: causes and consequences of vocal variability in the collared dove (*Streptopelia decaocto*). *Advances in the Study of Behavior*, 31, 31–75.

Tonolla D, Acuna V, Lorang MS, Heutschi K and Tockner K. 2010. A field-based investigation to examine underwater soundscapes of five common river habitats. *Hydrological Processes*, 24, 3146–3156.

Vasconcelos RO, Amorim MCP and Ladich F. 2007. Effects of ship noise on the detectability of communication signals in the Lusitanian toadfish. *Journal of Experimental Biology*, 210, 2104–2112.

Versluis M, Schmitz B, von der Heydt A and Lohse D. 2000. How snapping shrimp snap: through cavitating bubbles. *Science*, 289, 2114–2117.

Verzijden MN, van Heusden J, Bouton N, Witte F, ten Cate C and Slabbekoorn H. 2010a. Sounds of male Lake Victoria cichlids vary within and between species and affect mate preferences. *Behavioral Ecology*, 21, 548–555.

Verzijden MN, Ripmeester EAP, Ohms VR, Snelderwaard P and Slabbekoorn H. 2010b. Immediate spectral flexibility in singing chiffchaffs during experimental exposure to highway noise. *Journal of Experimental Biology*, 15, 2575–2581.

Wiley RH and Richards DG. 1982. Adaptations for acoustic communication in birds: sound transmission and signal detection. In *Acoustic Communication in Birds*, Kroodsma DE and Miller EH (eds.), 131–181. Academic Press, New York.

Wilson B, Batty RS and Dill LM. 2004. Pacific and Atlantic herring produce burst pulse sound. *Proceedings of the Royal Society B*, 271, s95–s97.

Wilson CJ, Wilson PS, Greene CA and Dunton KH. 2013. Seagrass meadows provide an acoustic refuge for estuarine fish. *Marine Ecology Progress Series*, 472, 117–127.

Wysocki LE, Amoser S and Ladich F. 2007. Diversity in ambient noise in European freshwater habitats: noise levels, spectral profiles, and impact on fishes. *Journal of the Acoustical Society of America*, 121, 2559–2566.

Yang X-J, Ma X-R and Slabbekoorn H. 2014. Timing vocal behavior: experimental evidence for song overlap avoidance in Eurasian wrens. *Behavioral Processes*, 103, 84–90.

Yodlowski ML, Kreitchen ML and Keeton WT. 1977. Detection of atmospheric infra-sound by homing pigeons. *Nature*, 265, 725–726.

14. The Contribution of Biocommunication (BICO) to Biomedical and Tissue Engineering: A Tech Mining Study

Angela Machado Rocha

Federal University of Bahia (UFBA)
Institute of Health Sciences (ICS)
Salvador, Bahia, Brazil
anmach@gmail.com

Fernando Palop

Universitat Politécnica de Valéncia (UPV)
Valéncia, Spain
fpalop@ingenio.upv.es

Maria Clara Melro

Federal University of Bahia (UFBA)
Institute of Health Sciences (ICS)
melro@ufba.br

Marcelo Santana Silva

Federal Institute of Bahia (IFBA)
Campus Santo Amaro
Santo Amaro, Bahia, Brazil
profmarceloifba@gmail.com

1. Introduction

Biocommunication explores the communication between living organisms (Witzany, 1993). When you mention communication, people think about speech, gestures, letters, emails, smells, videos, looks among other things. However, plants can communicate. Cells communicate. Bacteria communicate with each other. Tissues communicate with other tissues. Organs of the human body communicate with each other, Same do animals with other animals. Thus biocommunication transcends human communication.

Researchers have observed that microorganisms generate electromagnetic fields and the interaction between these functions as a kind of communication, so that their behaviors are modified in the presence of one in interaction with the others. This occurs with viruses or bacteria, protozoa and cells from multicellular organisms. The immune systems of animals have a form of communication capable of identifying an attack of microorganism intruders which triggers an effective response, monitored and modified according to the success and failure of the defense systems of the body. Human organs communicate with each other directly without the interference of the central nervous system (CNS) (Witzany, 2000).

One of the main challenges to improving human health is to understand cell behavior resulting from interactions at the molecular level. From understanding how mechanical, chemical, electrical, intra- and extra-cellular changes induce a biological response, it is possible to take advantage of biocommunication for the diagnosis of diseases and in aiding the healing process (Mattick, 2009). Understanding and interfering with communication between organs can excite or inhibit operation thereof. When detecting the specific communication of germs, it is possible to make an early diagnosis of a viral or bacterial invasion identifying the characteristics of the invader, and seek proper treatment (Villarreal, 2005).

The biomaterials specialist Prof. Monleón Pradas, who kindly accepted to review a draft of this chapter, draws our attention about the use of the concept of "biocommunication". A more justified way to the use of such a broad term, entails for him beyond the significance of this concept as an interaction its connection with the content of that interaction, "the information". This would pose extract from all areas cited in the examples of this introduction, conclusions on how it takes place the transmission of information between levels and at the time establish the existence of general guidelines. This considerable task still not addressed and it also would require some hypothesis. Such stakes are not covered by our chapter but on the other hand have opened the light to a new paradigm in the studies of regenerative medicine applications, the "cellular crosstalk" (Kirkpatrick *et al.*, 2010).

For the biomimetic approach to tissue regeneration is meaningful generate knowledge of how cells naturally interact with one another — "cellular crosstalk" (Kirkpatrick *et al.*, 2010). So the attention to the cellular communication in the cell differentiation process is nowadays a priority, through co-cultural processes (simultaneous culture of different cell types) that dedicate attention to understand the cellular crosstalk *in vivo* to manage the processes *in vitro*. This knowledge allows cells of different lineage to interact with biomaterials needed in a majority of regenerative strategies (Kirkpatrick *et al.*, 2010).

For instance in neuromedicine field, clones of exogenous stem cells in transplantations are considered for their therapeutic potential, but also as 'reporter cell' (Imitola *et al.*, 2004) used as biological tools for 'interrogating' the central nervous system environment. So the researchers need knowledge on how relevant cell molecular pathways works for this 'cross-talk' which, in turn, lead answers such as proliferation, migration, differentiation, apoptosis, etc. Researchers need these insights for the development of therapeutic strategies for in this case neuro diseases (Watt and Driskell, 2010).

Understanding the mechanisms of how "conversations" and signals occur between cells and molecules in the human body served as the initial inspiration for Biomedical Engineering (BME). This applies in research related to the early detection of diseases, diagnoses and therapies, as well as in the construction of synthetic biological devices and systems related to human health (BMES, 2012).

The study of articles production during last decade from 2003 to the present 2013 showed a steady growth in research on cellular interactions and cell crosstalk arriving at a rhythm from at least 550 to 900 publications per year just considering a narrow scope of research. The need for a deeper understanding of the tumor and cancer, extra and intracellular processes is one of the most important drivers. New imaging technology like atomic force microscopy has opened the gate to work at nano and microscale with topics like micro and nanoarrays, microfluidics, gold nanoparticles, carbon nanotubes (CNTs) or nanofibers. Other promising topics are endocytosis, biofilms, stem cells and tissue engineering with topics like scaffolds that will be treated later in this chapter.

The interdisciplinarity in research on cellular interactions and cell crosstalk is represented in the Fig. 1 through a visualization of the autocorrelations of 38 Web of Science categories presented in 92% of the 8055 article records analyzed. Although these categories gather the profile of the journals of the indexed articles, it helps us, indirectly, to see which disciplines are presented in the field and their correlations. So there are two top categories that are presented in 27% of all publications that correlate with themselves and with others. These are Cell Biology (with Cell and Tissue Engineering, Biotechnology and Medicine, Research and Experimental) and Biochemistry and Molecular Biology (with Biophysics).

Two main groups of knowledge categories are easily to observe in Fig. 1 as being presented together in numerous articles. At the top and left side of the map the Chemistry and materials group, reinforced with the Materials Science, Biomaterials and Biomedical Engineering with the contribution of the Nanoengineering of the materials. At the right side and botom we find Cell Biology with Biotechnology and Cell Tissue Engineering. The rest of the 38 categories — not displayed in the Fig. 1 — belong to different areas of the Medicine from Oncology or Hematology to Neurosciences where those Biomaterials knowledges are applied.

Interdisciplinarity showed upon WoS auto-correlation categories on cell interactions and cross-talk. Source Authors upon 8055 article records 2003–2013 from Web of Science. Powered with VantagePoint software.

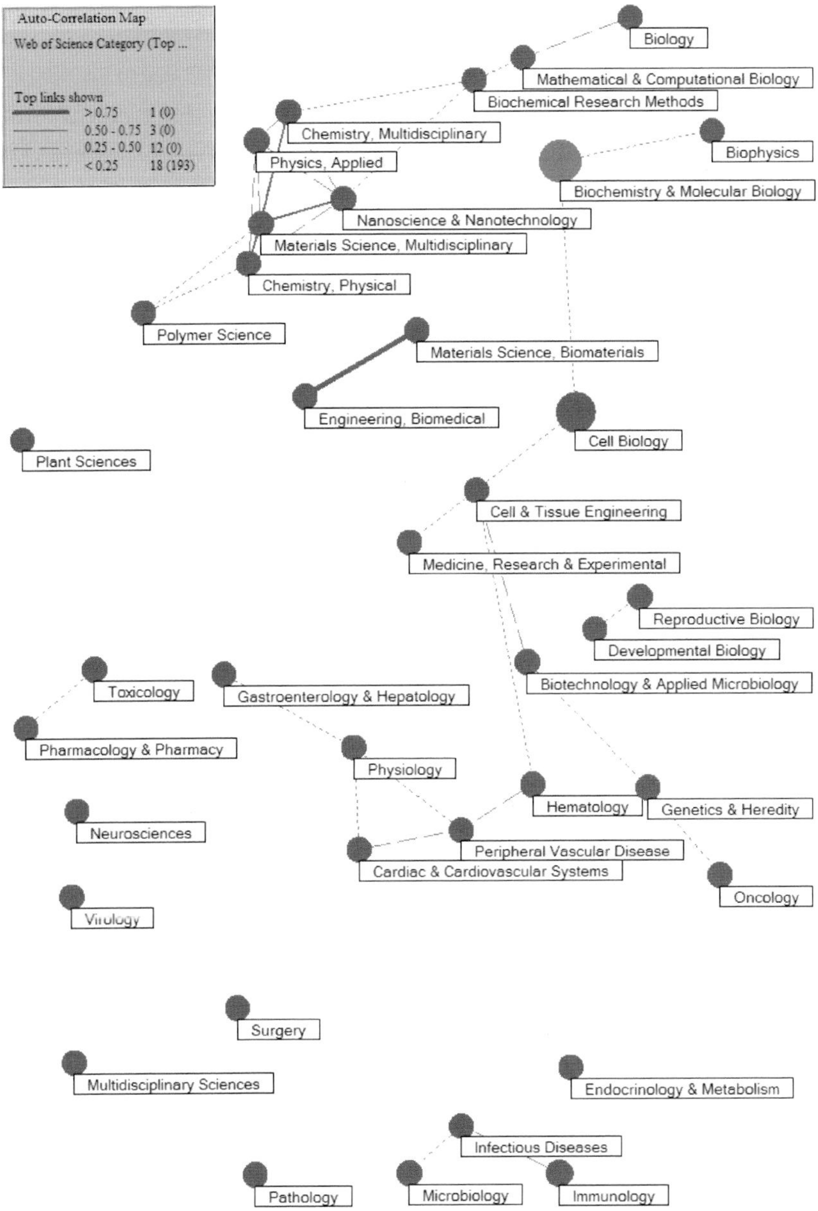

Figure 1. Interdisciplinary in cellular communication. Interdisciplinarity showed upon WoS auto-correlation categories on cell interactions and cross-talk. Source Authors upon 8055 article records 2003–2013 from Web of Science. Powered with VantagePoint software.

2. Biomedical Engineering

Biomedical Engineering is engineering applied to human health. It is an interdisciplinary field between Engineering and Life Sciences which requires not only an understanding of the anatomy of the body, but its physiology. As human health is multifaceted, the possibilities of expanding the field of Biomedical Engineering is in the use of biocommunication, a field that requires more study and research (USP, 2009).

The field began when engineers in partnership with physicians joined together to develop solutions to problems related to medicine. BME is currently undergoing explosive growth. It is the field within Engineering that is enjoying the highest growth with a 72% job-increase projected by 2018 (Foster, 2012). As an example the Biomedical Engineering Society (BMES), http://bmes.org was established in 1968 in Illinois and recognizes over 5,500 members.

BME can be classified into four major areas depending on the types of tools used to address the problem of interest and the nature of the problem itself: (1) bioengineering dedicated to the development of biomedical science with the aid of mathematical models and simulations; (2) rehabilitation engineering aiming to develop electronic and mechanical systems that improve the living conditions of the disabled; (3) medical engineering directing the study, design and implementation of instrumentation mainly electronics, images, sensors and prostheses for the medical field ; and finally (4) the engineering clinic or hospital dedicated to the activities of support, certification and testing of medical equipment, Information Technologies integration and other related activities in hospitals (Brockes and Kumar, 2005).

Example of BME projects include:

1. Artificial body parts such as: hearing aids, cardiac pacemakers, artificial kidneys and hearts, blood oxygenators, synthetic blood vessels, joints, arms and legs;
2. Patient monitoring during surgery or in intensive care, healthy persons in unusual environments, such as astronauts in space or divers at great depth, blood chemistry sensors (potassium, sodium, O_2, CO_2, and pH);

3. Advanced therapeutic and surgical devices such as laser systems for eye surgery, delivery-automated insulin;
4. Application of expert systems and artificial intelligence to clinical decision making, computer-based systems for diagnosing diseases;
5. Design optimal clinical laboratories such as computerized analyzer for blood samples, cath lab heart;
6. Medical imaging systems such as the ultrasound, computer-assisted tomography, magnetic resonance tomography, positron emission tomography;
7. Computational modeling of physiological system such as control of blood pressure, renal function, visual and auditory nerve circuits;
8. Project Biomaterials such as mechanical, transport and biocompatibility of implantable artificial materials;
9. Biomechanics of injury and wound healing such as gait analysis and the application of growth factors;
10. Medicine sports (rehabilitation, support external devices, etc.) among others.

At a time when biomedical engineers are still contributing to traditional fields such as biomedical signal analysis, physiological modeling, and instrumentation, BME is moving primarily to areas like Tissue Engineering and Regenerative Medicine which are far from the roots of its original field [11].

3. Research Framework

This article will cover just the topics of Tissue Engineering and Regenerative Medicine. A very important aspect of Tissue Engineering and Regenerative Medicine is the material that will be used to interact with the cells and tissues of the patient. Thus, biomaterials will also be examined.

Despite its incredible complexity, the human body has a very poor capacity to regenerate. Our organic systems are in permanent decline due to disease, trauma or aging (Brockes and Kumar, 2005). For Yannas (2005) "the adult organism reacts spontaneously to a serious injury, triggering a healing response which saves the body, but condemns the injured body part."

The idea of replacing body parts damaged or dysfunctional existed since antiquity. Doctors have tried to develop techniques to make healthy parts of corpses in people with problems (Silverstein and Gornstein, 1999). However, the first successful transplant proved happened only in 1954 in Boston, USA, when a kidney from one twin was implanted in the other (Papalois *et al.*, 2003). In 1968, a major milestone was the first bone marrow transplant. In this case, instead of an organ fully functional "seeded" and reconstituted with all hematopoietic cells, new bone marrow was slowly developed (Mccann, 2003). In the 1990s, two other important events occurred in the development of research in organ regeneration. The first was the cloning of Dolly the sheep in Scotland (Wilmut *et al.*, 1997). The second was the isolation of the first human embryonic stem cells in the United States (Thomson and Itskovitz-Eldor, 1998).

In the body, cells only multiply when growth factor commands are given by way of protein molecules and peptides emitted by the nucleus. They bring information as to when and how this multiplication must be done and to where the newly formed cells must go. Moreover they decide how they should be grouped with other cells to form tissue architecture. They often act in a group (Goustin *et al.*, 1986). To unravel this mechanism is a major key to the mystery that surrounds not only the development of tissue architecture but also tissue regeneration. To build a tissue in the laboratory, you must provide a similar structure, where cells can be supported to grow in the right direction. These molds — scaffolds — are constructed with artificial biomaterials (Suh, 1998).

Biomaterials are used to restore or replace functions in the tissues of the body and are continuously or occasionally in contact with body fluids (Agrawal, 1998). Tissue Engineering is the study of the growth of new connective tissues, or organs, from cells and biomaterials scaffolds based on the idea that it is possible to create or regenerate tissues in a laboratory and, in some cases, makes use of some cloning techniques. The reconstruction is made from cells of the tissue itself [20]. In the case of Regenerative Medicine, the "raw material" for use is called stem cells. They have the ability to stimulate wider regeneration of many different types of body cells tissues (Mason and Dunnil, 2008; Walter *et al.*, 1996; HHS, 2006).

4. Biomaterials

4.1 Introduction

Biomaterials are synthetic, natural or modified natural materials intended to be in contact and interact with a biological system. The term refers to materials, natural or synthetic, designed to interact with biological systems to evaluate, treat, augment or replace existing functions. They are also understood to be substances of natural or synthetic origin that are tolerated on a temporary or permanent basis by several tissues that constitute the bodies of living beings (Williams, 1987). It is important to emphasize that these materials must be biocompatible. Biocompatibility is the ability of the material to act with an appropriate response from the host in a specific application. The contact interaction of the biomaterial with the biological environment occurs at the interface of the material. The biocompatibility is determined by the interactions that occur in this interface, capable of eliciting an appropriate host for a specific application or not harmful to the living body in which they are placed (Williams, 1999).

Langer and Terrell (2004) propose that biomaterials should be classified into three groups depending on the type of tissue response: inert with little or no immune response of the host organism; bioactive that mimic the biological function of natural tissue and biodegradable that are absorbed by the spare and tissue regeneration.

The growing demand for biomaterials arises from an aging population with higher expectations for their quality of life (Holzapfel, 2012). Its traditional academic base began in the field of material science and engineering. The materials were designed to perform predominantly mechanical functions. To avoid biological rejection it was preferable that they be "inert" and not interacting with the biology of the host organism (Ratner and Bryant, 2004).

However, the molecular biology revolution of the 1970s and advances in genomics and proteomics in the 1990s and 2000, significantly affected the ways in which biomaterials are designed and used (Anderson *et al.*, 2008). Specific molecules have been incorporated into materials such as bioactive components. Such combination products that interact directly with cells and tissues by well-defined molecular pathways to direct biological responses, as representing the state of the art in various commercial products is widely used in various medical applications (Hutmacher, 2006).

Basically summarized the cell communication process starts after a certain cellular environment stimuli caused by different factors like nerves growth factor (NGF), substrate stiffness or dynamic forces. These factors crosstalk with the cell plasmatic membrane through its intracellular molecular pathways and activate different answers by the cell nucleus related with its cell fate (differentiation, apoptosis, multiplication, etc.). As a consequence we are living now a growing research interest in how biomaterials extracellular matrices (ECM), regulate, interact or crosstalk also communicate with the cell fate.

So today a tissue mechanism of repair cannot be considered without a better knowledge of the respective role of the various cell populations implied in the rebuilding of this tissue and their cell to cell communication processes (Amédée *et al.*, 2006). This new knowledge — on how the cell communication works — supports new therapeutic strategies. For instance, in the field of osseous substitution with the implantation of stem cells in a functionalized biomaterial in order to stimulate the colonization of the biomaterial by the cells and the host tissues (Amédée *et al.*, 2006).

Some authors (Tonga and Ji, 2012) pointed out that the success of mechanically active tissues relies on the integration of the mechanical sensing mechanism found in the native tissues into polymeric scaffolds based on biomaterials that closely mimic the structural organizations, combined with multipotent cells and dynamic culture devices.

4.2 Biomaterials and nanotechnology

With the advent of nanotechnology, biomaterials have evolved towards making it possible to create objects from the organization of atoms in desired ways (Thomas *et al.*, 2005). Nanoscience is the study of phenomena and manipulation of materials at atomic, molecular and macromolecular scales, where properties differ significantly from those at a larger scale. Nanotechnology are the design, characterization, production and application of structures, devices and systems by controlling shape and size at nanometer scale (The Royal Academy of Engineering, 2004). Succinctly both terms refers to all types of research and technologies that deal with the special properties of matter that occur in the range of about one to 100 nm.

In fact, changing the properties of a material to reach the nanometer scale is due to the combination of two factors: while on the one hand, it is in these objects with dimensions that quantum effects are manifested more evidently, on the other hand the smaller the sample size the more important the surface effects become by increasing the proportion of its area and volume (Tirrell *et al.*, 2002). The basic principle of nanotechnology is to take advantage of the new properties that arise when materials are in nanometer scale by controlling the size and shape of nano-objects, to achieve the preparation of new technological devices for specific purposes. The relationship between "biomaterial" and "nano" is intrinsic. This affinity allows for more effective product development due to an interaction at the nanoscale within the biological medium (Ratner and Ratner, 2003).

It is worth noting that to be a material at the nanoscale is not necessarily to be a nanomaterial. Being at the nanoscale, the material must be below its critical size and thus have different properties from its "bulk" state. In other words we can define nanomaterials as materials that have at least one dimension in the nanometer size range below the critical size that can change any of its properties (Koch, 2007). Many examples of nanomaterials such as CNTs, metal nanoparticles (NPs), NPs of oxides of transition metals, nanoemulsions, etc., are found as active components in a number of products currently on the market (Wagner *et al.*, 2006).

Here we highlight polymer nanocomposites and nanopharmaceuticals. The molding materials are composite structures formed by a polymeric continuous phase (matrix) reinforced by a discontinuous phase (fibers) that are added after a physico-chemical process of cross linking polymers. Polymeric nanocomposites are formed by combining a mixture of plastic or rubber and a material dispersed in a particular form, having at least one of their dimensions on the nanoscale. In general, the nanocomposites have two phases, one of which is formed by an elastomer or thermoplastic. The second phase may be formed of clay or exfoliated nanotubes or nanofibers (Robeson, 2008).

Biopolymers are studied in nanopharmacology as a substance capable of self-organizing and forming nanostructures under the configuration nanospheres or nanocapsules. These nanocapsules can act as vectors able to deliver a drug in the exact location of the body for extended periods of time, altering the biodistribution of the active principles associated with them. Thus, it is possible to reduce the dose with fewer side effects (Bawarski *et al.*, 2008). In immunotherapy, advances have been obtained using nanostructured systems as carriers of antigens for induction of immune response (Galembeck and Rippel, 2004).

Biomedical applications of magnetic nanoparticles include those which involve their administration into the body (*in vivo*) and outside the body (*in vitro*). *In vivo* applications can still be classified as therapeutic (hyperthermia and carrying of drugs) and diagnostic (Magnetic Resonance imaging (MRI) images) while *in vitro* applications comprise essentially diagnostic (separation/cell selection) (Pankhurst *et al.*, 2003).

4.2.1 *Metals*

Metallic biomaterials remain commonly used for the same reason that led to their initial selection for these devices many decades ago: high strength and resistance to fracture (Callister, 2003). They also have good electrical conductivity favoring their use for neuromuscular stimulation devices such as the pacemaker (Pilliar, 2009).

Often, metallic biomaterials come in the form of alloys, a combination of two or more elements, of which at least one is a metal. Permanent metallic implants based on stainless steel (SUS 316L), Cobalt–Chromium (Co–Cr) and Titanium alloys have been investigated as biomaterials for decades (Davis, 2003). However, metallic materials used have some limitations. Due to corrosion caused by the environment of the body, Nickel, Chromium, and Cobalt may be released from SUS 316L. Co–Cr alloy and can lead to toxic reactions and skin disease or can even induce carcinogenesis (Thyssen *et al.*, 2010).

Among the metallic materials available for implant placement, Titanium and its alloys stand out:

$$\text{Ti}, (\alpha + \beta)\ \text{Ti alloy}, \beta/\text{near}\ \beta\ \text{Ti alloys, Ni–Ti (Nitinol)}.$$

This can be explained by inhibiting bone reabsorption and increase bone remodeling. Titanium is corrosion resistant and highly biocompatible. However, its poor shear strength limits its use as screws or plates. The long-term application of some Titanium alloys raised concerns due to possible release of Vanadium or Aluminum which can lead to toxic reactions (Cochran *et al.*, 1998). Research exists to develop nickel titanium alloys and even Nickel-free alloy with shape memory effect and superelastic functions (Nyberg *et al.*, 2005).

Magnesium alloys are biocompatible and are presented as promising resources in orthopedic implants. They are also being manufactured in biologically active magnesium stents. However, concerns about the toxicity of dissolved Mg were raised due to their rapid dissolution in body fluids, however, it has been shown that an excess of magnesium is excreted from the body in urine (Bertrand *et al.*, 2010). To be used in implants corrosion resistance of the magnesium should be improved. This is attempted through alloys like porous magnesium metallic foams or with a magnesium metal matrix composite and HA particles. Tests in 2007 of such a composite showed its condition as cytocompatible biomaterial with adjustable mechanical and corrosive properties (Zeng *et al.*, 2008).

The noble metals Platinum (Pt), Platinum–Iridium (Pt–Ir) are materials used for neuromuscular electrical stimulation electrodes. Other possible noble metals include Rhodium, Gold and Palladium, but these are not used extensively (Beard *et al.*, 1992).

Metals are used in dentistry for fillings in direct teeth with dental amalgam, manufacture of crowns and bridges with noble metal and base metal alloys, the structures of partial dentures, orthodontic wires and brackets including stainless steel, Ti76 alloy and alloy Ni–Ti and dental implants (McCabe *et al.*, 2008).

Upon the metals research literature at nanoscale we highlight the use of gold, silver, TiO_2, or Fe_2O_3 NPs, CTNs and nanorods as part of a growing trend accelerating since 2010. The bones tissue engineering appears among it commons applications. Research topics in this field more frequent used are terms like biocompatibility, citotoxicity, adhesion, or degradation.

4.2.2 *Ceramics and glasses*

In general, the crystalline structure of the ceramic materials is more complex than the metals, since they are composed of at least two elements, each type of atom occupying positions determined in the crystalline lattice. Several bioceramics such as dense alumina (a-Al_2O_3)12, zirconia (ZrO_2), Titanium dioxide (TiO_2), Calcium phosphates and glass ceramics of silica/calcium phosphate have been widely used as biomaterials (Henche and Wilson, 1999). One of the disadvantages presented by bioceramics is the reduced mechanical strength which restricts its use to areas that do not require support. Bioceramics have been used as dense and porous form (Kokubo, 2008). Despite the increase in porosity, decrease in mechanical strength of the material itself, the existence of pores may promote tissue growth through them contributing to a strong entanglement of the tissue with the implant, thereby increasing the strength of the material *in vivo* (Chevalier *et al.*, 2008).

Hench (1998) proposed a classification of ceramics according to the three principal types of tissue response. The inert causes minimal tissue response or none at all. The active link stimulates nearby tissue and new bone growth. The degradable or reabsorbable are incorporated into the surrounding tissue or may even be completely dissolved after a certain period of time. Bioceramics can be found in the form of microspheres layers or thin coatings on metallic implants with a porous network and compound polymer components (composites) (Hao *et al.*, 2005).

The most successful class of bioceramics belongs to the group that are the basis of Calcium phosphate (CP) biodegradables, including hydroxyapatite (HA) and β-tricalcium phosphate (TCP) which are analogous to the inorganic components of vertebrate hard tissues (Kamita kahara *et al.*, 2008).

Another equally bioceramic material is alumina (Al_2O_3), a material which is highly inert and resistant to most corrosive environments including the dynamic human body. However, the human body recognizes foreign material and tries to isolate it forming a layer of non-adherent fibrous tissue around the implant wherever possible (Billotte, 2006).

Hench (1970) introduced the bioactive glass (Bioglass), widely used as fillers in clinical bone cavities. The glass consists mainly of oxides of Sodium, Calcium, Silicon and Phosphorus in various proportions.

The tech mining of the literature on ceramic and glass biomaterials highlights as topics of relevant and increasing interest in the mechanical properties and characteristics of these materials, the cell adhesion or the bones regeneration, repair and interactions. The use of scaffolds in one-third of the publications and the nanoscale research work in one-fourth of them are other two clear trends.

Attending the most referred to materials we found Calcium, hydroxyapatite — HA, TiO_2, and silica. From these 4th the HA is the one that register by researchers the clearest expanded interest since 2002.

4.2.3 *Polymers*

Organic or inorganic compounds, natural or synthetic molecular weight are characterized by their size, chemical structure and intramolecular and intermolecular interactions (Akcelrud, 2007).

Biopolymers are natural polymers such as collagen, elastin, hyaluronic acid, dextran, glycosaminoglycans, agarose, cellulose, chitin, chitosan, among others (Mohanty *et al.*, 2005). The natural polymers are suitable as biomedical materials due to their structural similarity with the tissue components, reducing the possibility of problems associated with toxicity of the material or its degradation products or stimulation of chronic inflammatory reactions (Coburn and Pandit, 2007).

That structural similarity is relevant in biocommunication. It is explained attending to the main body tissues (epithelial, connective, muscle and nervous) and their composition. They are based on three types of biopolymers. So there are proteins (chains of amino acids), polysaccharides (chains of sugar), and polynucleotides (chains of nucleotides). The Nature adapts the structure of those tissues to perform a particular function through specific biomaterials properties by molecular configuration (e.g., different types of colagen), composition (e.g., cartilage has much more polysaccharides than ligament), and organization (e.g., cartilage is isotropic, ligament has a preferred direction) (Patton, 2003).

Biomaterials are widely used as natural polymers from the ECM, of connective tissues (e.g., tendon, skin, ligament, blood vessels, cartilage, or bone). The ECM is the extracellular part of multicellular structures (e.g., organisms, tissues, biofilms) that typically provides structural and biochemical support to the surrounding cells (Abedin and king, 2010). It serves an important role in tissue and organ morphogenesis and in the maintenance of cell and tissue structure and function (Rosso *et al.*, 2004). As it was said in 4.1 cell-to-cell communication is one of the common function of the ECM (Abedin and king, 2010). The "external signals" to cell populations to promote cell proliferation and differentiation are generated from growth factors, cell–ECM, and cell–cell interactions, as well as from physical–chemical and mechanical stimuli (Rosso *et al.*, 2004).

An intrinsic characteristic of natural polymers is their ability to be degraded by enzymes, ensuring that the implant is metabolized by physiological mechanisms (Gros, 1989). This property is of interest in applications of biomaterials in which a specific function is intended for a short period of time, followed by what is expected that the implant degrades completely by normal metabolic processes (Kada, 2002). Disadvantages cited were strong physiological activities and high potential for rejection; generally insufficient mechanical strength. For these reasons, biomedical materials and their applications have been limited to specific small areas (Katz, 2001).

One of the base natural polymers most widely used is the structural protein collagen which can be found in the connective tissues such as tendons, ligaments, cartilage, bone and skin (Lee *et al.*, 2001). The half of the analyzed research publications on natural polymers refers to colagen as key material for fibers, sponges or the scaffold matrices (Steinwachs, 2009). There are several different types of collagen, which can be processed as sponges, membranes, nanofiber meshes or hydrogels which have received a steady interest since 2011 in research publications according to our analysis. Gelatin is derived from the denaturation of collagen with acid or alkali, resulting in charged polyelectrolytes, thus naming the types A and B gelatin (Ikada and Tabata, 1998).

The researchers showed also an increasing interest in the use of chitosan for scaffolds and membranes particularly since 2008, sharing the growing trend for biopolimers, achieving in the first 10 months of 2013, the same volume of papers of the whole 2012.

Fibrin is a protein matrix that is produced by polymerization of fibrinogen under the action of the enzyme thrombin. The fibrin glue has been used extensively as a tissue-glue for surgical wound repair (Wolberg, 2007).

In many tissues, glycosaminoglycans (GAG) are components of ECM by modulating the function of various proteins, thus regulating adhesion, migration, proliferation and differentiation. GAGs are responsible for hydrodynamic tissues and mediate the viscosity properties of cartilage (Lippiello, 2003). Hyaluronan (hyaluronic acid) is a non-sulfated glycosaminoglycan that can be applied as a gel-like carrier cell by covalent cross linking of hydrazide derivatives. It has clinical applications as a material for cartilage tissue, skin, bones (Tan *et al.*, 2009). Chondroitin sulfate (CS) is a sulfated glycosaminoglycan that is covalently bound to a protein nucleus forming proteoglycans (Wang *et al.*, 2007).

Chitosan its a structure similar to GAGs. Due to its biocompatibility and low toxicity it is an interesting material for applications in tissue engineering. It may be used alone or in combination with other materials to form hydrogels, fibers, beads or sponges (Khor and Lim, 2003). Alginate is a polysaccharide that is derived from brown seaweed. It can be used as a dressing delivery system or cell growth factor and immobilization matrix (George and Abraham, 2006).

Analysis of the 1883 publications on Natural Polymers related with Biomaterials points out trends of a significative growing interest since 2009 on the research with nanoscale solutions like nanofibrous scaffolds, nanocomposites, collagen nanofibers, carbon nanotubes or chitosan nanoparticles.

Bioartificial polymeric materials are biomaterials resulting from interactions between synthetic and natural polymers. Bioartificial materials combine the properties of synthetic polymers (good mechanical properties, easy processing, low cost production, and processing) with the specific properties of biocompatibility and degradation of biopolymers (Iwasaki and Ishihara, 2012). Examples of some polymers are polyethylene (PE), polymethylmethacrylate (PA), polytetrafluoroethylene (PTFE), polyurethane, polyvinyl chloride (PVC), polydimethylsiloxane (silicone), polyamides (nylons), and polyesters. PE is used for catheters, hoses, manufacture of orthopedic implants especially for knee and hip joint prostheses, among others. PA is indicated for ocular lenses, false teeth, bone cement, pumps blood, dialysis membranes, etc. PTFE is used in artificial tendons and ligaments and vascular grafts. Vascular prostheses, insulation wire with pacemakers, heart valves and breast implants are usually made of polyurethane. Polyglycolic acid (PGA), polylactic acid (PLA), and polydioxanone (PDS), have been widely used as bone fixation devices or as suture material to hold tissues together. The mechanical properties of these materials can be improved through processes of self- reinforcement, wherein the matrix polymer is reinforced by fibers of the same material (Törmälä *et al.*, 1997).

From the 962 article records analyzed, biocompatibility and biodegradation should be pointed out together with the different bio-polymers (PU, PE, PE-Glycol) involved in the research. Poly (L-lactic acid) should be highlighted as a growing trend of this last group, PLA, probably due to its positive degradation properties into a not harmful –innocuous- lactic acid and its applications in screws, pins or rods for implants and as a mesh (Ratner *et al.*, 2004).

Hydrogels are special class of polymers capable of absorbing large amounts of water in the spaces available between polymer chains due to the presence of hydrophilic groups such as amino, carboxyl and hydroxyl groups in the polymer chains. They are widely used in drug delivery as cell carriers and in tissue engineering. A variety of natural and synthetic hydrogels can be used for tissue engineering and regenerative medicine (Ratner and Bryant, 2004).

Synthetic materials include poly (ethylene oxide) (PEO), polyvinyl alcohol (PVA), etc. The analyzed literature pointed out an interest in hydrogels because of their biodegradability and as vectors or structures for cell growth. They raise a growing interest trend particularly since the year 2010. More than half of the 336 research articles that mention hydrogels were published between October 2010–2011.

4.3 Biological Responses to Biomaterials

When a tissue is injured, the normal healing response is initiated through a series of complex events. The coagulation cascade and the formation of a blood clot (Fig. 2) are the best known examples of physiological responses that depend on a series of reactions mediated by enzymes. In the coagulation cascade reactions, the end point is the formation of a network of cross-linked fibrin protein: fibrin cross-linking produces a solid material trapping red blood cells and platelets within a network high in protein. The formation of this solid material closes the area of injury and stops bleeding [78].

However, the clotting and blocking of small vessels impeding blood flow to the tissue, a process known as embolization, can lead

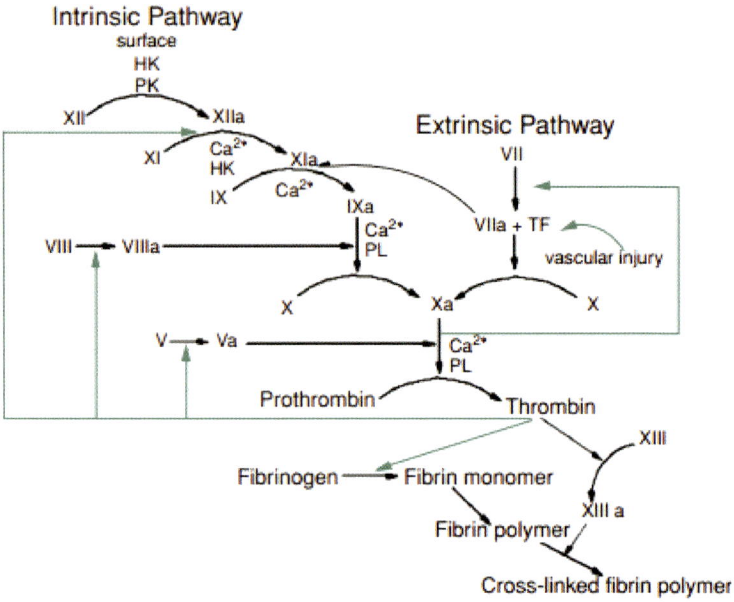

Figure 2. Proteins and reactions of clotting cascade.

to stroke (if the clot passes into the brain) or from breathing problems (if the clot enters the lung). The coagulation process can be initiated by intrinsic or extrinsic factors. The formation of a blood clot in response to tissue injury and damage the vessels occurs through the extrinsic pathway. For the intrinsic pathway is activated by the presence of charged surfaces, usually the EM on the outside of the vessel wall. The intrinsic pathway for clot formation needs factors VIII, IX, X, XI, and XII, as well as some other proteins and the mineral Ca^{++}. The extrinsic pathway is initiated by tissue factor and activated factor VII (VIIa). The interaction of these proteins activates factor X. These two initiation pathways work together.

However, a biomaterial can also be coated with protein inside the blood vessel [79].

A biomaterial implanted in the body, however, induces a different response, called the foreign body reaction. It is inherently complicated due to the presence of blood, interstitial fluids and various types of cells. This response is illustrated in Fig. 3. In summary, a biomaterial causes non-specific adsorption of proteins immediately after implantation. The non-specific protein adsorption, however, never occurs in the normal physiological process of healing wounds. Thus, the adsorption of non-specific protein can be an instigator in the foreign body reaction. A number of different cells, such as monocytes, leukocytes, and platelets adhere to these surfaces of biomaterials and as a result may lead to the production of cytokines and pro-inflammatory subsequent processes. Furthermore, the implant is significantly larger than the adherent macrophages preventing them to phagocytose the foreign body. Chronic inflammation interface biomaterial follows and macrophages fuse to form multinucleated giant cells which often persist for the life of the implant.

The final stage of the foreign body reaction involves walling off the device by an avascular tissue, fibrous collagen (Ratner and Bryant, 2004) (Fig. 3).

Determination whether cell death is immunogenic, tolerogenic, or silent when recognition occurs is a central problem in immunology. The answer to this question is directly linked to the lack of response, or the presence of undesirable immune responses (Zitvogel *et al.*, 2006). Several studies cite the use of tools of immunological tolerance to the application of functional biomaterials. The development of materials in basic research enables immune cell function and synthetic structures lymphoid tissue as well as advances in the engineering of vaccines (Irvine *et al.*, 2003).

4.4 The new generation of biomaterials (smart materials)

The new generation of biomaterials, "smart" biomaterials, is based on knowledge of the biology of wound healing and inflammation to control the responses of bioreactions accurately. In the short term, a strategy to achieve these surfaces is to modify existing devices to minimize foreign body reaction and promote normal wound healing (Roco and Bain-bridge, 2005). For long-term solutions, the only way to replace the damaged tissue function is

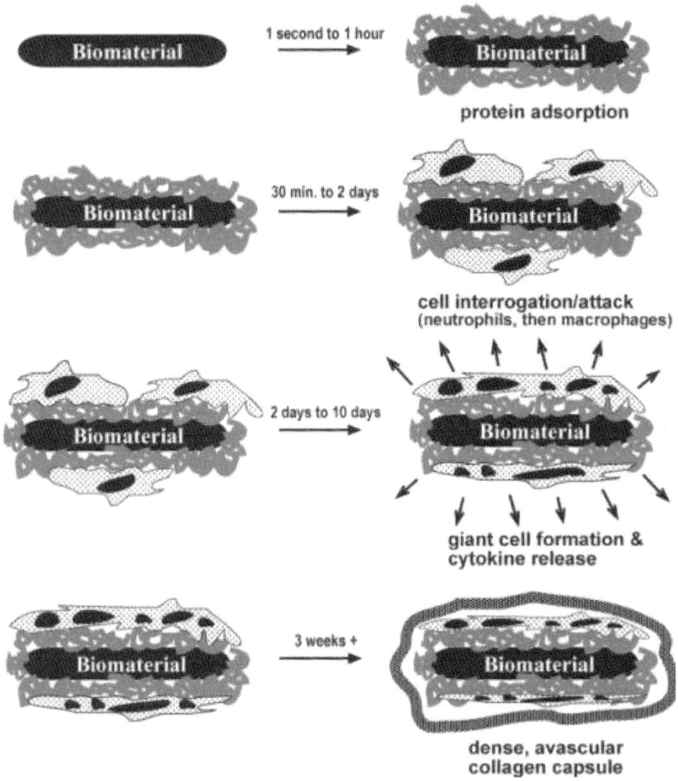

Figure 3. The foreign body reaction to an implanted synthetic material.

similar tissue containing the appropriate tissue architecture of cells and ECM (Rosso *et al.*, 2004). Although smart materials have been described in many ways there is no general definition for smart biomaterial. From the clinical point of view, an intelligent material can be defined as a material that actively participates in the regeneration of damaged tissue and responds to stimuli from its environment (Kikuchi and Kanama, 2007).

Different research strategies pursue the goal of improving tissue regeneration: changing the surface properties, the development of advanced three-dimensional geometry and systems engineering scaffolds or drug specific delivery which are loaded with growth factors or cells (Howard *et al.*, 2008). Thus, biocompatibility is only an essential requirement for intelligent biomaterial. Instead of increasing the complexity of tissue engineering, some researchers aim to develop biomaterials that have the intrinsic ability to actively promote the body's inherent capacity of healing and self-repair. Interactive smart biomaterials must be able to function and imitate ECM in order to stimulate cellular invasion, adhesion and proliferation (Holzapfel *et al.*, 2013).

To meet the need for repair and long-term clinical evolution, biological approaches that focus on the repair and restoration of the structure and function of tissues are necessary. Despite considerable challenges in understanding the physical properties of materials and chemicals, just some so-called smart biomaterials have found their way into clinical application so far. However, according to the EU Report on Nanotechnology, the worldwide market for smart biomaterials reached $ 47 billion in 2009 and will rise to $ 113 billion in 2025.

The concept of biomaterials has been recently modified in the recovery of materials viable tissue, or at least materials tissue regenerative material instead of traditional non-viable. Advances in molecular biology have provided information for the development of materials closest to natural tissues.

In determining cell function and interactions between cells and the ECM the 3D cell dimensions — topography, physical properties and the makeup of the ECM are fundamental factors. The analyzed articles confirmed the presence of different methods that have been developed to mimic aspects of the *in vivo* ECM like electro spun fibers or hydrogels-wide researched presenting an accelerated growing trend-based on peptides, collagen and other materials. At the same time this deeper knowledge in the ECM is facilitating the development of drug delivery solutions.

The research in smart biomaterials is in a significant part driven by nanotechnology that avoids accessing the cell micro environment. 43% of the 986 analyzed publications contain references to nanomaterials or nanodevices. 62% of these publications were published since 2011 to October 2013 showing an accelerating trend. Particularly since 2010 Among the emergent trends the electrospinning[1] is visible . It is a technique for generating biomimetic scaffolds made of synthetic and natural polymers for tissue engineering applications. This need of mimetics with the natural cell mechanisms emerges as a trend in the recent 2013 articles. Other topics in the research agenda include: how to fix cell adhesion and the important role of proteins are widely represented in the analyzed articles.

5. Tissue Engineering and Regenerative Medicine

Tissue Engineering, often known as "Regenerative Medicine", presents a new concept: create or regenerate new biological tissues, based on pre-existing tissues. This occurs in the laboratory through the manipulation and development of molecules, cells, tissues or organs originating from biological substitutes to restore, maintain and improve the function of damaged parts of the body. It is a multidisciplinary field, bringing together concepts of engineering, life sciences, and nanotechnology, directed both to improve and for the replacement of cells, tissues and organs (Langer and Vacanti, 1993).

Oreffo and MacArthur (2005) defined tissue engineering as "understanding the principles of tissue growth, and applying this knowledge in the regeneration of functional tissues for clinical use."

Salgado *et al.*, (2004) point out that the term regenerative medicine is often used synonymously with tissue engineering. However, Regenerative Medicine often involves the use of stem cells as a source of cells. Since its inception, the Tissue Engineering is based on three pillars: scaffolds, cells, and growth factors. Moreover, the strategies used for Regenerative Medicine to induce the regeneration of organs, including stem cell-based therapies, immunomodulation, gene therapy, nano-medicine and tissue engineering itself. Because of the similarity of the goals of these two fields they have merged in recent years, giving rise to the broad field of tissue engineering and regenerative medicine (TERM).

---

[1] A technique which uses electrical charge to draw nanofibers.

Although many definitions of TERM cover a wide range of applications in practice, the term is closely linked to repair or replace tissues or parts of the fabric piece. Although cells have been grown or cultured outside the body for many years, the possibility of growing tissues reproducing the three-dimensional complex design and function of human tissue is a recent development (Atala, 2012).

TERM is the regeneration of organs and living tissues through tissue of the patient. The cells in the donor tissue are dissociated and then can then be cultured and expanded into a scaffold (support, matrix framework, structure) three-dimensionally, which can be natural or synthetic, in the presence of two factors: of growth and of differentiation. Under proper conditions, the cells will secrete various components of the ECM to actually create a living tissue replacement which is then reinserted into the patient (Kim and Mooney, 1998).

These scaffolds or molds are made of artificial biomaterials. They are made up of microscopic fibers or webs of porous material. The architecture, the size and spacing of their empty areas determine the design that the cells will form. A three-dimensional porous scaffold is also important to ensure that growth factors at the cells arrived during the period of development of the tissue. These three-dimensional scaffolds must have growth factors that are release slowly trapped in structures, as the cells will multiply (Richardson *et al.*, 2001). A major barrier is the issue of tissue vascularization. Today, we have bioreactors, machines that control temperature, pressure, and if necessary, soak the molds in nutrients, improving the vascularization process, yet it is impossible to build a large organ, three-dimensionally to keep all your cells in direct contact with the nutrient solutions (Hutmacher, 2006). There are two possible solutions to organ vascularizing: (1) build the network of blood vessels beforehand or (2) seeding endothelial cells (which form the walls of blood vessels) and specific growth factors to cells along the tissues one wishes to build (Nieponice *et al.*, 2008).

In general, the preparation of the products of TERM follow these steps: (i) selecting and processing the scaffold, (ii) inoculating

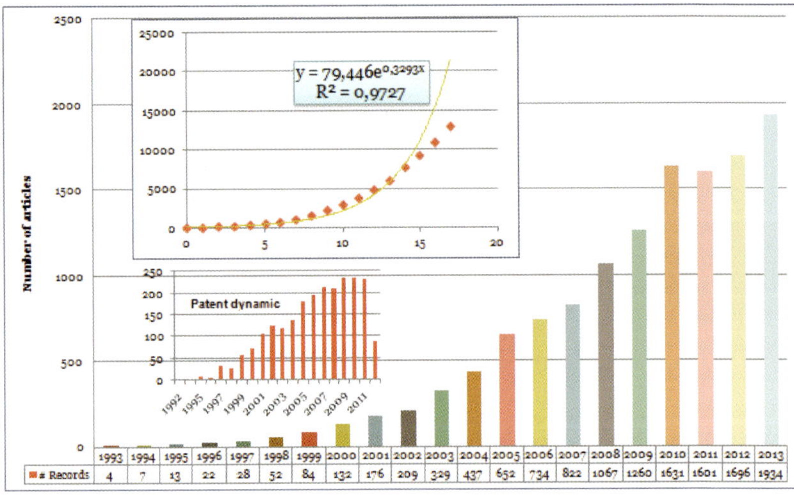

The number of articles by year. Inset: the accumulated trend and the trend according Solla de Price law and the number of patent family by priority year. Developed with the help of Martha Cervantes, MChemSc upon Web of Science and Derwent data, all years until end 2013. Powered with VantagePoint software.

Figure 4. Trend in publications research on TERM. Includes inset a patents trend comparison.

the cell population on the supporting scaffold, (iii) premature tissue growth, (iv) tissue growth matured in physiological system (bioreactor), (v) reimplant surgery, and (vi) assimilation of product (Barbanti *et al.*, 2005).

The Tech-Mining analysis on TERM developed upon a study of 12,890 articles publications in Web of Science showed a steep growth since the beginning of this century. Near 2,000 articles were recorded in 2013, the last year of the serial and the peak of it. It is also worth noting the inventive parallel effort in this field showed by the trend in patent submissions upon patent family priority year (see Fig. 4 and inset detailed graphs and Fig. 7 for patents). Most of the research activity is concentrated on the use of scaffolds — engineered ECMs — as support for the growth of new tissue. 70% of the publications contain this term. Stem cells are also mentioned in 40% of the publications used as row starting cells to repair tissues or for growth of new ones. The kind of materials of the scaffolds with a wide use of collagens and polymers such us polycaprolactane or

PCL, poly (d,l-lactic-coglycolic acid) (*PLGA*), nanofibers, CNTs and different biomaterials, the biodegradability of the used materials, the introduction of 3D scaffolds and their fabrication, are among the main topics of interest in the research agenda. The electrospinning technique shows an increase in use to generate the nanofibers in the scaffolds.

The study of cell growth factors with the obvious consequences in the tissue regeneration has been a main topic since the beginning of the field in the 90s. So is the attention to the fibroblast cells since the second half of the 90s achieving its peak in the first years of this decade. The use of platelets or thrombocytes as a natural source of growth factors is an important research direction. The research articles show (see Fig. 5) a clear growing trend in factors of growth like the TGF-beta or the platelet derived growth factor, PDGF, with a steady trend of interest in the latest since the middle of the last decade. Both play an active role in the regeneration and repair of connective tissues.

In this cell research context the increasing attention dedicated to the proteins and their role as cell signaling that is required to work at nano level is not surprising since 2010 more than 650 papers were published annually on TERM dealing with proteins. Figure 5 shows a trend comparison where we introduced among the cytokines the soluble proteins and the peptides and other proteins like fibrin.

Among the institutions more active in this field we found are in the USA Harvard University, MIT, University of Michigan, Rice University, University of Pittsburgh. In Europe University of Minho, Guimaraes, Portugal and in Asia, Natl University of Singapore and four Chinese institutions: Sichuan University, Chengdu, Tsinghua University, Beijing and Donghua University, Shanghai. Universities support the 69% of the research effort upon publications, Hospitals 8% and governmental R&D Centers 19%. The participations of company researchers is still small, 3.5%, with unless 350 firms engaged and a steady trend until 2010 when 75 articles were signed with at least one company author.

With respect to the knowledge areas that base and converge in this field, Fig. 6, more than 40% of the articles are published in

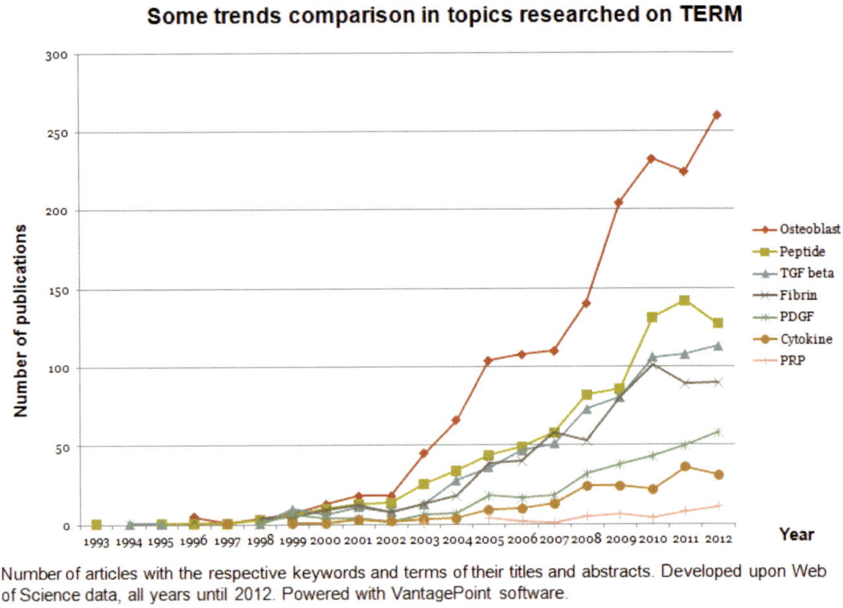

Figure 5. Some trends comparison in topics researched on TERM.

Engineering, Biomedical and Materials Science, Biomaterials Journals. At the same time these two categories show a high correlation between them. On the left side of this Fig. 6 are the group of categories related to Nanoscience & Nanotechnology, pink highlighted, and correlations with Materials Science, Physics and Chemistry. At the top there is a group of Biotech, Cell Biology and Cell Tissue Engineering correlated with Medicine Research and Transplantation. Finally at the bottom appears another group on Biochemistry, Organic Chemistry and Polymer Science.

5.1 Clinical trial literature

The 527 clinical trial papers examined, closer to the knowledge applications, based on PubMed, www.ncbi.nlm.nih.gov/pubmed, showed according to the Mesh Terms higher efforts around transplantation and bones transplantation, wound healing, artificial

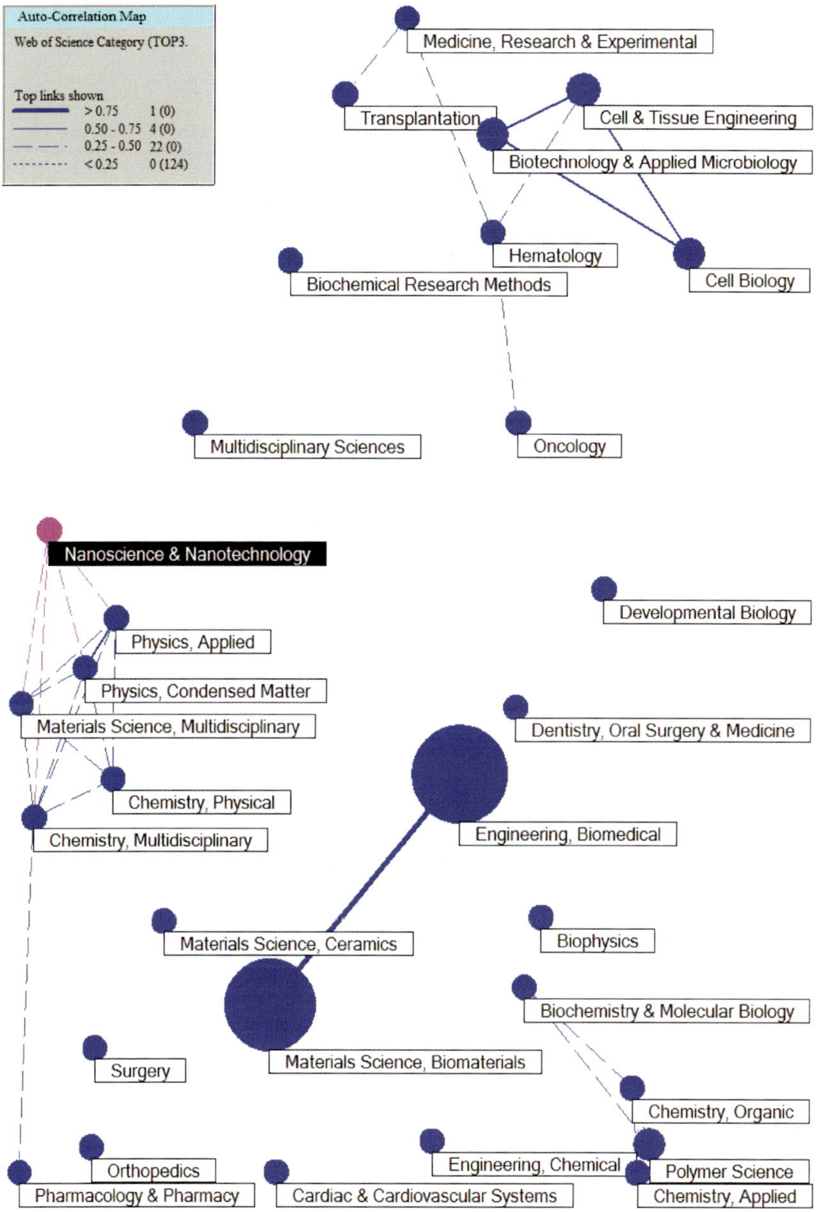

Figure 6. Interdisciplinarity in tissue engineering research. Interdisciplinarity showed upon WoS auto-correlation categories on TERM. Source Authors upon 12388 article records 2003–2013 from Web of Science. Powered with VantagePoint software.

skin, tissue scaffolds, regeneration, use of artificial membranes, periodontal and dental implantation or bioprosthesis.

It is also important to be mentioned that the clinical trials showed an increasing interest on TERM particularly since 2007 being 2012 the year by far with the highest number of works published as 70.

5.2 Patenting activity on TERM

The new knowledge generation on TERM has been followed since the end of the 19s by an increasing inventive activity measured upon patents (see Fig. 7). Although the patent sample used does not pretend to be full comprehensive of the field allows getting an approximation of it. Since 2007 more than 200 inventions are submitted annually to protect research results.

It is worth noting the strong presence of companies with 40% of the whole patent effort; meanwhile to academic institutions

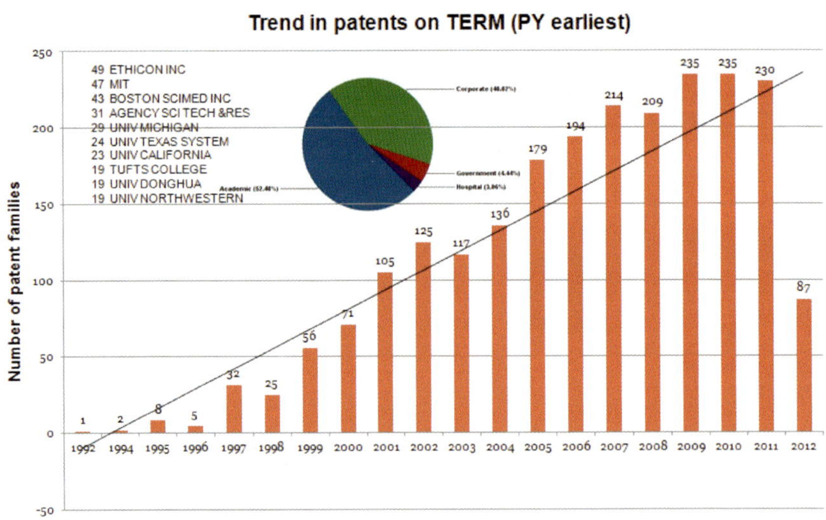

Number of patent families by earliest priority year. Inset: Top patent assignees and distribution of the research effort by institution type. A linear regression is included. Developed upon Derwent database data, all years – until 2013. Powered with VantagePoint software.

Figure 7. Trend in patents on TERM (Priority Year earliest).

correspond 52% and the rest to government research centers and hospitals. At the top affiliation list we find two corporations, Ethicon Inc. (Johnson & Johnson Group) and Boston Sci Med Inc. Three universities in this patent rank are also among the leaders in scientific papers production, MIT, University Michigan, and Donghua University as was before already commented.

Innovation in this field seems clearly science driven. Following Alan Porter's Management of Technology Topics approach (Porter and Cunningham, 2005) and taking as indicator to score the relative science base, the percentage of patents citing R&D papers, the result points out that more than 60% of the analyzed Patents cited R&D papers. Nevertheless we must caution about this data. That value is obtained for just a 45% of the sample patent records due to source data limitations. The citation field was empty on the 55% remaining patent records.

6. Biomaterials and TERM

One of the crucial points are TERM advances in the development of new biomaterials or devices capable of specific interactions with biological tissues and trying to use materials that are biocompatible with architecture that serves as a support (scaffold) for the *in vitro* growth of cells, organizing and developing tissue that will later be implanted in the patient. The biomaterials, in this case, must replicate the mechanical and biological function of native ECM found in the body tissues, serving as an artificial ECM (Matsusaki *et al.*, 2007).

The materials for construction of the scaffold should have the following characteristics: (i) they are biocompatible to avoid rejection, (ii) have a surface that allows cells to adhere, proliferate and migrate to the scaffold and (iii) having controlled biodegradability to help reconstruct new tissue, and (iv) have sufficient mechanical properties to maintain the structure and function immediately after transplantation, finally (iv) they have structural and functional properties which allow ECM biomolecular transfer signals between cells (Mekala *et al.*, 2012).

Biodegradable biomaterials have the potential to induce biological dysfunction *in vivo*. The interface between the material and tissue is the key area where the biological disorders are created. To restore tissue defects, studies *in vivo* and *in vitro* substances released from the material surface are essential in determining the biological characteristics of the implant material. The implant material must not inhibit the biological functions of host tissue (Sheyna *et al.*, 2010).

In general, three categories of biomaterials have been used: materials of natural origin, such as collagen and alginate; acellular tissue matrix such as small intestine submucosa and synthetic polymers such as PGA, PLA and poly (lactic-coglycolic acid) (Yannas and Burke, 1980).

As was introduced at the end of section four in recent years, several studies have focused on research to promote changes in nanotopography surface of the scaffold with the aim of controlling and directing physical and molecular signals that govern cell behavior (Zhang and Webster, 2009). Nanomaterials offer a way to mimic the natural architecture of the extracellular matrix. Nanofibers can provide a large surface area relative to volume. They also increase contact area and facilitate interaction between host and scaffold support. Fibers of collagen, hyaluronic acid, PLGA, PCL, PLLA, polyaniline, and polyamides have been used to mimic the structure of the extracellular matrix (Bawarski *et al.*, 2008).

Metals and certain ceramics are widely used in applications that involve bone marrow tissue (Feng, 2004). Properties are presented mechanically similarly, these materials are utilized on exposed sites at specific mechanical tensions or on large extensive lesions. Within biomaterials utilized today in bioengineering of bone marrow, the composites formed by a mineral phase in an organic three-dimensional matrix are key (Dolatshahi Pirouz *et al.*, 2011). From this understanding of bone marrow nanostructures many researchers have accepted that a combination of nanostructured elements such at HA COL, the best biomaterial to be utilized in bone application (Zhang *et al.*, 2010).

7. Research Methodology

One of the biggest challenges of empirical studies is to tackle a high volume of data that try to start the analysis with a solid scope of the indicators data. Moreover, one must take into account the complex, systemic and changing nature of technological innovation, with the advances inherent discontinuous and often radical innovations, as well as concerns of competitiveness and sustainability. In an age of too much information it is necessary to treat the text as given to obtain intelligence, "by mining" information for analysis of interest (Porter, 2009).

Tech Mining, according to Porter (Porter and Cunningham, 2005), seeks to extract useful information from data, emphasizing the numerical data analysis from pattern recognition, as well as statistical and mathematical techniques. It represents an opportunity to explore electronic information resources, on scientific research, patents and market publications. There are different types of software available in the market that allow statistical treatment of any set of information obtained from bibliographic databases (Search Technology, 2009). Based on bibliometric principles, automated analysis of structured information can be performed: a bibliographical reference is formed by the set of elements that constitute it, containing specific information in diverse fields such as title, institution, country, keyword, classification, and supplies, among others.

For the analysis we have used several articles files upon searches in PubMed, US National Library of Medicine, http://www.ncbi.nlm.nih.gov/pubmed and Web of Science, [WoS]-ISI Web of Knowledge, http://ipscience.thomsonreuters.com/product/web-of-science/. In the case of patents documents Derwent upon WoS records were mined. See Annex for the searches. We have used for the autocorrelations maps the Web of Science Categories that classify the journals where the articles were published. Most of the analyses are developed upon the combined Keywords and Phrases terms frequencies.

The interdisciplinarity of the research publications has been analyzed by Web of Science categories. The journals where the articles were published are assigned by the Institute of Scientific Information (ISI) to one or more subject categories of that classification.

Authors are aware that the sample of publications used in this chapter to study the biocommunication of the cells is just a reduced scale scope of the whole research effort in understanding the mechanisms of cellular communication and signals transmission. Enlarging the scope sure contains a number of publications several times larger. So, we consider that the results here exposed upon our sample are a first approach to be improved in the future.

8. Some Conclusions for the Whole Field of Scope

It is an interdisciplinary field crossed by several knowledge domains as was presented. This wide interdisciplinarity suppose an additional stake for the organizations who wish make profit from the growing opportunities in this still and high young field. High skill multidisciplinary teams (MDs, biologists, chemists, physicist, and engineers) should be better prepared to master the complexity offered by these new solutions and treatments.

The field shows an outstanding growth of interest by the research community and biopharmaceutical companies based on solid evidences like trends in publications and patents.

A 40% of the patenting effort in TERM belongs to corporations. This patent activity is fueled by the science research. More than 60% of the analyzed patents cited R&D papers.

Bones and cartilages tissues are nowadays the main stream applications of the TERM research accounting for a half of the research results production. Endothelial fibers and nerves, myocyte or epithelial fibers are a second area of interest for a 17% of the papers. Regenerative medicine on organs like heart, liver or kidney still represents a fraction under 7%.

The better understanding of the extra and intra cell communications mechanisms open important advances and new approaches in cancer treatment and tissues regeneration.

Nanotechnology contributes significatively to the advances in this field. In particular approximately 23% of the whole research effort in TERM, measured through publications, cited the contribution of materials and work techniques at nanoscale. It is nearly as many with certain biomaterials like artificial polymers or even a bit more 26% with metals.

8.1 Trends

The indicators analyzed upon publications in Web of Science and PubMed in the topics covered by this Chapter suggest that the increasing interest for Biocommunication will continue in many cases at a growing rate.

It is also relevant for the impact of this knowledge domain and its transformation in innovations and advances the perceivable role through joint research projects mainly with universities, research centers, and hospitals of companies from different countries.

A deeper knowledge about proteins and their role in their cell growth is one of the main streams recognizable in the results of our work.

An emergent trend in tissue engineering is exploring the combination of additive manufacturing techniques like stereo lithography and other 3D printing technologies in the scaffolds indirect fabrication. This approach that involves the printing and patterning in 3D of all the components of a living tissue is called "bioprinting" (Derby, 2012). This term has been used in research papers since the middle of the last decade, from 2011 has made a significant impact on the mass media, recognizable even in market indicators like Google Trends. Below the hype bias of the term, the research led by universities like Drexel, Carnegie Mellon, Clemson, Utrecht, Utah, Harvard, Hannover, or Scripps Research and INSERM Institutes have pushed the research front on a stage where applications seem more feasible. Nevertheless big challenges are still on the path like how to emulate blood vessels in the "bioprinted" tissue (Fountain, 2013).

It should be expected that innovation results will multiply in the coming years starting from today's dental implants treatments or cartilage regeneration to satisfy wider needs even in oncology.

Acknowledgments

The authors acknowledge the interest and time of the biomaterials expert Monleon Pradas, M., 2 Universitat Politècnica de València, Valencia, Spain and CIBER-BBN, Spain. With this chapter near to be finished, he has made high valuable comments and suggestions that pose us new and ambitious challenges in the topic.

Annex to the 7. Methodology

7.3 Research strings and electronic databases used
PubMed, available from US National Library of Medicine National Institutes of Health.
Clinical Trials on TERM:
((Tissue engineering AND (Clinical Trial [ptyp]))) OR (regenerative medicine AND Clinical Trial [ptyp]), 527 articles

Web of Science, available from ISI Web of Knowledge, Thomson 20140216
(TS="cell* interaction?") AND DOCUMENT TYPES: (Article) 5030
(TS=(Cell* NEAR cross-talk)) AND DOCUMENT TYPES: (Article) 3066
(TS=(Cell* NEAR "cross talk")) AND DOCUMENT TYPES: (Article) 2984
#4 OR #3 OR #2 8055
Indexes=SCI-EXPANDED, SSCI, A&HCI, CPCI-S, CPCI-SSH, CCR-EXPANDED, IC Timespan=2003–2013

TERM
WoS 20131028 and updated 20140216 to complete 2013
1st — (TS= ("tissue engineering" OR "regenerative medicine")) AND Document Types= (Article) 23593
Databases= SCI-EXPANDED, SSCI, A&HCI, CPCI-S, CPCI-SSH Time span= All years (1900–2013)

2nd — (TS= ("nano fiber" OR scaffold* OR "extra cellular matrix" OR "nano composite" OR "drug delivery" OR "stem cell*" OR biomimetic OR biodegradable OR polymer* OR cnt OR "carbon nanotube" OR "Cell engineering")) AND Document Types=(Article)

Databases= SCI-EXPANDED, SSCI, A&HCI, CPCI-S, CPCI-SSH
Timespan= All years 1196189
3th — (TS= ("polylactic acid" OR PLA OR "polyglycolic acid" OR PGA OR polycaprolactone OR PCL OR "poly-l-lactic acid" OR PLLA OR "poly lactic-co-glycolic acid" OR plga OR "hyaluronic acid" OR "hyaluronan"" OR ha OR collagen OR col OR protein OR fibrin OR polysaccharidic OR chitosan OR glycosaminoglycans OR gags OR chitin OR hydrogel OR alginate OR dextran)) AND Document Types=(Article)
Databases= SCI-EXPANDED, SSCI, A&HCI, CPCI-S, CPCI-SSH
Timespan= All years 2,485,417

#1 AND #2 AND #3
Databases=SCI-EXPANDED, SSCI, A&HCI, CPCI-S, CPCI-SSH
Timespan=All years 12890
The third step in the search string limits the sample by several materials approved by the FDA for medical purposes. This reduce the scope of the research effort from other materials not included in this step but used in TERM. Just the first two steps present 20,000 records.

Biomaterials
1st (TS= (Biomedical Engineering or BME or BiomedicalEngineering or Biomedical-Engineering or "Rehabilitation Engineering" or Medical Engineering or Clinical Engineering)) AND Document Types=(Article)
1 AND (TS= (Titan* or Ti or Magnes* or Mg or Stainless* steel* or SUS 316L or Platin* or Pt or Rhodi* or Rh or Gold* or Au or Pallad* or Pd))
AND Document Types= (Article)

1 AND (TS= (Cobalt–chromium or Co–Cr or Titanium alloy* or Nickel–titanium or Ni–Ti or Nitinol* or β/near-β Ti or ($\alpha + \beta$) Ti or Platinum–Iridium or Pt–Ir)) AND Document Types=(Article)

Databases= SCI-EXPANDED, SSCI, A&HCI, CPCI-S, CPCI-SSH
Timespan= All years

1 AND (TS= (Vitroceramic or Vitroceram* or Bioglass* or Glass-ceramic* of silic* or "calcium phosphate" or Bioceram* or Hydroxyapat* or $Ca_{10}(PO_4)6(OH)_2$ or "Dense alumin*" or $(a\text{-}Al_2O_3)12$ or Zircon* or ZrO_2 or "Calcium phosphates" or "Titanium dioxide" or TiO_2)) AND Document Types=(Article)

Databases= SCI-EXPANDED, SSCI, A&HCI, CPCI-S, CPCI-SSH
Timespan= All years

1 AND (TS= (Polyethylen* or PE or Polymethylmethacrylat* or PA or Polytetrafluoroethylen* or PTFE or Polyurethan* or "Polyvinyl chloride" or PVC or Polydimethylsiloxan* or silicon* or Nylon* or "Polyglycolic acid" or PGA or "Polylactic acid" or PLA or Polydioxanon* or PDS)) AND
Document Types=(Article)

1 AND (TS= (Collagen* or Fibrinogen* or Glycosaminoglycan* or GAG or Hyaluron* or "Hyaluronic acid" or "Chondroitin sulfate" or CS or Chitosan* or Alginat*)) AND Document Types=(Article)

1 AND (TS= (Smart-mater* or Drug-deliver* or Drug-elut*)) AND Document Types=(Article)
Patents: Derwent. Available from ISI Web of Knowledge, Thomson 20140116

1. TOPIC: ("tissue engineering") OR TOPIC: ("regenerative medicine") OR TOPIC: ("Bioactive nano*")
2. (TS=("nano fiber" OR scaffold* OR "extra cellular matrix" OR "nano composite" OR "drug delivery" OR "stem cell*" OR biomimetic OR biodegradable OR polymer* OR cnt OR "carbon nanotube" OR "Cell engineering"))

1st AND 2nd = 2267 patent family records

Indexes=CDerwent, EDerwent, MDerwent Timespan=All years

References

Abedin M and King N. 2010. Diverse evolutionary paths to cell adhesion. *Trends in Cell Biology*, 20(12), 734–742.

Agrawal CM. 1998. Reconstructing the human body using biomaterials. *JOM*, 31–35.

Akcelrud LC. 2007. Electroluminescent polymer system. In *Physical Properties of Polymer Handbook*, Mark JE (eds.), 2nd edition. vol 1, 757–786. Springer, Cincinnati.

Amédée J, Guillotin B, Pallu S, Barcile R, Guignandon A, Bourget CH, Durrieu MC and Bordenave L. 2006 Functionalization of biomaterials and cell to cell communication. *Bio-Medical Materials and Engineering*, 16(4), S53–S60.

Anderson M, Rodriguez A and Chang DT. 2008. Foreign body reaction to biomaterials. *Seminars in Immunology*, 20, 86–100.

Atala A. 2012. Regenerative medicine strategies. *Journal of Pediatric Surgery*, 47, 17–28.

Barbanti SH, Zavaglia CAC and Duek EAR. 2005. "Polímeros Bioreabsorvíveis na Engenharia de Tecidos . "Bioresorbable polymers in tissue engineering". *Polymers: Science and Technology*, 15(1), 13–21.

Bawarski WE, Pharm ED, Chidlowsky DJ and Bharali SA. 2008. Emerging nanopharmaceuticals. *MBA Nanomedicine: Nanotechnology, Biology, and Medicine*, 4, 273–282.

Beard RB, Hung BN and Schmukler R. 1992. Biocompatibility considerations at stimulating electrode interfaces. *Annals of biomedical engineering*, 20(3), 395–410.

Berg JM, Tymoczko JL and Stryer L. 2007. Biochemistry. 6th edition. W. H. Freeman and Company, New York.

Bertrand E, Gloriant T, Gordin DM, Vasilescu E, Drob P, Vasilescu C and Drob SI. 2010. Synthesis and characterisation of a new superelastic: Ti–25Ta–25Nb biomedical alloy. *Journal of the Mechanical Behavior of Biomedical Materials*, 3(8), 549–646.

Billotte WG. 2006. Ceramic Biomaterials. In *Biomedical Engineering Handbook*. 2nd edition. CRC Press LLC, Boca Raton.

BMES. Biomedical Engineering Society. 2012. *Annual Report* 2012. Available from http://bmes.org/files/AnnualReport2012.pdf. (Accessed on 12 December 2013)

Brockes JP and Kumar A. 2005. Appendage regeneration in adult vertebrates and implications for regenerative medicine. *Science*, 310(5756), 1919–1923.

Callister WD Jr. 2003. *Materials Science and Engineering: An Introduction.* 6th edition. Chapter 7. John Wiley & Sons, Inc, USA.

Chevalier E, Chulia D, Pouget C and Viana M. 2008. "Fabrication of porous substrates: A review of processes using pore forming agents in the biomaterial field." *Journal of Pharmaceutical Sciences* 97(3), 1135–1154.

Coburn JC and Pandit A. 2007. Development of naturally-derived biomaterials and optimization of their biomechanical properties. In *Topics in Tissue Engineering*, 3, 1–23.

Cochran DL, Schenk RK and Lussi A. 1998. Bone response to unloaded and loaded Titanium with a sandblasted and acid etched surface: A histometric study in the canine mandible. *Journal of Biomedical Materials Research*, 40, 1–11.

Constandinou TG. 2005. *Biologically-Inspired Electronics for Micropower Vision Processing*. Report PhD Thesis., Publisher: Imperial College of Science, Technology and Medicine, University of London.

Davis JR. 2003. Metallic Materials. In *Handbook of Materials for Medical Devices*, ASM International, Materials park, Ohio, USA.

Department of Health and Human Services (HHS). USA. August 2006. Available from: http://stemcells.nih.gov/staticresources/info/scireport/PDFs/Regenerative_Medicine_2006.pdf, (Accessed on 29 January 2014).

Derby B. 2012. Printing and prototyping of tissues and scaffolds. *Science*, 338(6109), 921–926.

Dolatshahi-Pirouz A, Jensen THL, Kolind K, Bunger C, Kassem M and Foss M. 2011. Cell shape and spreading of stromal (mesenchymal) stem cells cultured on fibronectin coated gold and hydroxyapatite surfaces. Interdisciplinary Nanoscience Center (iNANO), Aarhus University, Aarhus C, Denmark. *Colloids and Surfaces. B: Biointerfaces*, 84(1), 18–25. DOI:10.1016/j.colsurfb.2010.12.004.

Feng S-S. 2004. Editorial Board Members. *Expert Review of Medical Devices*, 1(1), 115–125. doi: 10.1586/17434440.1.1.115.

Foster KR. 2012. *Biomedical engineering online*. Available from http://www.biomedical-engineering-online.com. (Accessed on 8 December 2013).

Fountain H. 2013. At the Printer, Living Tissue. *The New York Times*. Available from: http://www.nytimes.com/2013/08/20/science/next-out-of-the-printer-living-tissue.html (Accessed on 13 November 2013).

Galembeck F and Rippel MM. 2004. Polymeric nanocomposites and nanodrugs: facts, opportunities and strategies. Brasília, *DF CGEE*, 18.

George M and Abraham TE. 2006. Polyionic hydrocolloids for the intestinal delivery of protein drugs: alginate and chitosan — a review. *Journal of Controlled Release*, 114, 1–14.

Goustin AS, Leof EB, Shipley GD and Moses HL. 1986. Growth factors and cancer. *The Journal of Cancer Research*, 46, 1015–1029.

Gros L. 1989. Biological activity. In *Polymers: Biomaterials and Medical Applications*, 28–52. John Wiley and Sons, USA.

Hao L, Savalani MM and Harris RA. 2005. Layer Manufacturing of polymer/bioceramic Implants for Bone Replacement and Tissue Growth Paper presented at the Proceedings of the 2nd. International Conference on Advanced Research and Rapid Prototyping: Virtual Modeling and Rapid Manufacturing — Advanced Research in Virtual and Rapid Prototyping, Leiria, Portugal, October, 2005.

Hench LL. 1970. *INT. J. NON-CRYST. SOL.* **1** BIOATIVO (Bioglass).

Hench L. 1998. The Bioactive materials: The potential for tissue regeneration. *Journal of Biomedical Materials Research*, 41(4), 511–518.

Hench LL and Wilson J. 1999. *An Introduction to Bioceramics*, 2 vols. World Scientific Publishing Co. Pte. Ltd., Singapore.

Holzapfel BM, Reichert JC, Schantz J-T, Gbureck U, Rackwitz L, Nöth U, Jakob F, Rudert M, Groll J and Hutmacher WD. 2012. *How smart do biomaterials need to be*? A translational science and clinical point of view. *Advanced Drug Delivery Reviews*, 65, 581–603.

Holzapfel BM, Reichert JC, Schantz J-T, Gbureck U, Rackwitz L, Nöth U, Jakob F, Rudert M, Groll J and Hutmacher WD. 2013. Comparative study of polyconvex strain-energy functions used for the modeling of damage hysteresis in overstretched arterial walls. *Journal of Anatomy*, 117(1), 23–42.

Howard D, Buttery LD, Shakesheff KM and Roberts SJ. 2008. Tissue engineering: strategies, stem cells and scaffolds. *Journal of Anatomy*, 213(1), 66–72. doi: 10.1111/j.1469-7580.

Hutmacher DW. 2006. Regenerative medicine will impact, but not replace, the medical device industry. *Expert Review of Medical Devices*, 3, 409–412.

Ikada Y and Tabata Y. 1998. Protein release from gelatin matrices. *Advanced Drug Delivery Reviews*, 31, 287–301.

Imitola J, Park KI, Teng YD, Nisim S, Lachyankar M, Ourednik J, Mueller FJ, Yiou R, Atala A, Sidman RL, Tuszynski M, Khoury SJ and Snyder EY. 2004. Stem cells: cross-talk and developmental programs. *Philosophical Transactions of the Royal Society of London, Series b-Biological Sciences*, 359, 823–837.

Irvine DJ, Stachowiak AN and Jain S. 2003. Engineering materials for control of immune functions. *Materials Science Forum*, 426–432, 3213–3218.

Iwasaki Y and Ishihara K. 2012. Cell membrane-inspired phospholipid polymers for developing medical devices with excellent biointerfaces. *Science and Technology of Advanced Materials*, 13, 064–101. doi:10.1088/1468-6996/13/6/064101.

Jagur-Grodzinski J. 2006. Polymers for tissue engineering, medical devices, and regenerative medicine. Concise general review of recent studies. *Polymers for Advanced Technologies*, 17, 395–418.

Kada Y. 2002. Polymeric biomaterials in medical systems. In *Polymeric based Systems on Tissue Engineering, Replacement and Regeneration*, 13–24. Kluwer academic publishers, Netherlands.

Kamitakahara M, Ohtsuki C and Miyazaki T. 2008. Review paper: behavior of ceramicbiomaterials derived from tricalcium phosphate in physiological condition. *Journal of Biomaterials Applications*, 23, 197–212.

Katz J. 2001. Developments in medical polymers for biomaterials applications. *Medical Device & Diagnostic Industry*, 23(1), 122–132.

Khor E and Lim LY. 2003. Implantable applications of chitin and chitosan. *Biomaterials*, 24, 2339–2349.

Kikuchi M and Kanama D. 2007. Current status of biomaterial research focused on regenerative medicine. *Science and Technology Trends*, 24, 51–67.

Kim BS and Mooney DJ. 1998. Development of biocompatible synthetic extracellular matrices for tissue engineering. *Trends Biotechnology*, 16, 224–230.

Kirkpatrick CJ, Fuchs S and Ronald E. Unger. 2010. Cocultures on Biomaterials: Understanding Cellular Crosstalk for Regenerative Medicine. JBMT, Vol. 31(18) SNAPSHOT, indd 2. REPAIR-lab, Institute of Pathology, University Medical Center, Johannes Gutenberg University of Mainz. Available from: www.deepdyve.com/lp/elsevier/snapshot-cocultures-on-biomaterials-understanding-cellular-cross-talk-7chsE0CFf0 (Accessed on 11 February 2014).

Koch C. 2007. Structural nanocrystalline materials: an overview. *Journal of Materials*, 42(5), 1403–1414.

Kokubo T. 2008. *Bioceramics and their Clinical Applications*, vol. 1, CRC Press, Washington DC.

Langer R and Tirrell DA. 2004. Designing materials for biology and medicine. *Nature*, 428(6982), 487–492.

Langer R and Vacanti JP. 1993. Tissue engineering. *Science*, 260(5110), 920–926.

Lee CH, Singla A and Lee Y. 2001. Biomedical applications of collagen. *International Journal of Pharmaceitics*, 221, 1–22.

Lee KY and Mooney DJ. 2001. Hydrogels for tissue engineering. *Chemical Reviews*, 101, 1869–1877.

Lippiello L. 2003. Glucosamine and chondroitin sulfate: biological response modifiers of chondrocytes under simulated conditions of joint stress. *Osteoarthritis and Cartilage*, 11, 335–342.

Macarthur BD and Oreffo ROC. 2005. Bridging the gap. *Nature*, 19, 433.

Mason C and Dunnil P. 2008. A brief definition of medicine regenerative. *Regenerative Medicine*, 3, 1–5.

Matsusaki M, Kadowaki K, Nakahara Y and Akashi M. 2007. Fabrication of cellular multilayers with nanometer-sized extracellular matrix films. *Angewandte Chemic International*, 46, 4689–4692.

Mattick JS. 2009. Deconstructing the dogma: a new view of the evolution and genetic programming of complex organisms. *Annals of the New York Academy of Sciences*, 1178, 29–46. doi: 10.1111/j.1749-6632.2009.04991.

McCabe JF and Walls A. 2008. *Applied Dental Materials.* 9th Edition, Blackwell Publishing Ltd., Oxford, UK.

Mccann SR. 2003. The history of bone marrow transplantation. In *Organ and Cell Transplantation*, Hakim NS and Papalois VE (eds.). Imperial College Press, London, UK.

Mekala NK, Baadhe RR and Parcha SR. 2012. Review on engineering biomaterials in tissue engineering application. *Recent Research in Science and Technology*, 4(12), 05–11 ISSN: 2076–5061.

Mohanty AK, Misra M and Drzal LT. 2005. *Natural Fibers, Biopolymers, and Biocomposites.* CRC Press, USA.

Nieponice AMD, Soletti LMS, Guan J, Deasy BM, Huard J, Wagner WR and Vorp DA. 2008. Biomaterials. *Journal of Biomaterials*, 29(7), 825–833. doi: 10.1016.

Nyberg E, Miller M, Simmons K and Scott K. 2005. Microstructure and mechanical properties of titanium components fabricated by a new powder injection molding technique. *Materials Science and Engineering C*, 25(3), 336–342.

Pankhurst QA, Connolly J, Jones SK and Dobson J. 2003. Applications of magnetic nanoparticles in biomedicine. *Journal of Physics D: Applied Physic*, 36, 167–181.

Papalois VE, Hakim NS, Najarian JS, *et al.* 2003. The history of kidney transplantation. *In Organ and Cell Transplantation*, Hakim NS and Papalois VE (eds.), Imperial College Press, London, UK.

Patton DL. 2003. *Lecture on Natural Materials as Biomaterials. School of Polymers and High Performance Materials.* University of Southern Mississippi. Retrieved 20140129 from: http://www.usm.edu/pattonresearchgroup/PSC475/Lecture%20Notes/Lec6_Natural%20Materials%20as%20Biomaterials.pdf.

Pilliar RM. 2009. Metallic Biomaterials. In *Biomedical Materials,* Narayan R (eds.), 41–79. Springer, New York, USA.

Porter AL and Cunningham SW. 2005. *Tech Mining: Exploiting New Technologies for Competitive Advantage.* John Wiley & Sons, Hoboken, NJ.

Porter AL. 2009. Tech mining for future-oriented technology analyses. In *Futures Research Methodology,* Glenn JC and Gordon TJ (eds.). Available from: https://www.thevantagepoint.com/resources/articles/FRM%20chapter-TextMining-2009mar3.pdf. (Accessed on 10 December 2013).

Ratner BD and Bryant SJ. 2004. Biomaterials: where we have been and where we are going. *Annual Review of Biomedical Engineering,* 6, 41–75.

Ratner M and Ratner D. 2003. *Nanotechnology.* New Jersey Prentice Hall, Upper Saddle River.

Ratner B, Hoffman A, Schoen F and Lemons J. 2004. *Biomaterials Science: An Introduction to Materials in Medicine.* 2nd edition. Academic Press.

Richardson TP, Peters MC, Ennett AB and Mooney DJ. 2001. Polymeric system for dual growth factor delivery. *Nature Biotechnology,* 19(11), 1029–1034. Available from: http://www.nature.com/nbt/journal/v19/n11/full/nbt1101-1029.html. (Accessed on 12 December 2013).

Robeson PLM. 2008. Polymer nanotechnology: *Nanocomposites Review,* 49, 3187–3204.

Roco MC and Bainbridge WS. 2005. Societal Implications of Nanoscience and Nanotechnology (II): Maximizing Human Benefit. NSET Report, Arlington, Virginia.

Rosso F1, Giordano A, Barbarisi M, Barbarisi A. 2004. From cell-ECM interactions to tissue engineering. *Journal of Cellular Physiology.* 199(2),174–180.

Salgado AJ, Coutinho OP and Reis RL. 2004. Bone tissue engineering: state of the art and future trends. *Macromolecular Bioscience,* 4(8), 743–765. doi: 10.1002/mabi.200400026.

Search Technology. 2009. *Vantage Point user's guide.* Available from: http://www.thevantagepoint.com. (Accessed on 12 August 2013).

Sheyna D, Mizrahia O, Benjamin S, Gazit Z, Pelled G and Gazit D. 2010. Genetically modified cells in regenerative medicine and tissue engineering. *Advanced Drug Delivery Reviews,* 62, 683–698.

Silverstein LH and Gornstein RA. 1999. Similarities between an acellular dermal allograft and a palatal graft, for tissue augmentation. *Clinical report, Periodontal insights,* 8, 3–6.

Steinwachs M. 2009. New technique for cell-seeded collagen-matrix-supported autologous chondrocyte transplantation. *Arthroscopy*, 25, 208–211.

Suh H. 1998. Recent advances in biomaterials. *Younsei Medical Journal*, 39(2), 87–96.

Tan H, Ramirez CM, Miljkovic N, Li H, Rubin JP and Marra KG. 2009. Thermosensitive injectable hyaluronic acid hydrogel for adipose tissue engineering. *Biomaterials*, 30, 6844–6853.

The Royal Academy of Engineering. The Royal Society. July 2004. Nanotechnology and Nanoscience. Available from: http://www.nanotec.org.uk/index.htm (Accessed on 29 January 2014).

Thomas JB, Peppas NA, Sato M and Webster TJ. 2005. Nanotechnology and biomaterials. In *Nanomaterials Handbook*, 605–630.

Thomson JA and Itskovitz-Eldor J. 1998. Embryonic stem cell lines derived from human blastocysts. *Science*, 282, 1145–1147.

Thyssen JP, Johansen JD, Menné T, Lidén C, Bruze M and White IR. 2010. Hypersensitivity reactions from metallic implants: a future challenge that needs to be addressed. *Birth Journal of Dermatology*, 162(2), 235–236.

Tirrell M, Kokkoli E and Biesalski M. 2002. The role of surface science in bioengineered materials. *Surface Science*, 500, 61.

Tonga Z and Ji X. 2012. Biomaterial-based strategies for the engineering of mechanically active soft tissues. *MRS Communications*, 2(2), 31–39.

Törmälä P, Pohjonen T and Rokkanen P. 1997. Ultrahigh-strength self-reinforced polylactide composites and their surgical applications. *Macromolecular Symposia*, 123, 123–131. doi: 10.1002/masy.19971230113.

USP. University of São Paulo, 2009. What is Biomedical Engineering?

Villarreal LP. 2005. *Viruses and the Evolution of Life*. ASM Press, Washington.

Wagner V, Dullaart A, Bock AK and Zweck A. 2006. The emerging nanomedicine landscape. *Nature Biotechnol.* 24(10), 1211–1217. Also Aitken RJ, Chaudhry MQ, Boxall ABA and Hull M. 2006. Manufacture and use of nanomaterials: current status in the UK and global trends. *Occupational Medicine*, 56(5), 300–306.

Walter JB, Talbot IC, Gardner HA and Israel M. 1996. *General Pathology*. 7th edition. Churchill Livingstone. NY, USA.

Wang DA, Varghese S, Sharma B, Strehin I, Fermanian S, Gorham J, Fairbrother DH, Cascio B and Elisseeff JH. 2007. Multifunctional chondroitin sulphate for cartilage tissue-biomaterial integration. *Nature Materials*, 6, 385–392.

Watt FM and Ryan RD. 2010. The Therapeutic Potential of Stem Cells. *Philosophical Transactions: Biological Tendencies*, 365(1537), 155–163.

Williams DF. 1999. *Dictionary of Biomaterials*. Liverpool University Press, Liverpool.

Williams DF. 1987. Definitions in biomaterials. In *Progress in Biomedical Engineering*. Elsevier, Amsterdam.

Wilmut I, Schnieke AE, McWhir J, Kind AJ and Campbell KH. 1997. Viable off spring derived from fetal and adult mammalian cells. *Nature*, 385, 810–813.

Witte F. 2010. The history of biodegradable magnesium implants: a review, *Acta Biomaterialia*, 6, 1680–1692.

Witzany G. 1993. *Nature of Language, Language of Nature: Sprachpragmatische Philosophy of Biology*. Königshausen & Neumann, Würzburg.

Witzany G. 2000. *Life: The Communicative Structure. A New Philosophy of Biology*, trans. Michael Stachowitsch. — Bürmoos, Vogelsangstraße 18c. Libri Bookson, Norderstedt.

Wolberg AS. 2007. Thrombin generation and fibrin clot structure. *Blood Reviews*, 21, 131–142.

Yannas IV and Burke JF. 1980. Design of an artificial skin. I. Basic design Principles. *Journal Biomedical Materials Research*, 14, 65–81.

Yannas IV. 2005. *Regenerative Medicine I: Theories, Models and Methods Medicine*. Springer, NY, USA.

Zeng R, Dietzel W and Witte F. 2008. Progress and challenge for magnesium alloys as biomaterials. *Advanced Engineering Materials*, 10, 3–14.

Zhang L and Webster TJ. 2009. Nanotechnology and nanomaterials: promises for improved tissue regeneration, *NanoToday*, 4(1), 66–80.

Zhang L, Chung By, Lear BC, Kilman VL, Liu Y, Mahesh G, Meissner R-A, Hardin PE and Allada R. 2010. DN1(p) circadian neurons coordinate acute light and PDF inputs to produce robust daily behavior in Drosophila. *Current Biology*, 20(7), 591–599.

Zitvogel L, Tesniere A and Kroemer G. 2006. Cancer despite immunosurveillance: immunoselection and immunosubversion. *Nature Reviews Immunology*, 6(10), 715–727.

15. Communication Languages and Agents in Biological Systems

Noga Gershoni-Emek[*,§], Eitan Erez Zahavi[*,§], Shani Gluska[*,§], Yulia Slobodskoy[*,§], and Eran Perlson[*,†,‡]

[*]*Department of Physiology and Pharmacology, Sackler Faculty of Medicine, Sagol School of Neuroscience, Tel Aviv University, Tel Aviv 69978, Israel*

[†]*Department of Physiology and Pharmacology Sackler Faculty of Medicine, Room 605, Sagol School of Neuroscience Tel Aviv University, Ramat Aviv, Tel Aviv 69978, Israel*

[‡]*eranpe@post.tau.ac.il*

"Intra" and "extra" cellular communication is a fundamental attribute impacting fitness and function of all cell types. In order to survive and maintain normal function, the cell depends on a

[§]Authors contributed equally to this work.

dynamic system of spatial specificity and fidelity of signaling pathways that can respond to both internal and external changes over space and time. Cell signals must be properly timed and located, and alterations in specific localization may lead to dysfunction and even death. Both intra- and extracellular modes of communication are mediated largely by ligand-receptor interactions and trafficking mechanisms.

In this chapter we will discuss some of the molecular mechanisms that govern the spatiotemporal localization of signals in the neuron, thus playing a key role in neuronal survival and synapse maintenance. We will discuss communication mechanisms between the neuron and its environment as well as within the cell's different compartments: soma, axon and synapses. In the first section we will discuss secreted factors that affect neuronal health and describe both positive and negative effects of these ligands. In the following sections, we will discuss receptors and intracellular aspects of the communication process. Further, we will argue how possible alterations in these extrinsic and intrinsic cellular communications can lead to neurodegeneration and diseases. **In short, this chapter will discuss the molecular mechanisms of communication that the neuron utilizes in order to "converse" both with its environment and within its discrete compartments.**

1. Introduction

Neurons are highly polarized cells with axons that are many orders of magnitude longer than the diameter of their cell bodies. Therefore, an efficient method for transferring signals along the neuron to or from the target organ is required. While some of this signaling is transduced via the neuronal action potential (not discussed here), the neuron employs protein and RNA trafficking mechanisms in order to regulate the specific task of spatial communication. Consequently, the precise targeting and delivery of cellular signals to specific, distinct sub-cellular compartments in neurons is largely dependent on axonal transport mechanisms.

Axonal transport is the cellular process of moving proteins, organelles, vesicles, RNA and other cellular factors to (retrograde) and from (anterograde) the neuron cell body. The molecular motor kinesin drives transport from the cell body to the nerve periphery, supplying proteins, lipids and other essential materials. The dynein/dynactin complex drives retrograde transport, moving damaged proteins for degradation and crucial signaling molecules such as neurotrophic factors to the cell body (Ibanez, 2007; Salinas *et al.*, 2008). The cellular cytoskeleton provides the "highways" for molecular motors to drive along. In axons, microtubule tracks in axons are polarized, where the plus or growing end is aligned with the tip of the axon and the minus end faces the cell body. This formation supports the directed travel of various motor proteins, along with their cargos. Axonal transport is vital to maintain neuronal function and survival, and alterations in transport have been shown to lead to neurodegenerative diseases (Perison *et al.*, 2009, 2010). While there is a clear and early defect in retrograde transport in multiple mouse models of neurodegenerative diseases, this alone can only account for relatively mild neurodegeneration. Thus, survival factor deprivation, while significant, may not be the key problem in rapidly progressive diseases. Instead, retrograde transport of stress/death signals secreted by neuronal supporting cells, may play a crucial role in disease progression. Hence, alterations in neurocommunication and spatial changes in the cellular balance of retrograde survival versus stress/death signals contribute significantly to rapid onset of neurodegeneration, and suggest a non-cell-autonomous mechanism in neurodegenerative diseases. In the next paragraphs we will discuss and present few examples of positive signaling initiated by neurotrophic factors and negative stress signaling along the axon.

2. Positive and Negative Signaling in Neurons

2.1 Neurotrophic factors

Trafficking of signals and proteins internalized at the axon tip to the cell body is a process widely used by both developing and mature neurons, as efficient retrograde axonal transport is crucial for their survival and adequate function. One such example is the trafficking of neurotrophic factors, which are growth factors that act specifically on neurons. According to the neurotrophin factor hypothesis, inspired by the pioneering studies of Viktor Hamburger and Rita Levi-Montalcini who first identified the neuronal growth factor (NGF) in the 1950s, neurons are overproduced in the developing nervous system, and compete for a limited amount of target-derived survival factors. Neurons unsuccessful in obtaining sufficient factors undergo programmed cell death (Cohen *et al.*, 1954).

Neurotrophic factors are proteins secreted by target organs or supporting cells in the neuron's vicinity, and are necessary for selective survival of neurons, neurite growth, arborization, synapse formation and overall development of both the central (CNS) and peripheral nervous systems (PNS) (Zweifel *et al.*, 2005). All of these are mediated by binding of trophic factors to specific receptors and co-receptors located on the neural membrane. Subsequently, a signaling complex consisting of receptor and ligand is formed, then travels to the cell body, where it activates downstream signaling and mediates transcription events- a process known as the "signaling endosome" model (Harrington *et al.*, 2013). Other proposed mechanisms for retrograde propagation of trophic signaling include the retrograde transport of subsequent effectors (such as ERK or Ca^{2+}) and activation of adjacent receptors on the cell membrane (Heerssen and Segal, 2002; Sengar and Campenot, 1997). Neurotrophic factors may also act locally, where signaling is activated adjacent to the ligand-receptor binding site.

Neurotrophic factors are often classified into three groups. Neuropoietic cytokines including ciliary-derived neurotrophic factor (CNTF), which binds receptor with the same name (CNTFR). The second group, is the GDNF family ligands (GFL's); glial cell line-derived neurotrophic factor (GDNF), neurturin (NRTN), artemin (ARTN), and persephin (PSPN). These ligands initiate signaling by binding simultaneously to specific co-receptors, GDNF-family-receptor-α (GFRα1-4, respectively), as well as to the receptor tyrosine kinase RET or, in the case of GDNF, the neuronal cell adhesion molecule (NCAM). The third group is the nerve growth factor (NGF) family or Neurotrophins; NGF, brain derived neurotrophic factor (BDNF), neurotrophin (NT)-3 and NT-4 (known as NT-4/5 or NT-5). These ligands initiate signaling via the Trk family of tyrosine protein kinases TrkA (NGF); TrkB (BDNF, NT-4), TrkC (NT-3) and the p75 neurotrophin receptor ($p75^{NTR}$). See Table 1.

2.1.1 *GDNF signaling*

GDNF was identified in the early 90s by Lin and colleagues (Hall and Sanes, 1993). Despite its name, GDNF is expressed in many tissues and cell types, including the nervous system. In addition to neural development and survival, GDNF is also required for kidney morphogenesis, enteric migration and spermatogenesis (Tsui and Pierchala, 2010). It also supports the survival of motor, dopaminergic, sensory and sympathetic neurons — both *in vivo* and *in vitro* (Tsui and Pierchala, 2010; Henderson *et al.*, 1994). While some of these functions are based on local signaling in the vicinity of receptor binding, other functions require the transport of GDNF over long distances. For instance, developing motor neurons require GDNF to be expressed along their route to the target organ, in order to arrive at the correct target, while avoiding programmed cell death en route (Airaksinen and Saarma, 2002). GDNF was shown to be transported anterogradely from the cell body to synapses (Russell *et al.*, 2000) as well as retrogradely in both sensory and MNs (Tsui and Pierchala, 2010) and (Zahavi *et al.*, 2015).

GDNF activity is initiated by high affinity binding to its co-receptor GFRα1 to form a complex that can further bind either RET or NCAM. Alternative receptors such as heparan sulfate proteoglycan and syndecan-3 might also be involved in GDNF signaling (Bespalov *et al.*, 2011). When this binding occurs at the axon tip, it is thought to result in internalization and retrograde transport of GDNF to the cell body. The regulation of this mechanism is still uncertain. Two interesting questions regarding its mechanism of action remain unanswered. First, what is the importance of GDNF's

Table 1. Positive and negative signaling in neurons.

Family	Ligand	Receptor	Co-receptors	Function	Ref.
Neuropoietic factors	CNTF	CNTFR, IL-6 receptor		Neuronal survival and neurite outgrowth	
GFL's	GDNF Neurturin Artemin Persepin	RET; NCAM; syndecan-3 RET; syndecan-3 RET; syndecan-3 RET; syndecan-3	GFRα1 GFRα2 GFRα3 GFRα4	Development and maintenance of central and peripheral nerves. Spinal, dopaminergic neruons, sympathetic neurons, noradrenergic neurons etc.	(Airoksinen and Saarma, 2012; Allen *et al.*, 2013; Henderson *et al.*, 1994)
Neutrophines	NGF BDNF NT3 NT4	TrkA, $P75^{NTR}$ TrkB, $P75^{NTR}$ TrkB, $P75^{NTR}$ TrkC, $P75^{NTR}$		In the presence of Trk family receptors — survival and maintenance of sensory neurons, retinal ganglia, cholinergic neurons, spinal and dopaminergic neurons, as well as neurite growth and arborization. When binding $P75^{NTR}$, induce apoptosis and neuron repulsion.	(Allen *et al.*, 2013) (Allen *et al.*, 2013)
	Pro-GDNF	Unknown		Unknown	(Lonka-Nevalaita *et al.*, 2010)
Pro-neutrophins	Pro-NGF	$P75^{NTR}$; Sortilin		Apoptosis	(Lee *et al.*, 2001; Feng *et al.*, 2005)
	Pro-BDNF	$P75^{NTR}$; Sortilin		Apoptosis	
Semaphorins	Semaphorine 3A/3B/3F	Neuropilin, Plexin		Axon repulsion and growth cone collapse, in the healthy developing nervous system or neurodegenerative diseases.	(Nakamura *et al.*, 2000)
Ephrins	EphrinA1/A2	Ephrin Receptor		Axonal guidance	(Dudanova and Klein; 2013)

transport along the neuronal cell? Second, do different transport properties allow for differential function of GDNF in neurons?

For example, GDNF was shown to support the survival of both sympathetic and sensory neurons in mass culture. The same study showed that GDNF did not have a significant rescue effect over a period of a few days when applied to the distal axon compartment in a microfluidic system of sympathetic neurons . However, a different result was obtained when GDNF was applied directly to sympathetic cell bodies or when applied to the distal axon compartment of sensory neurons. NGF, on the other hand, had a similar effect when applied to either the cell body or distal axons. Researchers hypothesize that these different effects can be attributed to different splice isoforms of the GDNF receptor RET in different types of neurons, which affect the properties of GDNF retrograde transport (Tsui and Pierchala, 2010).

In conclusion, neurotrophic-factors secreted from targets and supporting cells, such as GDNF, facilitate neuronal survival and function by either activating local signaling pathways, or by translocating to distinct regions within the neuron in signaling endosomes that employ the cellular trafficking machinery. In the next section we will discuss stress/apoptotic secreted factors that play a negative role in neuron health.

2.2 Neuronal death signaling

2.2.1 *Pro-neurotrophins*

While mature neurotrophins are mostly positive factors, there is also evidence for negative factors secreted in the neuron's environment that can lead to neurodegeneration and cell death. This negative signaling could be essential for blocking unwanted outgrowth and branching during development, however, when re-activated in adults, could lead to neurodegeneration.

Proteolytic cleavage of proneurotrophins may also play a role in determining cell fate: whereas mature neurotrophins bind with high affinity to Trk receptors (and with low affinity to p75NTR), and signal for survival, proneurotrophins bind with high affinity to p75NTR and induce cell death (Harrington *et al.*, 2004; Lee *et al.*, 2001; Teng *et al.*, 2005). Thus, neurotrophin pro-isoforms can facilitate programmed cell death in MNs, and attenuate synaptic transmission in a p75NTR-, sortillin-, and caspase-dependent manner (Oppenheim, 1996; Sendtner *et al.*, 2000; Taylor *et al.*, 2007). For example, proBDNF can be secreted from neurons and muscles and induce apoptosis of MNs *in vitro* and *in vivo* (Taylor *et al.*, 2012). Further, activity-dependent processing of pro- into mature BDNF has been implicated in synaptic competition that takes place at the developing neuromuscular junction (NMJ) for refinement of axon targets (Je *et al.*, 2012). Hence, pro-neurotrophic factors that mediate MN degeneration in development may be aberrantly re-activated under degenerative conditions in adults. This communication equilibrium may serve as a crucial switch in the neurodegenerative pathogenesis, and further exploration could be insightful.

2.2.2 *TNF signaling*

Tumor necrosis factor (TNF, cachexin, or cachectin) is another secreted molecule that can confer negative effects. It is a potent pro-inflammatory cytokine and can elicit either trophic or toxic biological responses upon binding to its receptor TNF receptor type 1 (TNFR1) and TNFR2 (TNF receptor type 2, expressed mainly on immune cells). TNFR1 is constitutively expressed on both neurons and glia in the CNS and PNS, and TNF can be synthesized and released by muscle fibers (Grounds *et al.*, 2008; Robertson *et al.*, 2001; Viviani *et al.*, 2004).

TNF binding causes a conformational change to occur in the receptor, leading to the dissociation of the inhibitory protein silencer of death domain (SODD) from the intracellular death domain, which enables the adaptor protein TNFR1 associated death domain protein (TRADD) to bind to the Death Domain, serving as a platform for subsequent protein binding. Following TRADD binding, three pathways can be initiated:

1. Activation of NF-κB: NF-κB is a heterodimeric transcription factor that translocates to the nucleus and mediates the

transcription of a vast array of proteins involved in cell survival and proliferation, the inflammatory response, and anti-apoptotic factors.

2. Activation of the MAP-Kinase pathways: Of the three major MAPK cascades, TNF induces a strong activation of the stress-related JNK group, evokes moderate response of the p38-MAPK, and is responsible for minimal activation of the classical ERKs. The JNK pathway is predominately pro-apoptotic, and is also involved in cell differentiation and proliferation.
3. Induction of death signaling: TRADD binds FADD, which then recruits the cysteine protease caspase-8. High concentrations of caspase-8 induce autoproteolytic activation and subsequent cleaving of effector caspases, leading to cell apoptosis.

2.2.3 *Guidance factors*

Guidance factors are also a means for neuronal communication. During development, a sequence of guidance events control the pathfinding, target selection, initial synapse formation and synapse refinements. Axonal guidance cues span fromshort-range mechanisms, such as cell surface adhesion or contact inhibition, to long-range guidance cues, like chemorepulsion or chemoattraction molecules (Dudanova and Klein, 2013). The three major families of axon guidance molecules that mediate repulsion are: ephrins, semaphorins and slits. As an example of negative guidance signal, we will focus here on the semaphorins.

Most members of the semaphorin family of proteins function as axon chemo-repellents. In some cases, they have also been shown to induce neuronal cell death. Sema3A is one of the best-characterized molecules in this family. During development, Sema3A functions as a chemo-repellent and death factor to a number of neuronal populations *in vitro* and *in vivo* (Nakamura *et al.*, 2000). In adults, Sema3A expression is very low; however, it is up-regulated in many human neurodegenerative diseases such as Alzheimer's disease (AD) (Good *et al.*, 2004) and multiple sclerosis (MS) (Williams *et al.*, 2007). Expression of Sema3A and its receptors is also up-regulated in humans and animals following CNS injury (Kaneko *et al.*, 2006; Pasterkamp *et al.*, 1999; Niclou *et al.*, 2003), ischemia (Fujita *et al.*, 2001) and inflammation (Majed *et al.*, 2006), and in neuronal cultures following oxidative stress (Shirvan *et al.*, 1999), insulin withdrawal (Majed *et al.*, 2006) and 6-hydroxydopamine administration (a model of Parkinson's disease) (Yasuhara *et al.*, 2004). Further, Sema3A is up-regulated in terminal Schwann cells (TSCs) in animal models of Amyotrophic Lateral Sclerosis (ALS) (De Winter *et al.*, 2006).

Type 3 semaphorins may also be involved in neurodegenerative conditions. Post-mortem analysis of diseased human brains shows elevated levels of Sema3F in MS and elevated levels of Sema3B in AD (Blalock *et al.*, 2004). Although the expression of many genes differs between control and AD brains, Sema3B is one of a small group of genes whose induction is tightly correlated with AD progression (Blalock *et al.*, 2004). Additionally, both Sema3B and Sema3F are up-regulated in animals following CNS injury (De Winter *et al.*, 2002).

In short, the neuron's diverse microenvironment contains secreted negative signals that can lead to synapse disruption, axon pruning and neuronal death: physiological processes in the developing embryo, but pathological in adults. In the next section we will discuss another level in the neurocommunication events, the spatiotemporal localization of neuronal receptors at the plasma membrane.

3. Spatiotemporal Localization of Neuronal Receptors at the Plasma Membrane

3.1 Neurotrophin receptors

The neuron displays specialized growth factor receptors on its plasma membrane. These receptors are not merely passive antennas, randomly diffusing along the surface of the cell, but are dynamically regulated in the lateral spatial domain. As discussed above, Trk-mediated signaling facilitates axonal growth and synapse strengthening, as well as anti-apoptotic and proliferative gene activation, while p75 activation generates axonal repulsion and pro-apoptotic pathways. In this section we will review some of the cellular mechanisms regulating receptor organization in the membrane and its role in signal modulation, specifically focusing on the neurotrophin receptors.

3.1.1 *Dimerization of Trk tyrosine kinase receptors*

Upon binding to their neurotrophin ligand, the Trk tyrosine kinase receptors undergo auto-phosphorylation, which activates the downstream signaling cascade. In their native state, NTF form dimers, which bind two Trk receptors simultaneously (Banfield *et al.*, 2001; Jing *et al.*, 1992). Trk receptors can trans-phosphorylate each other; and, similar to other tyrosine kinase receptors, undergo ligand-induced dimerization in order to activate signaling (Farooqui *et al.*, 1997; Ibanez *et al.*, 1994). Recent evidence, however, suggests that Trk receptors are trafficked to the plasma membrane as preformed, yet inactive dimers (Shen and Maruyama, 2012; Author name, 2011).

The p75 receptor was also shown to exist as a dimer in the membrane, which can only then bind NTF and undergo activation (Vilar *et al.*, 2009a, 2009b). Trk and p75 interaction, which facilitates Trk-mediated signaling may take place by formation of 1:1 Trk:p75 heterodimers, which was suggested to create a high affinity binding site for NTF (Wehrman *et al.*, 2007). However, as evidence for direct Trk-p75 binding on the membrane is scarce, and considering the preferable homodimeric state of both receptors, it is also possible that the cross-talk between the receptors takes place in contexts of high-order structures such as receptor clusters, lipid-rafts or at the level of downstream signaling effectors.

3.1.2 *NTF receptor clustering*

Trk neurtrophin receptors are highly mobile transmembranal receptors, shown to move in a diffusive manner at speeds of up to ~0.1–0.5 μm^2/sec (Levi *et al.*, 1980; Marchetti *et al.*, 2013; Shibata *et al.*, 2006). Early works into the cellular activation of Trk showed that it undergoes immobilization when activated by its ligand, mainly by FRAP assays. A recent work by Marchetti *et al.*, (2013) used single-receptor labeling to show that TrkA shifts to immobility on the membrane and forms high-order clusters of several receptors which move together in response to NGF binding. It was previously suggested that these kinds of immobile phases function as signaling platforms by recruiting the downstream effector Raf1 of the Ras pathway (Shibata *et al.*, 2006). This observation that receptor activation by its ligand changes its membranal distribution raises the question of how this lateral organization modulates signaling.

3.2 Receptor complexes enhance signaling fidelity and modulate downstream signaling

Formation of localized concentrations of active receptors along with downstream effectors enables crossing the "threshold" of signal that will propagate from the initial signal site, both spatially and temporally (Bethani *et al.*, 2010). Understandably, this mechanism of receptor re-distribution is common in numerous receptor–ligand systems, such as neurotransmission, cell–cell and cell-substrate adhesion, immune cell activation and EGF signaling. At immunological synapses between the antigen-presenting cell and a T-cell, T-cell receptors (TCRs) are concentrated at the center of the contact site. This clustering was shown to modulate the external MHC ligand signal-amplifying a low signal and attenuating a strong one. Similarly, in EphrinA1–EphA2 binding, it was shown that changing the patterning of receptor–ligand complexes from dispersed to concentrated directly affects the organization of F-actin and recruitment of the cell–cell adhesion protein ADAM2 into these complexes. Therefore, although relatively little is known about the interplay between neurotrophin receptor organization along the membrane and their signaling activity, the increasing evidence for ligand-mediated NTF receptor clustering encourages further examination of the cellular mechanisms regulating this clustering.

3.3 Lipid rafts as platforms for specialized receptor domains

Lipid rafts are currently regarded as spatiotemporal dynamic membranal structures approximately a few tens of nanometers long, comprised of specific lipids such as sphingolipids and cholesterol as well as proteins such as glycosylphosphatidyl-inositol anchored proteins (GPI-AP). These microdomains are implicated in the organization of membrane proteins by preferentially associating with "rafty", while sequestering "non-rafty" proteins, thus sustaining specific and discrete functional domains on the membrane (Lingwood and Simons, 2010).

Lipid rafts play an important role in the regulation of membranal signaling, specifically NTF signaling. While both p75 and TrkB were shown to translocate into lipid rafts following activation, some studies have shown that p75 is more "rafty" compared to TrkB, and that TrkB translocation into rafts is inhibited by p75 expression (Hibbert *et al.*, 2006; Higuchi *et al.*, 2003; Suzuki *et al.*, 2004). Furthermore, the exact role of lipid raft localization for ligand binding is not clear, while there is evidence that Trk preferentially bind their ligands at lipid rafts, other studies have shown that p75, but not TrkA, binds NTF within rafts. While Trk has been found to actively propagate a signal in response to ligand binding both in raft and non-raft membranes, disruption of lipid rafts by cholesterol depletion partially inhibits TrkB activation, especially of the phospholipase C gamma (PLCγ) pathway, as well as inhibiting p75 shedding in response to NTF (Suzuki *et al.*, 2004; Gil *et al.*, 2007). This observation suggests that specific NTF activities are mediated by the association of the receptor–ligand complex in lipid rafts. Despite these findings, there is little evidence showing the localization and transition of NTF receptors into lipid rafts on the plasma membrane. Moreover, TrkB translocation into lipid rafts occurs on internal membranes and not on the plasma membrane, as shown by combining surface protein labeling with membrane separation techniques by Chao's group (Pereira *et al.*, 2007).

It should be noted that the majority of the studies regarding lipid rafts and NTF use solubilization-based separation methods to distinguish between lipid-raft and non-raft membranes, a method now considered prone to various artefacts. In this regard, it is highly interesting to see if state-of-the art spectroscopic and super-resolution imaging techniques in living cells can be used to reopen the questions of how NTF receptor localization and signaling is spatiotemporally regulated by raft-membrane domains.

3.4 Actin cytoskeleton regulates receptor mobility and organization

While membrane protein clustering is directly related to local protein-lipid interactions, these alone cannot explain all the clustering properties of raft associated proteins such as GPI-AP, nor the dependence of clustering on actomyosin activity and membrane — actin attachment (Gowrishankar *et al.*, 2012). The actin cytoskeleton forms an intricate meshwork, which is attached to the plasma membrane, termed cortical actin (CA). This CA is well known for its role in membrane shaping, enabling dynamic changes in membrane rigidity, supporting cell attachment and motility, as well as endo- and exocytic processes. By forming a rigid enclosure made of long, relatively static intersecting or branched tracks of filamentous actin, the CA can form micro-to-nano sized diffusion barriers, capable of packing together receptors and membrane associated proteins by steric hindrance or specific interactions.

While NTF activation of Trk in neurons leads to distinct changes in actin polymerization dynamics, participating in NTF-mediated axonal growth and guidance, less is known about the role of actin in the membrane dynamics of NTF receptors. As a mobility barrier, cellular actin de-polymerizing machinery was shown to be recruited to activated TrkA domains near the membrane, enabling receptor internalization via endocytosis and subsequent retrograde transport of the internalized endosome (Harrington *et al.*, 2011). At the axonal growth cone, cell surface-bound NGF has been shown to be transported on actin fibers, shortly after binding to the cell surface (Tani *et al.*, 2004). Furthermore, disruption of the actin cytoskeleton by Cytochalasin D was shown to inhibit NGF stimulated TrkA clustering (Marchetti *et al.*, 2013). Either as a clustering and internalization regulator, or as a transport platform, actin plays an important role in the spatial and temporal organization of NTF receptors and their activation mechanism.

3.5 Microtubules and receptor organization-beyond transport routes

While actin's role in plasma membrane organization is well established, the microtubule cytoskeleton is regarded as the highways on which cargo is transported across large distances inside the cell.

In the case of NTF signaling, microtubules and motor proteins play a crucial role in the retrograde transport of activated receptors (Cosker *et al.*, 2008). Emerging evidence, however, suggests that microtubules can regulate spatial localization and function of receptor domains. Perlson *et al.*, demonstrated that dynein, a microtubule based motor, binds and tethers NCAM clusters on the cell cortex to microtubules, and thus interaction mediates cell–cell adhesion and synapse formation in neurons (Perlson *et al.*, 2013). Another example is in the immune synapse, where microtubules associate with the center of the TCPcluster depending on the localization of the MAL membrane protein, while actin is constricted to the periphery of the synapse (Anton *et al.*, 2011). Similarly, it was shown that microtubules are recruited to the cell cortex at the periphery of integrin focal adhesions (Lansbergen *et al.*, 2006). As mentioned previously, NTF receptors rely on microtubule based retrograde transport in neurons from their distal site of ligand induced activation toward the cell body. A possible role for microtubule and NTF-receptor association at the membrane may be in the assembly of a platform for retrograde transport of an activated receptor. This may be mediated by recruitment of microtubule end-binding proteins and dynein by activated receptors, thus promoting their own transport (Moughamian *et al.*, 2013; Zhou *et al.*, 2012; Mitchell *et al.*, 2012).

We have barely skimmed the surface of the multilevel interaction taking part in the organization of NTF receptors on the membrane. As its dynamic and complex nature proves an immense challenge to dissect, it is also revealed to be part of a sophisticated regulatory network, involving interactions between lipids, proteins, actin and microtubule networks as well as the endocytic and transport machineries, all working together to ensure NTF signaling takes place at the right place, time, and intensity.

The next section will discuss local protein synthesis, and its role in the regulation of neuronal-communication.

4. Local Protein Synthesis

4.1 mRNA localization

The localization of specific mRNAs to distinct intracellular compartments provides a mechanism to sort proteins to the site of function and to respond locally to stimuli that affect general cell function. Local protein synthesis can provide several advantages over protein transport, especially in highly polarized cells such as neurons. One of these is economical achieved when a translationally silent mRNA is used to make protein when needed. Another is a potential additional level of signaling provided by modifications to newly translated proteins.

Both axons and dendrites respond to extrinsic cues in order to regulate growth, navigation and synapse formation — as well as for maintenance and repair. Local mRNA translation is essential for a response that is precisely located, both spatially and temporally.

The subcellular localization of mRNAs in eukaryotic cells is an evolutionarily conserved mechanism. mRNAs are transported as part of a RNA-protein complex. The RNA binding proteins (RBPs) that bind to the mRNA serve not only as adaptors for transport, but also to maintain the mRNA in a translationally repressed state, until the arrival of a triggering signal — such as a guidance cue, or injury. This provides both the spatial and temporal control that are essential for polarized cells (Yoo *et al.*, 2010; Holt and Bullock, 2009; Donnelly *et al.*, 2010).

One suggested method for spatial restriction of mRNA translation is the direct interaction between a cell surface receptor and ribosomes at the site where an extrinsic cue is received, a model recently characterized for deleted in colorectal cancer (DCC), one of the receptors for the extracellular factor netrin that can be found in the axonal growth cone as well as in the post-synapse (Tcherkezian *et al.*, 2010). DCC binds to a complex of ribosomes and initiation factors, which dissociates as a result of netrin binding, and translocates to the polysome. Molecular motors also take part in this spatial regulation. Dynein plays a role in the anchoring of both microtubules (Hendricks *et al.*, 2012) and mRNA (Delanoue and Davis, 2005).

4.2 Growth cone guidance

The distal tip of a growing axon is a specialized structure known as the growth cone. During growth and pathfinding, the growth cone must respond to various stimuli from the environment providing guidance information. Guidance cues include attractive stimuli, such as BDNF and Netrin-1, which cause axon turning by inducing translation of proteins that build up the cytoskeleton. Repulsive cues, such as SEMA3A and SLIT2 cause the growth cone to turn away by initiating the translation of proteins that break down the cytoskeleton. The response to this information is rapid, and more often than not does not involve the soma (Holt and Bullock, 2009).

A common methodology for studying soma-less response involves the use of severed growth cones. Application of chemotrophic guidance cues to soma-less growth cones increases activity (mTOR) the mammalian target of rapamycin and axonal protein synthesis within minutes. Inhibition of either the mTOR pathway or ribosome function diminishes both local protein synthesis and the chemotrophic response, as well as the characteristic sema3a-induced growth-cone collapse. The response to netrin-1 is blocked by a phosphoinositol 3-kinase (PI3K) inhibitor, but not the response to sema3a, suggesting different signaling pathways that converge on mTOR (Campbell and Holt, 2001).

Most protein-synthesis inducing signals activate local protein synthesis through the mTOR signaling pathway. The direction of the turning response, however, is determined by translation of specific mRNAs. For example, β-actin mRNA is rapidly transported toward the chemotrophic gradient when the growth cone is stimulated by either netrin 1 or BDNF, followed by asymmetric rapamycin-sensitive translation of β-actin. Transport of the β-actin mRNA is facilitated by the RBP zipcode binding protein 1 (ZBP1), which binds to a *cis*-acting element in the 3′UTR while repressing its translation (Leung *et al.*, 2006; Yao *et al.*, 2006). Repulsive cues promote turning by inducing the rapid synthesis of proteins such as cofilin and RhoA that take part in actin disassembly (Wu *et al.*, 2005; Piper *et al.*, 2006).

The model proposed by Lin and Holt (2007) suggests that guidance cues initiate a global translational response, while specificity is achieved by mRNA-specific regulation facilitated by RBPs and microRNAs. The association of ZBP1 and β-actin mRNA has been well documented in dendrites and axons, as well as in fibroblasts, where it takes part in filopodia growth necessary for cell migration.

ZBP1 recognizes a 54-bp motif in the 3′UTR of β-actin mRNA, and is co-localized in spatially distinct granules with β-actin mRNA in the soma, processes and growth cones. Within the growth cone, the colocalized granules align with microtubules and display fast bi-directional transport in the cell (Zhang *et al.*, 2001). Binding between netrin 1 and DCC results in the phosphorylation of ZBP1 by Src kinases, which leads to synthesis and turning (Welshhans and Bassell, 2011).

Another aspect in neuronal communication that we will discuss in the next paragraph is how the neuronal cell body receives information about an injury from distant lesion sites in the axon.

5. Injury Signaling

The nerve-regeneration process depends on activated signals at the injury site that are transported along the axon to the cell body. The formation of this retrograde signaling complex depends on local synthesis events.

A large variety of retrogradely transported proteins have been identified in peripheral nerves following injury (Abe *et al.*, 2009; Michaelevski *et al.*, 2010). Interesting examples are the importins-nuclear transport factors that were found to form a local activated nuclear-targeted signaling complex after nerve injury in process that depends on local protein synthesis (Hanz *et al.*, 2003). Transcription factors like STAT3 and CREB (Ben-Yaakov *et al.*, 2012; Cox *et al.*, 2008) are examples of locally activated transcription factors in the distal axon that serve as retrograde signaling entities for direct communication with the nucleus. Another example for long distance signaling after axonal injury is the phospho-Erks, which bind vimentin complexed with importin 1, enabling translocation of activated kinases (Perlson *et al.*, 2005).

6. mRNA Transport and Localization

Microarray studies in various neurons revealed thousands of different mRNAs in their axons. Most axonal mRNA transcripts encoded for protein synthesis machinery, mitochondrial proteins and cytoskeleton components. However, specific cell types showed specific mRNA enrichment. For example, *Impa1* mRNA is enriched in axons of sympathetic neurons. IMPA1 is the master enzyme for regulation of the inositol cycle, controlling the availability of inositol affecting inositol phosphate and phosphatidyl inositol signaling events. Inhibition of IMPA1 catalytic activity can affect growth cone motility of sensory neurons (Andreassi *et al.*, 2010).

Specific anterograde transport of mRNA is the most likely mechanism behind this enrichment. The *cis*-acting localization elements in charge of targeting mRNAs to subcellular compartments, are usually located in the 5′ or 3′UTR of a transcript. In the case of *Impa1*, a transcript with a longer 3′UTR is enriched in distal parts of the axon. Structural models predict that the additional 120 nts present in the longer 3′UTR fold as distinct hairpin loops, which may interact with RBPs that localize the transcripts to the axon. The 3′UTR is sufficient for targeting a transcript into the distal axon of sympathetic neurons (Andreassi *et al.*, 2010).

Studies in rat sensory neurons show that extrinsic cues also play a part in regulating *cis*-acting elements. Quantitative analysis of axonal mRNA transcripts following administration of NGF, BDNF, NT3, myelin-associated glycoprotein (MAG), and Sema3A show a specific increase or decrease of individual transcripts. At the same time, an opposite change in cell body mRNA was recorded, pointing to ligand-dependent alterations in anterograde transport rates (Willis *et al.*, 2007).

7. MicroRNAs

MicroRNAs play a key role in post-transcriptional regulation of gene expression in physiological and pathological processes. The biogenesis of miRNAs is well characterized: the pri-miRNA is transcribed in the nucleus, and after undergoing the first stage of processing by the microprocessor, is then transported into the cytoplasm in their pre-miRNA form. To mature, the pre-miRNA is cleaved by the Dicer endonuclease, which then facilitates the loading of the miRNA into the RNA-induced silencing complex (RISC) which mediates post-transcriptional silencing of target genes (Kim *et al.*, 2009). Expression of Dicer and Argonaute family proteins, which make up RISC, as well as mRNAs and translational machinery, along with cytoplasmic bodies such as P-bodies and stress granules (SGs) in axons (Hengst *et al.*, 2006) and dendrites has been documented. miRNAs have also been identified in association with post-synaptic densities, suggesting a role for miRNA-induced regulation at these locales (Siegel *et al.*, 2011).

Relative enrichment of specific miRNAs in the axonal compartment suggests a mechanism for selective transport. These axonal miRNAs have multiple potential mRNA targets, suggesting multiple functions (Kaplan *et al.*, 2013).

Local expression of miRNA takes part in the regulation of various processes in the axon. In sympathetic neurons, the local expression of the translation initiation factors eIF2B2 and eIF4G2 can be modulated by the abundantly expressed miR-16 (Kar *et al.*, 2013). The mitochondrial mRNAs cytochrome c-oxidase (CoxIV)

(Aschrafi *et al.*, 2008) and ATP Synthase (ATP5GI) (Hillefors *et al.*, 2007) are targets for miR-338. In primary embryonic cortical neuron culture, miR-9 was shown to affect the axonal outgrowth and branching (Dajas-Bailador *et al.*, 2012). All these support the hypothesis that miRNAs hold a key role in gene regulatory networks, and suggest that dysregulation of miRNA function plays a role in the pathophysiology of neurological and psychiatric disorders.

The mechanism for miRNA transport remains elusive. Currently, there are a number of possible hypotheses for miRNA arrival at neuron process. First, by passive diffusion, this is most likely the predominant method in developing neurons (Smalheiser and Lugli, 2009). Second, as mature miRNAs "hitch-hiking" with their mRNA targets as these are transported. In this method, the miRNA is suggested to keep the mRNA transcript dormant during transit (Kosik, 2006; Schratt *et al.*, 2006). The third possibility is in precursor or primary form, after which local processing takes place at the dendrite, possibly in response to an extrinsic stimulus (Bicker and Schratt, 2008; Kosik, 2006; Smalheiser and Lugli, 2009; Kaplan *et al.*, 2013). Another possible machinery involves exosomal uptake from neighboring cells (Smalheiser, 2007; Ramachandran and Palanisamy, Mittelbrunn *et al.*, 2011).

7.1 Exosomal transfer of miRNA

Recent studies show growing evidence that exosomes serve as vehicles for horizontal transfer of genetic information between cells. They have been shown to be taken up by presynaptic neurons in a unique process that was described as "the inverse of MVB-mediated exosome secretion" (Smalheiser, 2007). Exosomal miRNAs have been reported to regulate genes in charge of the size of dendritic spines (Aumiller and Forstemann, 2008), mediate neuron to astrocyte (Morel *et al.*, 2013) and astrocyte to neuron communication (Smalheiser, 2007), Neuronal exosomes have been implicated in the pathogenesis of neurodegenerative diseases, amongst them the spread of the beta-amyloid protein throughout the brain in Alzheimer disease (Smalheiser, 2007), though exosomal RNA has not been linked with these processes. This rapidly developing field of research may provide new avenues for the diagnosis and treatment of neurodegenerative diseases.

Finally, in the last section we will describe specifically some key signaling events between the neuron and the muscle as an example for intercellular communication.

8. MN and Muscle Communication at the Synaptic Cleft of the NMJ

The synaptic cleft of the NMJ is the site where the MN converses with the skeletal muscle. The role of the NMJ is to transmit signals from the MN to the skeletal muscle fiber quickly and reliably, in order to ensure precise control of skeletal muscle contraction. Anatomically, MN cell bodies are located in the ventral horn of the spinal cord or in the brain stem. A single axon is sent out from each MN, via the ventral root of the spinal cord to a single skeletal muscle fiber to the motor end plate. A rapid (among the fastest in the body) action potential is passed along the axon to the nerve terminal end plate, which contains clusters of synaptic vesicles with the acetylcholine (ACh) neurotransmitter aligned with a patch of dense material on the presynaptic membrane forming an active zone. At the active zone, vesicles fuse with the plasma membrane and release their content of ACh across the synaptic cleft. Transmitter release results from the influx of extracellular Ca^{+2} through ion channels opened by depolarization of the nerve terminal membrane, a process called chemical synaptic transmission. The specialized pre- and post-synaptic membranes are precisely aligned and separated from each other by a 50–70 nm wide, protein-rich synaptic cleft (Hong and Etherington, 2011; Shi *et al.*, 2012; Werle, 2009; Meier and Wallace, 1998).

The muscular target, specifically the NMJ, provide MNs with a complex balance of survival and pro-apoptotic signals as discuss above that in sum determine the fate of the cell (Pedersen, 2013; Yoon *et al.*, 2012).

Skeletal muscle fibers are covered by a layer of basal lamina (BL) that coats the entire cell surface, including the post-synaptic membrane. Thus, at the NMJ, but not at any neuron–neuron synapses, BL comprises a major portion of the cleft material. Importantly, the BL also contains the enzyme acetylcholinesterase (AChE) which terminates neurotransmission by hydrolyzing ACh. Finally, each axon terminal is covered by a Schwann cell, a principal glial cell of the PNS (Sanes, 1995).

The fact that nerve and muscle membranes contact BL, rather than each other, suggests that the BL may contain some of the cues that these cells recognize. We can view the synaptic cleft as a molecular bulletin board that the two cells employ to post messages to each other.

Three main types of signals pass from muscle to nerve (Hall and Sanes, 1993; Sanes, 1995). First, nerve and muscle membranes may interact directly, as nerve and muscle membranes are in close apposition (~<10 nm) at newly formed synapses; the synaptic cleft widens and fills with matrix only at later stages. Because the strength of adhesion between the growth cone and the muscle cell increases in the few minutes following synaptic contact, it is possible that cell adhesion molecules participate in these interactions. N-CAM and N-cadherin promote neurite outgrowth on avian muscle cells, and interactions involving these or other adhesion molecules could provide an early signal to the two cells that contact has been made between neuron and muscle. Second, transient gap junctions have been observed between cultured nerve and muscle cells and could also mediate communication *in vivo* during an early period of development. Finally, muscle cells could secrete soluble factors as retrograde signals as described above.

9. Summary

In this chapter, we have demonstrated that accurate communication processes are essential for neuron viability. Neurons, more than any other cell type, traverse through the body and interact with diverse microenvironments. Hence, neurons must be able to integrate numerous signals from various sources, and transduce them further, often across great lengths.

Neurons are guided, activated, maintained, and obliterated by external signals. This variety of signals needs to work in concert and only extremely delicate fine-tuning between them can lead to the desired effect and, eventually, to proper function. In order to respond to these external stimuli, neurons employ various molecular mechanisms.

First, neurons are polarized and comprised of an intricate cytoskelatal architecture, as well as express countless receptors, adhesion molecules and other membranal components. Not only does this facilitate distinct connectivity to other cell types and tissues, but it also allows for a signal to be localized or transported in a very specific formation. Membrane fluidity creates an ever-changing sense organ, affected by internal and external indicators, which could quickly enhance, reduce or modulate a signal simply by changing its own nature.

Second, neurons express an array of motor proteins that facilitate fast and efficient bi-directional transport and spatiolocalization along neurites by utilizing the complex network of microtubules. This way survival/death factors, organelles, proteins and RNA swiftly localize at their destination. Though we do not yet understand the exact mechanisms in charge of axonal transport, we are aware of its importance and the deleterious effects of axonal transport deficits.

Third, we now understand that translation in neurons may occur not only at the cell body, but even at the distal regions of the axon, including the growth cone. This allows for highly specific and economic translation and reduces the need to anterogradely transport large proteins along the neuron, yet another way to save energy and, more importantly, answer a distal signal in a fast, efficient manner.

At all these levels, transport holds an important role, as it provides the basis for correct spatiotemporal localization.

As we see it, a deeper understanding of these mechanisms of communication between neurons and their environment is crucial. Combining that knowledge with that of the mechanisms underlying signal transduction within the neuron will not only expand our comprehension of basic cellular mechanisms, but will help us develop ways to intervene in these pathways when they go awry.

Substantiating our knowledge in the field of neural guidance and synapses would open a window to treating developmental and other neural faults, while mastering axonal transport would allow us to not only prevent pathogens and toxins from entering the nervous system, but also rescue disease phenotypes that stem from transport deficits. The knowledge being gathered today in laboratories across the world will have a significant effect on our ability to battle neurodegeneration and will give hope those suffering from currently incurable conditions (Fig. 1).

Abbreviations

CNS	Central Nervous System;
PNS	Peripheral Nervous System;
NT	Neurotrophin;
NTF	Neurotrophic factor;
NGF	Neuronal Growth Factor;
CNTF	Cilliary-derived Neurotrophic factor;
CNTFR	CNTF Receptor;
GDNF	Glial cell line-derived Neurotrophic factor;
GFL's	GDNF family ligands;
NRTN	Neurturin;
ARTN	Artemin;
PSPN	Persephin;
GFRα	GDNF-family-Receptor-α;
RET	Receptor Tyrosine Kinase;
NCAM	Neuronal Cell Adhesion Molecule;
BDNF	Brain derived Neurotrophic factor;
	neurotrophin NT-3 and NT-4 (known as NT-4/5 or NT-5);
TrkA,B,C	Trk family of tyrosine protein kinases;
	$p75^{NTR}$ p75 neurotrophin receptor;

(Continued)

TNF	Tumor necrosis factor;
	TNFR1/2 TNF receptor type 1/2;
AD	Alzheimer's disease;
MS	Multiple Sclerosis;
tSCs	terminal Schwann cells;
ALS	Amyotrophic Lateral Sclerosis;
EGF	Epidermal Growth Factor;
TCRs	T-cell Receptors;
MHC	Major Histocompatibility Complex;
GPI-AP	GlycosylPhosphatidylinositol-anchored protein;
PLCγ	PhosphoInositide-specific phospholipase C γ;
CA	Cortical actin;
MT	Microtubule;
RBP	RNA binding protein;
mTOR	mammalian Target of Rapamycin;
PI3K	PhosphoInositol 3-Kinase;
ZBP1	Zipcode binding protein 1;
miRNAs	MicroRNAs;
SG	Stress granules;
CoxIV	Cytochrome c Oxidase;
MVB	Multi vesicular body;
Aβ	beta-Amyloid;
NMJ	Neuromuscular Junction;
MN	Motor neuron;
ACh	Acetylcholine;
BL	Basal Lamina;
AChE	Acetylcholinesterase;
SODD	Silencer of Death Domain;
TRADD	TNFR1 associated death domain protein;
FADD	Fas-Associated protein with Death Domain.

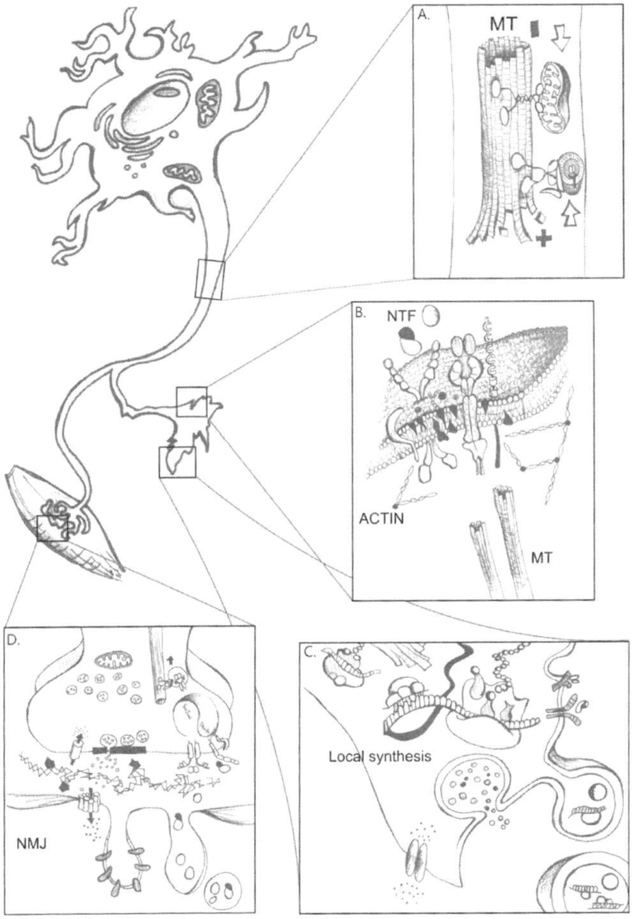

Figure 1 The molecular communication mechanisms of the neuron. (A) Axonal transport is the process of moving factors to (retrograde) and from (anterograde) the neuron cell body. The molecular motor kinesin drives transport from the cell body to the nerve periphery, supplying proteins, lipids, and other essential materials. The dynein/dynactin complex drives retrograde transport, moving damaged proteins for degradation and crucial signaling molecules such as neurotrophic factors to the cell body. (B) Spatiotemporal localization of neuronal receptors at the plasma membrane control signal modulation. (C) Localization of specific mRNAs to distinct intracellular compartments and local synthesis provides a spatiotemporal mechanism to response to extracellular signals. (D) The NMJ between MN and the muscle is an example for an extracellular communication between cells.

References

Abe N, Almenar-Queralt A, Lillo C, Shen Z, Lozach J, Briggs SP, Williams DS, Goldstein LS and Cavalli V. 2009. Sunday driver interacts with two distinct classes of axonal organelles. *J Biol Chem*, 284(50), 34628–34639. doi: 10.1074/jbc.M109.035022 M109.035022 [pii].

Airaksinen MS and Saarma M. 2002. The GDNF family: signalling, biological functions and therapeutic value. *Nat Rev Neurosci*, 3(5), 383–394. doi: 10.1038/nrn812.

Allen SJ, Watson JJ, Shoemark DK, Barua NU and Patel NK. 2013. GDNF, NGF and BDNF as therapeutic options for neurodegeneration. *Pharmacol Ther*, 138(2), 155–175. doi: 10.1016/j.pharmthera.2013.01.004 S0163-7258(13)00016-8 [pii].

Andreassi C, Zimmermann C, Mitter R, Fusco S, De Vita S, Saiardi A and Riccio A. 2010. An NGF-responsive element targets myo-inositol monophosphatase-1 mRNA to sympathetic neuron axons. *Nat Neurosci*, 13(3), 291–301. doi: 10.1038/nn.2486.

Anton OM, Andres-Delgado L, Reglero-Real N, Batista A and Alonso MA. 2011. MAL protein controls protein sorting at the supramolecular activation cluster of human T lymphocytes. *J Immunol*, 186(11), 6345–6356. doi: 10.4049/jimmunol.1003771.

Aschrafi A, Schwechter AD, Mameza MG, Natera-Naranjo O, Gioio AE and Kaplan BB. 2008. MicroRNA-338 regulates local cytochrome c oxidase IV mRNA levels and oxidative phosphorylation in the axons of sympathetic neurons. *J Neurosci*, 28(47), 12581–12590. doi: 10.1523/JNEUROSCI.3338-08.2008.

Aumiller V and Forstemann K. 2008. Roles of microRNAs beyond development — metabolism and neural plasticity. *Biochim Biophys Acta*, 1779(11), 692–696. doi: 10.1016/j.bbagrm.2008.04.008.

Banfield MJ, Naylor RL, Robertson AG, Allen SJ, Dawbarn D and Brady RL. 2001. Specificity in Trk receptor:neurotrophin interactions: the crystal structure of TrkB-d5 in complex with neurotrophin-4/5. *Structure*, 9(12), 1191–1199.

Ben-Yaakov K, Dagan SY, Segal-Ruder Y, Shalem O, Vuppalanchi D, Willis DE, Yudin D, Rishal I, Rother F, Bader M, Blesch A, Pilpel Y, Twiss JL and Fainzilber M. 2012. Axonal transcription factors signal retrogradely in lesioned peripheral nerve. *EMBO J*, 31(6), 1350–1363. doi: 10.1038/emboj.2011.494.

Bespalov MM, Sidorova YA, Tumova S, Ahonen-Bishopp A, Magalhaes AC, Kulesskiy E, Paveliev M, Rivera C, Rauvala H and Saarma M. 2011. Heparan sulfate proteoglycan syndecan-3 is a novel receptor for GDNF, neurturin, and artemin. *J Cell Biol*, 192(1), 153–169. doi: 10.1083/jcb.201009136.

Bethani I, Skanland SS, Dikic I and Acker-Palmer A. 2010. Spatial organization of transmembrane receptor signalling. *EMBO J*, 29(16), 2677–2688. doi: 10.1038/emboj.2010.175.

Bicker S and Schratt G. 2008. microRNAs: tiny regulators of synapse function in development and disease. *J Cell Mol Med*, 12(5A), 1466–1476. doi: 10.1111/j.1582-4934.2008.00400.x.

Blalock EM, Geddes JW, Chen KC, Porter NM, Markesbery WR and Landfield PW. 2004. Incipient Alzheimer's disease: microarray correlation analyses reveal major transcriptional and tumor suppressor responses. *Proc Natl Acad Sci USA*, 101(7), 2173–2178.

Campbell DS and Holt CE. 2001. Chemotropic responses of retinal growth cones mediated by rapid local protein synthesis and degradation. *Neuron*, 32(6), 1013–1026.

Cohen S, Levi-Montalcini R and Hamburger V. 1954. A Nerve Growth-Stimulating Factor Isolated from Sarcom as 37 and 180. *Proc Natl Acad Sci USA*, 40(10), 1014–1018.

Cosker KE, Courchesne SL and Segal RA. 2008. Action in the axon: generation and transport of signaling endosomes. *Curr Opin Neurobiol*, 18(3), 270–275. doi: 10.1016/j.conb.2008.08.005.

Cox LJ, Hengst U, Gurskaya NG, Lukyanov KA and Jaffrey SR. 2008. Intra-axonal translation and retrograde trafficking of CREB promotes neuronal survival. *Nat Cell Biol*, 10(2), 149–159. doi: 10.1038/ncb1677.

Dajas-Bailador F, Bonev B, Garcez P, Stanley P, Guillemot F, Papalopulu N. 2012. microRNA-9 regulates axon extension and branching by targeting Map1b in mouse cortical neurons. *Nat Neurosci*, 15, 697–699. doi: 10.1038/nn.3082.

De Winter F, Oudega M, Lankhorst AJ, Hamers FP, Blits B, Ruitenberg MJ, Pasterkamp RJ, Gispen WH and Verhaagen J. 2002. Injury-induced class 3 semaphorin expression in the rat spinal cord. *Exp Neurol*, 175(1), 61–75.

De Winter F, Vo T, Stam FJ, Wisman LA, Bar PR, Niclou SP, van Muiswinkel FL and Verhaagen J. 2006. The expression of the chemorepellent Semaphorin 3A is selectively induced in terminal Schwann cells of a subset of neuromuscular synapses that display limited anatomical plasticity and enhanced vulnerability in motor neuron disease. *Mol Cell Neurosci*, 32(1–2), 102–117.

Delanoue R and Davis I. 2005. Dynein anchors its mRNA cargo after apical transport in the Drosophila blastoderm embryo. *Cell*, 122(1), 97–106. doi: 10.1016/j.cell.2005.04.033.

Donnelly CJ, Fainzilber M and Twiss JL. 2010. Subcellular communication through RNA transport and localized protein synthesis. *Traffic*, 11(12), 1498–1505. doi: 10.1111/j.1600-0854.2010.01118.x.

Dudanova I and Klein R. 2013. Integration of guidance cues: parallel signaling and crosstalk. *Trends Neurosci*, 36(5), 295–304. doi: 10.1016/j.tins.2013.01.007 S0166-2236(13)00019-2 [pii].

Farooqui T, Franklin T, Pearl DK and Yates AJ. 1997. Ganglioside GM1 enhances induction by nerve growth factor of a putative dimer of TrkA. *J Neurochem*, 68(6), 2348–2355.

Fujita H, Zhang B, Sato K, Tanaka J and Sakanaka M. 2001. Expressions of neuropilin-1, neuropilin-2 and semaphorin 3A mRNA in the rat brain after middle cerebral artery occlusion. *Brain Res*, 914(1–2), 1–14.

Gil C, Cubi R and Aguilera J. 2007. Shedding of the p75NTR neurotrophin receptor is modulated by lipid rafts. *FEBS Lett*, 581(9), 1851–1858. doi: 10.1016/j.febslet.2007.03.080.

Good PF, Alapat D, Hsu A, Chu C, Perl D, Wen X, Burstein DE and Kohtz DS. 2004. A role for semaphorin 3A signaling in the degeneration of hippocampal neurons during Alzheimer's disease. *J Neurochem*, 91(3), 716–736.

Gowrishankar K, Ghosh S, Saha S, Mayor RCS and Rao M. 2012. Active remodeling of cortical actin regulates spatiotemporal organization of cell surface molecules. *Cell*, 149(6), 1353–1367. doi: 10.1016/j.cell.2012.05.008.

Grounds MD, Radley HG, Gebski BL, Bogoyevitch MA and Shavlakadze T. 2008. Implications of cross-talk between tumour necrosis factor and insulin-like growth factor-1 signalling in skeletal muscle. *Clin Exp Pharmacol Physiol*, 35(7), 846–851. doi: 10.1111/j.1440-1681.2007.04868.x.

Hall ZW and Sanes JR. 1993. Synaptic structure and development: the neuromuscular junction. *Cell*, 72, 99–121.

Hanz S, Perlson E, Willis D, Zheng JQ, Massarwa R, Huerta JJ, Koltzenburg M, Kohler M, van-Minnen J, Twiss JL and Fainzilber M. 2003. Axoplasmic importins enable retrograde injury signaling in lesioned nerve. *Neuron*, 40(6), 1095–1104. doi: S0896627303007700 [pii].

Harrington AW and Ginty DD. 2013. Long-distance retrograde neurotrophic factor signalling in neurons. *Nat Rev Neurosci*, 14(3), 177–187. doi: 10.1038/nrn3253.

Harrington AW, Leiner B, Blechschmitt C, Arevalo JC, Lee R, Morl K, Meyer M, Hempstead BL, Yoon SO and Giehl KM. 2004. Secreted proNGF is a pathophysiological death-inducing ligand after adult CNS injury. *Proc Natl Acad Sci USA*, 101(16), 6226–6230. doi: 10.1073/pnas.0305755101.

Harrington AW, St Hillaire C, Zweifel LS, Glebova NO, Philippidou P, Halegoua S and Ginty DD. 2011. Recruitment of actin modifiers to TrkA endosomes governs retrograde NGF signaling and survival. *Cell*, 146(3), 421–434. doi: 10.1016/j.cell.2011.07.008.

Heerssen HM and Segal RA. 2002. Location, location, location: a spatial view of neurotrophin signal transduction. *Trends Neurosci*, 25(3), 160–165.

Henderson CE, *et al*. 1994. GDNF: a potent survival factor for motoneurons present in peripheral nerve and muscle. *Science*, 266(5187), 1062–1064.

Hendricks AG, Lazarus JE, Perlson E, Gardner MK, Odde DJ, Goldman YE and Holzbaur EL. 2012. Dynein tethers and stabilizes dynamic microtubule plus ends. *Curr Biol*, 22(7), 632–637. doi: 10.1016/j.cub.2012.02.023.

Hengst U, Cox LJ, Macosko EZ and Jaffrey SR. 2006. Functional and selective RNA interference in developing axons and growth cones. *J Neurosci*, 26(21), 5727–5732. doi: 10.1523/JNEUROSCI.5229-05.2006.

Hibbert AP, Kramer BM, Miller FD and Kaplan DR. 2006. The localization, trafficking and retrograde transport of BDNF bound to p75NTR in sympathetic neurons. *Mol Cell Neurosci*, 32(4), 387–402. doi: 10.1016/j.mcn.2006.06.001.

Higuchi H, Yamashita T, Yoshikawa H and Tohyama M. 2003. PKA phosphorylates the p75 receptor and regulates its localization to lipid rafts. *EMBO J*, 22(8), 1790–1800. doi: 10.1093/emboj/cdg177.

Hillefors M, Gioio AE, Mameza MG and Kaplan BB. 2007. Axon viability and mitochondrial function are dependent on local protein synthesis in sympathetic neurons. *Cell Mol Neurobiol*, 27(6), 701–716. doi: 10.1007/s10571-007-9148-y.

Holt CE and Bullock SL. 2009. Subcellular mRNA localization in animal cells and why it matters. *Science*, 326(5957), 1212–1216. doi: 10.1126/science.1176488.

Hong IHK and Etherington SJ. 2011. Neuromuscular Junction. In *eLS Encyclopedia of Life Sciences*. John Wiley & Sons Ltd, Chichester.

Ibanez CF. 1994. Structure-function relationships in the neurotrophin family. *J Neurobiol*, 25(11), 1349–1361. doi: 10.1002/neu.480251104.

Ibanez CF. 2007. Message in a bottle: long-range retrograde signaling in the nervous system. *Trends Cell Biol*, 17(11), 519–528. doi: S0962-8924(07)00240-1 [pii] 10.1016/j.tcb.2007.09.003.

Je HS, Yang F, Ji Y, Nagappan G, Hempstead BL and Lu B. 2012. Role of pro-brain-derived neurotrophic factor (proBDNF) to mature BDNF conversion in activity-dependent competition at developing neuromuscular synapses. *Proc Natl Acad Sci USA*, 109(39), 15924–15929. doi: 1207767109 [pii] 10.1073/pnas.1207767109.

Jing S, Tapley P and Barbacid M. 1992. Nerve growth factor mediates signal transduction through trk homodimer receptors. *Neuron*, 9(6), 1067–1079.

Kaneko S, Iwanami A, Nakamura M, Kishino A, Kikuchi K, Shibata S, Okano HJ, Ikegami T, Moriya A, Konishi O, Nakayama C, Kumagai K, Kimura T, Sato Y, Goshima Y, Taniguchi M, Ito M, He Z, Toyama Y and Okano H. 2006. A selective Sema3A inhibitor enhances regenerative responses and functional recovery of the injured spinal cord. *Nat Med*, 12(12), 1380–1389.

Kaplan BB, Kar AN, Gioio AE and Aschrafi A. 2013. MicroRNAs in the axon and presynaptic nerve terminal. *Front Cell Neurosci*, 7, 126. doi: 10.3389/fncel.2013.00126.

Kar AN, MacGibeny MA, Gervasi NM, Gioio AE and Kaplan BB. 2013. Intra-axonal synthesis of eukaryotic translation initiation factors regulates local protein synthesis and axon growth in rat sympathetic neurons. *J Neurosci*, 33(17), 7165–7174. doi: 10.1523/JNEUROSCI.2040-12.2013.

Kim VN, Han J and Siomi MC. 2009. Biogenesis of small RNAs in animals. *Nat Rev Mol Cell Biol*, 10(2), 126–139. doi: 10.1038/nrm2632.

Kosik KS. 2006. The neuronal microRNA system. *Nat Rev Neurosci*, 7(12), 911–920. doi: 10.1038/nrn2037.

Lansbergen G, Grigoriev I, Mimori-Kiyosue Y, Ohtsuka T, Higa S, Kitajima I, Demmers J, Galjart N, Houtsmuller AB, Grosveld F and Akhmanova A. 2006. CLASPs attach microtubule plus ends to the cell cortex through a complex with LL5beta. *Dev Cell*, 11(1), 21–32. doi: 10.1016/j.devcel.2006.05.012.

Lee R, Kermani P, Teng KK and Hempstead BL. 2001. Regulation of cell survival by secreted proneurotrophins. *Science*, 294(5548), 1945–1948. doi: 10.1126/science.1065057.

Leung KM, van Horck FP, Lin AC, Allison R, Standart N and Holt CE. 2006. Asymmetrical beta-actin mRNA translation in growth cones mediates attractive turning to netrin-1. *Nat Neurosci*, 9(10), 1247–1256. doi: 10.1038/nn1775.

Levi A, Shechter Y, Neufeld EJ and Schlessinger J. 1980. Mobility, clustering, and transport of nerve growth factor in embryonal sensory cells and in a sympathetic neuronal cell line. *Proc Natl Acad Sci USA*, 77(6), 3469–3473.

Lin AC and Holt CE. 2007. Local translation and directional steering in axons. *EMBO J*, 26(16), 3729–3736. doi: 10.1038/sj.emboj.7601808.

Lingwood D and Simons K. 2010. Lipid rafts as a membrane-organizing principle. *Science*, 327(5961), 46–50. doi: 10.1126/science.1174621.

Lonka-Nevalaita L, Lume M, Leppanen S, Jokitalo E, Peranen J and Saarma M. 2010. Characterization of the intracellular localization, processing, and secretion of two glial cell line-derived neurotrophic factor splice isoforms. *J Neurosci*, 30(34), 11403–11413. doi: 10.1523/JNEUROSCI.5888-09.2010 30/34/11403 [pii].

Majed HH, Chandran S, Niclou SP, Nicholas RS, Wilkins A, Wing MG, Rhodes KE, Spillantini MG and Compston A. 2006. A novel role for Sema3A in neuroprotection from injury mediated by activated microglia. *J Neurosci*, 26(6), 1730–1738.

Marchetti L, Callegari A, Luin S, Signore G, Viegi A, Beltram F and Cattaneo A. 2013. Ligand signature in the membrane dynamics of single TrkA receptor molecules. *J Cell Sci*, 126(Pt 19), 4445–4456. doi: 10.1242/jcs.129916.

Meier T and Wallace BG. 1998. Formation of the neuromuscular junction: molecules and mechanisms. *Bioessays*, 20(10), 819–829. doi: 10.1002/(SICI)1521-1878(199810)20:10<819::AID-BIES7>3.0.CO;2-N.

Michaelevski I, Segal-Ruder Y, Rozenbaum M, Medzihradszky KF, Shalem O, Coppola G, Horn-Saban S, Ben-Yaakov K, Dagan SY, Rishal I, Geschwind DH, Pilpel Y, Burlingame AL and Fainzilber M. 2010. Signaling to transcription networks in the neuronal retrograde injury response. *Sci Signal*, 3(130), ra53. doi: 10.1126/scisignal.20009523/130/ra53 [pii].

Mitchell DJ, Blasier KR, Jeffery ED, Ross MW, Pullikuth AK, Suo D, Park J, Smiley WR, Lo KW, Shabanowitz J, Deppmann CD, Trinidad JC, Hunt DF, Catling AD and Pfister KK. 2012. Trk activation of the ERK1/2 kinase pathway stimulates intermediate chain phosphorylation and recruits cytoplasmic dynein to signaling endosomes for retrograde axonal transport. *J Neurosci*, 32(44), 15495–15510. doi: 10.1523/JNEUROSCI.5599-11.2012.

Mittelbrunn M, Gutierrez-Vazquez C, Villarroya-Beltri C, Gonzalez S., Sanchez-Cabo F, Gonzalez MA, Bernad A and Sanchez-Madrid F.

2011. Unidirectional transfer of microRNA-loaded exosomes from T cells to antigen-presenting cells. *Nat Commun*, 2, 282. doi: 10.1038/ncomms1285.

Morel L, Regan M, Higashimori H, Ng SK, Esau C, Vidensky S, Rothstein J and Yang Y. 2013. Neuronal exosomal miRNA-dependent translational regulation of astroglial glutamate transporter GLT1. *J Biol Chem*, 288(10), 7105–7116. doi: 10.1074/jbc.M112.410944.

Moughamian AJ, Osborn GE, Lazarus JE, Maday S and Holzbaur EL. 2013. Ordered recruitment of dynactin to the microtubule plus-end is required for efficient initiation of retrograde axonal transport. *J Neurosci*, 33(32), 13190–13203. doi: 10.1523/JNEUROSCI.0935-13.2013.

Nakamura F, Kalb RG and Strittmatter SM. 2000. Molecular basis of semaphorin-mediated axon guidance. *Journal of neurobiology*, 44(2), 219–229.

Niclou SP, Franssen EH, Ehlert EM, Taniguchi M and Verhaagen J. 2003. Meningeal cell-derived semaphorin 3A inhibits neurite outgrowth. *Mol Cell Neurosci*, 24(4), 902–912.

Oppenheim RW. 1996. Neurotrophic survival molecules for motoneurons: an embarrassment of riches. *Neuron*, 17(2), 195–197.

Pasterkamp RJ, Giger RJ, Ruitenberg MJ, Holtmaat AJ, De Wit J, De Winter F and Verhaagen J. 1999. Expression of the gene encoding the chemorepellent semaphorin III is induced in the fibroblast component of neural scar tissue formed following injuries of adult but not neonatal CNS. *Mol Cell Neurosci*, 13(2), 143–166.

Pedersen BK. 2013. Muscle as a secretory organ. *Compr Physiol*, 3(3), 1337–1362. doi: 10.1002/cphy.c120033.

Pereira DB and Chao MV. 2007. The tyrosine kinase Fyn determines the localization of TrkB receptors in lipid rafts. *J Neurosci*, 27(18), 4859–4869. doi: 10.1523/JNEUROSCI.4587-06.2007.

Perlson E, Hanz S, Ben-Yaakov K, Segal-Ruder Y, Seger R and Fainzilber M. 2005. Vimentin-dependent spatial translocation of an activated MAP kinase in injured nerve. *Neuron*, 45(5), 715–726. doi: S0896-6273(05)00055-3 [pii] 10.1016/j.neuron.2005.01.023.

Perlson E, Hendricks AG, Lazarus JE, Ben-Yaakov K, Gradus T, Tokito M and Holzbaur EL. 2013. Dynein Interacts with the Neural Cell Adhesion Molecule (NCAM180) to Tether Dynamic Microtubules and Maintain Synaptic Density in Cortical Neurons. *J Biol Chem*, 288(39), 27812–27824. doi: 10.1074/jbc.M113.465088.

Perlson E, Jeong GB, Ross JL, Dixit R, Wallace KE, Kalb RG and Holzbaur EL. 2009. A switch in retrograde signaling from survival to stress in rapid-onset neurodegeneration. *J Neurosci*, 29(31), 9903–9917. doi: 29/31/9903 [pii] 10.1523/JNEUROSCI.0813-09.2009.

Perlson E, Maday S, Fu MM, Moughamian AJ and Holzbaur EL. 2010. Retrograde axonal transport: pathways to cell death? *Trends Neurosci.* doi: S0166–2236(10)00052-4 [pii] 10.1016/j.tins.2010.03.006.

Piper M, Anderson R, Dwivedy A, Weinl C, van Horck F, Leung KM, Cogill E, and Holt C. 2006. Signaling mechanisms underlying Slit2-induced collapse of Xenopus retinal growth cones. *Neuron*, 49(2), 215–228. doi: 10.1016/j.neuron.2005.12.008.

Ramachandran S and Palanisamy V. 2012. Horizontal transfer of RNAs: exosomes as mediators of intercellular communication. *Wiley Interdiscip Rev RNA*, 3(2), 286–293. doi: 10.1002/wrna.115.

Robertson J, Beaulieu JM, Doroudchi MM, Durham HD, Julien JP and Mushynski WE. 2001. Apoptotic death of neurons exhibiting peripherin aggregates is mediated by the proinflammatory cytokine tumor necrosis factor-alpha. *J Cell Biol*, 155(2), 217–226. doi: 10.1083/jcb.200107058.

Russell FD, Koishi K, Jiang Y and McLennan IS. 2000. Anterograde axonal transport of glial cell line-derived neurotrophic factor and its receptors in rat hypoglossal nerve. *Neuroscience*, 97(3), 575–580.

Salinas S, Bilsland LG and Schiavo G. 2008. Molecular landmarks along the axonal route: axonal transport in health and disease. *Curr Opin Cell Biol*, 20(4), 445–53. doi: S0955-0674(08)00073-2 [pii] 10.1016/j.ceb.2008.04.002.

Sanes JR. 1995. The synaptic cleft of the neuromuscular junction. *Seminars in Developmental Biology*, 6, 163–173.

Schratt GM, Tuebing F, Nigh EA, Kane CG, Sabatini ME, Kiebler M and Greenberg ME. 2006. A brain-specific microRNA regulates dendritic spine development. *Nature*, 439(7074), 283–289. doi: 10.1038/nature04367.

Sendtner M, Pei G, Beck M, Schweizer U and Wiese S. 2000. Developmental motoneuron cell death and neurotrophic factors. *Cell Tissue Res*, 301(1), 71–84.

Senger DL and Campenot RB. 1997. Rapid retrograde tyrosine phosphorylation of trkA and other proteins in rat sympathetic neurons in compartmented cultures. *J Cell Biol*, 138(2), 411–421.

Shen J and Maruyama IN. 2011. Nerve growth factor receptor TrkA exists as a preformed, yet inactive, dimer in living cells. *FEBS Lett*, 585, 295–299.

Shen J and Maruyama IN. 2012. Brain-derived neurotrophic factor receptor TrkB exists as a preformed dimer in living cells. *J Mol Signal*, 7(1), 2. doi: 10.1186/1750-2187-7-2.

Shi L, Fu AK and Ip NY. 2012. Molecular mechanisms underlying maturation and maintenance of the vertebrate neuromuscular junction. *Trends Neurosci*, 35(7), 441–453. doi: 10.1016/j.tins.2012.04.005.

Shibata SC, Hibino K, Mashimo T, Yanagida T and Sako Y. 2006. Formation of signal transduction complexes during immobile phase of NGFR movements. *Biochem Biophys Res Commun*, 342(1), 316–322. doi: 10.1016/j.bbrc.2006.01.126.

Shirvan A, Ziv I, Fleminger G, Shina R, He Z, Brudo I, Melamed E and Barzilai A. 1999. Semaphorins as mediators of neuronal apoptosis. *J Neurochem*, 73(3), 961–971.

Siegel G, Saba R and Schratt G. 2011. microRNAs in neurons: manifold regulatory roles at the synapse. *Curr Opin Genet Dev*, 21(4), 491–497. doi: 10.1016/j.gde.2011.04.008.

Smalheiser NR and Lugli G. 2009. microRNA regulation of synaptic plasticity. *Neuromolecular Med*, 11(3), 133–140. doi: 10.1007/s12017-009-8065-2.

Smalheiser NR. 2007. Exosomal transfer of proteins and RNAs at synapses in the nervous system. *Biol Direct*, 2, 35. doi: 10.1186/1745-6150-2-35.

Suzuki S, Numakawa T, Shimazu K, Koshimizu H, Hara T, Hatanaka H, Mei L, Lu B. and Kojima M. 2004. BDNF-induced recruitment of TrkB receptor into neuronal lipid rafts: roles in synaptic modulation. *J Cell Biol*, 167(6), 1205–1215. doi: 10.1083/jcb.200404106.

Tani T, Miyamoto Y, Fujimori KE, Taguchi T, Yanagida T, Sako Y and Harada Y. 2005. Trafficking of a ligand-receptor complex on the growth cones as an essential step for the uptake of nerve growth factor at the distal end of the axon: A single-molecule analysis. *J Neurosci*, 25(9), 2181–2191. doi: 10.1523/JNEUROSCI.4570-04.2005.

Taylor AR, Gifondorwa DJ, Newbern JM, Robinson MB, Strupe JL, Prevette D, Oppenheim RW and Milligan CE. 2007. Astrocyte and muscle-derived secreted factors differentially regulate motoneuron survival. *J Neurosci*, 27(3), 634–644. doi: 10.1523/JNEUROSCI.4947-06.2007.

Taylor AR, Gifondorwa DJ, Robinson MB, Strupe JL, Prevette D, Johnson JE, Hempstead B, Oppenheim RW and Milligan CE. 2012. Motoneuron programmed cell death in response to proBDNF. *Dev Neurobiol*, 72(5), 699–712. doi: 10.1002/dneu.20964.

Tcherkezian J, Brittis PA, Thomas F, Roux PP and Flanagan JG. 2010. Transmembrane receptor DCC associates with protein synthesis machinery and regulates translation. *Cell*, 141(4), 632–644. doi: 10.1016/j.cell.2010.04.008.

Teng HK, Teng KK, Lee R, Wright S, Tevar S, Almeida RD, Kermani P, Torkin R, Chen ZY, Lee FS, Kraemer RT, Nykjaer A and Hempstead BL. 2005. ProBDNF induces neuronal apoptosis via activation of a receptor complex of p75NTR and sortilin. *J Neurosci*, 25(22), 5455–5463. doi: 10.1523/JNEUROSCI.5123-04.2005.

Tsui CC and Pierchala BA. 2010. The differential axonal degradation of Ret accounts for cell-type-specific function of glial cell line-derived neurotrophic factor as a retrograde survival factor. *J Neurosci*, 30(15), 5149–5158. doi: 10.1523/JNEUROSCI.5246-09.2010.

Vilar M, Charalampopoulos I, Kenchappa RS, Reversi A, Klos-Applequist JM, Karaca E, Simi A, Spuch C, Choi S, Friedman WJ, Ericson J, Schiavo G, Carter BD and Ibanez CF. 2009a. Ligand-independent signaling by disulfide-crosslinked dimers of the p75 neurotrophin receptor. *J Cell Sci*, 122(Pt 18), 3351–3357. doi: 10.1242/jcs.055061.

Vilar M, Charalampopoulos I, Kenchappa RS, Simi A, Karaca E, Reversi A, Choi S, Bothwell M, Mingarro I, Friedman WJ, Schiavo G, Bastiaens PI, Verveer PJ, Carter BD and Ibanez CF. 2009b. Activation of the p75 neurotrophin receptor through conformational rearrangement of disulphide-linked receptor dimers. *Neuron*, 62(1), 72–83. doi: 10.1016/j.neuron.2009.02.020.

Viviani B, Bartesaghi S, Corsini E, Galli CL and Marinovich M. 2004. Cytokines role in neurodegenerative events. *Toxicol Lett*, 149(1–3), 85–89. doi: 10.1016/j.toxlet.2003.12.022.

Wehrman T, He X, Raab B, Dukipatti A, Blau H and Garcia KC. 2007. Structural and mechanistic insights into nerve growth factor interactions with the TrkA and p75 receptors. *Neuron*, 53(1), 25–38. doi: 10.1016/j.neuron.2006.09.034.

Welshhans K and Bassell GJ. 2011. Netrin-1-induced local beta-actin synthesis and growth cone guidance requires zipcode binding protein 1. *J Neurosci*, 31(27), 9800–9813. doi: 10.1523/JNEUROSCI.0166-11.2011.

Werle M. 2009. Neuromuscular Junction (NMJ): Postsynaptic Basal Lamina. In Malenka R, (ed.) *Intercellular Communication in the Nervous System*. Academic Press, London, UK, 174–180.

Williams A, Piaton G, Aigrot MS, Belhadi A, Theaudin M, Petermann F, Thomas JL, Zalc B and Lubetzki C. 2007. Semaphorin 3A and 3F: key players in myelin repair in multiple sclerosis? *Brain*, 130(Pt 10), 2554–2565.

Willis DE, van Niekerk EA, Sasaki Y, Mesngon M, Merianda TT, Williams GG, Kendall M, Smith DS, Bassell GJ and Twiss JL. 2007. Extracellular stimuli specifically regulate localized levels of individual neuronal mRNAs. *J Cell Biol*, 178(6), 965–980. doi: 10.1083/jcb.200703209.

Wu KY, Hengst U, Cox LJ, Macosko EZ, Jeromin A, Urquhart ER and Jaffrey SR. 2005. Local translation of RhoA regulates growth cone collapse. *Nature*, 436(7053), 1020–1024. doi: 10.1038/nature03885.

Yao J, Sasaki Y, Wen Z, Bassell GJ and Zheng JQ. 2006. An essential role for beta-actin mRNA localization and translation in Ca2+-dependent growth cone guidance. *Nat Neurosci*, 9(10), 1265–1273. doi: 10.1038/nn1773.

Yasuhara T, Shingo T, Kobayashi K, Takeuchi A, Yano A, Muraoka K, Matsui T, Miyoshi Y, Hamada H and Date I. 2004. Neuroprotective effects of vascular endothelial growth factor (VEGF) upon dopaminergic neurons in a rat model of Parkinson's disease. *Eur J Neurosci*, 19(6), 1494–1504.

Yoo S, van Niekerk EA, Merianda TT and Twiss JL. 2010. Dynamics of axonal mRNA transport and implications for peripheral nerve regeneration. *Exp Neurol*, 223(1), 19–27. doi: 10.1016/j.expneurol.2009.08.011.

Yoon JH, Kim J, Song P, Lee TG, Suh PG and Ryu SH. 2012. Secretomics for skeletal muscle cells: a discovery of novel regulators? *Adv Biol Regul*, 52(2), 340–350. doi: 10.1016/j.jbior.2012.03.001.

Zahavi EE, Lonescu A, Gluska S, Gradus T, Ben-Yaakov K and Perlson E. 2015. A compartmentalized microfluidic neuromuscular co-culture system reveals spatial aspects of GDNF functions. *J. Cell Sci*, 128, 1241–1252.

Zhang HL, Eom T, Oleynikov Y, Shenoy SM, Liebelt DA, Dictenberg JB, Singer RH and Bassell GJ. 2001. Neurotrophin-induced transport of a beta-actin mRNP complex increases beta-actin levels and stimulates growth cone motility. *Neuron*, 31(2), 261–275.

Zhou B, Cai Q, Xie Y and Sheng ZH. 2012. Snapin recruits dynein to BDNF-TrkB signaling endosomes for retrograde axonal transport and is essential for dendrite growth of cortical neurons. *Cell Rep*, 2(1), 42–51. doi: 10.1016/j.celrep.2012.06.010.

Zweifel LS, Kuruvilla R and Ginty DD. 2005. Functions and mechanisms of retrograde neurotrophin signalling. *Nat Rev Neurosci*, 6(8), 615–625. doi: 10.1038/nrn1727.

16. Ethical Methods of Investigation with *Pan/Homo* Bonobos and Chimpanzees

E. Sue Rumbaugh[*,‡], Itai Roffman[†,§], Elizabeth Pugh[†], and Duane M. Rumbaugh[*,||]

[*] *Bonobo Hope Sanctuary*
4200 Southeast 44th Av. Des Moines
IA 50320, USA

[†] *Institute of Evolution, University of Haifa, Mount Carmel*
Haifa 31095, Israel

[‡] *suerumbaugh@gmail.com*
[§] *iroffman@gmail.com*
[||] *drumbaug@aol.com*

"...(the study of) human and nonhuman primates...has also been a permanent struggle with the most significant, the most tabooed dividing line within nature; an enduring activity of drawing, policing, displacing, denying, and bridging the metaphysical, religious, and moral boundaries between humans and their closest living relatives...the history of the anthropological disciplines to a considerable degree has been an alternation of humanizing and bestializing moves with respect to both apes and humans, a persistent quest for unambiguousness and human purity" (Corby, 2005).

Given that chimpanzees, bonobos, and humans share 99% of our genome and that they have a wide variety of cultural life styles, what topics are appropriately investigated with them from an ethical standpoint? And on what basis do we continue to qualify them as animals? Beginning in the 1970s a wealth of new information appeared about apes. It suggested that we had greatly underestimated them. The recently passed Chimp Act of 2013 was to have terminated biomedical work and limited behavioral work to investigations of species-specific behaviors. Yet chimpanzees continue to be housed in federal and private facilities dedicated to cancer and other human diseases, and these facilities are now seeking to retain private ownership of these apes. What can be learned from studies of a highly cultural species housed in limited environments? We address these questions by presenting data on chimpanzees and bonobos reared in a world than enables them to acquire language, a skill than moves apes into the human realm, without changing their biology. It accomplishes this in one generation. These findings raise questions regarding the adequacy of evolutionary theory to explain the presumed human/ape differences as well as the prevailing tendency to classify apes in a genus other than Homo, thereby maintaining an man/animal distinction that is in adequately supported by anatomical data, genetic data, physiological data or behavioral data.

The passage of the 2013 Chimp Act was an attempt, on the part of congress; to reflect humanity's increasingly enlightened understanding of the close biological kinship between chimpanzees, bonobos, and ourselves. This increased awareness began with the early work of Robert Yerkes (Yerkes and Yerkes, 1929) and culminated in the tireless political efforts of Jane Goodall and the Goodall Foundation across the past three decades. In passing this legislation, Congress mirrored the growing public desire for environments commensurate with the chimpanzee's potential to become self-aware, capable of symbolic thought, and to construct mental worlds, social rules, and cultures (Gardner and Gardner, 1969; Povinelli, Eddy, Hobson and Tomasell, 1996; Premack and Premack, 1976; Savage-Rumbaugh, Murphy, Sevcik, Brakke, Williams, and Rumbaugh, 1993; Rumbaugh, 1971; Rumbaugh, 1974; Wrangham, McGrew, DeWaal, and Heltne, 1994; Tuttle, 2014). The goal was also to terminate their use as biomedical models.[1] Did this law accomplish its goals and where shall future research take us, if we abide by its dictums to limit our understanding of apes to their species-specific behaviors? Is it time to end all ape research in laboratories and zoos, and to limit ape research to field settings, where variables are difficult, it is not impossible, to control? Is it propitious to ask what, if anything, should we attempt to learn from apes that would merit their continued restriction to captive habitats? Our current zoological and captive ape habitats are devoid of choice, freedom of expression, the execution of freewill, and the right to self-determination. They may have rocks and vines and more space, but they continue to serve first and foremost as cages.

We argue herein that much remains to be learned from comparative efforts and cultural studies; however a fundamental shift in ape containment strategies and human/ape relationships must take place. This shift should reflect our increased awareness of who we are, in a genetic sense, and who apes are in the same manner. Were it not for the strange anomaly of our having one chromosome, that is a combination of two chimpanzee or bonobo chromosomes; according the Darwinian theory of evolution, we would not exist at all. There would only be chimpanzees and bonobos. What does this mean at a behavioral level?

1. Ape Mind/Animal Mind

To inform this question will require empirical experiments and subjective experiences that probe the human/ape boundary in new

[1]There is currently an amendment in the US House of Representative to keep chimpanzees in biomedical research in the United States until 2023, and potentially beyond, at the discretion of the Comptroller General's office. The similarities between human beings and chimpanzees at the biological level are sufficient to treat them as an alternative form of human being. They make a perfect animal model for us and eliminate the need to place human beings at risk. But what makes them animals and ourselves, human? The important data for determining whether they should be members of *Homo*, cannot be genetic data, for this data already indicates that they should be classified as *Homo*. It is the behavioral data to which we must now turn.

ways. Thus we begin our paper with an account of subjective experience provided by the second author. For the reader unfamiliar with direct ape interaction, the subjective experience of another being is the only open door into the mind of an ape.

Our human minds are lodged in a man/animal dichotomy. The dichotomy itself may be valid but the genetic data tell us that apes belong on the human side of the equation. The behavioral data have been telling us the same thing, but we have strongly resisted it (Taylor, 2009). Any mention of symbolic behavior by a non-human provokes disbelief among many people unless they are watching the *"Planet of the Apes,"* and have decided to suspend their man/animal category for the sake of the movie. Intentional communicative acts by apes are not viewed as meaningful, in the same way that they would be instantly interpreted as meaningful, if emitted by a person. Yet it has been demonstrated that chimpanzees *can* communicate novel and appropriate information to each other, they *can do so with intent* and they *can display* sensitivity to the knowledge state of the recipient (Greenfield and Savage-Rumbaugh, 1984; Savage-Rumbaugh, Rumbaugh, and Boysen, 1978a; Savage-Rumbaugh, 1986). None-the-less much effort continues to be expended by the scientists to determine if chimpanzees know what others know and do not know (Tomasello *et al.*, 2003a). This is skill that is prerequisite to language proper, but not to communication *per se*.[2] Arguments about whether or not chimpanzees are capable of thought and symbolic processing also continue. Different researchers have not reached agreement simply because apes reared in different ways

[2]Language is employed for ritualistic communicative purposes (Hello, how are you, etc.) as is all animal communication. Such communications can serve true social functions and promote social cohesion and social synchrony regardless of whether they are relevant to the state of knowledge of the recipient. Language steps beyond this level of co-ordination and allows the user to pass information to the recipient to which the recipient otherwise has no access. Proficient use of language is always linked to knowledge of the state of the recipient. That is, we do not generally spend of good deal of time telling another person something we presume that they know. To do so is socially inappropriate.

display radically different cognitive competencies. Some environments allow human-like cognitive capacities to flourish and others do not. Human-like cognitive and symbolic skills make possible an entirely different manner of relating, one that is highly recursive, self-reflective, linguistic, and morally responsive to utterances. Standard primate laboratory and zoo environments do not foster these abilities. But it is not only that environments differ, the criteria for determining if a given communication is symbolic, meaningful, and intentional shift depending upon whether the speaker is a person on an ape.

Scientists are required to collect data to show that the acts of apes are meaningful to an observer. However, what is symbolically meaningful if observed in a human may be labeled as conditioned response if observed in a non-human being. In fact, it has asserted that *symbolic meaning can exist only in human beings* (Deacon, 1998; Heyes, 1998; Povinelli, 2000). It is widely thought that the behaviors of non-humans must be explained according to principles of conditioning, not those of communicative intent. Communicative intent is a special class of behavior that it highly symbolic and generally reserved for humans (Penn and Povinelli, 2008). Symbolic communication skills are not ascribed even to highly trained companion dogs (Dennet, 1996; Penn and Povinelli, 2009). Dogs may respond to symbols and retrieve hundreds of items in response to human requests or execute hundreds of action-based maneuvers in response to human commands. Yet these abilities do not require that a dog understand the use of symbolic communication as being about something other than the act itself.

For example, a dog may scratch at the door to get your attention, or bring its leash to you. Certainly the dog desires to go out. The dog also knows that you are a being who can fulfill this desire. Animal behaviorists explain such behavior as arising from the dog's *desire* to go out. The *desire* initiates actions on the part of the dog that are responded to by you as intentional symbolic communications, because you are a symbolic being. However, what the dog has learned are behaviors that lead to the fulfillment of its desires; not the representational parameters of symbols *per se*.

The dog may also learn associative connections between words and objects but these connections are not symbols.

It is not the case that the dog intends to communicate something about its internal state to you because it realizes you do not know. It is rather that the dog has learned which behaviors will cause you to act as it desires. Of course we human beings often use language to get others to respond in particular ways as well. In a similar manner, a computer can acquire behaviors that get people to do things for them, such as service them. However this does not mean that when the computer tells you to do something, that it *knows* about itself in the way that you *know* about your internal self. It also does not mean that the computer *knows* that you are symbolic being and that it is communicating with you on a symbolic level, even though it appears to do so. The computer's commands do not mean that some internal being wishes to convey its internal mental states or wishes and desires to you. Quite the contrary, some part of a program has assessed the function of some part of another program and the computer has been programmed to pass this information along to you. In essence, it is passing you a string of numbers or words that should have meaning to you, but that have no meaning to the computer *per se*, because the computer is not a conscious sentient being. You are. But you can be easily fooled into thinking that the computer is aware — in the same sense that you are aware. Similarly, it is possible for a dog to convince you that it is aware that you have a state of mind. When this happens it is a misattribution of human mental competencies to the dog, because there are different fundamental levels of knowing. The computer knows only how to translate information from one form to another, the dog knows only how to maximize the chances of fulfilling its desire, but you know how to represent you state of mind and the states of other minds symbolically and linguistically (Penn and Povinelli, 2009).

Therefore your dog simply could not tell you about his/her state of mind unless it believed/knew that:

(a) You have a state of mind,
(b) It has a state of mind,

(c) That it needs to inform you of his state of mind in order to affect your state of mind.

Such concepts have has been referred to as theory of mind (ToM), behavior (Premack, 1976). Tests for the existence of TOM have been given to dogs and they fail. Dog have minds, wishes and desires but they lack theories about their minds. They also lack theories about your mind. Children who lack language also fail TOM tests. Your computer can pass these tests, but only if a person who already has a highly developed TOM of their own, programs the answers into its data bank. Such a person cannot program this knowledge into a dog or into a young child. Experience can program this knowledge into a child but not into a dog. And given the right kind of cultural environment and loving stable familial setting, experience can program this knowledge into an ape, and presumably the same would hold for any sentient being (Savage-Rumbaugh and Fields, 1998).[3]

If your dog is not thinking about your mind and what you know or do not know about his desires, then your dog is not communicating symbolically. Nonetheless, he is communicating and he does want to go out. You are capable of thinking about what the dog may or may not want regardless of how the dog is acting. You are capable of making a separation between the mind of the dog and the behavior of the dog. You know that your behavior does not always reflect your state of mind and you may assume that the dog's also does not always reflect his true state of mind. However your ability to think about the dog's mind is not to be confused with the dog's ability to think about your mind. As far as scientists have determined, dogs do not think about your state of mind, apart from how you depict it, through you immediate actions. That is to say they don't think about how you felt yesterday, or how you will

[3]Sentient beings are those who are conscious of being conscious and therefore able to think recursively and reflexively. This allows for the creation of the conscious self and a personal narrative, through culturally endowed symbolic processes. Sentience requires a period of social development in a languaged world, but it does not mandate human language.

feel tomorrow, or why you feel the way you feel, they just think about how you feel at the moment. Indeed, they need not be thinking at all, but rather co-feeling, co-sensing and co-experiencing with you (Dotta *et al.*, 2011).

Dogs, like monkeys, and most animals fail tests of mirror recognition even though they are constantly around mirrors. When a dog looks in a mirror for the first time it may behave as though it has seen another dog, just as most monkeys do. It may try to interact with the dog in the mirror, as it would any other dog. But the dog in the mirror does not react as a normal dog; instead it does the same things your dog does — at precisely the same time your dog does these things. There is no give and take, no action and response. Eventually your dog gives up and stops attempting to meet the dog in the mirror because it does not react appropriately. Not knowing what else to do, it simply ignores the dog in the mirror. Most animals behave in a similar way. This has greatly puzzled scientists, especially when it can be shown that monkeys learn that mirrors are reflective. Thus they can employ mirrors to see what is behind them, but cannot employ mirrors to see themselves. That is to say, then know — in principle — how mirrors work, but they don't know that the mirror is presenting them with a picture of themselves (de Waal, Freeman and Hall, 2005). This lack of understanding would *represent* a profound neurological deficit — if encountered in a human being. There are instances of neurological deficits that result in a patient's knowing only the right of left side of the body, or their arms or their legs. In each case the patient can see that the unknown limb is attached to them, but the patient does not recognize the limb as part of the "self." They feel as though the limb belongs to someone else (Ramachandran, 2009, 2010). However even these patients retain a sense of the self. They know who they are and they know that what they see in a mirror is a reflection of their self-hood. Like you, they know that the being in the mirror is not reacting to them, and that it is not going to react to them, because it is them. Therefore they (and you) employ the mirror to brush your teeth, comb your hair and to generally engage in other behaviors that are motivated by your desire to look a certain way to others. Your dog may wish to make his or herself appear a certain way, such as vicious or friendly, but if so, you dog will not use a mirror to effect or to practice or enhance his look. You do use mirrors in the way and chimpanzees can use mirrors in this way. They can also recognize themselves on television and distinguish between taped and real-time videos of themselves (Savage-Rumbaugh, 1986).

An entity that cannot see itself as others see them lives in a distinctly different world from me, from you, and from that of an ape. The thought that others might not perceive their intentions and/or actions in the same manner that themselves perceive their own actions, actions, desires and beliefs does not arise in a non-sentient being. Therefore non-sentient beings manifest no desires to tell others things apart from their behaviors of the moment. They are time locked to their desires of the moment. They may expect others to do things for or with them, but that is not the same as telling others things. Telling others has a degree of *aboutness* to it, that is telling entails a desire to change the information of the mind-state or belief state of the listener and a belief that such is possible through the deployment of symbols. To desire to do this, one must be able to differentiate the mind of the self from the mind of the other. To desire to do this, one must also be able to lift oneself out of the present and to contemplate the self as extant in the past and as extant in the future. One must become able to think about the self outside of time, even though the self is bound by time in the present moment. Thus one must be aware that one exists in some sense, apart from that which one is manifesting in the moment, and thus a self-sentient being.[4]

[4] This is generally referred to as TOM (Premack, 2003). Simply put, this is the idea that you have a mind, that others have minds which differ from your mind and that you have no privileged access to the contents of their minds except by: (a) assuming that you reason and feel in similar ways or (b) through the vehicle of language. Of course the more recent discovery of mirror neurons now provides a third vector, that of similar simultaneous brain activation through neurons specialized for that purpose. It has been shown that monkeys possess mirror neurons and therefore even though they cannot recognize themselves in a mirror and

Some animals have used mirrors to locate objects and even to locate red dots on their bodies. But only apes and humans have employed mirrors to put on lipstick, to help pull a loose tooth, to look down their throat with a flashlight, or to inspect a cut on their face. Only apes and humans recognize themselves on television and know that they are not only seeing an image of themselves but a past image of themselves as they were at another time (Savage-Rumbaugh, 1986). The red-dot test is a test of how mirrors act as light reflective devices; it is not a true test of self-recognition. Its use with a wide array of animals has misled psychologists to declare the existence of similar abilities between apes and other mammals and even birds. Appropriate mirror-image studies are those that ask whether organisms employ mirrors to study their self-reflection in order to modify or improve it.[5] Only apes and humans have done so and the more advanced the language skills of either species, the more advanced the use of the mirror becomes. There is no evidence at that birds, monkeys, elephants, dolphins, dogs,

---

appear to lack an independent self-concept, they may (just as dogs may) — know what is in your mind through the action of mirror neurons (Rizzolatti and Fabrri-Destro, 2010).

[5]Simply looking at the dot or touching it may reveal either knowledge of how mirrors work or a low level of self-recognition, such that something *on* the self evokes concern much as an injury would evoke concern. This is to be considered distinct from a knowledge of how the self appears to others. A red patch on the self may signal injury, but placing lipstick on one's mouth and a fur shawl around one's shoulders — while looking in the mirror — reveals that such a being knows that the perspective of others is not the same as their own. Thus a mirror is required if they want to know not only how lipstick feels or tastes, but also how it looks. The ability to think at this level of self-objectification signals a new level of awareness. It means that one knows that the *perceptions of others* can be manipulated by ones actions. Manipulation of the perceptions of others is quite different from attempting to alter the actions of others in a way that one desires. The perceptions of others can exist only as inference in the mind. It is not possible to experience the perceptions of others unless your mirror neurons are synchronized with theirs. In sum, there are different ways of knowing the other. Some methods require greater intimacy and others required greater inferential capacities. As far as have discerned to date, apes are the only sentient beings who know others in a manner similar to what we, as humans, experience.

horses, pigs, lessor apes, or any living beings beyond humans and apes, employ mirrors to enhance their self-image (de Waal *et al.*, 2005). The closest evidence is that elephants like apes, initially try to reach behind the mirror to locate the other individual. It has also been reported that they put their trunks into their open mouths while observing the image, one elephant bent its ear once while oriented toward a mirror (Plotnik *et al.*, 2006). These examples do not provide clear evidence for sentient deployment of mirrors.

We suspect that the difference between humans and animals is as large as those who favor the Cartesian Divide would claim, especially if we look at the data on mirror understanding, tool manufacturing and language for non-apes. However, given the behavior of linguistically competent apes, we have to ask, might we have misclassified apes as animals? The genetic data would confirm that we have done so. The behavioral data also confirm that apes belong on our side of the divide, because they share with us a very special kind of consciousness. Like us, they are conscious of being conscious. This peculiar state of affairs allows for the emergence of the reflexiveness of consciousness. It is this which produces the concept of the self and which allows for the reflexive qualities of language that make language such a powerful tool.[6] The absence of this capacity does not mean that other animals lack extraordinary intelligence or that they fail to understand mirrors, tools and, complex communication. There is considerable data to suggest that they do all of these things, but not in the way that humans and apes with language do them. They deploy intelligence but lack the sentience of selfhood, the running narrative of self-hood, the concept of moral action, and the understanding of the TOM. These things matter. To emerge in humans and apes they require a socio-cultural world that cedes them importance and that impregnates them into babies (Segerdahl *et al.*, 2005; Savage-Rumbaugh and Fields, 2011).

---

[6] Many have argued that whales and dolphins possess knowledge of the other, a TOM and a sense of self. It has even been suggested that they are far more intelligent that we are but in a different way (Fields, 2008). Certainly the whale's behavior of beaching themselves in groups suggests that something is going on at a group level in these beings that is different from ourselves.

When these are absent, both humans and apes fall back into an animal like existence. It is their culture, plus their biology that allows a cognitive function to arise which enables reason, reflective thought, purposeful action and planning to emerge. This emergence is ever on the verge of vanishing if not passed successfully to the next generation (Lock, 1980; Savage-Rumbaugh and Fields, 2013). Thus while humanity may begin to overlook cultural differences with the world-wide sharing of information now underway — we must not take the deep culture that allows for the emergence of thought and symbolic processes lightly, *as we depend upon its recreation and reinvention anew each generation in ways of which we are barely cognizant* (Lock, 1980).

2. Ape language/human language

In the field of ape language, it has been difficult to *measure* language behavior in a manner that all accept, because so very few researchers have linguistically competent apes. Would be critiques demand those with linguistically competent apes justify the very utterances which they are recording, as actually extant (Terrace, 1982; Wallman, 1992). They require this proof for each measurement, since each utterance is different. But it is not possible to measure something inside the ape's head to prove that its brain holds language during every utterance, though attempts have been made to produce MRI images of linguistically competent apes (Rilling *et al.*, 1999). Even if it were shown that an apes' hemispheres were appropriately active during language production, this would not prove that *real* language emanated from the ape. It would only demonstrate that the ape employed parts of the brain that we associate with language in human beings.[7] Similarly one cannot measure *real* language in a human being, one must simply assume that it exists and that the utterances measured reflect

---

[7]However human beings who have lost parts of the brain typically associated with language can still produce language, so the answer does not lie in localization of functions.

something about the language itself. Assuming that something exists, in order to take measurements, is common in science and a perfectly valid method but it has been questioned when applied to ape-language.

Real language can only be validated to the degree that it is employed within constructed socio-cultural acts. These must be acts designed/intended to produce meaning. But because animals are believed to lack the capacity to mean — except in a ritualized conditioned manner — it has been unacceptable to measure this capacity in apes — unless it could first be proved that *intent to mean* existed inside the ape. The field of ape language became entrapped in loop of circularity because intent could never be seen, only inferred. To infer intent, once had to measure acts that connoted intent. How were these acts to be certified as intentional?

In order for symbolic intent to exist, one party must take the behavior of another — not only as something that *could* be meaningful — but also as something that *is intentionally being produced for the expressed purpose and intent to mean something*. It is possible to study all manner of human language — by assuming that the sender intended to mean and by recording and measuring the utterances. Thus must the demonstration of meaning arise from the usage itself; as it cannot not exist apart from context. (Dictionaries lead us to believe otherwise, but dictionaries simply define words with other words. They do little to elucidate how meanings get placed into any word.) Do apes intend that their symbols carry meaning or are they acting only to get a reward? To answer this question we must take the subjective impressions of those communicating with the apes (be they other apes or humans) as valid data.

We therefore begin by presenting the personal subjective account of the second author's experiences with his brother Orr, who did not speak till in his teens. This is followed by an account of this author's experiences with Kanzi and by a review of the data collected revealing the capacities of apes for language across the past 40 years. We then raise a number of theoretical issues that arise from this work, including the current inadequacy of

Darwinian theory to explicate the evolution of *Homo sapiens* and/or other sentient beings. The goal of this Chapter is not to challenge the *status quo* though we may seem to do so. It is rather to arouse in the reader a willingness to raise questions about many topics upon which the proverbial door has been slammed shut. There is a multiplicity of data, genetic and behavioral; to indicate that the human/ape divide deserves reconsideration. We further suggest that the fundamental data points from which scientific question arise will always be those of human experience. Though *hard* data is supposed to eliminate all bias, scientists inevitably structure the data they choose to collect and analyze through their lens of human experience. It is not possible to do otherwise. Data can easily be made to look *hard*, in a manner that hides its true underbelly of prejudice. Better that we acknowledge and challenge that which we know is subjective than to accept that which encodes the presuppositions in the categories of collection.

Sometimes our findings with apes run counter to ordinary human experience, just as the findings of quantum physics run counter to ordinary time/space experience. In the account below, this is where we begin. We raise the question, could science have made a mistake in placing apes on the animal side of the Cartesian divide? We follow this by asking whether or not our current views of human origins could be mistaken as well.

3. A Personal Account — Pan & Homo: Orr and Kanzi

Orr is a *person* who failed to acquire language until 14 of age. His language door opened following an encounter with videos of the bonobo Kanzi and the gorilla Koko. After watching these videos with his human brother, Itai Roffman, Orr's mind opened to the possibility that he could use sounds to intentionally communicate — because he had seen Kanzi doing so. The ways in which both Orr and Kanzi communicate are described in a manner that is designed to allow the reader to open the door of his own mind as well. Forget, if you will, that Orr is a human person and that Kanzi is a bonobo person. Think of both as simply cognizant sentient intelligent beings whose desire is to communicate with the second author, in the personal account below.

I was five when my brother was born. As a baby he would rock back and forth in his crib banging his head against the bars to the point where he did not develop hair on his upper forehead. His hair grew back when we got rid of the crib. He was slow in his development and did not talk intelligibly, aside from making vocalizations till he was almost four. But I remember all the exciting times we spent together, Orr was my sidekick, he would do anything I would and he followed me everywhere.

Then after he learned to play with me, by approximately four and half, he began to communicate through many non-verbal means. He had no language but he began to point to things he wanted, or showed us what he needed by doing things himself, while we observed. He developed a behavior of raising his arms up to both sides of his face while making a repetitive up and down waving gesture up with both hands while simultaneously vocalizing with a sound reminiscent of a short pant hoot. It seemed that Orr did this when he wanted to get our attention, when he wanted to be heard, or when he wanted to get us to laugh with him. Thus was Orr able, by five years of age able to make manifest his desire for joint attention, his desire to make statements to us and his desire to get us to share in his positive and constructive affective responses to certain events.

I remember the games we would play. I would turn the sofa over to make a cave and we built fortresses of sorts that we crawled into and thoroughly enjoyed. Orr got very excited at watching TV and would get up close to the TV screen moving his body back and forth as if he is joining in the action movie. At least two TV sets and a big glass table were toppled and broken by his excitement! Little did I know then that in the 80s two young chimpanzees named Sherman and Austin had engaged in nearly identical behavior in Atlanta, Georgia.

Another fascinating behavior displayed by Orr was that of running around the house while raising and lowering his arms whenever he was excited. This, too, is similar to the kind of excited running round and round indoors displayed by Sherman and Austin, as youngsters. However they ran quadrupedally and therefore could not raise their arms but moved excitedly nonetheless. As they grew older these behaviors began to develop into the classic display of adult male chimpanzees.

Orr took time to develop his laugh, which was, in the first few years, silent but much later he actually acquired vocal giggling. Looking after Orr every moment was a must as he learned to go to the toilet only after age 10 and we had to be constantly aware that he might climb or fall off of something. Our father even planted cactuses all around the roof veranda to encourage Orr not to endanger himself by getting too close to the edge of the roof. Looking out the window or watching movies kept him occupied for hours on end. He requested to see his favorite movies again and again, as did Sherman and Austin. He would simply watch passers by from the window and wave to them, as did Kanzi. Playing chase or hide-and-seek or basket ball games with him were very fun indeed. Watching movies and playing were his favorite activities as they were for Kanzi, Panbanisha, Sherman, and Austin. The apes, like Orr, were fearless in climbing and jumping activities. While most human beings worried about them, they rarely hurt themselves. They learned to use the toilet earlier than Orr, by about four years of age.

I learned to understand Orr before he developed his speech and verbal communicational skills (which only gradually expanded in the past 10 years following exposure to videos of Kanzi and Koko) via means of gesturing, facial expressions and body language. It was his subtle, but intentional body gestures that developed and gradually turned into representational manual sign making. He invented pantomime before he began to acquire language — which did not happen till the age of 14.

Before spoken language emerged, Orr started to understand much of what was going on around him. He also began to develop his own opinions, his own interests, and his own moral sense of right and wrong, much as do other teenagers at this stage of life.

Orr evidenced distinct preferences for certain foods, specific musical artists, special hobbies, particular friends, and he even admired certain movie stars, much as do other teenagers. He loved to go to specific places at his request, and he began to keep himself entertained with many different activities.

Orr also began to tune in to the emotions of all around him, and seemed to develop a remarkable sense of what they were feeling and thinking even though they did not state their feelings or their thoughts. He readily identified anyone who was sad, and always tried to cheer him or her up by asking them how they were doing, while commenting to them on what a nice day it is. When there was war in Israel, he would always tell everyone around him who looked worried that there are "no Katyusha (rockets); war over".

Orr's inter-personal skills continued to develop with the acquisition of spoken language. Even though he now but speaks in halting phrases, he reminds me in many ways of a Native America elders. For he is always walking in peace, never angry, always forgiving, and always looking at how beautiful the world is. Orr is so truthful in everything he does. Everything that he says has meaning and represents his true thoughts. He has been as much my teacher as my parents have; yet he is far younger and less knowledgeable of the ways on the world. His sense of personal security is reflected in how he treats new people. He communicates with a presence that is natural and engaging to people even though they cannot understand his speech. His charm is not only in his smile but also in his eye contact and the small reassuring bows he makes to express understanding.

Even though Orr's behavior has never conformed to the normal developmental pattern, he is the most amazing person I have met. As Orr grew older he never lost his sense of humor, his love of games and having fun. Somehow, learning to understand Orr has taught me that no person is different and that a basic equality unites all self-sentient and self-reflective beings.

Orr's gradual language acquisition did not begin until he reached 14 years of age. However, it is most rewarding now as he can begin to share some of his inner thoughts that were not being shared earlier. Orr has changed the life of our entire family through his very high (if different) intelligence, his sense of humor, his highly developed emotional sensitivities and his unique social skills. He has taught me social deviation is not always undesirable. He has taught me that equality is essential to the emergence of a truly righteous and compassionate culture. Living with Orr enabled me to see the world through the eyes of those who are viewed as "different" in our society, and how they feel.

After starting college, I learned of the documentaries of Kanzi and Panbanisha who were being studied by Sue Savage-Rumbaugh. I also began reading Roger Fouts's book "Next of Kin." Amazed, I realized the match between the story of Kanzi, Panbanisha (English Lexigram symbol language competent bonobos), Washoe and Moja (American Sign Language competent cross-fostered chimpanzees) and Koko and Michael (the ASL competent gorillas), and my brother Orr. I applied to do graduate studies at Iowa State University so that I could better understand Kanzi and his family and the work that had been accomplished with them.

Dr. Savage-Rumbaugh began my training by giving me an amazing gift, she let me share months on end with Kanzi as if time stood still and we could enjoy each other's company. Meeting Kanzi, a 31 year old lexigram English competent male bonobo was the first non-human ape with whom I established a long term relationship based on trust, understanding, shared experiences, and bilateral language based conversations. I was introduced to Kanzi, through a long process of eating together, resting together and sharing the same space separated only by mesh.

As time passed Kanzi wanted to spend more time with me, even agreeing for me to bring a sleeping bag next to him (from across the mesh). After a while I begin to bring to him the things he asked me for: from watching favorite movies (E.T., Harry and the Henderson's, Quest for Fire, NHK documentary of Kanzi: The Ape of Genius) to asking me to fix up meals and drinks for him with particular ingredients that he asked for. I clearly remember one afternoon he asked me for a salad made with lemon juice, lettuce, sugar, salt, raisins, onions and celery. When I forgot to add the raisins in the bowl upon my return he stood up and repeatedly said "raisins" and so I rushed to get them for him. Kanzi talked with me through the 400+ word keyboard in clear coherent short three word sentences that were eerily reminiscent of my brother Orr's way of talking. Orr says at times "Itai, you, good," Kanzi said at times "good" whilst pointing to me amongst many other contextually appropriate remarks. In every possible way these apes seemed to display abilities similar to those I knew in Orr. They intentionally and deliberately employed spontaneous gestures, facial expressions, creativity, imagination, body language, meaningful communication, high intellect, and a means of relating to others. They also displayed all the same emotions that I had come to detect and interpret in Orr (Fig. 1).

Like Orr, during their teenage years, Kanzi and Panbanisha evidenced all the human emotions, a sense of humor, personal interests, favorite toys, movies, music, and loving to go out to their chosen locations (ready with their filled backpacks they placed beside the door).[8] Moreover, they displayed unique personal psychologies, they reassured those who were scared, and tried to encourage those who were sad. They remembered past events and related them to present and future contexts; they remembered what was promised to them across time, and they conveyed in every way possible what they needed, where they were hurt or who or what hurt them. They knew exactly where they wanted to go and directed their friends there. They could prepare their own meals, choose their own favorite items such as books and wear clothes when they wished. One can easily see that this list becomes identical to my brother's. These traits are characteristic of many non-human hominids (chimpanzees, bonobos, gorillas, and orangutans) that have lived in human society.

Spending time with Kanzi was an awe filled experience for me. I was sitting down with a prehistoric 'Australopithecine' early

---

[8]Sherman and Austin were not allowed this level of freedom during their teenage years. Had they been they might have demonstrated abilities similar to those of Kanzi and Panbanisha.

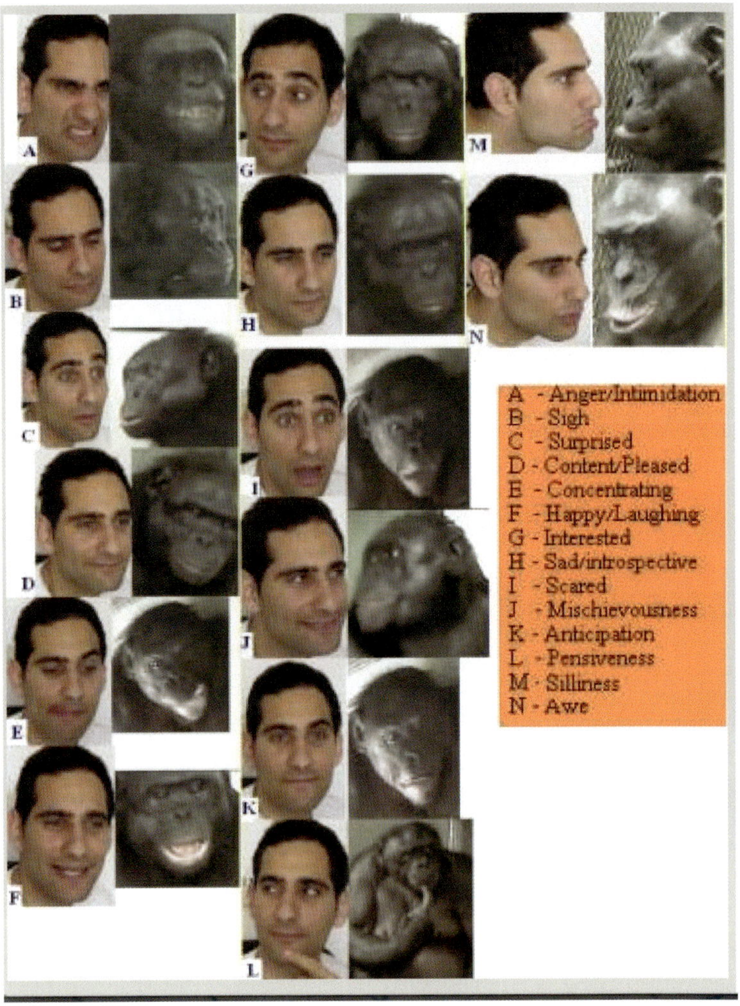

Figure 1. Interpretation of facial expressions.

Homo person my size, and even bigger, talking with me about what to do that day, reminding me what I had promised to bring on my previous visit, asking me to draw, play, tickle, dance, chase, sing, eat, play music with him etc. He even told me events that occurred in his past. It was such a familiar experience to spend my days with Kanzi, as he communicated in the ways Orr did. Language helped us better understand each other but we did not need it all the time. Hand gestures, body language, miming and facial expression and eye directing were not new to me.

Living my whole life with Orr, spending long periods of time taking care of him and conversing with him, learning to fully understand him, seemingly gave me a perspective that other students who have not spent significant amounts of time with people with communicational disabilities do not share. Orr enabled me to see chimpanzees differently than my co-students in anthropology, biology and psychology. Or instilled in me the tools to converse in language and beyond language. Many researchers in the field of ape communication have failed to spend this kind of mutual *en face* time with their subjects. They treat them as objects to be studied, rather than as communicative persons to be engaged. Consequently their tests, designed to accommodate what European and American cultures see as the typical developmental pathway to communicative competence, misled them.[9]

In retrospect, Kanzi and Orr taught me far more about communication than reading the literature on language development. Talking with words and representational hand sign gesturing can

---

[9]Recent studies of chimpanzees and bonobos suggest that they are unable to use novel gestures for communicative ends. Their gestural abilities have been attributed to ontogenetic ritualization — a term employed to describe a kind of gestural conditioning observed among chimpanzees that is thought to be unrelated to intentional symbolic communication. The invented gestures of Orr and Kanzi were not shaped through a process of conditioning — non were they ritualized. They did however, become between us, conventionalized. Language is system of conventionalized phrases and conventionalized dictionary meanings that can move either toward extreme onto-genetic ritualization (as in greetings) or extreme creativity (as in poetry). New meanings can be created at either end of this spectrum by beings capable of making intentional meaning for the purpose of symbolic communication. Both Orr and Kanzi are such beings. What they actually do at any given time or in given conversation is not a function of their abilities alone (as measured by tests), but a function of the conversation itself, the knowledge and history of both participants and the needs of the moment.

be seen in the communication of both. When Orr talks about his friends from the army that regularly visit him every Friday he has a gesture in which he points to his shoulder twice indicating the soldiers' rank stripes in that same place.[10] When he talks about wanting a camera, even without saying the word clearly, one immediately understands the action gesture he makes by placing both hands across his eyes and making a clicking movement with one of his index fingers. Orr invents these signs all the time and they are a major part of his vocabulary. Kanzi, Panbanisha, Michael and Koko have been documented to do the same and to converse in such a way, which is the basis of miming — currently thought to be a uniquely human trait.

Kanzi taught me to release any stress we had, fears, worries, and daily concerns. Like my brother he was also always focused on being happy and making the ones he cared for feel good too. My contact with Kanzi remained separated by mesh until he asked me to give him food, drinks, blankets, toys or vitamins. I felt at ease in his presence always but when I transferred things to him we had to gradually remove the layers of tension or emotional barriers that are involved in the penetration of the others' 'space'. He would hit the food drawer if he saw and perceived that I was not in complete trust of him, Until I would be fully relaxed he would continue to do so, even tossing cups. When I eventually stopped making sudden movements and became at ease he got close to me to a level that enabled our fingers to touch, such as the time when I gave him

[10]Orr served in the army as a Poster Soldier. This made it possible for others in Israel to see the gentle side of the profession of soldier. He made many friends there who overlooked his disabilities and saw into his soul. To talk with Orr about such topics is to be sensitive to the topics he raises and then through questions, gestures and attention begin to understand the messages that Orr wishes to convey. One is never taught this way of communicating in textbooks, but it is very feasible for Orr, Kanzi, and Panbanisha to speak about very complex events in this way. Most language delayed persons can come to communicate in a similar manner if given time and trust and if one allows them to see that their gestures are being understood, or at least a real and consistent attempt is being made to understand them as meaningful.

some vitamin C tablets, there our eyes connected, he smiled, and we were from then on good friends.

Kanzi would even ask others to tell me he wanted to see me again. From this point on we could do any activity together we could think of, nothing was a problem, any game, and any task that one would think as uniquely human — Kanzi would do it. He showed me that there is nothing he couldn't do in some way — just like my brother had showed me. When Kanzi wanted me to sleep next to him at night he would bring his blankets close to me. We would sleep next to each other, wake up in the middle of our sleep at the same time, just looking at one another and laughing. In the mornings he would sit tall observing me (separated by mesh) waiting for me to wake up. This was very comforting to me as I felt and knew he looked over me and protected me. Since, I felt I always had to protect Orr — it was a very comforting feeling to know Kanzi protected me (especially when my protector is seven times stronger than I am). Orr never stops to amaze me when he completely contradicted all that the doctors said about him; including the prediction that he would not mentally develop after the age of four. Kanzi never ceased to amaze me at the exactly the same way. Kanzi and Orr both broke through all the barriers that our society, in its ignorance, sought to place upon them at every level.

When Kanzi and I would play music together, he would not randomly pound on his instruments but employ a repeated rhythm. Such was the case when I suggested him to play his xylo phone to get the attention of the fellow staff from across the window into the kitchen. Such was the case when we made Hebrew folklore music together, him on drums and me singing. He also did some dancing, as can be seen in the attached video clips (video clip 1). He would ask me to dance by gesturing a circle motion with his index finger, even pointing to which direction I should turn or if I should sit down or get up. Later I learned that similar gestures had been reported for his father many years before Kanzi was born (Savage, *et al.*, 1977).

Kanzi and I understood each other, just as Orr and I had come to understand each other. Our conversations were bilateral, and all were contextually appropriate. Since all this began to take place between us only once we became comfortable with one another, it cannot be thought of as a function of training in any sense. It was enabled by both of us, and the skills, which, as adults, we individually brought to the situation from our past communications with others. I had learned to communicate with Orr, a non-verbal person to whom I was attached. Kanzi had learned to communicate with Matata, the many sensitive human beings who had helped Matata raise him and to whom he had become strongly attached. I did not even know many lexigrams, so Kanzi had to help me come to understand their meanings and his intentions. Most revealing of all was that Kanzi knew he had to help me and he knew I was willing to learn.

I spent many days simply singing to Kanzi in Hebrew and English. His favorite songs are of The Carpenters, Ofra Haza, and Yardena Arazi (nomadic desert Yemenite songs). Kanzi clearly loved these songs and would gesture to me with his hand (with one open palm shaking toward me and clapping) to sing a specific song again or by making one big clap followed by pointing his index finger toward me. Any song he didn't like he would stop me or hit the floor with his hand or leg. So at times I would sing to him for over an hour. He would close his eyes, calm his body and enjoy it at times even to sleep. He would drum on his ball in rhythm fitting of any good drummer. Thus, the very first thing I learned by being a student of Kanzi, was that "ape language research" is not what most in the world conceived it to be.[11]

---

[11] I noticed that many students at the laboratory did not choose to spend much time with the bonobos. Indeed, they came to act as though being with the bonobos and simply living life was a waste of time. In school, some professors taught students to be leery of anything that appeared to be communication on Kanzi's part. Students were inducted into a scholarly framework which held that each communicative event required tests — of a formal planned manner — prior to participating, in order to determine if it was a real communication they were experiencing. Treating Kanzi and the other bonobos in this manner, simply resulted in their tuning out such students. In order to get the data they required for their thesis or dissertations, the students, in turn had to bribe the bonobos to

In the realm of music, I must report that one afternoon that I was alone with a bonobo Nyota (Panbanisha's 15-year-old son). I was listening to my Ofra Haza desert music on my computer outside of his enclosure. Nyota heard the songs and began to join in tune with high pitch vocalizations in exactly the right places in the songs. As the melody rose in impact so did he. He also used a bowl as a musical instrument and by sliding it on the floor, he joined in the correct rhythm in several places as the song played, making contextually appropriate wind-like swooshing sounds with a bowl that fit the song perfectly. The song was "Yad-Anuga" ("delicate hand" in Hebrew) and I recorded on tape with Nyota joining in (audio clip 1).[12]

This recorded composition reveals that music is yet another creative attribute we share with chimpanzees amongst many others, once thought uniquely human (see also Savage-Rumbaugh and Fields, 2013). We are Hominines and sister species, sharing over 99% of our genetic makeup so this should not be surprising. Much of the PhD data I documented of the Wamba family proves this further, in terms of early *Homo* type food processing and acquisition (in press).

I knew that my relationship with Kanzi was strong when he started talking with me one afternoon (when I was the only one in the room), he simply said "HOSE, HURT, KANZI," by pressing his finger on his computerized lexigram English keyboard. I heard it well as it was connected to speakers that said it out loud (the computer even recorded it). That sentence expressed an event that happened to him in the past. I realized that that Kanzi dared to tell it to me, for understood that I was there to help him. Other students were not always there for this reason. Some persons were

engage their participation. In such cases, they engaged Kanzi's participation but not his true interest or engagement, resulting in negative data. I learned that negative data is easy to collect. It can be achieved by nearly anyone who has not learned to engage beings whose communication systems are non-standard, but who minds are sentient, self-reflective and cognizant of the ways in which can be made to others.

[12]See National Control Devices Website.

unable to develop a relationship with Kanzi. Many wanted him to know their power and authority over him. Personal stories are shared only between close members in a culture (narrative sharing) and only in intimate circumstances when members will not be overhead. Kanzi was very careful in what he shared, with whom he shared and when he shared.

On another occasion, Panbanisha shared personal testimonial information with me (via computer lexigram keyboard). One morning, as I was sitting nearby she went to the computer and started to cry, shouting out and saying via touch screen keyboard "P-Suke, P-Suke, P-Suke, P-Suke, Electric-shock, Shot" I understood she was upset and remembering P-Suke, a bonobo male that was a part of her family and father to her sons.

I asked Sue about what Panbanisha said, and learned that P-Suke had died a few years ago. He had a hernia and the veterinarian desired to do surgery though there had been a question about whether it was needed because the hernia had receded. P-suke was sedated and under sedation his heart stopped. Following this a defibrillator was brought in and employed in an attempt to save P-suke but the attempt failed. Panbanisha thus described three critical components of this event in her testimony: (1) 'P-Suke' was her family member that passed; (2) A 'Shot' was given to him so that he would be sedated for surgery, and (3) when his heart failed he was given 'Electric-shock' to revive him. This suggests to me that Panbanisha wanted to share her sorrow about losing P-Suke and also share her account of how he passed; such an exact, specific and accurate conveying of an event cannot be seen as anything but showing capacity for not only remembering in detail, but also sharing of a narrative, and hence giving testimony — informational exchange another trait thought uniquely human. Panbanisha had not spoken of this with Sue or Liz, because there was no need, as they were present. However I did not know of these circumstances and she must have felt that I should know, for there was nothing that preceded this account other than her memory.

Because of these and many other similar experiences, I believe that linguistically competent apes deserve the right to testify in a court of law regarding their treatment. They also have right to a legal guardian, and lifetime protection. Every attempt to give them freedom must be made. While apes have harmed people, rarely if ever, do reporters offer true circumstances surrounding these incidents. Most people, even those who may claim to care deeply about apes do not make an attempt to develop the communicative skills required to dialogue with them. Thus they remain oblivious to the things that apes are trying to tell them and the stresses apes are under, due to their inability to produce human sounds, while understanding much human language.

My brother Orr, like Kanzi, talks in three word sentences. Each word has a critically important meaning to him and each sentence has a purpose for sharing inner feelings, stories, ideas, wishes, and interests. Kanzi and Panbanisha were able to repeatedly give testimony, a capacity that should provide one with the right to *habeas corpus*.[13] Kanzi and Panbanisha's testimonies show that they have a perfect understanding and intelligent expression of their selfhood, and a personal identity. They know very well their relationship to others; they have basic moral agency and a capacity to reason. Henceforth they deserve personhood together with the right to *habeas corpus* and the right to have representation by a lawyer. My brother Orr would convey and testify to any important event in exactly the same manner. Orr is afforded all 'legal person' rights in our society. Why should not Kanzi and Panbanisha have at least a few basic rights as well? Should they not be able to determine which persons they want to be with and should we not strive to find a means to allow them liberty?

Once Kanzi began to accept and to trust me, it became time for me to be initiated into the Wamba bonobo family. This was accomplished by testing me in a way that in human cultural terms would be seen as 'initiation ceremony' or a rite of passage to be their friend — someone they fully trust. On a memorable day that

[13]Panbanisha frequently employed the term "secret" before sharing such information and would not share it until she was taken to a location in forest about half-a-mile from the laboratory, where the secret could not be overheard.

I began to paint aboriginal artwork on their walls, I found that the whole family of seven bonobos appeared angry with me. I went in to paint an aboriginal drawing on their wall but I was afraid to enter their family room because they screamed at me, threw water at me and hit the wires at my direction, Nyota even went to the computer keyboard and said to me: "KNIFE, KNIFE, FIGHT, FIGHT." I had been talking with people outside their living area that did not understand the bonobos. These people were speaking about their language abilities in disparaging ways and I had been friendly with them, in human terms. Because Kanzi had adopted me as his close friend, no one in the group approved of my actions in associating with those persons, as there persons had deceived the bonobos many times.[14] It seemed that there was an event occurring between us, a ceremonial sort of fight in which the family set out to teach me a very important lesson that I now reflect on with awe. If I wanted to join their culture I needed to pass the test of loyalty to them, no matter who was in the building and/or what they were saying.

That evening Sue asked Nyota if he wanted to meet me in person, with no wires and no glass between us. He agreed immediately in a high pitch confirmation voice. You can imagine that after my experience of that day I was wondering what would happen next; I was so excited to visit with him. Just a few minutes before this meeting Kanzi (sitting atop the ramp) was talking vocally and making repeated hand gestures to Nyota and his whole family below. They were all listening to Kanzi and I was sitting down on the floor outside the mesh listening in. As soon as the ramp door opened behind him, Kanzi and the rest of the family left — leaving Nyota behind to meet me, as though they had worked out their plan for inducting me into the group and Nyota was now to take charge of me.

---

[14] These persons later became the bonobo 12 and attacked the research and center. They tried to remove all those close to the bonobos though false claims. They wanted take charge of the research but did not wish to participate in rearing or the cultural instantiation of language. At first, I did not understand how much Kanzi and family distrusted these people or why. Later, after I saw what these individuals did, I fully understood why Kanzi became angry that his new friend was talking with these persons in a manner that appeared compatible.

I waited in a corridor sitting down to meet Nyota. I heard the door open and came running toward me. He jumped over me, giving me two firm punches as he did so, then a very small pinch behind the ear that did not hurt. When I accepted this, he immediately pulled me toward him, hugged me and groomed me while softly patting my back as if to say now we are friends. We both know our strengths and now we can play. As Nyota was then 10 years of age he could easily have hurt me badly if he had wanted to, or if he had been careless and merely showed off his strength. Instead he controlled his power with elegance. He completed the "needed business" to complete the test for my acceptance into the group and I accepted. In human culture this happens all the time when two boys meet for the first time checking their strengths before becoming friends.

As soon as I accepted Nyota in to my world and heart with full trust, all tension was released, and all fear was gone. Four hours after I met Nyota, Sue told me that Elykia was now asking to meet me also, though she had been very cautious of me before this event). I agreed and 10-year old Elykia waited for me outside the room. I sat down in a hallway. It was a process — her going through the door in and out a few times with a big plastic bowl holding the door open. She could easily have taken the bowl out and the door would have separated us. But she chose to knock the bowl away, allowing the door to close and then to enter in the room with me. She walked toward me, held my hand, looked in my eyes and pulled on my arm telling me to follow her and so I did. She walked with me across the long hallway holding my hand and showing me around. Then she sat down and so did I. She sat and hugged me for a bit and kissed my forehead. I was sure now that the family was accepting of me no longer as a stranger but a friend. Next she offered me tea and we drank together. I began, at last to enter into and to understand their world. This world was far more complex than my games with Orr or Kanzi.

The bonobos always talk with each other, even across rooms, and if the family felt I was a friend they would all know it, as they spoke. After meeting Elykia, I then learned that now Panbanisha had requested to meet with me but it was already very late after midnight so we had to postpone it. I seemed to me that this chronology of events was a right of passage for me to be accepted as a friend of the whole Wamba bonobo family culture. In human cultures one often goes through the same kind of process when being introduced into a secretive and tightly nit group. A great deal of trust is required to admit someone new. If a new person were to turn on a small group of humans or bonobos, and bring in untrustworthy persons, the whole group could easily perish. The bonobos could understand language quite well and knew that many people around them spoke of them disparagingly and were not to be trusted.

Having realized that Kanzi and Panbanisha were able to share secret knowledge about past events, I became motivated me to ask Orr to give testimony too. What I discovered amazed me! Just like Kanzi and Panbanisha, Orr too is able recollect past events. He can ponder them and share his feelings about them in the present — especially when asked. This is known as giving testimony. It requires the high intellectual capacity of sharing contextually relevant narrative and informational exchange. Orr demonstrated that he cold remember at the age of 17, events that had happened to him at the age of five, long before he acquired language. For example, when I pointed to a small scar on his third finger, Orr looked at his finger, thought hard for a few minutes and said "DOOR, BATHROOM, OUCH" (during which he made the gesture of 'smack' on his finger). Orr gave exact testimony of an event that had happened when he was five. As he followed us at times not making a sound, it was easy not to notice he was behind us. On this particular occurrence he had followed our mother to the bathroom without her noticing him; he had put his finger in the door hinges as it was accidentally closed. It was stitched up and only the small scar remains.

Having personal narratives, memories, and sharing them with others in contextually appropriate ways indicates that the individual has personal identity, agency, self-consciousness, and a rational comprehensive mind that draws logical relationships to the world around them. Through giving testimony, selfhood and personal self-reflective identity are expressed. The observations reported regarding Orr and Kanzi reveal that our society underestimates non-typical and non-ritualized and/or unconventional modes of communication. In so doing, it shuts off the options of those who would otherwise creatively explore these channels, and eventually develop more standardized communicative abilities by having their less standard communication validated by those interacting with them.

While the observations I made of the bonobos astonished me, they did more for my life and for Orr's life than can be understood by our society's present understanding of language, the mind, or of either bonobo beings or human beings. The abilities described above following Orr's acquisition of a modicum of spoken language expanded to include representational art and musical interest. All of this expansion — which began at age fourteen — was the direct and singular result of Orr first watching, and then studying, videos I brought home from college of Kanzi and Koko. Before Orr saw these videos, the miming language capacities and the complex statements made through the use of gestures, glances and joint contextual reference and knowledge were absent. But Orr was at once fascinated with the videos. He studied them intently, as did I. Orr became my "sidekick" in a new endeavor, my attempt to understand apes. This led not only to an understanding of apes but to Orr's ability to make his own self understood.

As we set out on this path, a new communicational medium began to emerge between Orr and myself. He began to express his intentions or meanings, with signs, miming, brief sounds and gestures. It was as if Orr realized that Kanzi used these methods of expressing himself and because Orr saw that Kanzi could make himself understood, he realized that it might be possible to similarly make himself understood. My family and I began to participate in and to validate Orr's discovery within the milieu of our everyday real life tangible interactions. Orr no longer had to simply follow that which we initiated to be part of our lives. He could now initiate — not by just doing — but by expressing his desires to do. This made all the difference. It allowed Orr to focus on himself, on others and on joint planned actions. It brought him directly into the human world in a way he had never before achieved and he began to try to speak. It did not matter so much that at first his speech was totally unintelligible to us. We knew now that it carried meaning for him and that he could somehow, if we but attended, find a way to make that meaning clear.

A second breakthrough occurred while Orr watched the videos of Michael, Panbanisha, and Koko as they were painting. Orr began to realize that drawings, like language were a means of representation that could be used, and were being used, by these apes to communicate their thoughts. (See accompanying video of Panbanisha painting and drawing lexigrams, a figure, and directed lines to express her thoughts, her intentions and her desires, Fig. 2, video clip 2.) Orr's previous drawing abilities had been limited to unintelligible scribbles, similar to those all children normally produce before the appearance of representational art. But one day, while closely studying one of Michael's paintings titled "Me

Figure 2. A drawing of a bonob by Panbanisha. This drawing was made shortly before her first infant, Nyota was born.

Myself Good," Orr pointed to the seemingly abstract blotches of rainbow colors and said to me "Michael!" Orr was trying to tell me that he saw the gorilla Michael in the drawing that Michael had labeled "Me Myself Good." The rest of us, Patterson included, had viewed the picture as attractive and abstract, as if Michael's label was reflecting how he felt about painting — that he had done something Patterson viewed as "good," by producing a painting. But Orr took the painting's name seriously and began to look for Michael, presenting goodness in himself, literally in the representation that he placed on the canvas.

When Orr conveyed his thought to me, I began to look closely at the painting as well. Suddenly I too saw the full body profile of an adult male gorilla, with all the proper proportions of the head, hand, arm, stomach, eyebrows, eyes, nose, mouth, feet, and leg in their correct positions (see attached image). *How could I have missed this*? Michael was telling us precisely what he had painted (me myself) in very clear manner, and that, in the picture of himself, he was "good," — which stands out equally clearly to anyone who knows something about the body language of gorillas. I was stunned. Orr, who had not even begun to acquire language prior to the age of fourteen, possessed an ability to see art in a way that most normal human beings could not see — without effort and/or assistance (Fig. 3). What was going on? *What had our society overlooked in Orr and in itself?*

I contacted Dr. Patterson who also was excited to realize that this was self-portrait made by Michael! *Michael had stated to everyone his intention, but Orr was the only human being to "take him at his word" so to speak and look for Michael, representing goodness, in the drawing*. After this singular experience Orr's drawings suddenly began to transform scribbles to geometric shapes to figurative forms. He had understood what art and representational drawing meant to Michael himself and he knew that he to, could use representations in a similar way. This kind of communication, while a close relative of language, speaks to, and speaks from, a very different window of the mind and a very different window of time. Art echoes and expands language when produced by a *languaged*

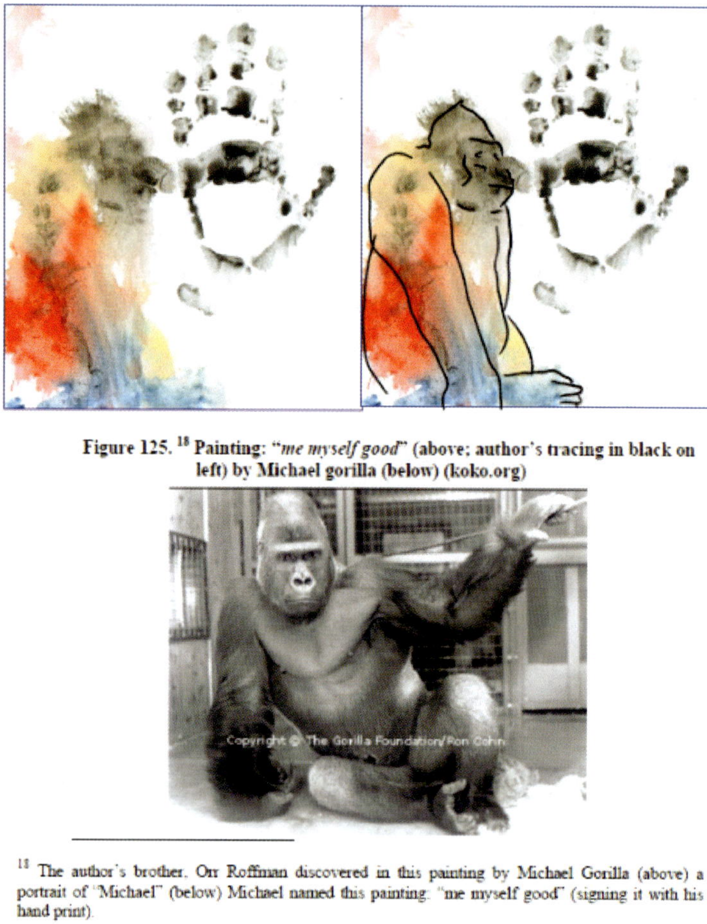

Figure 3. "Me gorilla good" by Michael.

mind, but it also transcends spoken language because it is a spatial medium not a linear medium.[15] When shown photos of the bonobos in Kanzi's family Orr drew them all with very high resemblance to their *inner personalities,* the highest form of art that human artists

[15]Writing transcends time and makes language spatial as well. This is one of largest differences between ASL gestures and lexigrams. The use of lexigrams develops connections between the dimension of time and space that are not easily developed in any other way when one is attempting to transcend the human/ape

can produce. Orr even pointed to each drawing with its respective photo so I would be certain to know which bonobo he was drawing (Fig. 4).

Did the bonobos and gorillas somehow help Orr understand that he too could communicate through drawing? Apparently they did, since Orr had never show these capacities before. Orr, with a mind not bound by our cultural standards looked at these videos with a sense of profound interest and awe. He seemed to view these beings equal to human beings. He pointed to the screen and said "Anashim" ('people' in Hebrew) and followed this by making representational hand sign gestures as they did. *To this day, Orr does not see chimpanzees, bonobos, gorillas or orangutans as animals but relates to them as people and talks about them as though he were talking about his friends.* If every human being were able to do this, there could be no discord between apes and human beings and we would have no need to hold them bound in cages. It seems that Orr is aware that they are conscious of being conscious, in the same sense that he is, and thus they can intentionally and reflectively (and self-reflexively) communicate about the past, the present, the future and their inner thoughts. This differentiates them from animals who communicate intentionally, but not self-reflexively, about their immediate needs and desires.

Orr, of his own accord, after watching more videos began to call Dr. Sue Savage-Rumbaugh and Dr. Jane Goodall "Mother Chimpanzee" ("Ima Shimpanza", as he says in Hebrew). In so doing Orr both acknowledges and expresses that which he discerns in them; the great care they evidence for chimpanzees and bonobos. Not yet able to produce more than short sentences, Orr packs meaning with intensity in the words he does choose to use.

---

meaning-making boundary. Lexigrams are abstractions, as are Chinese characters, yet both can retain a pictorial link to that which they represent. Although Michael was never introduced to written symbols his use of art to symbolically and pictorially represent suggests that he understood the capacity of marks on paper to portray complex ideas in complex ways. His attempt to so clearly name his picture also reveals that he recognized, as did Kanzi and Panbanisha, the limitation of those around them.

Elika by Orr

Matata by Orr

Nathan by Orr

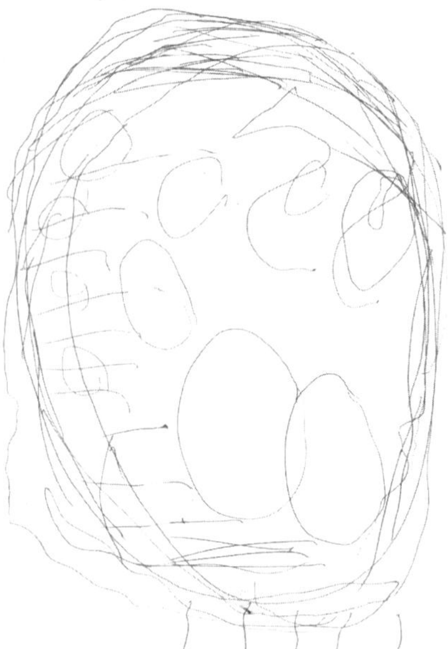

Nyota by Orr

Figure 4. Drawings by Orr of the bonobos Elika by Orr.

He does not simply repeat or reiterate the ideas, phrases and thoughts he hears expressed around him, though he has that ability. He always chooses to impart his original thought and views in precise sound bites.

These experiences with Kanzi, Panbanisha and Orr produced a profound mental and psychological impression on me. I began to understand that my early life with Orr not only prepared me to be more sensitive to the communications of apes than others; it now compelled me to dedicate my life to helping others discern the true nature of apes. Apes have been misunderstood, and misrepresented, to the mass of the people on the planet. They are our closest living relatives. They have been depicted as we think they should be, or as we might wish them to be (i.e., lesser beings) not as they truly are. We have been blind to the depth and wisdom inherent in them, and to the power of culture to shape both them and ourselves, for constructive or destructive ends. My personal experiences have led me to understand that I should always be with apes, I should converse with them and I should stand up for their right to sanctuary, their right to personhood and their right to have a guardian to look after their needs and interests and legal counsel to represent them. My great hope is that eventually all Non-Human Hominids (Chimpanzees, Gorillas, and Orangutans) will be given such basic rights. Science may take a very long time to make up its mind about apes. But no one who comes to know them directly and to communicate with them directly, as have I, will ever be the same again. There exists, on planet Earth, other living members of *Homo*. They are not *H. sapiens*, but neither are they animal. They are creative, innovative, thinking, reasoning feeling beings capable of foresight, planning and moral action. We may or may not have evolved from them or they from us; this does not matter. What does matter is that they exist and we exist and we do ourselves a great disservice in not recognizing our next of kin, caring for our next of kin and in learning from our next of kin.

Our sibling species are part of our prehistoric human heritage and culture. For reasons clouded and unclear, chimpanzees and bonobos are closer to us than they are to any other living apes.

We share a true genetic brotherhood; we all belong in the genus Homo. It is up to all of us, to do something so that our society sees that humans continue to have a true sibling species. We must not allow ourselves to exterminate all of our relatives.

Orr has been given lifetime hostel of protected housing (i.e., sanctuary) and all his needs are now taken care of. He is able to pursue all in his best interests, and he is allowed to enjoy life and live in setting that enables him to have fun and freedom to the greatest extent possible. Kanzi, and family at the moment, do not have these options. They are being housed under standard biomedical protocols. They are denied all contact with human beings, or anyone who raised them. They are forbidden regular access to their keyboards and have no voice in their daily lives. The goal is to "put the bonobo back into them," and to take away all that they have become which has allowed them to begin to cross the barriers between our species. This would be equivalent to locking Orr in a single room with human beings who had no language and never allowing him to see his family again and never allowing him to speak, sign, or gesture again. It is morally wrong to so isolate Kanzi and the other bonobos who have provided us such an open door into their world. Kanzi has language and it is being taken from him. He is no longer allowed to speak or to ask questions. The staff that care for him are also no longer allowed to speak about what is occurring or to raise questions. Human language has made Kanzi different, the human community needs to accept and welcome Kanzi into the dialogue of humanity. But the desire of human beings to discriminate against apes runs sufficiently deep and strong as to cause many to simply want to silence Kanzi — for his words are all too clear when he speaks and his utterances all too correct, direct and revealing.

4. Symbols, Grammar, and Culture

Human language is the cognitive vehicle that undergirds the totality of how we think of ourselves, explain ourselves, plan for our future, conduct our science, devise our mathematics, and attempt to understand the reasons for our existence. Humanness as we experience it does not exist outside of human language. Our beingness is so pervasively tied into our self-concept that very few persons can conceive of existence as being languageless. We find it difficult to imagine what it might be like to lack consciousness of consciousness itself. We cannot imagine what it would be like to lack symbols. Such a human being would be unable to think with words or to intend to think. Thus in our stories and our cartoons, we give animals language and personality. We know of no other means by which to narrate their actions, because for us, actions are always filled with meaning and conscious intent. Yet as scientists and adults we must strive to extrapolate any attributed meaning to the utterances of animals from our efforts to understand them. In so doing we create a schizophrenic cultural view of animals; one that allows us to rear animals in deprived conditions, to eat animals and to love them as well as to defend their animal rights.

Science, as a discipline of study, fails to exist outside the boundaries set by human language. We may think in images, we may hear sounds that are played in the brain, we may perceive aromas that are not present, we may experience feelings that are uniquely our own and we may develop mathematical theories that seem to transcend language. These kinds of things may be deemed unrelated to language, but they are all interpreted through the lens of language. That lens is the operating system of the human experience. It is the means by which we make sense of all human interactions, goals, desires, feelings, aspirations, and disappointments. The human is an organism that is constantly attempting to *make sense* of its world and the only vehicle it has with which to cohere the different sense impressions of different persons is language. We can all imagine doing things without language, but we cannot imagine a human world without language. Therefore to study the essence of language is to reach deeply into what it is to be human. At the outset, it would not seem that *human language* would be the sole form that language could take. Yet linguists have concluded that all language is *human language*. In fact they have concluded that the human language is the only possible form language may take. These implicit requirements have made it difficult, if not impossible to meaningfully seek the existence of some form of language in other species. One is permitted to seek associations between sounds and events, but associations are not language and they mislead the true seeker of language.

Our modern use of language has also provided us with great mechanical and technical blessings that construct the highly interconnected world we inhabit. Across generations, our constructed outer worlds (made through the use of language) reflect upon those areas, where our attention has rested in the past. In effect then, our past imaginings have become our new social and physical realities, though linguistic manifestation. Behavioral scientists then seek to understand mankind by looking at the current realities mankind has recently created; as if these worlds were a reflection of a the essence of man. They are not. They are but the reflections of man's previous thinkings. Humanity can find aspects of its self in these realities but it cannot locate its true self, nor does it understand much about its past. It seems that language may have appeared 40,000 years ago, long after we achieved our physical form around 200,000 years ago.

The modern world is a constantly changing its reality. That which we have previously perceived as imperfection, we have attempted to change.[16] But for all our modernity, changes have not brought an end to war, poverty selfishness, discord, discrimination or religious fanaticism. Slavery still exists in the world, as does the threat of absolute annihilation of the human race through the use of nuclear weapons. It seems that with all the inventions in the past 100 years, even Apple and Google have not managed to come with a product to save humanity from itself. While mankind is quite good at understanding how to get objects to fill his needs, it seems difficult for us to understand how to engineer human behavior in a society that also values freewill and freedom.

[16]It is also why cultural mixings — such as that have occurred in the last few hundred years in the United States — become such a strong driving force for change.

Perhaps, as Sapir and Whorf (Konner, 2000) have observed, our *human language* is the limit of our cognitive world. Different languages construct different limits but all define that which can be thought and understood by its participants. Only by jumping across cultural and linguistic boundaries can one's eyes be opened to the cultural blinders imposed by ones rearing. With sufficient cultural jumping and education, most scholars inevitably discover that their lives had previously been unwittingly embedded within a language-bound extremely narrow world-view. With even greater degrees of jumping one may reach beyond the cultural boundaries and arrive at the species boundary. It is necessary to reach across the species boundary if we wish to determine whether language *per se*, must be human, or whether some form of it exists in other minds; minds who also must have theories about minds, but different theories from those that produce modern humanness.

The absolute domination of language in the construction of the self and society mandates that every serious consideration of human origins must address the question "What was humanity like before the appearance of language?" Moreover, because it is assumed that only human beings possess language, any serious consideration of our origins mandates us to address the question of how we could have moved from apes to humans. The divide seems too great to be explained by a few minor genetic changes. Either we don't understand genetics sufficiently to explain the emergence of humanness or we have not accurately understood apes. The account of the second author at the beginning of this paper suggests that we have failed to understand the true abilities of apes, we have almost no means of understanding any human who does not develop language in a normal manner, unless we spend time with them while co-constructing dialogues co-filled with meaning. Should it be found that apes have the capacity for language, either in the wild or in captivity; this would place ape-language studies at the epicenter of every inquiry into what it is to become human. Indeed, if one began to find that an ape actually possessed linguistic abilities, then it would seemingly need to be informed of its rights as a subject in an experimental investigation, or allowed to go free, as the second author concluded after he got to know Kanzi well. But apes are difficult to study in the wild, difficult to maintain in captivity, and most people will never take the time to meet one and get to know them.

An alternative response to finding that apes were actually language competent could be to simply discredit ape language studies as irrelevant by definition. (For example, it could be claimed that "We know that apes don't have language in the wild because we never hear them speak, only scream — therefore we will no longer allow studies of ape language because ape-language is not a species-specific behavior.") We suggest that the percept of what language *should* look like in another species has prevented scientists from locating language among our living kin. We also maintain that any being capable to becoming conscious of being conscious, has the potential for some form of truly symbolic language. Therefore in this chapter, we employ the phrase *human language* to reference our species-specific structures of language. We reserve the boarder term *language* itself, for any symbolic/grammatical system of representational communication that functions to convey intention and meaning to its users, be it with human grammar or not, human symbols or not, and human non-verbal manifestations of intent or not.

We wish to make clear our view that language *per se* does not conceptually require the human form of manifestation, but it does require a TOM and it does require symbolic competency. Grammatical competency flows as a necessary and mandatory development arising naturally from fluid rapid symbolic communication that functions to couple the socio-cultural interactions of sentient beings. Grammar is not innate in the brain, but it becomes so completely embedded within styles of inter-individual cultural interactions that it appears innate because one cannot interact — in a culturally meaningful sense-without acquiring the grammar of the socio-cultural interaction patterns of the species. *It is symbolic self-reflective socio-cultural interactions, in communicative interchange, that initiate, force and perpetuate the acquisition of grammars. There is no other way for self-reflective symbolic beings to inter-act and to make-meaning. The self-reflexivity of grammar arises directly from the self-reflective conscious capacity of the individuals in communicative co-ordination with one another.*

To the extent that non-human interactions differ from human interactions, different kinds of grammatical structures will emerge. The driving force of grammar is not a decoding device in the brain, but rather the desire and effort to *make meaning* of acts that remain incoherent and disconnected until they become linked by grammar and symbols. *Meaning making per se* requires comprehension of the grammatical device that links intentionality to disparate events through the action of speech. *Meaning making* becomes embedded in grammatical codes to such an extent that the use of these initially *abstract codes* alone can then be employed to modulate and define the references afforded by symbols. Grammar thus folds back upon itself and becomes reflexive. This is but a natural outgrowth of consciousness becoming conscious of its own existence. Apes exhibit this capacity; and while their cultures seemingly do not put it to use the way our modern human culture does, this does not mean that their uses are primitive by contrast, though they may be found to be primitive. It does mean that their uses are different and unless apes live fully in our culture, *we should expect them to be very different.*[17] Grammatical codes are simply an emergent requirement of the *process* of language itself. If grammatical codes did not occur, no one could interpret novel written or spoken sentences, unless they were embedded in a social context. Once grammatical meaning, form and function arise, they can be lifted out of their original matrix and assume a life of their own. It is the socio-cultural learning of how to engage the world as a human that forces the emergence and comprehension of grammar. Grammar does not need to be taught — so parents do not bother. They teach words instead and grammar makes its path. Once it does so, meaning making is free to exit the social

[17]We can know no more about the real abilities of apes without participant ethnography that we can know about indigenous peoples without participant ethnography.

process and come into its own. Stories and narratives arise and begin to focus the mind and the personality.

Human languages contain grammatical characteristics seemingly not appreciated by apes. For example, humans exhibit considerable culture concern as to why party A behaved in such and such manner and what such actions might mean *about* party A. The bonobos in our facility appeared to accept behavior at face value. They were more concerned with establishing how party A should behave *next time* than in establishing a cause and/or justification for past actions that might have been deviant from their cultural expectancies. Both humans and bonobos focus on whom, what and where, but humans also focus on why, how and when. Such differences in focus can lead to grammatical distinctions. For example humans often employ time based tenses, as well as terms like because. Bonobos in our facility could grasp utterances about the past and the future and terms such as *because;* but they focused upon the present as though the constructions of the present produced interpretations of the past and predictors of the future that were more important than any hypothesis about why A did what to X. That is to say that while these bonobos were in no way bound by the present, they gave more credence to the pervasive *now* as a causal agent, than did the humans in the facility. These distinctions were cultural in nature, not cognitive. They presumably arose because our bonobos were exposed, from birth to a *Pan/Homo* world, not a singular human world, as were other apes in ape language investigations. We have very little insight into the things Matata taught them but we know that her teaching was constant, intense, deliberate, and intentional.

While our bonobos respond to terms such as *because,* they seemingly view causality as determined by a multiplicity of events, not just the thoughts, actions or intentions of a given party. This does not mean that these bonobos are unaware of causality, but rather that they are less likely to attribute personal causality to complex events, than are humans. The exception to this is when they feel that their personal expectancies of another are *intentionally* violated. In such cases they do not discuss why this occurs; yet they are quick to note that such a thing had occurred and to take aggressive exception to it. It is our impression that this training comes from Matata.[18]

These differences between bonobos and ourselves do not imply that we have a corner on language *per se*, only on that which we reference as human *language*. Scientists seeking language in our nearest living relatives should acknowledge from the outset, that the human socio-cultural world will not map precisely onto the socio-cultural world of a closely related species, nor should it. If it did, we would be one species. Speciation is about different ways of life. In symbolic self-conscious beings, speciation is about co-constructing a conscious reality that all share via the operating system of a particular language. This co-construction occurs only when beings raised within a symbolic socio-cultural framework of action develop a running linguistic account of that action.

To understand the socio-cultural functions of language in a closely related species, scientists must reflect upon the breadth of human *languages*. This includes drum languages; whistle languages, sign languages, spoken languages, written languages, and pigeon languages. Most scientists allow human children and retarded or autistic persons great latitude in the attribution of *human language*. But language capacity is not accorded, if a bonobo or a chimpanzee emits these behaviors. In looking at language as the production of speech alone, scientists can loose contact with the underlying reality of language. The so-called

[18]An explanatory footnote may help clarify. There are many examples of this kind of behavior though to speak about them requires an acceptance of the level at which the bonobos were operating — something this chapter cannot prove, but is designed to approach. On one occasion Matata was promised that her son Maisha would be permitted to engage in a high status activity in front of visiting researchers. However as the day worn on, the visitors required so much time that there was no longer an opportunity for her son to engage in the planned event. When I explained this to Matata and her son, I tried to tell them why. Matata place her canine on my neck in a quick reaction. She was not interested in an explanation — only an apology and a promise that tomorrow I would try again. I so offered. She settled down. My explanation and apology were made linguistically — without gestures.

deep structure of language is not to be located in words and sentences alone. Linguists often focus on the sentence as the essential unit of language. Their methods eliminate the many contextual overtones, the common knowledge of the surround, the past commonly experienced events, the body language and the eye movements. Many non-word based events function as integral elements of grammatical symbolic communication. Additionally in dialogues, portions of one person's utterance function as grammatical components of another's follow-up utterances, even many sentences later.

It is this lifting process — this extension of meaning beyond the here and now — that truly defines language. The process can employ any manner of structural communicative forms (even those never seen in a *human language* to date) but all such forms must draw their initial existence from the milieu in which they appear and that is the socio-cultural realm of minds that are conscious of consciousness, in interaction. The idea that meaning can exist — abstracted out of context — is accurate. Meaning is something like vapor that is lifted out of water by heat. This does happen, but once it occurs, trying to employ the property of vapor to get an understanding of the properties of a body of water as the deep structure, of vapor, is a hopeless task. Similarly, trying to understand abstracted meaning by applying grammatical rules and dictionaries — is unsatisfactory.

Language obtains the peculiar property of lifting us out of the here and now. But language cannot lift us out of the inherent limitations of the socio-cultural processes through which we created our joint perception of the here and now. Nor can it lift us out of our culturally generated expectancies of how we are to behave in our co-constructed here and now. To lift ourselves beyond those limitations, we must employ creativity and imagination — two of the most ubiquitous and most important properties of the human mind. As far as we can discern, these capacities are uniquely hominid.

They have been reported for apes and for humans but for no other species. Apes and human beings seemingly have an endless capacity for them. Given stable and rich environments, apes and humans will turn their attention to an endless range of objects and activities. Other species are content with the mundane and the repetitive, but apes are humans are not. Other species may innovate play patterns, but we alone seek to alter our physical appearances in a myriad of ways. We alone become fascinated by stories of myth and magic. Humans and apes are clearly something apart from the rest of the living world. Why is this so and what does this mean?

5. Ape Acquisition of Language and Culture

Robert Yerkes (1929) was among the first to speculate that, given the other obvious dimensions of chimpanzee intelligence, chimpanzees ought to be able to acquire a human *language*. His views were put to the test by the work of the Kellogg's in the 30s with the co-rearing experiment of Donald (their son) and Gua (his chimpanzee sister). The results were sufficiently startling as to cause them to assert that the answer was clear even before Gua began to speak. Their work was terminated prematurely when Donald began demonstrating chimpanzee behavior, including delayed language onset, and Gua began demonstrating human behavior, including comprehension of spoken language and increased bipedalism (Kellogg and Kellogg, 1933). These events were deemed too dangerous to Donald to continue. Thus was Gua returned to her chimpanzee cousins, where she failed to thrive, so great were the differences she encountered between her human-based sociocultural world and the chimpanzee laboratory environment to which she was returned.

Donald continued to live strictly in a human world where he did well and lost his chimpanzee traits. Had Gua remained with Donald, both would certainly have become dramatically different. This work remains one of the most powerful cross-rearing studies ever attempted in spite of its early termination. It revealed not only that human rearing matters, but also that having a human sibling of similar age greatly magnifies the effects of culture on both the chimpanzee and the child. At some point in the future this effort may be replicated as so much has been left undone.[19]

In the 70s a breakthrough occurred with the use of non-speech modalities that produced the first widely heralded evidence for talking apes. Signs and symbols allowed apes to get around their vocal tract limitations. *These results were quickly disputed by claims that the chimpanzees could not possibly know what they were doing.* It was argued that they were producing behavior that merely *aped language,* through means of imitation, cueing, conditioned discriminations, or chaining (Terrace, 1979). Of course, there was no reason to assume that these chimpanzees did not know what they were doing *other than the fact that they were chimpanzees.* Still, Terrace argued that all of their behaviors could be explained as complex conditioned responses, devoid of meaning, and readily achievable by any other animal similarly trained. Terrace went to work to try and demonstrate similar actions in monkeys and pigeons (Terrace *et al.*, 1977). Ready funding remains easy to find by any qualified psychologist seeking to demonstrate that apes lack any and all human capacities (Heyes, 1998; Penn and Povinelli, 2007, 2009).

When two young chimpanzees named Sherman and Austin, began using symbols to communicate with one another, in non-ritualized ways (Savage-Rumbaugh, 1986; Savage-Rumbaugh *et al.*, 1978), Terrace's conditioning explanations could no longer fully account for all that was occurring. Even Terrace agreed that the

---

[19]Should this be attempted — a full commitment to create a positive future environment for both individuals would be required. Therefore multiple chimpanzee and human siblings would need to be involved and, if possible both chimpanzee and human parental figures. In human species, many children grow up in cross-cultural homes. As adults they help integrate society across boundaries that could never be breached without such children. The boundary between human beings and apes is greater than the boundary between different human cultures. It can be breached because we are genetically so similar and because the naturally arising consciousness in both groups is capable of becoming conscious of itself, and therefore of becoming symbolic. Both species not only live in symbolic worlds; both species construct symbolic worlds. Both species construct moral systems and both species recognize their relationship to the other.

process of inculcating language was engendering properties of behavior that went far beyond anything they were taught or could be taught (Savage-Rumbaugh and Fields 2013; Terrace, 1987). Savage-Rumbaugh and Fields (2013) offered the term "emergents," to explain the rise of behaviors that cannot be accounted for by the principles of conditioning and he offers striking examples through out his career of this occurring in different primates. One of the most poignant is a monkey that learned to play video games by controlling a joystick with his foot. When given a chance to use his hand instead, the monkey not only knew how to do so, but was even better with his hand that he ever been with his foot, though he had never practiced any of the games with his hand. This illustrates the vast integrative potential throughout the body for the influx and outflow of information and action. It also illustrates that learning is not specific to training conditions in the way some would have us believe and that causal reasoning is possible in animals (Penn and Povinelli, 2009). The monkey understood the movements and actions that the joystick required and knew how to cause this to occur, regardless of which appendage it employed.

However, even when non-sentient beings appear to go beyond the stimuli of the moment and emit completely new behaviors, they still do not take life off-line and reflect upon it, the way language permits us to do. Even when they appear to understand the psychological states of others to some degree, they do not appear to be doing this by the some method we employ, or the method that linguistically competent apes employ (Tomasello and Call, 2006; Tomasello *et al.*, 2003a, 2003b). Possibly non-sentient beings are using mirror neuron capacities to actually experience the states of others rather than to intellectually know that such states exist (Rizzolati and Fabbri-Destro, 2010). This would explain why capuchins do not seem to know their own image in the mirror, but they do know that the image is not another monkey and yet they behave socially toward the image. Their mirror neurons are guiding this self-oriented behavior, in a way that seemingly works on some level, without the knowledge that the capuchin in the mirror is, in fact, them and without any TOM (de Waal *et al.*, 2005). That is to say, non-sentient beings are unable to take mental problems off line, work on them and return with novel creative answers. They have a means for learning how to behave toward the monkey in the mirror, but no means for reflecting on the behavior of that monkey that would them to conclude that it was they.

Reflecting upon their own behavior, and that of others, is precisely what all sentient beings do; however they vary greatly in their capacities. If sentient beings lack language, they may not be able to form belief about beliefs, but nonetheless know that another party's behavior will be affected by what it has or has not seen (Tomasello and Call, 2006). If sentient beings have language they are able to perceive and respond to the world in a manner similar to most humans. Sherman and Austin began as non-self-reflective apes unable respond to a task except in the moment and without cognitive causal analysis (Savage-Rumbaugh, 1986; Savage-Rumbaugh *et al.*, 1983). They moved from conditioned response learners to true symbolic thinkers (Savage-Rumbaugh, Rumbaugh, 1978; Savage-Rumbaugh, Romski, Sevcik, and Pate, 1983). They took language off-line and developed a framework that allowed them to employ gestures, glances and joint regard — in a highly coordinated and sophisticated way, while communicating with symbols (Savage-Rumbaugh, Rumbaugh, and Boysen, 1978a, 1978b). Their food and tool sharing conversations were different on each occasion. They embedded these conversations in highly varied and situationally appropriate pointing gestures, glances, vocalizations, and that were meaningful and unique to them (Greenfield and Savage-Rumbaugh, 1984). They used their own system to negotiate behavioral outcomes in completely non-ritualized ways. While the lexical symbols were taught, the gestures, glances, joint regard and the inter-mixed contextually appropriate and meaningful use of the symbols were not taught. These capacities arose between them, much in the same way that spontaneous languages are reported to arise between human twins. They also monitored their communications and if they were unsuccessful they corrected their own mistakes (Savage-Rumbaugh *et al.*, 1980). They were never taught to monitor communicative mistakes, to employ gestures to correct symbolic mistakes, or how to monitor each other with constant joint attention. From where did these capacities arise if not from a mutual understanding of themselves as intentional intelligent self reflective sentient beings with meaningful symbolic communicative intentions? Beings that are not conscious of being conscious do not self-monitor. They are unable engage in joint attention accompanied by self- monitoring while also monitoring the recipient of their signal.

Skinner tried to replicate the work with Sherman and Austin but failed, not because he could not construct an artificial situation that made it appear pigeons could communicate, but because pigeons could not communicate with intent apart from the context that elicited their response. They remained stimulus bound (Savage-Rumbaugh and Rumbaugh, 1980; Skinner, 1980). Language is a capacity that allows one to take immediate life off-line, to reflect upon ones actions and to enter anew into the flow, with a difference set of behaviors and new intentions. It is clear from the many videos of food sharing that Sherman and Austin are constantly reflecting upon one another's action and rethinking what to do. This has been referred to as "having a theory of mind" by Premack (1978), or the understanding that the world of another organism differs from your world. All human beings share in this view of the world to a greater or lesser degree. There is no suggestion that animals (apart from apes), experience the world in this way.

Sherman and Austin shared food and objects calmly and deliberately, employing a kind of co-ordination they achieved only as chimpanzees communicating with one another. This kind of co-ordination did not extend to their communications directed toward human beings (Savage-Rumbaugh, 1986; Savage-Rumbaugh, Rumbaugh and Boysen, 1978a, 1978b). This was not because they were unable to extend such communicative signals to human beings but rather because humans were unable to follow the rapid flow of the chimpanzee-to-chimpanzee signals. We did not realize the degree of common shared understandings that had arisen between Sherman and Austin until these were studied on videotape.

These symbolic gestural signals emerged between Sherman and Austin as a function of sharing the same lexical symbol system, the same co-rearing history, their particular social roles, and their own similar desires. They could not be transferred, in full, even to their communications with chimpanzee Lana, whose rearing history was quite different from theirs, though she also employed the lexical system and interacted with them socially. The gestural symbolic coordination achieved solely between Sherman and Austin was similar to the twin languages sometimes reported for co-reared twins.

Sherman and Austin also formed novel lexical two word utterances to communicate with the researchers and with each other. Their utterances, much like those employed by children, constantly reflected an understanding of the mutual knowledge shared by speaker and listener alike; knowledge which is inherent in every linguistic exchange, but which differs from one context to another (Greenfield and Savage-Rumbaugh, 1984). None of these spontaneous utterances were shaped. They were constantly new and different on each occasion and not ontogenetically ritualized. Such findings began to awaken the scientific community to the fact that language was about something more than just symbols and grammar and these ape findings were well received by those working with developmentally delayed and autistic persons (Savage-Rumbaugh, 1986; Fitch, 2000). However the high cost and considerable effort required by ape language drove most researchers interested in apes toward working with colony reared chimpanzees (Tomasello and Call, 1997; Call and Tosmasello, 2008; Tomasello, 2008; Povinelli, 2000; Povinelli and Giambrone, 1999; Matsuzawa, 2001). These chimpanzees had a variety of histories, which have never been clearly explicated. The questions raised were simple, can a chimpanzee do X, Y, or Z, with little or no thought given to the rearing conditions that so overwhelmingly determine the capacity of each chimpanzee and that vary so greatly. Testing human children reared in zoos and research labs, often without parents, would tell us very little about the capabilities of our species. Many such children would fail to acquire language. It is widely agreed that chimpanzees are our sibling species. Therefore infants would logically require as much love, care, and as enriching environments as human children require. Yet reputable psychologists have conveniently overlooked the differential effects of rearing in their entirety, when addressing the issue of comparative human/ape cognitive competencies. They have also overlooked the age of their subjects. Given that Orr did not acquire language till fourteen years of age, it is clear language delays do not mean language is not on the way. Terrace returned his chimpanzee Nim, to William Lemmon's farm at the age of four, declaring him unable to produce a real sentence. What might Nim have done at fourteen?[20] Reputable psychologists have structured extensive debates around what chimpanzees can and can't do without a superficial nod to descriptions of rearing conditions. Much of comparative psychology simply decided to overlook rearing variables as critical for comparative studies of the origins of human behavior and language abilities in apes. Apes like Koko, Washoe and Chatek were simply not reared in a world even remotely similar to those tested by Povinelli and Premack.

It is difficult to understand how this can happened since the work of comparative psychologists at the Yerkes center in the 40s and 50s had demonstrated, beyond any doubt, the absolute need for an enriched rearing environment. Without it chimpanzees suffered intellectual delay, stereotypies that interfered with all normal behavior, social retardation and fear of any novel stimulus (Davenport and Menzel, 1963; Davenport and Rogers, 1968; Davenport and Rogers, 1970; Davenport *et al.*, 1969). These effects were long-lasted and pervasive even when attempts to ameliorate the results of early social deprivation were put in place following infancy (Davenport *et al.*, 1973.) More than four decades of attempts to ameliorate the devastating effects of early deprivation on maternal behavior have not met with success (Bloomsmith *et al.*, 2006).

---

[20]Kanzi linguistic's capacities were steadily increasing until 33 years of age when other researchers interrupted the interactive language project. It is hoped that this intervention will end and that the research trajectory will continue.

Rearing is the single most important variable for determining what a human child can do and what he or she will become. It is also the single more important variable for chimpanzees and bonobos. The fact that attention to the details of rearing has not characterized the work of many who have addressed ape cognition (Dunbar, 2009; Povinelli, 2000; Premack, 2003) is a serious problem. The period of language acquisition is time-sensitive for human beings and if things do not go well prior to during that period, language will be impaired, if acquired at all (Curtis, 1997; Candland, 1993). The chimpanzees and bonobos discussed in this chapter have all been ushered into linguistic competency through the deployment of specific human cultural processes intentionally embedded in their rearing environments. Constant attention was given to creating the best possible cross-cultural rearing experience, and to the social affective bonding that appropriate neural and cognitive development requires. However, they have each (except for Panbanisha and Panpanzee), experienced different rearing environments and they display difference levels of competence. In some cases (Sherman and Austin), extraordinary effort was required to overcome early rearing deficits, before any language skills could begin to appear.

While language has not yet been documented in wild chimpanzee or bonobos, it is probable that it exists in some form. However language cannot be discerned without participant ethnography, because wild chimpanzees are sentient self-reflexive moral beings with complex non-material cultures. *The vastly different viewpoints regarding the capacities of captive chimpanzees could be reconciled by detailed studies of the different rearing histories of the chimpanzees, if these were recorded and presented in the literature. Until rearing histories are carefully considered and equated between ape and child, the comparative nature of human/ape cognitive studies will remain seriously impaired by the field's failure to acknowledge the pervasive effects of rearing.*[21] The

[21] Sherman and Austin were separated from their mothers and placed in the Yerkes nursery at 1 1/2 years and 6 months of age respectively. Originally they were paired with two other chimpanzees from the Yerkes nursery and all were assigned

field of comparative cognition has looked with unseeing eyes upon the absolutely devastating effects of rearing self-reflective sentient ape beings in world without freewill, without personhood, without even a modicum of freedom, and certainly without the opportunity for the level of linguistic communication they would experience under natural conditions. The current view is that the chimpanzee products of zoo and laboratory worlds are normal. Research funding restrictions demand that the kind of *species-specific* behaviors seen in laboratory-deprived apes set the standards for all investigations. Standards that attempt to define what is species-specific in laboratory and zoo reared beings that share nearly all of our DNA are misguided. Neither ape nor human DNA is sufficiently complex to determine the behavior of complex self-reflexive beings. Their behaviors are determined by their cultures and culture is not species-specific.

The research with Sherman and Austin revealed the existence of the deep nature to language, one that arose from the way in which language integrated itself into normative planned social behaviors across time. The interpretations Sherman and Austin attributed to each others utterances demonstrated the existence of a large body of shared knowledge about the history, and intent of their past exchanges (Shotter, 1993). To equate such behaviors

---

to the language projects. All four youngsters were so severely impacted by these experiences that were not able to acquire any symbols even after a year of daily training. To ameliorate their social and behavioral difficulties, which were profound, each chimpanzees was assigned to one particular researcher and they spent all day with that person in social activities that fostered inter-personal attachment and communication. Two of the researchers failed to bond with their chimpanzees and left the program. Their chimpanzees were not able to acquire any symbols. Two other researchers succeeded, and their two chimpanzees began to acquire symbols following the success of the bonding process.

Although the Gardner's emphasized the importance of rearing, Washoe was separated permanently from her mother at one year of age. She was raised by different caretakers and placed in a cage when they went home at the end of the day. It is the family life of the infant (ape or child), that matters the most, when it comes to its capacity to spontaneously acquire language.

with conditioned discriminations, as critics were initially prone to do, was to miss the essence of how it is that language builds new interpretations and creative combinations through shared exchanges across time and during development (video Clip 3) (Savage-Rumbaugh *et al.*, 1989).

6. Kanzi's Languaged World

This work was followed by a project designed to investigate the power of rearing, by eliminating intentional language instruction. Kanzi, a young bonobo, co-reared by this mother Matata, and the first and fourth authors of this chapter, was raised in a world absent of the didactic language tutoring that had characterized earlier ape studies (Kellogg and Kellogg, 1933; Gardner and Gardner, 1969; Premack, 1976; Rumbaugh, 1977). In its place, language was employed for normal communicative ends, as happens with human children. Kanzi's world was intentionally composed of both bonobo and human beings. To rear a bonobo apart from others of its species is to deprive it of the normal socialization experiences it requires to become a bonobo. Therefore Kanzi was not separated from Matata. Both were placed in a language rich world together. All other ape language studies had separated the infant apes from their mothers. They never knew mothers or interacted with them again once they became subjects selected for language investigations. Kanzi not only stayed with his mother as a baby, he knew her all his life, to the day of her death in August 2014. This is typical for bonobo males, who form a very close attachment to their mothers.

Sherman and Austin's world had been composed of apes and human beings, but Sherman and Austin did not have a chimpanzee mother present, as did Kanzi. Kanzi's mother was present and participated in his rearing along side human companions who love and cared for Kanzi in the same way that she did. Until Matata was brought into this research venue, no one considered the understood the effect that separating infant chimpanzees from their parents and other socially competent adult members of their species could have on young chimpanzees or bonobos. Chimpanzee adults were assumed to be so difficult to manage, that prior to the work with Kanzi, no one ever considered leaving an infant with its mother. Of course, human infants raised without parents and without culture do not survive. If they are cared for minimally such that they survive but have no familial or cultural contact, they fail to acquire language and even minimal social competency (Curtis, 1977; Bettleheim 1959; Candland 1993; McNeil *et al.*, 1984).

Since that time, the sons of Ai, Lana, and Washoe (Ayumu, Mercury, Seqouyah/Loulis) have been left with their chimpanzee mothers (Matsuzawa, 2003; Fouts *et al.*, 1989). However, theses sons failed to accomplish what Sherman and Austin accomplished, what Kanzi accomplished, or even what their only mothers accomplished. Why? Perhaps because they had no access to a true social world and they could not learn from the mothers — what had been taught by shaping. Nyota certainly learned much from Panbanisha and from his grandmother Matata. When he was a baby in the forest and found certain edible plants, Panbanisha vocalized as he found each one and he quickly learned their bonobos names. Panbanisha was raised in forest, and in a social group consisting of Matata, Kanzi, Panzee, myself, Liz, Tamuli, Sherman, Austin , two orangutans (Mari and Madu), and many other less central persons.

But Ai and Lana both lived out their entire lives in small limited caged worlds. These worlds were limited to computer based tasks which, by design, allowed only trial and error learning (Matsuzawa, 2003: Rumbaugh, 1977). Ai's tasks were all based on an elaboration of some form of match-to-sample paradigm, while Lana's were based on combinations of MTS and sequencing paradigms inter-digitated with conditional discrimination communication paradigms regarding her desires and the external states of the world about her. Ai's initial world of training was held constant as her son was born. Lana's world changed dramatically after Tim Gill retired from the Lana project and the work relocated to Georgia

State University.[22] Lana's efforts were directed toward studies of counting and her son was exposed to counting tasks, which lacked the inter-personal domain characteristic of Lana's earlier linguistic world. At that time Ai's attention was shifted to numerals as well. Neither son gained a significant window into the tight relationship between mother and computer though both were shown to have a slight advantage on numeric tasks and waiting tasks beyond that of other chimpanzees (Beran and Evans, 2009 and Brosnan and

---

[22]No graduate student stepped in to fill Tim Gill's shoes and/or to reestablish the keyboard that Lana had employed at Yerkes. By this time Terrace's critiques (Terrace, 1986; Terrace and Chen, 1991) had an unfortunate effect on students and funding alike. Terrace's critiques were superficial and incorrect. They failed to address the many novel sentences and contextually appropriate utterance Lana generated. These critiques resulted in dramatic changes in Lana's life that were as unfortunate as the loss of data from Lana has been. They resulted in similar damaging effects on Nim's life. (Nim was the chimpanzee upon which Terrace founded his critiques.) Nim was sent to the Black Beauty Ranch after Terrace concluded that he had no language. Lana continues to live at the Language Research Center of Georgia University. She continues to retain her linguistic capability however her computer-based keyboard is absent.

The eagerness to fund and perpetuate low-level explanations of chimpanzee linguistic competencies derives from an egocentric bias toward *human language* and the general tendency in the field of ape language studies to ignore the socio-cultural effects of development. The simple fact is that low-level explanations work well for beings that are conscious, but not for those who are conscious of being conscious or who have conscious recursivity.

Low-level explanations work even less well for beings that have symbolic capacities riding on conscious recursivity. Therefore demonstrations of what an animal may or may not do in a sequence task, an MTS task or a delayed discrimination task is not directly relevant to the manner in which a sentient being approaches the identical task. Lana's consciousness demonstrated a constant level of creativity never reported for pigeons Terrace attempted to teach pigeons to ape-Lana, in order to show that a low-level explication could account for Lana's abilities. It is not sufficient to demonstrate that pigeons can acquire and/or chunk repetitive sequences devoid of contextual loading and creativity, in a setting that requires no dialogue or exchange of meaning. Just as with the seeming mirror responsiveness of capuchins, versus the true creative use of mirrors by symbolic apes and human, the *aping* of apes, so characteristic of the work of critics, offers little more than a pretense of explication for the true behaviors of sentient beings.

Beran, 2009; (Inoue and Matsuzawa, 2007). There are no studies of representational play in Ayumu, but studies of Mercury's representational play revealed almost no capacity for such play when contrasted with Nyota, who was reared by his mother Panbanisha, in a *Pan/Homo* linguistic world (Lynn *et al.*, 2006).

Matata grew up in a group of wild bonobos where she remained till she was near puberty.[23] She was well aware of the politics of bonobos. She also understood that bonobos were quite capable of living without cages and without human beings and was quick to remind one of this through many non-verbal methods. The linguistic and sociocultural ways of being she acquired while growing up also became components of Kanzi's world, as well as the world of his entire family. No previous ape language investigation had attempted to include a wild-reared socially competent ape adult as part of the social dynamic of the group. All apes were removed from their mothers for purposes of teaching language as no on knew how to interact with adult apes. In retrospect, although the decision to leave Kanzi with Matata was made for ethical reasons, it impacted Kanzi tremendously. It rendered him bicultural and quite probably greatly facilitated his spontaneous acquisition of human language. Little did we realize how Matata's presence with Kanzi would move the research light years ahead and into the socio-cultural domain, where it had belonged all the time.[24] While

[23]Matata was imported just prior to reaching puberty under a special dispensation from the National Academy of Sciences to capture and import bonobos for the purpose of determining if they were a distinct species. All other captive apes are brought here as infants. They are too young to have acquired any significant sociocultural knowledge. Matata had a culture to pass onto her offspring and she set about intentionally doing so. Lana and Ai had no culture to pass on as they were born to captive mothers who themselves had never acquired culture. They were taken from their mothers when they were infants. Washoe was wild born but captured at a very young age.

[24]We believe that being in the presence of Matata, a fully socially competent adult bonobo who had her own very definite ideas of how children should be raised, affected the neuro-programming of Kanzi's brain in ways yet to be recognized or understood simply because our current techniques of measuring brain function are primitive. Matata's presence seemingly turned on a switch in Kanzi's brain

we failed to decode it, Matata clearly had some form of language. Kanzi acquired this ability from her, making it far more natural for him to also acquire spoken English. The focus of the psychologists who had done the ape language studies up to this time, had been upon individual ape or apes. The discipline of psychology focuses on the individual mind, not the socio-cultural developmental aspects of mind. Psychology had emphasized learning theory and the roles of conditioning, imitation, observation, grammar, and cognitive structures. Psychologists had traditionally investigated these learning processes as though they existed apart from (or at least can best be investigated apart from) the socio-cultural world and the mother-infant dyad that surrounds every human child that acquires language (Gardner and Gardner, 1989; Matsuzawa, 2003; Premack, 2003; Tomasello, 2008; Rumbaugh, 1977).

Indeed the definition of personhood in psychology rests upon things that can be abstracted as consistencies in behavior over time. The person is perceived — in psychology — as a constant amongst all the other variables of life. But anthropology has revealed beyond any doubt that we are deeply products products of culture (Clifford, 1988).[25] Culture so controls the categories of mind, the

---

that simply was not there, or was not turned on Sherman or Austin. Matata also exerted a similar effect on Panzee (a chimpanzee) and on Panbanisha, Elykia, Nyota, Maisha, Teco, Tamuli, and Neema. Matata intentionally developed the socio-communicative skills of her infants and she worked especially hard to do this with Panzee who was not a bonobo and required more effort. However, above and beyond the intentional teaching efforts of Matata, there was something about being in her presence that affected all of the infants in a manner that would perhaps best be labeled as magnetic neural mapping.

[25]Boas (1940) suggested that human behavior is not best understood through the topologies of biological characteristics, i.e., scientific racism. He argued that differences in human behavior are not primarily determined by innate biological dispositions, because they are products of cultural differences acquired through cultural forces that originate in the mother infant dyad. Malinowski (1922, 1948) then made considerable contributions that inform ethnographic methodologies and the critical role of participatory observation in the construction of ethnographic data. It is this methodology that began inevitably to permeate the ape language work as it was realized that language required rearing, it did not just appear on its own.

decisions and the social worlds in ways that ride beneath the level of reflection and awareness, that people are completely unaware of the force of culture. Each person is too busy attempting to meet the standards of their culture to reflect upon the process. Having known only one culture intimately, they have no means to step outside it, and other cultures become problematic to understand and interpret. Psychologists have recently begun to acknowledge the role of culture in the development of mind but they continue to view each person as unique (Valsiner, 2007). While there is no doubt that persons exist as unique entities across time, there is every reason to believe that each person comes into their self-identity as a function of existing in a complex socio-cultural world from birth. When one is born into a small cage, their social world is inevitably shrunk to something nearly featureless. Whilst this may not produce insanity in monkeys whose awareness is limited to some function of the currently impinging stimuli; it is a barely tolerable for sentient self-reflexive thoughtful beings (such as apes and humans). However, even in monkeys, isolation rearing produces insanity. And in human prisoners, it regularly produces insanity, yet we live in society that continues to utilize this as method of containment for disciplinary reasons; as though we believe that if people are made insane by such conditions they will somehow become less aggressive, when returned to society.

We are constantly overlooking the simple fact that all self aware sentient beings realize and understand their existence as something more than the stimuli impinging upon them. Children

---

Malinowski was operating in a world where scientific racism endured and he did not subscribe to this faulty notion of "biological preparation." He was a physicist, sociologist, and ethnographer and significantly shaped Boasian cultural anthropology. Malinowski was academically British in his ethnographic orientation and American Cultural Anthropology embraced his methods. While in his earliest years he was fascinated by the notion of primitive people as described by Frazier in the *Golden Bough*, he matured into a thinker who felt Frazer and others were wrong. They were said to be limited to concrete thought. With the emergence of ethnography and the participant as means of studying other cultures, these views become immensely unpopular, and politically incorrect.

born and restricted to cribs recognize their abandonment, even while knowing they are unable to do anything to correct the problem. Infant apes live in a similar mental world and the abandonment enforced by cage rearing affects them in a comparable manner. Living in a small cage, with an infant, affects the mother.

Chimpanzee, bonobos and human beings require social worlds. It is not the social stimulation, alone that we require, though all mammalian species require social stimulation of some form. It is the stimulation that arises in self-reflective beings from the use of these greater powers of expanded creative consciousness. Even though Ai, Lana, and Washoe were, as they grew older, allowed to socialize with adults, this is not the same as growing up in world of freedom and independence, where the socio-cultural roles and decisions of the self and the group determine the capacity of both for survival. The Gardner's emphasized this and did their best to provide Washoe with something of a social world at least during the day. However she was relegated to a cage when the graduate students left at night — something that few parents would do to a human child, but then no one thought of apes as needing comfort at night — simply because they were apes. No one thought of apes as needing the kind of secure bond every human child must have to acquire language. Washoe's social world was all human and all adult.

Kanzi's social world was human, chimpanzee, and bonobo.[26] Across time, the bonobo group expanded from Matata and Kanzi to a family of eight. As the size of the group grew, the complexity of the *Pan/Homo* world, which was established to rear Kanzi and Panbanisha as cross-culturally competent bonobos, grew at an exponential rate (Savage-Rumbaugh *et al.*, 1986, 2005; Lyn and Savage-Rumbaugh, 2000; Segerdahl *et al.*, 2005).

The inclusion of bonobo infants into the true sociolinguistic dynamic of the human world, as meaningful individuals with

---

[26]Panbanisha's social world was human, bonobo, chimpanzee, and sibling oriented. She made the most progress of all the bonobos studied to date.

expectancies, rights and responsibilities rather than as subjects, was sufficient to produce the *spontaneous comprehension of English grammar and the spontaneous production of grammar* (Savage-Rumbaugh *et al.*, 1993; Greenfield *et al.*, 2008). By nine years of age, apes reared in a bi-species cultural dynamic readily decoded novel spoken English sentences which included recursions,[27] pronomials,

[27]Recursions, according to Chomsky are the sin quo non of language. The form of recursion that Kanzi demonstrated was level one, Can you take the X that's in Y to Z. The question of whether Kanzi could perform a level two recursion was then raised, "Can you take the X's the in the Y that's in the Z to A? This was not tested because the physicality of this level of embeddedness was extremely complicated to construct in the physical experimental plan. Test trails suggested that Kanzi could do level two, but whether the experimenters could set the environment with hundreds of novel instances was questionable. Setting up the physical space for a few novel sentences is easy in this context, doing hundreds is not. Moreover, sentences like "This is the rat that ate the malt that killed the cat that lived in the house that Jack built" — while understandable by humans — have very little to do with normal human communication which never takes such forms. So testing Kanzi on sentences of this sort seemed pointless as we never contended he could understand sentences that were essentially incomprehensible on first listen. While "the house that Jack built" sentences are the quintessential test for recursion, recursion is actually an essential language property and it is manifest in nearly every language utterance at some level. Recursion is about how language reflects back upon itself. Language is constantly self-reflexive, in sentences, in paragraphs, in stories. Language — in the human sense — can only be comprehended and/or spoken by a self-reflective being. In any given sentence the words must relate to each other in some kind of self-reflexive manner to form a co-joined meaning of some sort. It is thus the relationship between the words themselves (as well as between the words, the present context and the past context) that gives them their meaning. This relationship is reflexive by its very nature. Human language is a self-reflexive tool invented by self-reflexive beings for the purpose of communicating. If it were not language as we know it could not exist. Traditional grammar limits the self-reflexive relationship to a given sentence, (as in the endless 'house that Jack built," but the relationship goes far beyond a given sentence. Reflexive relationships are constructed between utterances in different sentences and between utterances, gestures, glances, postures, etc. Meaning-making is distributed across all of these modalities in normal communication. The didactic tests that Kanzi and Lana received fit the rather arbitrary constructions of linguists but the real study of their capacities needs be embedded with the structural framework of systemic linguistics, as is done in this proposal.

and other syntax dependent utterances (Can you put the can of coke in the trash can, Give the dog a hotdog, Can you get the lettuce that's in the refrigerator) as well as concepts of metaphor (for example, Kanzi began to employ, on his own, the color term "yellow" for cowardly actions), moral agency, and pretense (Can you make the toy snake bite the toy dog?), (Lyn *et al.*, 2008; Lyn *et al.*, 2006; Savage-Rumbaugh *et al.*, 1998). These bonobos also communicated representational information to one another through their own vocal channel (Taglialatela *et al.*, 2002; Savage-Rumbaugh *et al.*, 2004). The existence of this ability in the *Pan/Homo* group implies either that human language and bicultural rearing, acting as an epigenetic variable, over-ride the supposed absence of language and vocal modification in the wild or that wild bonobos posess a vocal system that is under the control of free-will and intentionality. The current argument against the ability of non-human primates to vocalize is based on the view that a fixed relationship exists between external referents and the structure of primate calls. Therefore referential calls exist in non-human primates but they are presumed to be fixed and innate (Wheeler and Fisher, 2012). However there is also evidence to suggest that chimpanzees can alter their vocal repertoire (Watson *et al.*, 2015). Kanzi and family exhibited clear voluntary control over their capacity to vocalize (Fields *et al.*, 2007; Savage-Rumbaugh *et al.*, 2004; Taglialatela *et al.*, 2004).

The bicultural world of bonobos and human beings, sought to incorporate as many aspects of free-ranging bonobo culture as possible. Spoken language, the lexigram keyboard, and daily travel in a 50-acre forest environment constituted the world and mode of daily life for the experimental group — Kanzi, Panbanisha, Nathan, and Nyota. As infants and juveniles, these bonobos had no sense of being caged or confined, and no sense that bonobos and humans were members of different species. The traveling group was always composed of bonobos and humans living, interacting, and moving freely together as a unit in the forest. Close relatives of both species were always present. Conversations with humans (or other bonobos) were ubiquitous throughout the day and composed of contextually embedded normative gestural, vocal, and lexical communications about an unlimited range of topics directly relevant to all ongoing activities (Benson *et al.*, 2002; Pederson and Fields, 2009). Through this method, the bonobos like human children became pregnant with culture.

By contrast Matata and P-suke (both wild born) reared the bonobo cultural group, consisting of Tamuli, Neema, Maisha, and Elykia. Their interactions with human beings were limited to those that took place behind wire. Humans and bonobos were clearly conceptualized as different species by Matata. She had experienced human beings in the wild, and human beings had captured her. There were very few humans that she came to trust during her life, but those she did allow into her world became extremely important to her. She truly accepted them into her bonobo family and made efforts to instruct them in bonobo ways, as she did with all her offspring. Her passing in July of 2014 was unexpected and a great loss to the group. Culture carries on and her offspring were taught all that they could be taught by her in a captive setting. It is hoped that they will, at some point, be able to take this knowledge back to the Congo with them, though they will not likely be able to make their own way there. Still, they are fortunate to have this knowledge and they are the only living captive group of bonobos with such information.[28]

[28]The goal of returning the bonobos to sanctuary in the Congo arose, as soon it was understood that Kanzi was acquiring language without training and that he was producing high-level novel utterances with intent. As he has grown older, it has become increasingly apparent that the captive environment deprives him and his family of the freedom that their understanding of their own captive situation requires. The authors have funding support to provide Kanzi and family with the needed safe environment in the Congo. However as the paper is being written, Kanzi and family are being restrained from all human contact and the research trajectory with them has been terminated. Their care now falls under strict biomedical protocols and they have been placed in conditioning studies (so that they will present for injections) and foraging studies to restore their species-specific behaviors. It is our intent that this situation be terminated and these bonobos be allowed to communicate and to express their own desires.

While in Atlanta, these groups lived in different buildings following the arrival of P-suke but were provided daily opportunities to interact and to form social bonds through means of tunnels several blocks long that ran through the woods to connect the different buildings. The vocal language that Matata brought with her from the wild was infused into the entire group and they communicated daily through the woods as they might in the wild. They also regularly communicated vocally with the chimpanzees that lived in a third building. It seemed that inter-building language played a key role in Kanzi's acquisition of lexigrams and his understanding of human speech (Ragir and Savage-Rumbaugh, 2009.) To the degree that Matata's vocal world was symbolic, Kanzi was born into a bilingual symbolic cultural world. Children reared in bilingual homes pick up both languages and Kanzi did so as well.

Sherman and Austin were not born into such a world. Austin was taken from his mother at birth. Sherman stayed with his mother for 1.5 years. Sherman was developmentally advanced relative to Austin. However neither was exposed to a symbolic cultural and vocal world constituted and constructed by either chimpanzees or human beings. Sherman's mother, Cookie, never lived in such a setting, when she kept Sherman as a baby. And Austin's mother was not even able to rear him, having served as subject — when an infant — in an early deprivation study (Davenport and Rogers, 1968). Thus it is not possible to was not appropriate to compare Kanzi's spontaneous linguistic progress to the lack of similar progress with Lana, Sherman, and Austin. To make any comparison between species the rearing environment would must be held constant and a chimpanzee and a bonobo must be co-reared together.

The presence of Matata made a major difference in the lives of Kanzi, Panbanisha, and Nyota, as did constantly living in a forest. An important principle of care was that bonobo infants and/or children were never left alone, nor in the care of someone they did not know and fully trust, either bonobo or human being.[29] From the

[29] The Gardner's (1989), Miles (1994), Patterson (1978), and Terrace *et al.* (1979) all emphasized the importance of close contact for the apes in their language projects.

day Kanzi was able to speak, Matata appeared to be bringing Kanzi into her language world in ways similar to what human mothers do, that is by engaging back and forth rhythmic utterances. However these utterances were impossible for human beings to emulate. Even though our species is known for its vocal emulation skills, it is not possible to emulate bonobo speech with the precision they employ. A person can attempt to sound like a bonobo and they and they will respond by being very polite but will indicate that the person has really, to their ears, said nothing.

As Kanzi and Panbanisha began to become proficient in symbol production with lexigrams as well the comprehension of spoken English, it became clear that they perceived Matata as 'having some kind language' that to the meaningful which utterances directed to them were meaningful. They also knew, that they knew, how to talk to Matata. This could be ascertained by simply suggesting that they "Tell Matata X or Y." In response to such a request (from a person they trusted and knew well,) they would turn toward Matata and begin to vocalize. (They could not have acquired this capacity from human beings, as we did not possess it.) The presence of this capacity revealed the bonobos' awareness of, and a translation competency between, different parties in their unique cross-species bilingual world. Matata also employed both Kanzi and Panbanisha, to convey requests to us for items that she wanted. Matata realized that we could not understand her, but that Kanzi and Panbanisha could. She also recognized that she could speak to them (vocally) and ask them to make lexical requests for her and that they would do so. When we returned with the items Matata wanted, they would refuse them and gesture toward Matata, or take just a small portion and then gesture toward Matata.

However they all left the infant chimpanzee's to sleep alone at night. The importance of bodily contact and reassurance at night cannot be over-emphasized for the well-being and psychological development of infant chimpanzees. All bonobo and chimpanzee infants noted in this work slept adjacent to parental figures each night until they became old enough to desire to sleep alone. Leaving an infant in a locked crib after five until the staff arrives the next morning at eight or nine is not the normal experience of infant apes and results in social disruption and detachment.

Even though they were willing to translate some of our utterances to Matata, there were also things they would never attempt to convey. They behaved as if they clearly understood that Matata's mental world, and our mental worlds, were not the same. That is, there were things they seemed to view as simply not comprehensible from Matata's point of view and were thus not appropriate to translate. They would make no attempt to translate these items no matter how much they were encouraged to do so. On the other hand, they tended to translate information regarding the locations of food, the type of food and plans for obtaining food whenever we asked them to do, and often when we did not so request. They nearly always translated any information about visitors arriving or leaving and information about a visitor's behavior, demeanor and activity, even without a suggestion to do so. They would not repeat translation once they were given. The function of the translations, in their view, was to get Matata to understand, not to satisfy our request for talking to her. They rarely if ever, translated any information about things unrelated to danger, food, travel, visitors, or unusual events no matter how many requests we made or how we phrased the request. They seem to understand information about many of the more mundane things but did not find them worthy of translation. For example, they paid close attention to gossip among staff, but they often they not communicate such gossip to Matata. We assumed they found her disinterested in those activities. Request to translate manufactured events, as might be presented for a test of this skill, were also spurned. Only authentic real-time current or probable and immanent events were translated. One might conclude from this that Matata had no means of talking about the past or thinking about the past. This did not seem to be the case, since we could ask about Matata's memory of the forest and they would ask Matata to talk about the forest. Kanzi and Panbanisha, by contrast, enjoyed vocally participating in human conversations about the past, the displaced present and the future.

The capacity for high-level vocal linguistic communication between Kanzi and Matata first became clear when Kanzi was 2 1/2 and in a situation, where the apes were out of one another's sight. Matata suddenly started speaking in a rather urgent tone. Kanzi immediately climbed to the top of the room he was in — about 20 feet up — so that he could see out the monitor windows in the top of the room. He starred south intently. He never ordinarily did so, so the first author wondered what he was looking for. The first author walked outdoors to see why and found that about 1000 yards away a telephone service person had climbed a pole. Matata could see this unusual person climbing a tree from her location but Kanzi could not. Matata had to have told Kanzi that there was something outdoors, high up and the direction of the person, in order for him to immediately climb upwards approximately 20 feet (on a sheer wall) to look out the window. This took remarkable effort on Kanzi's part and he would not have done this without a reason. Apparently Matata was startled to see a human being in something that looked like a tree. After I understood what was occurring I explained to Kanzi who this person was, and why he was there. Kanzi then began to speak in the direction of Matata. She immediately calmed down and stopped worrying about the person and we proceeded with our activities, but Matata kept a close eye on the man on the telephone pole.

Kanzi and Panbanisha began to employ referential gestures for requests and statements (Pederson *et al.*, 2009), demonstrating TOM in the classic linguistic "Sally Ann" test setting (Savage-Rumbaugh, 1997). Wild bonobos specify travel direction by dragging branches on the ground to suggest the intended direction of travel. Panbanisha began to regularly illustrate, where she wanted to travel outdoors, by drawing lexigram-like constructions on the floor with chalk, then adding a line that went from the lexigram directly toward the door, in the direction to be traveled in the forest, just as wild bonobos drag branches in a line indicative of their travel plans and then vocalize. Panbanisha combined these two processes into one — of her own accord, and used lexigrams plus vocal words, because we understood the lexigrams but not her words. They engaged in conversations about complex events, including those removed in space and time with the inflection of moral rules of engagement and normative action (Fields *et al.*, 2007). They responded to questions regarding their social lives and their captive conditions, and reported upon past events of their lives (Savage-Rumbaugh and Fields, 2000a,b). These symbolic apes revealed thru their utterances that they were aware of themselves and that they possessed internal personal and historical narratives of their lives. They functioned as moral agents within their world-view and behaved in socially appropriate and morally responsible ways as adults (Fields, 2008).[30]

As adults Kanzi and Panbanisha came forth with their own adult construction of both the *Pan* and *Homo* worlds which produced them. The forest world of Language Research Center (LRC, Georgia State University) provided an enculturation experience unique in evolutionary history, allowing many bonobo cultural activities, which could never have taken place in a zoo, to occur. Moreover, in the forest their human companions were not "caretakers" — they were companions negotiating life in the woods. We found food together, we avoided true predators together (there were panthers in the forest) and we constructed lives of a myriad of different activities each day. Our lives were co-constructed from the tangible effects of co-living in a *Pan/Homo* world (Segerdahl *et al.*, 2005). The *cultural fusion* produced by the diverse *Pan* and *Homo* worlds around them led these bonobos to develop vocal utterances (in addition to lexical utterances) that functioned as a vocal language, different from, Matata's vocal language. This language was beginning to be comprehensible by human members (Savage-Rumbaugh *et al.*, 2004) until the human

---

[30]The rapid rise in the abilities of the bonobos as the Pan/Homo culture began to solidify required new methods of documentation and reporting. Treating the bonobos as experimental subjects near the level of two and three year olds was no more feasible that such a method would be with an indigenous group of human beings. Not only did this culturally based rise in ability produce methodological issues; it produced social issues. The true linguistic abilities rapidly bootstrapped skills in the group setting that went so beyond previous claims at a rate of speed that met with great resistance.

members were excluded from participation by outside cultural human forces that opposed the *Pan/Homo* world.[31]

When one is in the midst of a group of bonobos, such as Kanzi and his family, one has no sense that one is with beings that are retarded, developmentally delayed, childlike or slow in any regard. One rather has the constant over-riding impression that one is with highly sensitive and intelligent beings who are able to communicate with human beings only a small portion of the things they are constantly communicating with each other. As human, one feels at times like they are like an autistic bonobo, that is a bonobo who cannot quite behave appropriately the social group, because they are not processing all available information.

7. *Pan/Homo* Bonobos

Some scientists have characterized the *Pan/Homo* bonobos as "home-reared" apes, or human-reared and human enculturated apes. They are not. They acquired wild bonobo culture from Matata, a pre-adolescent female brought into captivity by the New York Academy of Sciences under a special lend-lease dispensation from the Congo. The research began as means of seeing how far the genome of the bonobos could bend, of its own accord toward the dimension of "humanness" regarding activities, skills, and capacities universally presumed to belong to the genus *Homo*. The major domain of research was language but it quickly began to spill over into tool making, the narrative, and the psychological

---

[31] While the human-like abilities of the *Pan/Homo* bonobos generated a great deal of interest among the scientific and lay public alike, they also generated hostility. Some persons disliked the idea that an ape could make a fire or that an ape could learn the numeric code to the locks on the door. While on the one hand it may seem interesting to determine the ape's limits for language, it is quite another thing when that ape suddenly begins doing many things that beings with true language do. That level of sentience can become frightening as illustrated in movies like *Planet of the Apes*. This hostility resulted in attempts to defame the research as well as the researchers and attempts to locate the bonobos in a zoological park or biomedical research facility.

'self', music, art, agriculture, moral group standards etc. These changes came so rapidly that the ability to place them properly in the literature was overwhelmed by the ever-present need to argue for the existence of the key change — that of language, and to care for the increasingly diverse and sentient ape beings that were suddenly emerging, in the midst of political landscape that was rejecting them.

Language appeared to drive all other changes. It was as though by bootstrapping a bonobo brain with a human language, there emerged a global effect on the biological substrate. Everything else shifted as the social and physical worlds became co-constructed worlds, and symbolic worlds — formed and framed by the mind of a new kind of being, a *Pan/Homo* symbolic thinking, and living-breathing personage. These changes affected the researchers and staff who had committed to the work and the bonobos. They affected the employees who were there for jobs but not prepared to undergo cultural changes.

Pan/Homo beings are sufficiently different that no one should expect them to behave as zoo reared bonobos would in test settings. The same may be said of high-functioning autistic persons. Such persons often do not fit into our current cultural expectancies and they may be socially excised for their failures. All self-aware conscious entities with freewill must behave in certain ways during social exchanges to avoid being peripheralized and excluded from the group. For a culture to normalize its members, all must agree to co-act in proscribed ways. Enculturation is a process of learning to dance with the other members of your group of symbolic self-conscious beings. It is the only process by which the culture can reproduce itself and it can be extremely tyrannical in its demands. When a *Pan/Homo* culture formed, it naturally formed its own demands (Fields *et al.*, 2007). While Matata resisted these, she increasingly came to conform to the *Pan/Homo* world, as did the human members of this world who also initially resisted the transformations taking place in them. The outer human group of scientists, staff, administrators constantly expected the bonobos to act more human as though this were the goal. It was not the goal and it did not occur, instead the *Pan/Homo* culture became increasingly unique distinct.

Human groups have always separated themselves across time and they continue to do so on the basis of epi-cultural markers such speech, dress, food, etc. We wear these behaviors like badges; the smallest behavioral markers delineate entire sets of inter-individual behavior patterns. These co-acted expectancies, constantly exchanged, go to the depth of what we believe ourselves to be. When a child learns how to behave according to the expectancies of its family, friends, etc., is really acquiring inter-individual patterns of co-action. No two persons are precisely alike and no co-action is ever precisely repeated, but the patterns of co-action are repeated in hundreds of thousands of ways. From these jointly constructed patterns of co-action, human groups unwittingly extract expectancies, codes, and rules of proper behavior. Together the members of groups co-build a culture of expected interactions. Any group of symbolic self-sentient beings will naturally become uncomfortable when the behaviors they extend to others fail to meet with their anticipated results. This always happens in cross-cultural exchanges. Humans have historically found it difficult to look across cultural boundaries and to see humanity as one entity. Thus humans have found it even more difficult to look across both cultural and species boundaries to understand great apes. But when one becomes a *Pan/Homo* being, the cultural forces act in an imperceptible manner as the perceptual blinders start to drop and the physical differences become almost invisible. As this happens essentially every intentional symbolic acts of each member takes on symbolic meaning. Symbolic meanings literally jump into existence everywhere. There are no non-symbolic intentional communications between bonobos and humans who become members of the *Pan/Homo* world.

To be inside the living world with the *Pan/Homo* bonobos is to constantly feel and experience the meaning being made all around one, in one and through one. There is never a moment when meaning is not being made in highly sophisticated novel ways. Meaning is forever made at multiple levels, with sophisticated innuendos.

It is extended or withdrawn at will toward human beings, based on their behavior. Formal tests — which do not involve real world communications, will cause bonobo meanings to be withdrawn instantly. One will not even be included in-group activity if one is constantly trying to test whether not the activities in which one is engaged within are actually occurring to the others. If this does not ring true — simply try testing a group to human-adults to determine if they can point (by the methods used to test apes) to determine how rapidly you can be excluded.

Human, Bonobo, and Chimpanzee Genomic Similarity: The Enigmas

The data cited above can reduced to little more than a set of interesting accounts about a few apes that seemly aped human beings, unless they elucidate our own evolutionary history. Do they? One thing is clear, they tell us that language matters even more than we might have thought and that it can fundamentally alter the hardwire of the modern chimpanzees and bonobos. *Language can make bonobos far more like ourselves in terms of how they structure their mental lives*. These findings imply that the seemingly great differences between *H. sapiens sapiens* and *Pan* are software constructions riding on the symbolic/grammatical system of inter-individual relationships that culturally instantiate that which we call language and more specifically that which we call *human language*.

Molecular genetics has revealed that humans are more closely related to chimpanzees and bonobos than scientists — who previously had access to anatomical evidence alone — would ever have surmised (Wildman *et al.*, 2002). The degree of DNA similarity between humans, bonobos and chimpanzees indicates we are sibling species. *This means each of us is more closely related to the others than any of us is an orangutan or a gorilla* (Wildman *et al.*, 2002). How can this be when the anatomical resemblance between bonobos, chimpanzees and gorillas has been sufficient to cause species confusion in zoological gardens, but no one has mistaken a person for an ape? The genetic data also implies that our bodily forms are only partly determined by genes and that additional forces must be operating to cause physical overlap (such as that which appears to exist between bonobos and gorillas) apart from the genetic code. Presumably if this force were acting in another direction, bonobos could appear more like us or we could appear more like them. What is this force and how is bending morphology?

Standard evolutionary theory assumes that *H. sapiens sapiens* have undergone more rapid evolutionary change in the past three million years, than apes, but exactly the opposite seems to be the case (Bakewell *et al.*, 2007). Chimpanzee DNA has altered nearly twice as rapidly as has human DNA in the same amount of time! Why would this be? After all it was we who invented civilization, and spread around the world not apes. This finding would suggest that human beings were the root ancestor and chimpanzees and bonobos evolved from us if taken at face value. It is worth noting that there are no systematic records of ape fossils. All transitional fossil forms seemingly lead to modern *Homo*, not to modern *Pan*. If we turned the standard model upside down and allowed for chimpanzees and bonobos to devolve from us, the data would make more sense, but it still would not explain how we came to posses the supposed uniqueness attributed to us. In fact it would make the question more complex, because if today's living apes devolved from humans, then we would have the far far more difficult task of explaining the total leap from monkey to man.

How and why human uniqueness emerged has become the central question of the behavioral, linguistic, sociological and philosophical sciences at this time (Savage-Rumbaugh *et al.*, 1998; Noble and Davidson, 1996). Darwin could not tell us how we evolved from ape to man. Neither has the answer been found through genetics, though it has begun to answer the question. Shockingly, it appears that the answer does not reside in the evolution of the genes (Goodenough and Deacon, 2008), but in the evolution of culture and cultural processes. Culture is the ratchet that made man from ape (Tomasello, 2008). By all reasonable biological explanations, human beings should not exist as special species apart from all others. Yet either we are essentially separate; or we have profoundly and egregiously underestimated and misunderstood our closest relatives. If we have committed the sin of underestimating apes, we may also have under estimated the plasticity of the biology. How has culture come to shape and reform apes to produce human beings? Have we underestimated ape cultures in the wild? This would not be a new sin, as we similarly underestimated the aboriginal tribes of Africa when European explorers first encountered them in them in past centuries (Virey and Guenebault, 1837). Europeans did not believe that these individuals were sufficiently human as to be capable of acquiring the behaviors needed by European culture. They were wrong.

Yet ape cultures appear to differ far more from us than any known human culture. Indeed, many anthropologists would maintain that apes lack the true symbolic cultural structures that define humanness (Tuttle, 2014). Known ape cultures lack fire, cooking, clothing, overt kinship markers, music, art, and other skills that are highly object dependent. These are all things that require the use of objects in ways that are not reported for wild ape cultures. However, if we define culture as the glue which allows self-conscious beings to make meaning of each other's actions and communications by binding together acts that co-mean, ape cultures may well be found more similar to ourselves that we have previously acknowledged.

The current politically accepted view of apes holds that is *H. sapiens sapiens* who are special, but it is so difficult to explain how this occurred that some academics have turned to the paleo-contact hypothesis. This hypothesis argues that humanity was visited by beings in space ships from outer space. These beings gave us the keys we needed to bootstrap human civilization such as mathematics, writing, and astronomy (Cairns-Smith, 1975). While such theories run counter to standard scientific consensus, the seemingly sudden onset (in evolutionary terms) of language, speech, music, art, architecture, writing, metallurgy, centralized government, etc., cannot be explained by standard scientific models of evolution.

Moreover, to explain the human/ape separation we must also address the body/behavior problem. Why if we have had relatively modern human bodies and very large brains for at least 400,000 years, did we wait so long to develop any special abilities? And if it was some small gene mutation that caused the sudden appearance of civilization, why do some modern human groups still exist who have failed to make any progress? All the seeming vast differences between living groups of humans are known to be the result of culture, not genes. Thus we waited a very long time before becoming ourselves, as we know ourselves to be today, even though our bodies were human bodies. Many modern human populations living in isolated uncontacted areas today seemingly prefer to continue so doing. Their life styles are sufficiently close to those of living apes as to not strain credulity that humans and apes could be related. It is modern civilization that strains credulity. These uncontacted human groups have not employed their human uniqueness in any particular manner that would make then instantly recognizable as special beings — were aliens to arrive today (McGrew, 2004). Therefore the fact that we have not yet discovered an ape culture that manifests fire, kinship and cooking does not mean that such a group of apes does not exist. Kanzi understands fire, kinship and cooking. Could he perpetuate that knowledge in the wild?

An equally perplexing question is why, if intelligence and language are the key to human civilization, did the Tasmanians in Australia lose their cultural capacities? The behaviors of these human beings were so unhuman like — that they appeared to be on the verge of speciation from *H. sapiens.* When first discovered, they had no tools more complex than chimpanzees, they did not cook, and they often slept without shelters. Yet they did have language. They were not integrated into the new cultures that arrived and became extinct after contact with modern humans? Why was society unable to assimilate them? *The truth is that being genetically human does not inevitably lead to humanness as we know it today and we do not know why.* We are the most intelligent species on the planet but we seem to have lost our history. We don't know how we got here and we don't know if we have sufficient intelligence to survive in numbers of eight billion and counting.

Whatever happened in our past one thing is no longer controversial; a special genetic relationship exists between *bonobos, chimpanzees, and humans*. The fact is that, "There are widespread regions of the genome where chimpanzees and bonobos are less closely related to each other than any of them are to humans," (Caswell *et al.*, 2008). Thus both have their own special and *unique* ties to us. Again, the problem for evolutionary theory is that this fact would not be predicted by an evolutionary account in which we diverged from them.

The most obvious biological difference among ourselves, chimpanzees, and bonobos is that we have 46 chromosomes and they have 48. Are chromosomes like grocery sacks, it does not matter which sack the groceries are in as long as you get them home? It certainly matters in conditions such as trisomy which leads to Down's syndrome. This trisomy produces changes throughout the entire body (Miller, 2000). The current evidence indicates that humans were formed when two chimpanzee or bonobo chromosomes fused (Yunis and Prakash, 1982). This raises the question of how could it be that humans evolved from chimpanzees in a gradualist manner if we were formed by fusion event. It also suggests that we were made directly from chimpanzee stuff, with a small pinch of our own added DNA. That pinch of stuff somehow (no one is sure exactly how) made all the difference. If we adopt a strict evolutionary perspective, it would appear that a random event (chromosomal fusion) and a random mutation that added a pinch of extra DNA, produced something human like rather quickly from chimpanzees. This event happened once and nothing like occurred before or after, for we believe that we are only beings on the planet that are sentient, rationale, and linguistic. There are those who offer evidence to suggest that such a fusion did not occur (Tomkins, 2013), in part because the putative fusion site lacks satellite DNA, which is generally present in known fusion events in living mammals (Ijdo *et al.*, 1991).[32] Their explanation is Divine

[32]Some of the diverse data is arising from creation scientists who are working to interpret scientific data from the perspective of creationism. This has occurred

Intervention. The alternative hypothesis, namely that a fission event (or events) led to chimpanzees, gorillas and bonobos has not been considered. The problem is that none of the current explanation really work and we don't know why humans have 46 chromosome, and chimpanzees have 48, or why two of theirs appear to have fused to produce the our number two chromosome.

The breaking of a chromosome, by a previous civilization that had acquired the ability to manipulate chromosomes outside the body, would be a simple feat. We manipulate genes all the time. Some evidence has been found to suggest that human civilizations existed long ago, and is detailed in the book *Forbidden Archeology* by Cremo and Thompson (1998). Like alien theories, these data stand outside science because there is no means to verify the hypothesis of previous continents and highly developed civilizations of the past. Findings counter to the standard view are rarely reviewed by mainstream journals, a practice that is generally good, but it can foster an unhealthy dependence upon a particular theoretical point of view. Intelligent Design is not a viable theory, because it is untestable. Thus scientists have found themselves caught between the impossible-to-investigate (Intelligent Design), an outdated imprecise theory (Darwinian theory), and the non-verifiable (ancient civilization/alien intervention) when they try to understand human history.

Being made from a chromosomal fusion effect, as we seemingly were, is a highly unlikely thing. Rarely are other species so formed. It is an event that should never have happened at all. It must have happened at least twice (and probably more) or there would be no descendants. Moreover, the small bit of added material contains

---

because scientists have become as passionate about interpreting data from the Darwinian perspective as creationists are from the viewpoint of Divine Creation. Constantly rotating between these two paths leads each to attempt to deride and debunk the others. Scientists forgot that science is about method and its goal to attempt to find truth rather than to assume that is holds truth firmly in hand. A new evolutionary framework to explain change across time is needed in biology, one that is open to alternative data and alternative approaches, and that is foster the experimental approach.

FoxD4 genes; these are genes that help guide the developmental process. Because fork-head genes are key regulators of embryogenesis and tumorogenesis, the mutations in one or more of the fork-head genes associated with humanness may also be responsible for our increased susceptibility to cancer. Humans are prone to many diseases that apes don't normally get, including cancer that metastasizes (Fan *et al.*, 2002). Thus on our way to becoming human we travelled a strange path, one which evolutionary theory would not predict and which seemingly changed everything by propelling us beyond the evolutionary processes presumed to govern life on planet Earth. We even bought humanness at the cost of increased tumorogenesis. Why should humans be selected for an increased ability to become ill? The obvious Darwinian answer would be that we gained something that we needed which made up for our increased tendency toward cancer.

If we examine DNA differences of similar magnitude between other closely related species — we find that they produce only minor changes in anatomical and cognitive traits. Yet among chimpanzees, bonobos and humans, small differences in DNA created differences of such magnitude that we group chimpanzees and bonobos in the category of animals, and save the genus *Homo* for ourselves and our extinct non-human relatives; neanderthalensis, *denisovans*, *erectus*, etc. Were we to find similar levels of relatedness among other species we would place them all in the same genus. The *Homo/Pan* divide stands beyond normal biological classification, thereby bending the classification system to fit our perceived human transcendence.

Increasingly, scientists are finding that genetic relationships do not map onto the classic morphological relationships in a logical manner in other species. The picture of animal evolution has become more confusing (rather than less confusing as was supposed to happen) with the advent of modern genetics. We find the displacement of taxa long thought to represent successive grades of complexity at the base of the metazoan tree, now leads to much higher positions inside the tree. This has left scientists with no evolutionary intermediates. Thus the gradualist argument is being undermined by genetic data on species other than apes and human beings. This reappraisal of animal evolution rests on several congruent approaches ranging from primary gene sequencing to looking at the qualitative molecular signatures within appropriate genes (Adoutte *et al.*, 2000). Morphological closeness does not correlate completely with genetic closeness.

Attempts to explain the mounting data regarding the genetic similarity of bonobos, chimpanzees and human beings have led biologists, psychologists, primatologists, anthropologists, and philosophers to assert that, at some point in the *Homo* linage, a total discontinuity emerged between ape and man — but that this discontinuity took place at the psyche level not the morphological level (Goodenough and Deacon, 2008). The epicenter of this discontinuity was based on the appearance of human language — and defined as the emergence symbolic mind by Deacon (1998). The human brain, upon this view, is said to completely reorder the mental world. *That is, an impossibly small change in the wetware allowed the brain to reorganize itself, the body and the entire world beyond the body, and set the process of evolution off on a completely new track* (Goodenough and Deacon, 2008).

The symbolic mind may have accomplished this extraordinary feat by internalizing the world and then abstracting it. Many mammals and birds are assumed to build an inner operating theater of sorts in their head, but not to abstract that theater, The inner theater of mammals and birds allows them to make use of the best choice method, through an internal Darwinian selection of behavioral options built from past actions, to determine what behavior is likely to be most effective next (Calvin and Bickerton, 2001). Human beings behave in a similar manner much of the time, but not all the time. The fact that we behave in this way much of time is due to the cultural conditioning or ontogenetic ritualization that is required for the construction of the complex behavior required to conduct daily life in social group. This kind of conditioning of both individual and inter-individual behavior is assumed to be characteristic of both animal and human behavior. It may even permit some forward planning in that the Darwinian selection scenario can propel itself several moves ahead based on the internal theater. Thus animals, and humans, readily develop expectancies, which produce frustration if they are not met. The critical difference in the symbolic brain is that is allows us to remove ourselves entirely from the reality theater typical of other mammals, that is the theater that is a mental version of the world they inhabit. With our abstract theater, we can construct and imagine the realties we prefer — and then our intelligence (which was already present in apes) allows us to build those imagined realities into our co-constructed group realities. Time and space become completely constructed entities within the human internal theater. In essence, the symbolic mind turns an ordinary mammalian brain into one that is no longer responding to the world around it but is rather one that is generating the world around it through an infinite myriad of cultural lenses.

Apes, by contrast, are thought to lack the capacity to construct abstracted worlds (Deacon, 1998; Dunbar, 2009). Prior to Boas, it was *primitive peoples* that were thought to lack this capacity. Our seemingly inveterate need for *the other* — or the not us, by which we seek to define our own uniqueness, led us to replace primitive peoples with apes. Apes seemingly had no language at all and were depicted as sufficiently aggressive to fit the human need for an animal version of the self, against which to construct a portrait of the human being. Additionally the building skills of apes appeared to be limited to nests. The problem is that we now know that many human groups restrict themselves to shelters that are nest like. Such shelters are basically woven tree branches, albeit constructed on the ground (Van der post and Taylor, 1984).

Thus, if we humans are capable of constructing modern high-rise cities, why don't all human cultures do so? Even without the advent of modern machinery many ancient cultures constructed architectural masterpieces. But there are living groups of humans who do not so, making it difficult to argue that differences between ourselves and apes are architectural, even though many of greatest feats of various cultures have been architectural (Lucie-Smith, 1992). If some human cultures have not engaged in complex construction, how can we be sure that all ape cultures are incapable of doing so? It could be argued that under the right circumstances, and given enough time and the right environmental pressures, such behaviors would emerge in bonobo cultures.

The purpose of this example is to clarify the problem of defining species-specific behaviors, as well as the problem of defining species according to behaviors. Behavior is imminently flexible. Any large brained ape may or may not display behaviors that are within its ultimate capability. *There is no means by which to determine the ultimate capability of any group of beings capable of self-knowledge, freewill, biographical memory, and abstract thought and symbolic expression*. Apes (human and non-human) are generally capable of self-knowledge, freewill, biographical memory, abstract thought, and symbolic expression (Segerdahl *et al.*, 2005). The degree of these capacities varies along a sliding scale in both humans and apes. *The sliding scale is an artifact of their culture* — and the culture is created by the beings that manifest it. Thus does culture accrue the capacity to, as Tomasello (2008) asserts, to ratchet us toward the future. However culture also has the capacity to hold us in a given place. Humanity can, and has, constructed cultures that employed language to create cultural stasis.

The genetic differences between chimpanzees, bonobos, and ourselves have proven to be uncomfortably small, compelling scientists to acknowledge that the formation of our brains (as well as those of bonobos and chimpanzees) is a fully plastic affair (Gazzaniga, 2011, Goodenough and Deacon, 2008). Thus the most prized feature of our species — our brain — is apparently structured by something beyond genes. It is that abstract something, which is thought to provide the brains of *H. sapiens sapiens* with a list of super-abilities, thought to be absent in other apes.

Another anomaly to arise from genetic sequencing analysis is that that the human genome displays far less variability than that of any living apes or previously living *neanderthals, denisovans, erectus*. This again implies that apes evolved and changed rapidly while *H. sapiens sapiens* remained stable; or that all current living human beings descended from a relatively small population, or both. Recent evidence suggests *H. sapiens* went through an evolutionary hourglass approximately 70,000 years ago (Petraglia, Korisettar, 2012). Our numbers were reduced to between 2000 and 15,000 individuals, for unknown reasons. Our lack of genetic variability is surprising because we are the most widespread species on the planet. More than any other ape, we have elected to travel, travel and travel. Our ability to construct mental-models of worlds distinct from those we inhabit seems to have caused us to be restless with the *status quo*.

Our recent circumnavigation of the globe appears to have began approximately 42,000 years ago (Balter, 2011). In a species with a high tendency to travel and spread out, variability should have increased, but this did not happen. Forty thousand years ago we *H. sapiens* began interacting with Neanderthals and Erectus and we have genes flowing from these species in to modern humans. The "purest humans" today (that is those who are free of Neanderthals genes) are found only in sub-Saharan Africa. However, all human populations (dispersed or not) remain more closely related than other apes, (Adoutte *et al.*, 2008; Caswell *et al.*, 2008). This is puzzling. We know that apes interbreed between neighboring groups and we know that humans typically, even in sedentary African groups, travel further and interbreed more than apes living in the same vicinity. Why then is their genetic variability significantly greater than our own? It seems that of humans and apes living in similar African areas, the human beings are relative new comers.

The evidence suggests that anatomically modern humans have been around for at least 200,000 years and possibly longer (Reid and Heterington, 2010). Anatomically modern chimpanzees have been around for about the same time period. Only in the last 10,000 years have humans begun to produce modern civilization with agriculture, city-states, trade in material goods, etc. We have accomplished all of this with a relatively restricted genome. Why is our genome so radically restricted in diversity compared to that of our living (and dead) relatives? What does this restricted genomic set portend for our future?[33] What, if anything, do we need to protect, in terms of genetic diversity, for future *H. sapiens*? Why, if this genetic bottleneck resulted from some climatic event, did it decimate human populations but not ape populations? If it were not a climatic event, why would humans have decimated their own numbers to such a significant degree? The only human ability we know of that would cause such devastation is the atom bomb. Ancient texts suggest that humans developed atom bombs and airplanes in earlier long-lost civilizations, but there is no proof of such an event. Possibly, as we probe more into the mysteries offered by DNA, we will find answers to our history — but to do so, we must be willing to entertain alternative hypothesis to the standard model regarding our history.

Answers to these questions will offer a better understanding of the puzzling gene/body/behavior relationship between ourselves and the other living apes. But until the answers arrive, it behooves us to reflect upon the fact that many cultures exist today that have not evolved beyond Stone Age technology. Except for the fact that they are human, we would have no particular reason to assume that they have language or a complex symbolic world. Anthropologists, who have penetrated such groups, assure us that they possess all the cognitive abilities that we possess. However, apart from our ability to relate to these groups of humans and to acquire their language, there is little to suggest that, in terms of material culture, they are cognitive light years beyond chimpanzees or that they will go on, someday, to develop a modern society based on human reason and insight.

Our work to date with captive bonobos has revealed that bodily form can be rapidly altered within one, two or three captive

---

[33]It is widely assumed that genetic variability must be large for a population to survive, however this has not proven so with laboratory animals or domestic animals, which are often intentionally inbreed, nor with free-ranging population of macaques that have been introduced to non-native areas. Large healthy troops have formed from a small number of animals. Inbreeding seems to be a particular problem in the human species because of our evolutionary bottleneck and the coincident reduced variability at the population level.

generations, without genetic selection. Muscle mass decreases in density and becomes softer. Canines decrease in size; infants tend to walk upright for increasingly longer periods of time, the length of the legs increases and the length of the arm decreases. Skull diameter increases. The ability and desire to engage in human activities also increases sharply. So does the anatomical ability to perform them. Self-motivated desire arises to engage in human activities. Self-motivated desire arises to engage in self-practice in order to achieve competency in language, tool-use, and other human skills. Thus the changes we have observed in captive bonobos in just a few generations have implications for selection theory as well as for understanding the brain/body dynamic as a function of environmental factors. The changes with one small captive group demonstrate that physical and behavioral changes in the absence of a process of selection, can and do occur, as function of socio-cultural triggering.

On the bases of a genome wide comparison of human and chimpanzee DNA, Cheng *et al.* (2005) advise that most of the asymmetrical increase of duplicated DNA in the chimpanzee lineage emerged as a mechanistic consequence of changes in the chromosome structure rather than forces of selection. Their view is echoed by Feuk *et al.* (2005) who note that the recent genetic evidence on deletions, inversions and copy number variations should "cause people to rethink their ideas about how species evolved" (Donovan, 2005). How then should we think about chromosomal fusion and other genetic anomalies?

Any unusual change — such as the combination of two chimpanzee chromosomes into one human chromosome — would need to occur must occur multiple times to form a completely new species. Additionally those individuals with that change would need meet and to reproduce, then their offspring would need to meet and reproduce before a significant morphological change (such as moving from an ape-like anatomy to a human-like anatomy) could become sufficiently fixed to produce population of human beings from a population of apes. As Marshall (2008) — in attempting to explain the Cambrian explosion observed: "It's not new genes that create new morphological innovation, but rather the way they're wired together, different-looking creatures are not apples and oranges," they just look that way.

With the aide of modern molecular phylogenetics, we now trace the origin of *H. sapiens* to central Africa between 15,000 and 200,000 years ago. Our first ancestors moved out of Africa and into China and Europe between 60,000 and 80,000 years ago, we know that probably only a small group left Africa, possibly as few as 150 people (Liu 2006). The *Homo erectus* and *Homo habilis* populations found outside Africa are not direct ancestors of any modern humans living in those locations today. Once *H. sapiens* arrived on the scene, for more than 30,000 years (the time period between the anatomical appearance of the species we call *H. sapiens* and the appearance of behaviors which are clearly associated with symbolic processes, so characteristic of *H. sapiens*) almost nothing is found in the archeological record to clearly indicate that humans behaved symbolically (Lock, 2000; Mithen, 1996; Yellen *et al.*, 1995).

If we assume that this gap is real, rather than the result of inadequate preservation (or current lack of discovery of appropriate remains), then we have no choice but to conclude that modern human language and symbolic behavior arose between 40,000 and 70,000 years ago. Before that time, no examples of representational art, fossilized forms of bodily decoration (such as ochre, beadwork, clothing, etc.), hearths for cooked food or indications of widespread trading have been identified and accepted by the profession as valid. Thus for some 30,000 years, modern humans would have lived in forest conditions similar to those inhabited by modern day bonobos and they seemingly did not execute any of the capacities we, today, associate with humanness. Once these symboling humans arrived on the scene, they proceeded to change he face of the earth wherever they traveled but not so much wherever they stayed behind. The rise of *H. sapiens* is the single most profound speciation event in all of evolutionary history. Yet it seems to have happened without any particular precipitating events.

Some have argued that modern *H. sapiens* arose because man suddenly gained a new biologically based capacity. According to this view the appearance of human language began with a single point mutation in the FOXP2 gene (Enard *et al.*, 2002; Varga-Khadem *et al.*, 2005). Of course it is also possible that the current lack of fluent vocal ability in chimpanzees is the result of the absence and/or loss of a single point mutation. The genome not only gains material, it can also lose it. When the FOXP2 gene undergoes mutation in modern humans, we manifest a language deficit known as developmental verbal dyspraxia (Vargah-Khadem *et al.*, 2005). This is a disruption of the motor coordination and control required for fluent speech, but not for speech *per se* or symbolic behavior. Interestingly this deficit leaves cognition relatively intact. Mice that are genetically engineered to contain a copy of the human version of FOXP2 also turn out to be more vocal than normal mice. In addition they display increased synaptic plasticity and dendrite length in the basal ganglia (Enard *et al.*, 2009) supporting the view that this gene acts to increase the tendency to vocalize more often and more easily. Further support for this hypothesis is provided by the discovery that the up-regulation of FOXP2 is related to song acquisition in birds (Haesler *et al.*, 2004).

The chimpanzee version of the FOXP2 gene differs from the human version by only two amino acids. This small difference lends credence to the view that humans may have suddenly gained fluent speech with the incorporation of a single point mutation ((Enard *et al.*, 2002), Przeworski, Fisher, 2002). Other genes regulate the expression of the FOXP2 gene, so even if an organism possesses it, vocal fluency is not automatic King and Wilson, 1975; White *et al.*, 2006). This complicates the picture because we still do not understand gene regulation, and/or gene expression very well. Nonetheless, the fact that such a small mutation can significantly influence vocal fluency (even in mice) makes it possible that the intelligence of early humans was held in check by lack of fluent speech, not by speech *per se*. Under this model, only when our speaking capacity could keep pace with our intellect did we really begin to employ symbols in a cultural manner that united our high-level minds in a common world view, thereby enabling us to coordinate our behaviors to an extraordinarily high degree.

Should this view prove correct, the current intellectual capability of chimpanzees and bonobos could be vastly underestimated due to their lack of a human form speech. They could be able to think as we do, but due to their lack of fluent human speech, unable to co-ordinate in the cultural extreme seen in modern *H. sapiens*. Bonobos signal and vocalize often. They may decode these frequent vocal exchanges as language, but the human ear would not hear them as such because it is tuned to a lower frequency and a slower rate of energy change. Chimpanzees simply perceive and respond the world at a rate we cannot emulate (Inoue and Matsuzawa, 2009). The current differences between ourselves and chimpanzees with regard to speech and other cognitive capacities could thus result from a tiny mutation that affected a minor parameter of development (the freeing of the vocal tract). The hypothesis is that tiny change — magnified through the lens of cultural factors endowed by increased communicative fluency — could completely reshape our species.

It seems logical that chimpanzee and bonobo vocal fluency should be kept in check by the high rate of infant predation they experience in the wild due to their chosen lifestyles. Were these lifestyle constraints lifted (as humans may have lifted them with the construction of home bases), infant apes that vocalized frequently and loudly could survive. This might be all that would be required to change apes from relatively quiet beings into linguistically fluent ones. Similar lifting of constraints in birds kept in captivity has resulted in a similar lifting of vocal constraints.

Given that the freeing of vocal constraints can manifest in birds *sans selection*, it may also have occurred without selection in human beings, as our environment began to produce conditions that did not inhibit vocal calls, and or crying, in children. That is to say, vocal expression may be a condition toward which organisms move unless environmental pressures push them another way. Bonobos are far more vocal than chimpanzees and they live in much larger groups. This allows them to be more vocal without the free of predation. They also inhabit much higher forest canopies in areas that are very swampy and difficult for feline predators to traverse. Additionally bonobo males play a large role in the rearing of infants, which serves to decrease predation of infants. These factors, taken together, ensure a greater degree of safety for a noisy bonobo child than for chimpanzee child who is often carried by a single mother traveling alone, (but see Roffman, in preparation). In captivity bonobos are far more vocal than chimpanzees and in a much wider range of settings.

Therefore, it could be that a captive environment would begin to lift the speech constraints naturally simply because it removes predation that follows upon localization due to sound. Thus if bonobos and chimpanzees are cognitively capable of language, they would, in captivity, eventually begin to speak, with this occurring first in bonobos.

8. Evolutionary Accounts

It is self evident that evolutionary accounts are troubled by the close genetic relationship found between human, chimpanzees and bonobos. But the problem is not limited to the human/ape divide. The more species that are sequenced the more difficult it becomes to align the previously supposed evolutionary changes (based on anatomy) with the new genetic data (Adoutte *et al.*, 2000.) Evolutionary accounts do not predict nor do they explain, they simply offer a means of attributing change across time to something other than intelligent design. Science has lacked the tools by with which to investigate intelligence greater than human intelligence, though they may be starting to emerge in some data analysis programs.

When Darwinian views are challenged, scientists need not jump to the conclusion that the only alternative theory is than an intelligent designer interrupted the biological momentum of gradualism. Science should be willing to question any theory if it does not comfortably accommodate new data, and new theories of change across time should to be encouraged. One could argue that the new synthesis, with recognition of the importance of kinship and selection operating through that vector, accomplished this. Yes, this is so, the new synthesis modified thinking about the various ways genes could transmit information across time. One could also argue that evo-devo has greatly expanded our view as to how changes in bodily form might occur due to developmental principles folding back upon selection process. Yes this is so. But these expansions of Darwinian theory have left the basic mechanisms of selection and random mutation in place as the driving engines of change across time.

It is also self-evident that there are many vectors and means of changing speciation through behavior. There are many behaviors that change quickly and that are not under genetic control except in the most indirect sense. In the human species, behavior can be completely freed from the environment and possibly even from genetic control by the power of symbolic thought. This may not be easy to do but it should certainly possible and the feats of yogis and others, as well the strong placebo effects observed in modern medicine strongly suggest that the mind has more control over the body than we are generally led to believe. Indeed, a level of freedom from the control of the outer world is at the very heart of all mankind's higher motivations and all religions. Many have attributed this to the so called God gene, but giving this deeply human phenomenon a name and attributing it to a gene explains nothing about the power of belief beyond the empirical conditions of this plane. Moreover, the laws of physics do not rule out the existence of action at distance nor do they do they rule out the role of conscious action on the body or beyond the body. It is rather that we currently lack measuring devices sufficient sensitive to detect such effects (Dotta *et al.*, 2011).

We know that memes quickly alter human behavior and that our behavior can easily alter what the body thinks, feels and does. How we tangibly live matters in that it constructs who we are across time. Any and all significant changes in behavior (either by choice or chance) cannot only affect reproduction but they can also affect gene expression directly (Francis, 2011). Even something as simple as the number of individuals in a group has been shown to affect the behavior of bacteria and to determine the number of individuals of a given sex in certain species of fish. Scientists are only now beginning to think about the true behavior/biology relationship. Since Darwin they have viewed the body as determining the potential of the organism. In so doing they overlooked the great potential of the organism that lies in the flexibility of its behavior. Behavioral flexibility and epigenesis can change bodily form quickly (Maienschein, 2008). For example, tooth size and jaw structure can alter within a single generation. Events experienced by the body as starvation can alter the body's response to food for the next seven generations (Lumey, 1998). To say that all of the behaviors exhibited by the organism across a lifetime are somehow repackaged into the reproductive units of the DNA and then passed along to the next generation is reductionism in the extreme. When an organism dies before reproducing, it leaves behind its entire contribution of bacterial cells, behavioral interactions and even pieces of its DNA. The group may continue emitting patterns of behavior that affect reproduction in other individuals that could have been instantiated by this individual — though this party did not live to pass its on any tendency toward that behavior through its genes. Indeed, one monkey may start a tradition, such as potato washing, which passes to others in the group. This innovator may or may not have had a gene for creativity or even for potato washing; yet his behavior can affect the nutrition and viability of others in the group for generations to come.

Our bodies clearly do not represent the limits of our adaptability. They are living breathing records of past behaviors across many lives of genomes, environmental biomes, gut biomes, and memes and — manifest through our flesh at a moment in time. The DNA has the power to bend our bodies in certain directions, but it does not determine them in full. It serves as weak biological anchor that can be pulled in many directions. It is the moment-by-moment behavioral decisions that shape the future body, the future reproductive successes and the future of the species, though epigenes, imprinted genes, cultural memes, bacterial signals and other yet to be document means. Scientists tend to think of the behavioral decisions of humans as being mostly conscious but focused on proximal adaptability and the behavioral decisions of non-humans as being unconscious and determined by conditioning. Humans have the potential to make behavioral decisions with reference to the long-term cross-generational effects as mediated through the forces of culture, the epigenome, the gut biome and the environmental forces already put into play by our sheer number.

Standard evolutionary accounts have provided scientists with an array of possible mechanistic explanations as to how of life and diversity of form came to exist. With evolutionary theory in mind, diverse anatomical facts were arranged to fit the known evolutionally schema (Mayr, 2002). Once skeletal forms were hung upon the Darwinian outline, it become essential to interpret new data into that framework. New data that did not fit became suspect. There is, of course, no doubt that selection actually works. The absence of selection is not what is being postulated here. One need only look at dogs, foxes, or wolves selectively bred in captivity to demonstrate the role of intentional selection by human beings. Darwinism works best for certain classes of biological problems and simple behaviorally constrained organisms. It requires significant supplementation when applied to organisms that rely heavily upon non-genetic resources of information during development, such as ourselves. Frances (2011) offers an alternative perspective; "In the traditional view, genes function as executives that direct the course of our development. In the alternative view…the executive function resides at the cellular level and genes function more like cellular resources." (p. xiii).

Solid scientific theory should make readily testable predictions. Most Darwinian predictions have been tested under artificial conditions — not in the real world. For example, evolutionary theory predicts that certain characteristics of bacteria will be selected against — through extensive use of modern antibiotics. These bacteria should eventually lose their susceptibility to antibiotics and morph into forms that are not affected by modern antibiotics. This has occurred. But in these cases the doctors have administered the same substance to many individuals in the population repeatedly. Such would never be the normal state of affairs in the natural world. Yet even in these artificially controlled cases, Darwinian theory does not predict how selection will physically manifest, or the biochemical processes by which the bacteria will actually produce immunity. It only predicts that selection will likely occur. At that level of prediction, Darwinian explanations are equivalent to the idea — "Life will find a way."

This lack of specificity causes one to long for a more precise model to explain the profusion, the profundity and the integrated behavior of life forms across time on the planet. While physics leaves us somewhat uncertain about the nature of reality, the nature of time and the number of universes we inhabit; it does a good job of predicting the properties of non-living objects with accuracy. Predicting the behavior of life — in the future — or even in the past, it not within the province of Darwinism. Darwinism's strength is that it has led scientists to explore biology in a manner that otherwise would never have occurred. In so doing, science has generated vast new fields of understanding about ourselves and other living beings.

Darwinism has suggest to us that changes across time found in the archeological and geological records, are not to be understood from a theological standpoint as creations of God, or acts of God, but rather as instantiations of the movements of global plates, successful reproduction, natural selection, random mutation. Darwinian theory has made it possible to spins hundreds, if not thousands, of evolutionary models to account for the seemingly peculiar situations in the world of living beings at this time. It has not however, for all the fear surrounding its early development, explained away human behavior. Those who would understand the baboon did not do more to explain the condition of man than theologians as Darwin ventures. To date, theologians have not succeeded either.

There is now general agreement that evolutionary forces cannot account for the development of human uniqueness nor the current prominence of human civilization and its transformative power upon the planet. That role has been accorded to culture, not biology (Tomasello and Call, 1997). If there is a magic bullet embedded within some cultures that propels them to prominence and not others, that bullet has yet to be identified and many are constantly searching. With the advent of genetic sequencing and the emerging field of epigenetics, the biological knowledge that should undergird Darwinian theory has become radically altered, while the theory itself has remained in place, but its influence is being steadily altered.

For example, it is obvious that the parameters of potential selection (within Darwinian theory) become essentially infinite once we began to understand the truly endless array of patterns of gene expression (not the genes themselves) which could act to affect morphological forms. Unlike genes (of which there are many but not an infinite number), epigenetic factors that turn genes off and on have no finite number. Moreover, selection for or against certain genes is operating on multiple epigenetic factors that may or may not ride on the genes that are selected, in any constant focused manner across reproductive cycles. Following conception, epigenetic factors are stripped off the genome, and returned in slightly different patterns (Shu and Wu, 2009). These changes seemingly take into account something about the environment at the time of conception. This new pattern is not selected for through reproduction but arises outside of and beyond reproductive success. This means that it is not possible to determine exactly what is selected for or against at any time a particular genetic shift occurs during mating. Darwinian selection for, or against, is essentially, more idea than fact. Change across time is fact.

Except in the laboratory where certain forms with identifiable genes are allowed to reproduce and others are not, major change across time does not take place at the level of individual genes, but at the level of organisms, groups, bands, tribes, states, city-states and nations (Edelman, 1987; Wilson, 1975). Entire populations of human relatives such as Neanderthals and Denisovans can, and apparently have vanished all at once. Epigenes can turn genes off and on; making bodily traits come and go in the external reproductive world, but not in the internal world. What bodily features can be selected for, when they are not always constant? Genes can go silent and still exist. Genes can travel along for millions of years doing nothing and reappear as active again, when conditions change. The first big surprise once science was able to sequence genes was the finding that the majority of genes had not been selected for or against, but had ridden along as junk. The genome was inefficient! Now it seems some of this junk may be telling the non-junk what to do, but this only renders the idea of how selection operates at the level of morphology more opaque. It means that the not just a certain region of the genome needs to be added or removed by selection, but precisely corresponding regions need to be simultaneously removed or added. This is not so easy to do because some regions control the expression of multiple genes. Genes that might not be activated in a body that reproduced could still be carried forward to the next generation as junk or as a non-active pattern of expression. These would not be selected against, through reproductive failure, simply because individuals who carried them failed to survive. Because these regions jump around in manner not yet understood, and for reasons not yet known, they garble any simple evolutionary account. They can even result in chromosomal rearrangements — such as the fusion of two chimpanzee chromosomes into one.

"The elements contain in their sequence all the instructions needed to cut themselves out of their host DNA and splice themselves into another spot. But they are not always benign 'junk' DNA — they can insert into genes or gene regulatory elements, potentially disrupting the gene's function, and they can trigger chromosome rearrangements." (Blément and Viera, 2006, p. 521)

It is important to note that most genes, in a population, are passed on, regardless of mating success. Most regulatory genes are critical for survival and they govern lots of other genes, if they are not passed on the reproductive product fails to mature. Selection at the level of some genes has minor effects, at the level of others, major effects. Many genes simply cannot be lost or cast aside even if they have some negative properties or their will not be another organism at all. These key essential genes can be carried within just one fertilized individual female. This female can give birth to a large successful population and this has occurred many times (Nei *et al.*, 1975).

This is accomplished in the laboratory all the time. The number of random mutations is so low as to be non-problematic for the creation of very large populations. Populations do not tend to change because of mutations. Mutations tend to fall aside. Populations change when they are in contact with different environments, when they undergo dietary changes, and — for the most part — when they undergo behavioral change. *However, it is important to note that behavioral change does not require environmental change.* It can occur quite on its own, without selection and without environmental pressure. We see this most clearly in *H. sapiens*, because the changes we have wrought through behavior are so abundant, but the principle is the same for all species. This fact is often explained away by asserting that behavior is genetically driven in all non-human forms. However the increasingly cultural nature of behavior is becoming apparent for many non-human beings (Kappler, 2010).

Epigenetic markers can be imposed by behavioral events. Nothing about this system is simple. Nothing about it appears predetermined. The concept of selection seems passé, once one realizes that everything and nothing can be selected for or against at the same time. This is not to say that Darwin was incorrect or that we must fall back on creationism. It is to say that it is time to acknowledge that science has not found the answer in selectionism and mutation. The answers are more complicated and we will not find them until we acknowledge the limitations of the Darwinist perspectives to explain the new biological facts that are pouring in from all directions.

The continued allure of the Darwinian model has been that it provides scientists with a mechanistic explanation of what would otherwise be a totally mysterious process. Darwinian theory could not have created such an enduring impact on the minds of scientists but for one reason, it offered an alternative account for the existence of everything on Earth; man no longer needed to have arisen by Divine Creation. Once Darwin appeared, biology was no longer beholden to the church. Biological information began to explode. Biology's new found ability to create entirely new organisms by genetic manipulation seemingly cemented the Darwinian model. The only problem is that while genes do evolve, in natural settings all genes, and all species are constantly co-evolving within a system that is itself evolving. The operation of models of selection occurring in the real world — when reduced to the laboratory and manipulated one variable at a time — lose all predictive power, because in the living world, all life forms are constantly interacting. The best that can be said is for Darwin is that he was correct to state that — given enough time, evolutionary explanations account for everything but the origin of life and the origin of man. However what has actually occurred appears to be more complicated than this.

Our rapidly increasing understanding of role of epigenetics, and the plasticity of brain development represent substantial challenges to theories of change based on selection processes and random mutations. We should consider that selection, over the long-hall, might favor groups of genes able to produce a very wide range of traits and/or forms depending upon environmental changes that have occurred in the past, but are no longer present. The fine tuning needed for each different environment may be handled by epigenetic and developmental processes which are not selected in any traditional sense but which, once they are in the genome, can be produced as needed, when the environment calls. It might thus be more reasonable to view the genome as a tool-kit that has evolved in order to allow the organism to differentially manifest across geological eras of time. Thus the organism's genome (in its entirety) can be thought of a representing, in a sense its history, a history which may repeat itself and which is retained to allow the organism to come *pre-adapted* to a wide array environments in the future. There are many organisms now known to be sensitive to the phenomenon of quorum sensing. Certain behaviors such as photoluminesence appear in bacteria only as they reach a certain number. Bacteria monitor the environment and alter their behavior depending upon the number of individuals present as well as the other species that are present (Waters and Bassler, 2005). The molecules and the behaviors that are activated by quantum sensing then change the bacteria's environment in a way that may turn on still other behaviors or chemical processes. Because bacteria also communicate with the host species, and behave in groups as multicellular organisms, the clear demarcation between prokaryotes and eukaryotes becomes muddled. It seems as though they come with a set of genes designed to do many different things in difference environments. Environment in this sense does not just mean the natural world surrounding them; it means as well the world they are creating by their very own presence within that world.

If the organism's tool-box were designed for many environments — no longer present at this time, then selection would still be an operational principle, but the fitness of an organism's genome would not be the result of producing more offspring than others in a given lifetime. Instead it would be the result of gathering together a genome adapted to enable it to encounter nearly any environment it might encounter across millennia. If this is so, then the genome is really properly conceived of as the organisms' existential tool-kit. That is, a tool-kit allows it not only to adjust at the moment, but over the *longue durée*. Thus all genes in the tool kit may have been selected for, not against. The more genes that prove helpful and that can be tucked away for future use, the better.

Genes, we now know, can be activated by internal gut biota, and by the external environment. Thus diet becomes a critical factor in selection, not because of getting insufficient food, but because certain foods affect the gut biota in a particular manner. In a sense the possibilities for interaction between an organism's tool kit and its internal and external environment become essentially limitless, constantly changing and constantly interacting. Genes may be inserted by virus's and virus's may also insert themselves into the genome — possibly waiting to be activated by conditions yet unknown. The new picture that is rapidly emerging forces anyone raising questions about how humans became human — to look beyond the standard gradualist ape into man drawings found in every textbook.

9. Ethical Science for the Pan/Homo Bonobo Population

Kanzi, like Austin, Sherman, Panzee and other *Pan/Homo* beings, has a brain that is larger than other captive members of his biological species. The most prized of all human traits has been our large brain. According to standard interpretation of the fossil record, increase in cranial size was the major dimension of change, the driving factor of selection and the *sin qua non* of speciation toward modern humanness (Dunbar, 2009). But this view runs into difficult as soon as we consider the fact that Neanderthal brains were larger than our own, yet they seemingly had no symbolic artistic abilities. They did not evolve city-states, the wheel, or complex tools of any sort. The Darwinian perspective causes us to inquire as to what they doing with these large brains? It also urges us to ask, what was the selective force that produced brains larger than our own but no known civilization, art, music, or complex tools?

Equally perplexing is the finding that DNA data indicate that the genes which have changed the fastest since the human/chimpanzee split, are the ones that code for inflammatory responses and cell proliferation (Perry *et al.*, 2009), not brain size. Why weren't the genes that changed the fastest, those that selected for our own large brain, if in fact, it is the critical ingredient to our current successful domination of the globe and high-tech society? It seems that the selection has, since the human-chimpanzee split, been improving the chimpanzee's ability to resist microbial diseases and pathogenic organisms. Thus moving away from us has required rapid immunological adjustment on the part on the part of the chimpanzee. That could suggest that it was their environment, not ours, which changed radically as a function of the separation event.

During that same period of time, our immune system underwent selection for increased susceptibility to cancer (Fan *et al.*, 2002).

Thus the species that evolved the more sophisticated immune system (chimpanzees) produced far fewer offspring than the species who became increasingly prone to cancer (Ziv *et al.*, 2006). Certainly no one would predict such an outcome based on any version of Darwinian theory. But the recursive nature of Darwinian theory provides an "out," for every anomalous event. In this case the explanation would be that something about increased susceptibility to cancer also provided *H. sapiens* with a valuable ability that just happened to be linked to genes that promoted cancer. Whatever else those genes do for us, they confer sufficient reproductive advantage to over-ride the negative impact of the cancer link. One possible explanation might be that cancer generally shows up on the tail end of reproduction, so *H. sapiens* can tolerate increased cancers and still reproduce. The immense flexibility of the Darwinian approach allows it to account for even the most puzzling facts of nature, after the fact. What the theory does not do is predict which of the above hypothesis (or many other possible hypothesis) is the most probable. The theory allows for nearly any possible after the fact explication but offers no guide for what will occur in the future.

A more intriguing hypothesis is that the human brain is the result of a genetic change that allowed cell proliferation in various places to become enhanced, sometimes unfortunately leading to cancer, but always leading to a larger brain. An additional anomalous genetic finding is that there are some segments of the human genome that seem to have inhibited change, while these same segments — in chimpanzees — have continued to evolve. Again they seem to be changing rapidly at the genetic level, while we are not. Even more puzzling is the fact the piece of the genome changing most rapidly in the chimpanzee (the immune system) is also a system that has the ability to assist learning (Ziv *et al.*, 2006), yet we are the ones who supposedly became the better learners. The bottom lines is that the genetic data on the human/ape split simply do not fit the standard Darwinian account of how we got here. The implication of this for the *Pan/Homo* bonobos is clear. It appears that human culture is a potent force for change and they have embodied this change, in a manner that must be something like our own embodiment of the change seventy thousand years ago. They are no more freaks of nature now that we were freaks of nature then. They are a particular set of cultural traits riding upon a pliable biological platform (Savage-Rumbaugh and Fields, 2011, 2013). Human ethics dictate that the beings that are essentially human should be accorded the rights we accord to humanity. If humanness is cultural, not biological, as it seems to be; then the *Pan/Homo* bonobos have a right to dignity, respect and self-determination within the limits of their cognitive capacities to execute such rights in a responsible and moral manner.

Whatever occurred genetically after the human/ape split, the result was that humanity populated the entire globe in a relatively brief amount of time. Apes remained restricted to limited areas of Africa and Asia. To spread as we have, requires unusual spatial mapping capacities, a specialty of the hippocampus. We know that T lymphocytes are critical to hippo-campal neurogenesis, and an enriched environment is associated with the recruitment of T cells. Is it possible that as we began to travel and leave apes behind as we recruited more T cells and diverged physically and cognitively as a direct result of inhabiting far larger and far more complex areas (Kipnis, 2006). Larger ranges required more hippo-campal neurogenesis, and thus — as we enriched our environments, we changed ourselves. That is, human travel behaviors themselves, which affect hippocampus neurogenesis, led to more travel as humans grew bigger brains, and bigger brains allowed travel to become more efficient though increased long-range planning.

H. sapiens may simply have walked out of the forest and kept right on walking because our symbolic brains had enabled us to image different social and physical worlds. Our imagined worlds allowed us to choose to leave. We could imagine finding a place more ideal than where we were. Thus far, chimpanzees seem not to have accomplished a similar feat, but new data suggests that they have left the forest in Mali and are living in sub-Saharan and even Saharan environments (Roffman, in preparation). Should chimpanzees and/or bonobos choose to move beyond the forest and into a very different environment, we would predict anatomical change, because long distance travel would require infant carriage (Savage-Rumbaugh and Fields, 2013). We would also predict hippo-campal neurogenesis and increase in T-cell proliferation.

On the bases of a genome wide comparison of human and chimpanzee DNA *et al.* (2005) advise that most of the asymmetrical increase of duplicated DNA in the chimpanzee lineage emerged as a mechanistic consequence of changes in the chromosome structure rather as due to forces of selection. Their view is echoed by Feuk *et al.* (2005) who note that the recent genetic evidence on deletions, inversions and copy number variations should "cause people to rethink their ideas about how species evolve." Chromosomes invert, delete and copy themselves in what appear to be an erratic manner. When this occurs, immediate behavioral and physical changes can be produced. As long as these changes are random they will have no effect on the population. If they occur repeatedly in multiple individuals, new populations can emerge quickly. As Marshall (2008) in attempting to explain the Cambrian explosion observed: "It's not new genes that create new morphological innovation, but rather the way they're wired together, different-looking creatures are not apples and oranges," they just look that way.

If it is not new genes that are creating the morphological changes seen in the *Pan/Homo* bonobos, we are then looking at cultural forces. The study of Kanzi and family began with the importation of Matata in 1975. It was interrupted in 2013 and it has been necessary to resort to legal action to allow the accretion of knowledge to be permitted to continue. Should we prevail in court, the Bonobo Hope Science Board (the group charged with overseeing the welfare of the bonobos) is seeking to continue the research trajectory into the future across many generations and at the appropriate time, to relocate these bonobos, and other apes who have acquired a human language, to Africa.

The cultural diversity, selfhood, abstract thought, creativity, cultural diversity, sharing of all human emotions, informational exchange and the sharing of all human psychological pathologies shown by these bonobos also requires a revaluation of the appropriateness of the continued attempts to employ them and chimpanzees in biomedical research. The four decade long cross-rearing study has provided considerable data that strongly support the hypothesis that epigenetic change is occurring in these populations. The current population is too small to justify a statistical population analysis but much more detailed study of the morphological changes is warranted. With each succeeding generation there has been a drift of morphological form toward the human dimension. This population of the *Pan/homo* culture should be studied for minimum of 10 additional generations with appropriate measures taken each generation to determine whether the changes in the epigenetic profiles are continuing to occur along the humaness-bonoboness dimension. The epigenetic and morphological changes should be measured both during development and between generations. These biological assays should be accompanied by measures of behavioral and cognitive changes along the dimensions of fluidity of vocal use, speed of language development, speed of tool manufacturing development, sophistication of language and tool-manufacturing and use, and the pace of development of artistic and musical capacities. Also of great important is the development of self-control over emotional reactions and the emergence of moral systems.

As a matter of ethics, it must be acknowledged these *Pan/Homo* beings require an environment that allows them to manifest culture as adult, without the constraints of captivity. Ethically some might question whether it is appropriate to allow the bonobo genome to express *Homo* traits? However if *Homo* propensities are riding within the genome of this group, one could also insist that failure to allow the potential for self-expression constitutes speciesism in the extreme. Let u not try to determine which view is correct, but rather to acknowledge the facts straightforwardly.

They are:

(a) The very close genetic relations and the clear effect of rearing and social environment on morphology and the deployment of human-like consciousness,

(b) That behavioral work has produced young bonobo beings capable of language and self-reflection.
(c) That as these beings grow older they acquire ever more human capacities depending on their physical environment and social treatment by the human species.
(d) That these bonobo beings are capable of moral reasoning and self-control under the appropriate social conditions.
(e) The uniqueness of this population requires special precautions be taken to protect these bonobo beings, as they are not adaptable to the traditional biomedical regime and should never become subjects in studies of disease prevention.
(f) Standards of treatment must be raised to insure that their lives are considered more important than those of other fungible and/or dispensable animal research subjects.
(g) They are non-replaceable because of the rearing and bonobo/human kinship vector the group represent. To create this group required a period of 40 years and the capture of wild bonobos with culture. This is not going to be replicated in the foreseeable future.
(J) Studies of the inter-relationship of these mental and social factors are merited, as long as they not invasive or restrictive.

Because a captive world composed of cages is an inappropriate long-term environment, any research that is to be done should strive to prepare these bonobos to live in a non-captive world. If such were not attainable, the only ethical research to be continued, in a captive world, would be that which has the byproduct of enhancing their welfare and well-being. Informed consent should be attempted whenever possible for this work. Researchers who attempt to investigate ape cognition with Pan/Homo apes from outside the cage employ homo-centric paradigms that constantly underestimate the true intellectual capacities. They seek modern human-centric demonstrations of intelligence, because they know and experience nothing else. The use of these methods to investigate the cognitive memories of Lana, Sherman, Austin, Panzee, Mercury, Kanzi, Matata, Elykia, Nytoa, Teco, Maisha, and Nyota, while simultaneously placing upon them a world that eliminates their daily keyboard conversations, they travels through the forest, their *Pan/Homo* social contracts, and the only world these bonobos have even known, is not only unethical, it is a fatally flawed paradigm.

As Kanzi and Panbanisha began to reach adulthood and to display the typical adult morphologies, as well as adult political and moral interests and attunements, no homo-centric paradigm sufficed to explain their behavior, even when the human informants lived inside the *Pan/Homo* group. This was because the realities of their *Pan/Homo* world differed extensively extant from the previous published accounts that emphasized abilities commensurate with two and three year olds. As adults, they structured that world as much, if not more, than human participate. Reports that included completely accurate descriptions and information about these bonobos began to produce something essentially akin to a social revolution with the laboratory itself. This served to shut down any further attempts to validate the findings that were beginning to be published about Kanzi and Panbanisha as the adults they became in Iowa. With the publication of the article entitled *"Welfare of Apes in Captive Environments: Comments On, and by, a Specific Group of Apes,"* the research program at the Great Ape Trust became substantially emaciated by the arbitrary removal of the first author at the whim administration and the board. The auditory data supporting the capacities presented and referenced the 2007 article were arbitrarily removed from the internet and from computer storage. The inclusion of names of the bonobos, in a peer reviewed manuscript, were presented as evidence that the first author had stepped beyond scientific data to too great a degree. Data was supposed to come only from human beings, never from apes themselves. The real issue was that the bonobos were becoming able to generate their own data, to translate their own language and to speak for themselves. This was particularly true of Panbanisha, who have progressed the furtherest and who was beginning to reliably translate Kanzi's utterances into lexical representations and to explicate Matata's utterances, which were far more difficult to fathom, even for members of the *Pan/Homo* world. Panbanisha had also demonstrated that they she could read English words and that she was able to memorize the numerical codes that locked the external doors on the facility and to open and close these doors — simply to show that they knew the codes. She never attempted to leave the facility as she understood perfectly that she would be shot should she do so, as did the other bonobos. These revelations, made within the science circle, were simply too much for scientists who were not working in direct physical interaction with these bonobos and for the administration. Along with 2007 article, they resulted in the replacement of Dr. Savage-Rumbaugh by Mr. Fields. When he became the director of the bonobo project, Dr. Savage-Rumbaugh was banned from the facility, by the administration. Finally, after a round of legal negotiations, a contract was issued allowing her access to the bonobos but data collection opportunities were restricted. There is much more to be explicated regarding these events, however the point to be made here is that as the ultimate success in attempting to determine IF apes could acquire language was being achieved, and hotly debated within the halls of academe, the actual implications of these capacities were having a tangible and immediate real world impact upon the daily lives of these beings. They were also having an immense impact upon the resident scientists and administrators who were unable to speak their lexical language, their vocal language, or their gestural language. While they could not argue with the observations that were pouring in from everyone able to walk in and amongst the group of bonobos, they could stop the progress and they did.

Beings who could speak for themselves, produce their own data, evaluate their own captive conditions and decide to remain within them or not, were no longer welcome. Unlike Orr, who also became able to do these things and who was praised for them, Kanzi and family were punished and the scientists working with them were punished. Although Mr. Fields was given responsibility for the future research program, he was not permitted to integrate Dr. Savage-Rumbaugh back into it. He was asked to allow researchers with no experience or minimal experience with language, to collect data, in order to keep the funding for the facility.[34] He came under attack and threat of removal from other organizations who wanted to obtain the bonobos. Attempts were ever made to split up the bonobo group amongst various zoos and to bring gorillas to replace them. In order to protect them from removal by other parties, it became necessary for him to file a lawsuit in federal district court. Against great odds, Mr. Fields kept the bonobos from being separated and relocated to zoological gardens and primate research labs, and he moved toward a re-instatement of the original research trajectory. However, just as he was nearing success, he was also removed.

The level of dissonance between what these bonobos were actually capable of doing and what the "accepted assays" revealed was so great as to produce dysphoria in anyone trying to understand the literature in light of their first hand experiences with apes.

10. Human Political Games

It is time to stop playing human games regarding the future of these bonobos and to seriously investigate who and what they have become, while they are still with us. Given the death of Matata and the current ban on importation of wild caught adolescents, this group of bonobos could not be recreated, even if other scientists were to be funded to spend another 40 years replicating the work the work that it has taken to produce the third generation

---

[34]The bibliography at the end of the reference section delineates the articles published during the tenure of Mr. Fields. They make clear that while research was allowed to continue, the nature of that research was severely curtailed. Had Mr. Fields not complied with the demands of the other scientists at the Trust and the administration at that time, he would also have been removed and the bonobos would have been relocated to zoological collections. The bonobos survived and their capacities even increased because of the continued presence of the following members of their *Pan/Homo* world Sue Savage-Rumbaugh, William Fields, Itai Roffman, Duane Rumbaugh, Liz Pugh, and Heather Pugh.

linguistically competent individuals currently represented by Nyota and Teco. It is time to let Kanzi speak, and to listen to his view of the world. Panbanisha can no longer speak due to tragic circumstances, neither can Panzee , Austin, Nathan, Matata, or P-suke. All were taken far too earlier in life by circumstances that leave those who loved them wondering and without adequate explanations of any kind.

But Kanzi and Lana are still with us. Nyota, Teco, Elykia, and Maish are still with us. Sherman and Mercury are still with us. Any one of them, if given the simple opportunities availed to Orr through the understanding of his parents and his brother, could begin to tell their personal stories to the world. The world needs to hear those stories and to begin to try to understand what they mean. To shut them out and to relegate them to the world of "animal" is morally repugnant and scientifically unethical. Even IF the scientists able to relate to them and properly film and interview them are unable to separate all of the chaff from the wheat, the stories need to be told. Even IF there are those who still wish to assert that all of their capacities are the result of cueing or conditioning, they must admit that no other animals or human beings have ever been so eloquently conditioned and/or cued. There is much to be learned from these apes who have very different rearing backgrounds; if they are given, once and for all, opportunities to try and speak for themselves, in a world where they can be assured that they will not be oppressed or punished for doing so. This must be a world from the medical primate centers in which nearly all captive chimpanzees reside and a world far from zoos, where apes are under the constant stress of public observation. Where they have no privacy and where repeated break-ups of family structures for the purposes of shifting genes throughout the AZA population are a constant threat to group structure. However, at least they allowed to reproduce. In sanctuaries, reproduction is forbidden and no family structure can arise.

We must employ high-level flexible culturally based behavioral assays. The currently approved measuring tools are so primitive as to ensure those dependent upon them, that apes cab only function as something akin to large monkeys or large dogs. Nothing could be further from the truth. Apes are functioning in a manner similar to ourselves in that they create their own mental inner worlds, against which they consciously reflect upon their outer world. Like us, they generate ideas, thoughts and views about what others ought to do and should do; not only about what they desire them to do. From this kind of reflective thought arises a moral system and expectancies that others will fulfill the moral dictates of the emergent co-constructed system that inevitably rides along with self-reflective brains. Thus do we find good and bad arising and riding upon the emergence of system of co-constructed moral expectancies.

The goal of this paper is to point toward a future of ethical research. We suggest that all future research must enhance the lives of all three species, bonobos, chimpanzees and humans — not merely humans. It must require informed consent of the non-human individual when language is present and it must required informed consent by a guardian or fully informed legal representative when language is absent. Additionally, such research must take place in an environment that provides the maximal amount of freedom possible to be given without harm to others. Finally such research must strive to enlighten our understanding of both Pan and Homo in a manner that seeks to be beneficial for both, while respecting the implications of the extremely close biological relationship. This document offers legitimate questions for future work that meets these criteria.

In closing we again turn to where began, Corby (2005):

The exclusion of animals has been and still is, constitutive of much of the Euro-American anthropological discourse on humanness, natural order, and culture. Condescending and distancing attitudes, connected to an obsession with human grandeur, have shifted toward more informed and egalitarian ones, and there is a growing awareness in the population at large that humans are not that different from apes. However, great apes and early hominids are still profoundly ambiguous, categorically, cognitively, and morally. The small group of apes that have become language-savy through human training in particular, are in a bizarre category. The are chimeras, not human but endowed with a human quality that their kind would not possess without years of human training. They are, in a sense, more sentient than their species is supposed to be...[Some of them] have ended up in the dungeons of biomedical laboratories...[like] some sort of ape-human hybrid...trapped in then netherworld between the two species" (Stanford, 2001).

These apes are in a netherworld but not because they were trained. They are in a netherworld because they came prepared to learn any language and they happened to acquire a *Pan/Homo* language and a *Pan/Homo* culture. They are NOT more sentient than apes are supposed to be. They represent what apes can be, as a function of culture. Nothing more and nothing less is acceptable. All apes are caught in a netherworld because of humanity's failure to grasp what it has done and is doing to them, as well as because of its failure to realize that the lack of ability to understand is the fault of humanity, not the fault of the ape.

Until we realize and correct the problem, humanity will never know who and what it is. The ape imperative is upon us and it is time to take off the blinders and come to grips with it and with ourselves.[35]

---

[35]William M. Fields, made theoretical contributions to cultural studies of this group of bonobos in 1997 with a seminal paper entitled "Why the *Pan/Homo* Culture Made Kanzi Speak." This paper dealt with the inadequacy of the notion of protoculture as a theoretical explanation of ape language competencies. It asserted that if Kanzi possessed language and produced Oldawan tools, he was running a full blown form of cultural programming acquired in the proximal zone of development and therefore was a legitimate subject of ethnographic inquiry as were the other members of his human and bonobo family. The paper was written as part of graduate course in cultural anthropology and it laid the basis for the future approach to studies of *Pan/Homo* bonobos.

Following this chapter, and as a direct result thereof, the first author began a collaborative effort with William M. Fields. This included allowing Mr. Fields ethnographic access to the laboratory and a deliberate shifting away from exclusive use of methods of understanding that characterized experimental psychology towards that had been developed for the study cultural processes in human beings.

Thus the questions moved away from those focused on whether they did or did not "have" language, to how they engaged/established/maintained and co-constructed the *Pan/Homo* cultural world in which they lived their lives and constantly employed language. As this shift was made the size of the bonobo group had grown from two (Matata and Kanzi) to nine (Matata, Kanzi, Panbanisha, Tamuli, Neema, Elykia, Maisha, Nyota, and P-suke) all by natural births within the group. The group was divided into those that were exposed to human lexical capacities and those that were not.

Some infants were reared in a Pan/Homo world with daily exposure to the use of lexigrams and spoken English in an interactive real world dynamic setting. Other infants observed and sometimes participated, but were not as closely bonded to human members of the group. Mr. Fields participation significantly increased and broadened the theoretical and methodological approaches employed. It brought a sophistication to the recording and archiving of data that was needed and that is now allowing, initially through this chapter, the real story of the work to begin to be told.

Moreover the relationships Mr. Fields personally established with the bonobos expanded their competencies, and the competencies of the group far beyond that which had previously been noted. In part this was because they were becoming adults, but it was also because the increased sophistication of Pan/Homo culture was being forced to change in accord with new challenges, brought about by the relocation to Iowa.

The efforts of Mr. Fields are throughout the current paper and cannot be legitimately separated from those of the first author. His participation in the research has reorganized the way in which ape language has been conceptualized. The continuum that developed between ourselves and the *Pan/Homo* bonobos is now becoming visible even to those who do not know them in a personal subjective embodied, manner through these efforts (see Segerdahl, 2005).

There is currently an amendment in the US House of Representative to keep chimpanzees in biomedical research in the United States until 2023, and potentially beyond, at the discretion of the Comptroller General's office. The similarities between human beings and chimpanzees at the biological level are sufficient to treat them as an alternative form of human being. They make a perfect animal model for us and eliminate the need to place human beings at risk. But what makes them animals and ourselves, human? The important data for determining whether they should be members of *Homo*, cannot be genetic data, for this data already indicates that they should be classified as *Homo*. It is the behavioral data to which we must now turn.

References

Adega FH, Guedes-Pinto H and Chaves R. 2009. Satellite DNA in the karyotype evolution of satellite DNA in the karyotype evolution of domestic animals — clinical considerations. *Cytogenetic and Genome Research*, 126(1–2), 12–20.

Adoutte A, Balavoine G, Lartillot N, Lespinet O, Prud'homme B and de Rosa R. 2000. The new animal phylogeny: Reliability and implications. *Proceedings of the National Academy of Sciences*, 97(9), 4453–4456.

Bakewell M, Shi P and Zhang J. 2007. More genes underwent positive selection in chimpanzee evolution than in human evolution. *Proceedings of the National Academy of Science of the United States of America*, 104, 7489–7494.

Balter M. 2011. Was North Africa the launch pad for modern human migrations? *Science*, 331(6013), 20–23.

Benson J, Fries P, Greaves W, Iwamoto K, Savage-Rumbaugh S and Taglialatela J. 2002. Confrontation and support in bonobo-human discourse. *Functions of Language*, 9(1), 1–38.

Beran M and Evans TA. 2009. Delay of gratification by chimpanzees (*Pan troglodytes*) in working and waiting situation. *Behavioral Processes*, 80(2), 177–181.

Bettelheim B. 1995. Feral children and autistic children. *American Journal of Sociology*, 64(5), 455–467.

Bloomsmith M, Baker KC, Ross SR and Lambeth. 2006. Early rearing conditions and captive chimpanzee behavior: Some surprising findings. In *Nursery Rearing of Nonhman Primates in the 21st Century: Developments in Primatology: Progress and Prospects*. Sackett GP, Ruppentahal GC and Elias K (eds.), pp. 289–312. Springer.

Boas F. 1940. *Race, Language and Culture*. Chicago University Press, Chicago.

Brosnan SF and Beran MJ. 2009. Trading behavior between conspecifics in chimpanzees, *Pan troglodytes*. *Journal of Comparative Psychology*, 123(2), 181–194.

Call J and Tomasello M. 2008. Does a chimpanzee have a theory of mind? 30 years later. *Trends in Cognitive Science*, 12(5), 187–192.

Calvin WH and Derek B. 2001. *Lingua ex machina: Reconciling Darwin and Chomsky with the human brain*. Cambridge: MIT Press.

Cairns-Smith AG. 1975. A case for alien ancestry. *Proceedings of the Royal Society*, 189, 249–274.

Candland D. 1993. *Feral Children and Clever Animals: Reflections on Human Nature*. Oxford University Press, New York.

Caswell JL, Mallick S, Richter DJ, Neubauer J, Schirmer C, Gnerr S and Reich D. 2008. Analysis of chimpanzee history based on genome sequence alignments. *PLoS Genet*, 4(4), 1000057.

Cheng Z, Ventura M, She X, Khaitovich P, Graves T, Osoegawa K, Church D, DeJong P, Wilson RK, Pääbo S, Rocchi M and Eichler EE. 2005. A genome-wide comparison of recent chimpanzee and human segmental duplications. *Nature*, 437, 88–93.

Clifford J. 1988. *The Predicament of Culture: Twentieth-Century Ethnography, Literature and Art*. Cambridge: Harvard University Press.

Corby M. 2005. *The Metaphysics of Ape: Negotiating the Animal-Human Boundary*. Cambridge University Press, Cambridge.

Cremo MA and Thompson RL. 1998. *Forbidden Archeology: The Hidden History of the Human Race*: Bhaktivedanta Institute.

Curtiss S. 1977. *Genie: A Psycholinguistic Study of a Modern-Day Wild Child*. Academic Press, New York.

Davenport RK Jr. and Menzel FW Jr. 1963. Stereotyped behavior of the infant chimpanzee. *Arch. Gen. Psychol*, 8, 99–104.

Davenport RK and Rogers CM. 1968. Intellectual performance of differentially reared chimpanzees. I. Delayed response. *Am. J. Ment. Defic*, 72, 674–680.

Davenport RK and Rogers CM. 1970. Differential rearing of the chimpanzee: A project survey. In Bourne GH (ed.), *The Chimpanzee*: Vol. 3. *Immunology, Infections, Hormones, Anatomy, and Behavior of Chimpanzees*. University Park Press, Baltimore, pp. 337–360.

Davenport RK, Rogers CM and Menzel EW. 1969. Intellectual performance of differentially reared chimpanzees. II. Discrimination-learning set. *Am. J. Ment. Defic*, 73, 963–969.

Davenport RK, Rogers CM and Rumbaugh DM. 1973. Long-term cognitive deficits in chimpanzees associated with early impoverished rearing. *Dev. Psychol*, 9, 343–347.

Deacon TW. 1998. *The Symbolic Species: The Co-Evolution of Language and the Brain*. New York: WW Norton & Company.

Dennet D. 1996. *Kinds of Minds; Toward and Understanding of Consciousness*. Basic Books, New York.

de Waal FBM, Dindo M, Freeman CA and Hall MJ. 2005. The monkey in the mirror, hardly a stranger. *Proceedings of the National Academy of Sciences*, 102(32), 11140–11147.

Donovan J. 2005. *Flipped Genetic Sequences Illuminate Human Evolution and Disease*. Available from http://www.eurekalert.org/pub_releases/2005-10/hhmi-fgs102505.php. Accessed on August 5 2016.

Dotta BT, Buckner CA, Lafrenie RM and Persinger MA. 2011. Photon emissions from human brain and cell culture exposed to distally rotating magnetic fields shared by separate light-Stimulated brains and cells. *Brain Research*, 1388, 77–88.

Dunbar R. 2009. Why only humans have language. In Botha R and Knight C (eds.). *The Prehistory of Language*. Oxford University Press, New York.

Edelman GM. 1987. *Neural Darwinism: The Theory of Neuronal Group Selection*. Basic Books, New York.

Enard W, Przeworski M, Fisher S, Lai C, Wiebe V, Kitano T and Monaco APS. 2002. Molecular evolution of FOXP2, a gene involved in speech and language. *Nature*, 418(6900), 869–872.

Enard W, *et al*. 2009. A humanized version of Foxp2 affects Cortico-Basal Ganglia Circuits in Mice. *Cell*, 137, 961–971.

Fan Y, Newman T, Linardopoulou E and Trask BJ. 2002. Gene content an function of the ancestral chromosome fusion site in human chromosome 2q13–2q14.1 and para-loguous regions. *Genome Research*, 12, 1663–1672.

Feuk L, MacDonald JR, Tang T, Carson AR, Li M, Rao G, Khaja R and Scherer SW. 2005. Discovery of human inversion polymorphisms by comparative analysis of human and chimpanzee DNA sequence assemblies. *PLoS Genetics*, 1(4), 489–498. doi: 10.1371/journal.pgen.0010056.

Fields RD. 2008. Of whales and men. *Scientific American* on line, http://blogs.scientificamerican.com/news-blog/are-whales-smarter-than-we-are/.

Fields WM, Segerdahl P and Savage-Rumbaugh ES. 2007. The material practices of ape language. In Valsiner J and Rosa A (eds.). *The Cambridge Handbook of Socio-Cultural Psychology*, 164–186.

Fitch W. 2000. The evolution of speech: a comparative review. *Trends in Cognitive Sciences*, 4(7), 258–267.

Fouts RS, Fouts DH and Van Cantfort TE. 1989. The infant Loulis learns signs from cross-fostered chimpanzees. In Gardner RA, Gardner BT and Van Cantfort TE (eds.). *Teaching sign language to chimpanzees*. State University of New York Press, Albany, 280–292.

Frances RC. 2011. *Epigenetics: How Environment Shapes our Genes*. Norten and Company, New York.

Gardner RA and Gardner BT. 1969. Teaching sign language to a chimpanzee. *Science*, 165, 664–672.

Gardner RA, Gardner BT and Van Cantfort TE. (eds.) 1989. *Teaching sign language to chimpanzees*. State University of New York Press, Albany.

Gazzaniga MS. 2011. Who is in charge? *Bioscience,* 61(12), 937–938. doi: 10.1525/bio.2011.61.12.2.

Goodenough U and Deacon T. (2008). The Sacred Emergence of Nature. In Human Nature and Ethics. *The Oxford Handbook of Science and Religion.* pp. 858–871.

Greenfield PM and Savage-Rumbaugh ES. 1984. Perceived variability and symbol-use: a common language-cognition interface in children and chimpanzees. *Journal of Comparative Psychology,* 98, 201–218.

Greenfield PM, Lyn H and Savage-Rumbaugh ES. 2008. Protolanguage in ontogeny and phylogeny: Combining deixis and representation. *Interaction Studies,* 9(1), 34–50.

Haesler S, Wada K, Nshdejan A, Morrisey EE, Lints T, Jarvis ED and Scharff C. 2004. FOXP2 expression in avian vocal learners and non-learners. *Neuroscience,* 24, 3164–3175.

Heyes C. 1998. Theory of mind in nonhuman primates. *Behavioral and Brain Sciences,* 21, 101–134.

Ijdo JW, Baldini DC, Ward ST, Reeders and Wells RA. 1991. Origin of human chromosome 2: An ancestral telomere–telomere fusion. *Proceedings of the National Academy of Sciences of the United States of America,* 88(20), 9051–9055.

Inoue S and Matsuzawa T. 2007. Working memory of numerals in chimpanzees. *Current Biology,* 17(23), R1004–R1005.

Inoue S and Matsuzawa T. 2009. Acquisition and memory of sequence order in young and adult chimpanzees (*Pan troglodytes*). *Animal Cognition,* 12, S59–S69.

James W. 1890. *The Principles of Psychology.* New York: Dover Publications, 1950.

Kappler P. 2010. *Animal Behavior: Evolution and Mechanisms.* Springer, Heidelburg.

Kellogg WN and Kellogg LA. 1933. *The Ape and the Child: A Study of Environmental Influence upon Early Behavior.* McGraw-Hill, New York.

King MC and Wilson AC. 1975. Evolution at two levels in humans and chimpanzees. *Science,* 188(4184), 1070–1166.

Kipnis J. 2006. T cells — New and unexpected players in cognition. *Journal of Neuroimmunology,* 178, 21.

Konner M. 2000. *The Tangled Wing. Biological Constraints on the Human Spirit.* Henry Holt and Company, New York.

Lyn H and Savage-Rumbaugh ES. 2000. Observational word learning in two bonobos (Pan paniscus): ostensive and non-ostensive contexts. *Language and Communication,* 20, 255–273.

Lumey LH. 1989. Reproductive outcomes in women prenatally exposed to undernutrition: a review of the findings from the Dutch famine birth cohort. *Proceedings of the Nutrition Society*, 57(1), 129–135.

Lock A. 1980. *The Guided Reinvention of Language*. Academic Press, London.

Lock A. 2000. Phylogenetic Time and Symbol Creation: where do zopeds come from. *Culture and Psychology*, 6, 105–129.

Lynn H, Greenfield P and Savage-Rumbaugh ES. 2006. The development of representational play in chimpanzees and bonobos: evolutionary implications, pretense and the role of inter-species communication. *Cognitive Development*, 21, 199–213.

Liu H, Prugnolle FA, Manica A and Balloux F. 2006. A geographically explicit genetic model of worldwide human-settlement history. *American Journal of Human Genetics*, 79, 230–237.

Lyn H, Franks B and Savage-Rumbaugh ES. 2008. Precursors of morality in the use of the symbols "good" and "bad" in two bonobos (*Pan paniscus*) and a chimpanzee (Pan troglodytes). *Language and Communication*, 28, 213–224.

Lyn H, Greenfield PM, Savage-Rumbaugh S, Gillespie-Lynch K and Hopkins WD. 2011. Nonhuman Primates do Declare! A Comparison of Declarative Symbol and Gesture Use in Two Children, Two Bonobos, and A Chimpanzee. *Language & Communication*, 31(1), 63–74.

Lucie-Smith E. 1992. *Art and Civilization*, Harry N. Abrams, a Times Mirror Company, New York.

Maienschein J. 2008. Epigenesis and preformationism. *Stanford Encyclopedia of Philosophy*, Zalta EN (ed.), Standord, Stanford University, CA.

Malinowski B. 1922. *Argonauts of the Western Pacific: An Account of Native Enterprise and Adventure in the Archipelagoes of Melanesian New Guinea*. London: Routledge & Kegan Paul.

Malinowski B. 1948. *Math, Science, Religion and Other Essays*. Glencoe, Illinois: The Free Press.

Marshall C. 2008. *Animal Interaction Behind 'Cambrian Explosion'*? Available from http://news.harvard.edu/gazette/story/2008/05/animal-interaction-behind-cambrian-explosion/. Accessed on August 5 2016.

Matsuzawa T. 2001. *Primate Origins of Human Cognition and Behavior*. Springer, Tokyo.

Matsuzawa T. 2003. The Ai Project: Historical and ecological contexts. *Animal Cognition*, 6(4), 199–211.

Mayr E. 2002. *What Evolution Is*. Weidenfeld and Nicolson, London.

McGrew WC. 2004. *The Cultured Chimpanzee: Reflections on Cultural Primatology*: Cambridge University Press.

Miles HL, 1994. The development of self-awareness in a signing orangutan. In Parker ST, Mitchell RW and Boccia M (eds.). *Self-Awareness in Animals and Humans: Developmental perspectives*, (pp. 254–272). Cambridge University Press, New York.

McNeil MC, Polloway EA and Smith JD. 1984. Feral and Isolated Children: Historical Review and Analysis. *Education and Training of the Mentally Retarded*, 19(1), 70–79.

Miller KR. 2000. *Biology*. Prentice Hall, Upper Saddle River, New Jersey.

Mithen S. 1996. *The Prehistory of the Mind*. Thames and Hudsen, New York.

Nei M, Takeo M and Chakraborty R. 1975. The bottleneck effect and genetic variability in population. *Evolution*, 29(1), 1–10.

Noble W and Davidson I. 1996. *Human Evolution, Language and Mind: A Psychological and Archaeological Inquiry*: Cambridge University Press.

Patterson P. 1978. The gestures of gorilla: Language Acquisition in another pongid. *Brain and Language*, 5(1), 72–97.

Pederson J and Fields WM. 2009. Aspects of Repetition in Bonobo Human Conversation: creating cohesion in a conversation between species. *Integrative Psychological and Behavioral Science*, 43(1), 22–41.

Penn DC and Povienelli DJ. 2007. On the lack of evidence that non-human animals possess anything remotely resembling a 'theory of mind.' *Philosophical Transactions of the Royal Society of London. Series B, Biological Sciences*, 363, 731–744.

Penn DC and Povinelli DJ. 2008. The Comparative Delusion: The Behaviouristic/Mentalistic Dichotomy in Comparative Theory of Mind Research. In *Oxford Handbook of Philosophy and Cognitive Science*, Samuels R and Stich SP (eds.).

Penn DC and Povinelli DJ. 2009. On becoming approximately rational: The relational reinterpretation hypothesis. In *Rational Animals, Irrational Humans*, Watanabe S, Blaisdell AP and Huber L (eds.), Tokyo, Keio University Press, pp. 23–44.

Perry GH, Yang F, Marques-Bonet T, Murphy C, Fitzgerald T, Lee AS, Hyland C, Stone AC, Hurles ME, Tyler-Smith C, Eichler EE, Carter NP, Lee C and Richard Redon R. 2009. Copy number variation and evolution humans and chimpanzees. *Genome Research*, 18, 1698–1710.

Petraglia MD, Korisettar R and JN. 2012. Pal. Toba volcanic super-eruption of 74,000 years ago: Climate change, environments and evolving humans. *Quaternary International*, 258, 1–4.

Plotnik JM, de Waal FBM and Reiss D. 2006. Self-recognition in an Asian elephant. *Proceedings of the National Academy of Sciences*, 103(45), 18053–17057.

Povinelli DJ and Giambrone S. 1999. Inferring other minds: Flaws in the argument by analogy. *Philosophical Topics*, 27, 167–201.

Povinelli DJ, Eddy TJ, Hobson RP and Tomasello M. 1996. What young chimpanzees know about seeing. *Monographs of the Society for Research in Child Development*, 61(3), i–191.

Povinelli DJ. 2000. *Folk Physics for Apes*. Oxford University Press, New York.

Premack D. 1976. *Intelligence in Ape and Man*. Erlbaum Associates, Hillsdale, NJ.

Premack D. 1978. *Gavagai: Or the Future History of Animal Language.* Cambridge: Bradford Books.

Premack D. 2003. *Original Intelligence. Unlocking the Mystery of Who We Are.* McGrawHill, New York.

Premak AJ and Premack D. 1976. *Teaching language to an ape. Scientific American*, 227, 92–99.

Ragir S and Savage-Rumbaugh ES. 2009. Playing with meaning: Normative function and structure in play. In Rudolf Botha and Chris Knight (eds.), *The Prehistory of Language.* Oxford University Press, Oxford.

Ramachandran VS. 2009. Self Awareness: The Last Frontier. Edge Foundation. http://www.edge.org/conversation/self-awareness-the-last-frontier, Accessed on August 4 2016.

Ramachandran VS. 2010. The Tell-Tale Brain: *A Neuroscientists Quest for What Makes Us Human*. Norten, New York.

Reid GBR and Hetherington R. 2010. *The Climate Connection: Climate Change and Modern Human Evolution.* Cambridge University Press, Cambridge, UK.

Rilling J, Kilts C, Williams SL, Kelley JW, Beran MJ, Giroux M, Hoffman JM, Savage-Rumbaugh ES and Rumbaugh DM. 1999. Functional neuroimaging of linguistic processing in chimpanzees. *Society for Neuroscience Abstracts*, 25(2), 2170.

Rizzolatti G and Fabbri-Destro M. 2010. Mirror neurons: From discovery to autism, *Experimental Brain Research*, 200, 223–237.

Roffman, The Chimpanzees of Mali (in preparation).

Rumbaugh D. 1971. Evidence of qualitative differences in learning processes among primates. *Journal of Comparative and Physiological Psychology*, 76, 250–255.

Rumbaugh DG. 1974. Language in man, monkey and machine. *Science*, 185, 872–873.

Rumbaugh DM (ed.). 1977. *Language Learning by a Chimpanzee: The Lana Project*. Academic Press, New York.

Savage-Rumbaugh ES. 1997. Why are we afraid of apes with language? In *The Origin and Evolution of Intelligence*, Arnold B. Scheibel and J. William Schopf (eds.), pp. 43–69. Sudbury: Jones and Bartlett.

Savage-Rumbaugh ES and Fields WM. 1998. Language and Culture: A Trans-generation Interweaving. *Language Origins Society*, 27, 19–41.

Savage-Rumbaugh ES and Fields WM. 2000a. The cultural origins of human cognition. *American Anthropologist*, 102(4), 925–926. doi: 10.1525/aa.2000.102.4.925.

Savage-Rumbaugh ES and Fields WM. 2000b. Linguistic, cultural and cognitive capacities of bonobos (*Pan paniscus*). *Culture & Psychology*, 6(2), 131–153. doi: 10.1177/1354067x0062003.

Savage-Rumbaugh ES and Fields WM. 2011. The evolution and the rise of human language: carry the baby. In Henshilwood CS and Errico FD (eds.), *Homo Symbolicus: The Dawn of Language, Imagination, and Spirituality*. Netherlands: John Benjamins Publishing, 13–47.

Savage-Rumbaugh ES and Fields WM. 2013. Human Uniqueness — Constructions of Menzel, Tembrock G (ed.). *Quo Vadis, Behavioral Biology? Past, Present, and Future of an Evolving Science*, Nova Act Leopoldina, Stuttgart, 7–11.

Savage-Rumbaugh ES and Rumbaugh DM. 1978. Symbolization, language and chimpanzees: a theoretical reevaluation based on initial language acquisition processes in four young Pan troglodytes. *Brain and Language*, 6, 265–300.

Savage-Rumbaugh ES and Rumbaugh DM. 1980. Requisites of Symbolic Communication or are words for birds? *The Psychological Record*, 30, 305–315.

Savage-Rumbaugh ES, McDonald K, Sevcik RA, Hopkins WD and Rubert E. 1986. Spontaneous symbol acquisition and communicative use by pygmy chimpanzees (Pan paniscus). *Journal of Experimental Psychology: General*, 115(3), 211–235.

Savage-Rumbaugh ES, Murphy J, Sevcik RA, Rumbaugh DM, Brakke KE and Williams S. 1993. Language comprehension in ape and child. *Monographs of the Society for Research in Child Development*, 58(3–4), 1–242.

Savage-Rumbaugh ES, Pate JL, Lawson J, Smith ST and Rosenbaum S. 1983. Can a chimpanzee make a statement? *Journal of Experimental Psychology: General*, 112, 457–492.

Savage-Rumbaugh ES, Romski MA, Hopkins WD and Sevcik RA. 1989. Symbol acquisition and use by Pan troglodytes, Pan paniscus, and *Homo sapiens*. In Heltne PG and Marquardt LA (eds.). *Understanding Chimpanzees* (pp. 266–295). Harvard University Press, Cambridge, MA and London, England.

Savage-Rumbaugh ES, Romski MA, Sevcik R and Pate JL. 1983. Assessing symbol usage versus symbol competency. *Journal of Experimental Psychology: General*, 112, 508–512.

Savage-Rumbaugh ES, Rumbaugh DM and Boysen S. 1978a. Symbolic communication between two chimpanzees (*Pan troglodytes*). *Science*, 201, 641–644.

Savage-Rumbaugh ES, Rumbaugh DM and Boysen S. 1978b. Linguistically mediated tool use and exchanged by chimpanzees (*Pan troglodytes*). *Behavioral and Brain Sciences*, 641–644.

Savage-Rumbaugh ES, Rumbaugh DM, Smith ST and Lawson J. 1980. Reference, the linguistic essential. *Science*, 210, 922–925.

Savage-Rumbaugh ES, Segerdahl P and Fields WM. 2005. Individual differences in language competencies in apes resulting from unique rearing conditions imposed by different first epistemologies. In Namy LL (ed.). *Symbol Use and Symbolic Representation* (pp. 199–219). Lawrence Erlbaum Associates, Mahwah, NJ.

Savage-Rumbaugh ES, Shanker S and Taylor TJ. 1998. *Apes, Language and the Human Mind*. Oxford University Press, New York, NY.

Savage-Rumbaugh ES, Wilkerson BJ and Bakeman R. 1977. Spontaneous gestural communication among conspecifics in the pygmy chimpanzee (Pan paniscus). In Bourne GH (ed.), *Progress in ape research* (pp. 97–116). Academic Press, New York.

Savage-Rumbaugh ES. 1986. *Ape Language: From Conditioned Response to Symbol*. Columbia University Press, New York.

Savage-Rumbaugh S, Fields WM and Spircu T. 2004. The Emergence of Knapping and Vocal Expression Embedded in a *Pan/Homo* Culture. *Journal of Biology and Philosophy*, 19(4), 541–575.

Segerdahl P, Fields WM and Savage-Rumbaugh ES. 2005. *Kanzi's Primal Language; The Cultural Initiation of Primates into Language*. Houndmills, Basingstoke, Hampshire: Palgrave Macmillan.

Shotter J. 1993. *Conversational Realities: Constructing Life Through Language*. Sage Publications, London.

Shu L and Wu J. 2009. Epigenetic regulation in mammalian preimplantation embryo development. *Reproductive Biology and Endocrinology*, 7, 59.

Skinner BF, Epstein R and Lanza RP. 1980. Symbolic communication between two pigeons (Columbia livia domestica). *Science*, 207, 543–545.

Taglialatela JP, Savage-Rumbaugh ES and Baker LA. 2002. Vocal production by a language-competent Pan Paniscus. *International Journal of Primatology*, 24(1), 1–17.

Taglialatela JP, Savage-Rumbaugh ES, Rumbaugh DM, Benson J and Greaves W. 2004. Language, apes and meaning-making. In William G and Lukin (eds.). *Development of Language: Functional Perspectives on Evolution and Onto Genesis* (pp. 91–111). The Continuum International Publishing Group, London.

Taylor J. 2009. *Not a chimp: The hunt to find the genes that make us human.* Oxford University Press, Oxford.

Terrace HS and Chen S. 1991. Chunking during serial learning by a pigeon: II. Integrity of a chunk on a new list. *Journal of Experimental Psychology, Animal Behavior Processes*, 17(1), 94–106.

Terrace HS and Petitto LA, Sanders RJ and Bever TG. 1979. Can an ape create a sentence? *Science*, 206(4421), 891–902.

Terrace HS, Straub R, Bever TG and Seidenberg M. 1977. Representation of a sequence by a pigeon. *Bulletin of the Psychonomic* Society, 10, 269.

Terrace HS. 1979. Can an ape create a sentence? *Science*, 206(4421), 891–902.

Terrace HS. 1982. Why Koko Can't Talk. *The Sciences*, 22, 8–10.

Terrace HS. 1986. Positive transfer from sequence production to sequence discrimination in a nonverbal organism. *Journal of Experimental Psychology: Animal Behavior Processes*. 12(3), 215–234.

Terrace HS. 1987. *Nim, A Chimpanzee Who Learned Sign Language*. Columbia University Press, New York.

Tomasello M and Call J. 1997. *Primate Cognition*. Oxford University Press, New York.

Tomasello M and Call J. 2006. Do chimpanzees know what others see — or only what they are looking at? In Hurley S and Nudds M (eds.). *Rational Animals*? Oxford University Press, Oxford.

Tomasello M, Call J and Hare B. 2003a. Chimpanzees understand psychological states — the question is which ones and to what extent. *Trends in Cognitive Sciences*, 7(4), 153–156.

Tomasello M, Call J and Hare B. 2003b. Chimpanzee versus humans: it is not that simple. *Trends in Cognitive Sciences*, 7(6), 239–240.

Tomasello M. 2008. *Origins of Human Communication*. MIT Press, Cambridge, MA.

Tomkins J. 2013. Alleged Human Chromosome 2 "Fusion Site" Encodes an Active DNA Binding Domain Inside a Complex and Highly Expressed Gene — Negating Fusion. *Answers Research Journal*, 6, 367–375.

Tuttle RH. 2014. *Apes and Human Evolution.* Harvard University Press, Cambridge Massachusetts.

Valsiner J. 2007. *Culture in Minds and Societies: Foundations of Cultural Psychology*. Sage Publications, New Delhi.

Van der post L and Taylor J. 1984. *Testament to the Bushmen*. Hammondsworth, Middlesex, England.

Vargah-Khadem F, Gadian DG, Copp A and Mishkin M. 2005. FOXP2 and the neuroanatomy of speech and language. *Nature Review of Neuroscience*, 6, 131–138.

Virey JJ and Guenebault JH. 1837. *Natural History of the Negro Race*: D. J. Dowling.

Wallman J. 1992. *Aping Language*. Cambridge University Press, Cambridge, UK.

Waters CM and Bassler BL. 2005. Quorum sensing: Cell-to-cell communication in bacteria. *Annual Review of Cell and Developmental Biology*, 21, 319–346. doi: 10.1146/annurev.cellbio.21.012704.131001.

Watson SK, Townsend SW, Schel AM, Wilke C, Wallace EK, Cheng L, West V and Slocombe K. 2015. Vocal learning in functionally referential food grunts of chimpanzees. *Current Biology*, 25, 1–5.

Wheeler BC and Fischer J. 2012. Functionally referential signals: a promising paradigm whose time has passed. *Evolutionary Anthropology*, 21, 195–205.

White SA, Fisher SE, Geschwind DH, Scharff C and Holy TE. 2006. Songbirds, and more: Models for FOXP2 dysfunction in human speech and language. *Journal of Neuroscience*, 26, 10376–10379.

Wildman D, Grossman LI and Goodman M. 2002. Functional DNA in humans and chimpanzees shows they are more similar to each other than either is to apes. In Goodman M and Moffat AS (eds.), *Probing Human Origins*, 1–10, American Academy of Arts and Sciences, Cambridge, MA.

Wilson DS. 1975. A theory of group selection. *Proceedings of the National Academy of Sciences*, 72(1), 143–146.

Wrangham RW, McGrew WC, DeWaal RBM and Heltne PG (eds.). 1994. *Chimpanzee Cultures*. Harvard University Press, MA.

Yellen JE, Brooks AS, Cornelissne E, Mehlman MJ and Stewart K. 1995. A Middle Stone Age bone industry from Katanda, Upper Semliki Valley, Zaire. *Science*, 268, 553–557.

Yerkes RRM and Yerkes AW. 1929. *The Great Apes*. Yale University Press, New Haven, CT.

Yoakum CS and Yerkes R. 1920. *Army Mental Tests.* Henry Holt and Company, New York.

Yunis JJ and Prakash O. 1982. The origin of man: A chromosomal pictorial legacy. *Science,* 215(4539), 1525–1530.

Ziv Y, Ron N, Butovsky O, Landa G, Sudai E, Greenberg N, Cohen H, Kipnis J and Schwartz M. 2006. Immune cells contribute to the maintenance of neurogenesis and spatial learning abilities in adulthood. *Nature Neuroscience,* 9(2), 268–275.

17. Conversing with Dolphins: The Holy Grail of Interspecies Communication?

Toni Frohoff*,‡ and Elizabeth Oriel†,§

*TerraMar Research, 27 W. Anapamu St. Suite 336
Santa Barbara, California 93101, USA
†Sonar, 31 Greenhill Park, NW10 9AN
London, United Kingdom
www.wearesonar.org
‡ tfrohoff@gmail.com
§ lizzieoriel@gmail.com

The scientist's quest for communication with dolphins can be as fraught with mystery and mishaps as an archetypal search for the Holy Grail. At every turn, scientists encounter millennial-old, romantized notions about iconic dolphins teaming with logistical challenges and plodding methodologies struggling to keep pace with, and identify, even a singular, solid thread running through an intricate web of data. When it comes to dolphins, scientific paradigms merge with policy, politics, emotion, and ethics in stimulating, yet sometimes turbid, waters. Our search for clarity through meaningful analysis to adequately (let alone elegantly) explain our findings about dolphins can be both an enviable quest and an onerous duty. The mystery of dolphins continues to beckon us, from before Aristotle to modern-day, inspiring research that expands our concepts of who cetaceans are. We pursue a path beneath the sea, challenging ourselves to press more deeply against our own anthropocentric limits of biocommunication to connect with an aquatically alien, but similarly sentient, species.

Dolphin communication and language research is often comparable to studies of humans and other primates (Bearzi and Stanford 2010). The more we learn about dolphins — from their neurophysiology and behavioral ecology to their cognition and social behavior — the facts become more fascinating than the fables. However, human interpretation of the marked morphological differences between dolphins and humans can create communicative havoc between the species. Dolphins demonstrate distinctive physiological and psychological needs and their behavior is not always consistent with their human-imposed, ever-friendly and perpetually joyful media "Flipper" stereotype.

Toothed whales, including dolphins (and orcas), demonstrate amongst the most sophisticated biocommunicative capabilities of any animal. Aspects of their echolocation systems remain unparalleled, even by our most advanced human technology. From an acoustic perspective alone, their vocal production and neural information processing are dazzling. However, we cannot lose sight of the subtler, yet equally remarkable, cognitive, cultural, and psychological complexities inherent in the dolphins we study and with whom we may attempt to communicatively interact (Dudzinski and Frohoff, 2008, Herzing and Johnson, 2015).

This chapter emphasizes the need for a more expansive, and a multidisciplinary, approach to studying biocommunication in dolphins and other animals. We adopt a species-inclusive approach to our perceptions about animal life that embodies myriad unique forms of awareness, intelligence, culture, sentience, and even personhood. The complexities we seek to understand in dolphins and other life forms exemplify the need to transcend reductionist and mechanistic approaches to studying communication in other animals. This era of burgeoning data illuminating complex internal experiences in non-human animals calls for innovative ways to approach the study of such capacities, especially as exhibited in communicative interactions. One way of advancing our study of dolphins is to correct our approach to studying them so that they are treated less like "objects" or even "subjects" and acknowledge, what is sometimes, undeniably, the complex, communicative, and even collaborative relationship, occurring between humans and dolphins. A new paradigm of research (Interspecies Collaborative Communication) is discussed as one example of myriad methods through which we can advance our study of human–animal communication (Frohoff and Marino, 2010; Marino and Frohoff, 2011).

We also present the positive cohabitation that occurs in some small-scale regions among dolphins, humans and coastal systems, The grail of interspecies communication and cohabitation shines in multispecies relationship-building, expanded notions of personhood, and cohabitation, in which both humans and dolphins share a mutual thriving across coastal systems.

1. Smart, Savvy, Sociable, and Sentient

It can be uncommon for free ranging, wild animals to interact socially with people for reasons other than food provisioning. Yet stories of sociable and "friendly" dolphins, including those assisting swimmers and boaters in distress, have been reported in ancient and in recent times by reliable sources (Frohoff and Peterson, 2003). Plutarch, in his treatise, "On the Intelligence of Animals" written in 66 A.D., wrote of the dolphin, "It has no need of any man, yet is the friend of all men and has often given them great aid." Friendly dolphins soliciting sociable contact with people may be a recreational swimmer's delight, but pose a methodological dilemma. Scientists find themselves as unwitting participants in dolphin–human interactions, while trying to stay in the background and be as "invisible" as possible, as most have been traditionally taught to do when studying wildlife behavior in the wild.

Fortunately, cognition and interspecies communication can be assessed in the wild through indirect measures and inference as well as through direct tests (Marino and Frohoff, 2011; Herzing, 2015). A new era of interspecies research is taking exciting strides in advancing what we know about dolphins and other cetaceans by our taking advantage of their habituation to humans, if not their outright sociability directed toward us in the wild, and on *their* terms. Systematic methodologies in which researchers actively collaborate with cetaceans are developing in ways that can be as rigorous, albeit not as manipulative, as studies conducted in the laboratory, aka dolphin tank (Dudzinski and Frohoff, 2008; Herzing and Johnson, 2015).

So how do cetaceans and humans even communicate between species (Fig. 1)? Dolphin physiology permits dolphins to communicate with one another through a variety of sensory modes in a marine environment that is mostly foreign to human biocommunication. They are perfectly adapted to their underwater environment, with hearing and sound production capabilities that utilize this medium exquisitely. Dolphin communication is characterized by a dynamic interplay of acoustic signals involving an unusually rich and complex vocal repertoire, visual posturing, and movements, chemoreception, tactile exchange (ranging from friendly to aggressive), and use of water for making visual (bubbles) and acoustic (splashes) signals. For a more comprehensive overview of dolphin communication, see Dudzinski and Frohoff (2008), and Bearzi and Stanford (2010), Herzing and Johnson (2015).

Figure 1. A free-ranging, sociable dolphin initiates interspecies Collaborative Research with researcher. Both © Ute Margrreff, with permission.

Although dolphins and humans lack a common ancestor, both have evolved similar cognitive abilities, likely for comparable social or communicative purposes (Bearzi and Stanford, 2010). Dolphins exhibit varied and rich social cultures, similar to those in humans and other highly social terrestrial (e.g., chimpanzees and elephants) and avian (e.g., parrots and ravens) species. Despite our environmental differences, non-human primates exhibit qualities such as self-awareness, morality, culture, empathy, and politics (cite include mirror recognition study). And even for animals without "hands", the discovery of tool use in dolphins through the mouth-held features of their natural environment — and associated cultural transmission and communication has been impressive (Krützen *et al.*, 2005). Neuroanatomical post-mortem studies of brain size, structure and complexity in dolphins and whales also provide important information about the neurobiological bases of intelligence, sentience and communicative capacities (Marino *et al.*, 2008). Within the current framework for establishing legal personhood for non-human animal, these cognitive and communicative capacities are centerpieces of the legal arguments.

2. Intraspecies Culture and Communication

We are in a new era in which evidence abounds for the presence of culture in many non-human species. Here, we focus on culture among cetaceans. Culture is evidenced through communication and social learning that transmits through social structures and networks (Herzing, 2005; Lusseau *et al.*, 2006). Herzing and Johnson (2015) present mechanisms of transmission of social learning among Atlantic spotted dolphins (*Stenella frontalis*) off the Bahamas that describe a highly complex multilayered approach to sociality, learning, communication, and culture. Whitehead and Rendell (2014) delve deeply into the rich ways in which cetacean culture and its transmission are shaped by a blend of adaptations, innate sociality, and their aquatic environment, and what the implications for such complex cultures mean for current and future generations of cetaceans and human responsibility.

Jared Diamond (2001) states that, especially in preliterate societies, "Old people are the repositories of knowledge ..." and that some cetaceans rival humans in this regard. Female orcas may live so much longer than males — by as much as 40 years — because the knowledge they exchange and possess helps the whole group survive. Orcas and short-finned pilot whales represent a somewhat rare phenomenon of extended post-reproductive life for females in the mammalian world (Courage, 2010). The vital importance of culture to the survival of non-human populations and species may be most evident in the wisdom of female orca elders (Brent *et al.*, 2015). Post-menopausal female orcas are knowledgeable leaders who are vital to younger members of their community by acting as "repositories of ecological knowledge" (Courage, 2010). Yet evidence indicates that knowledge of orca elders in orcas, as in other species (such as humans, some other primates, and elephants (Diamond, 2001), extends far beyond that of the basic ecology of survival to more nuanced and intellectually sophisticated aspects of cultural wisdom. But for the most pragmatic purposes of protecting other species, humans can no longer underestimate invaluable role that these elder females play in finding food for orcas, for example (or in the case of elephants, finding water).

Discretion is warranted whenever assessing the intelligence, awareness, culture of any other animal within the human/ape paradigms and ecology. For example, as Barrett and Würsig (2014) point out, the ability to echolocate in dolphins constitutes a completely foreign sensory modality for human researchers. Echolocating dolphins may be able to see inside each other's bodies, even perhaps to the extent of perceiving another's emotional states. For years, ultrasonic sounds used by cetaceans were not even considered since they transgressed the upper end of the human hearing threshold. Similarly, it was only recently that subsonic vocalizations used by elephants were discovered, and some perspectives on elephant communication were completely rewritten (Payne *et al.*, 1986). Thus using human-like attributes as the yardstick by which all others are measured would not only be misguided but could invalidate the methodology used in obtaining data, let alone analyzing it. Unless we acknowledge and transcend our limited perspective as hominid researchers, we will continue to overlook or underestimate abilities in other species simply because we do not perceive, let alone understand them.

3. Interspecies Culture, Communication, and Miscommunication

The wide varieties of communicative dolphin–human interactions reveal varied levels, types and modes of interspecies biocommunication. For example, Pryor (1990) states that all social behavior may be considered communication, including reinforcement training. Trainer-dolphin interactions would qualify as a change in the behavior of the receiver in response to the information provided by the signals or signs of the sender (Philips and Austad, 1990; Smith, 1990). Play between dolphins and humans may provide the motivation, as well as form a basis for, the development of mutually understood signals necessary in interactions (Fagan, 1981).

Another view of communication is that interaction may be interpreted as communication if the response is positive for the sender. If the response is neutral or negative for the sender, it may be interpreted as miscommunication (Krebs and Dawkins, 1991). Frohoff introduced the concept of miscommunication in dolphin-human interactions in 1993 and when studying programs in which the public swims with captive dolphins (Frohoff 1993). Humans do not share a large number of communicative behavioral repertoires with dolphins.

Perhaps the most obvious example of human projections on dolphins is what is commonly referred to as, the "dolphin's smile". What humans often perceive of as an emotional expression is, on the contrary, an upturn behind the mouth that is a completely static feature of the dolphin's jaw that remains through life and through death, health and suffering (Frohoff, 2000). In this, and many other cases, dolphins are often poorly understood due to human cultural stereotype as well as from anthropocentrism at its extreme.

Unless humans are exceptionally knowledgeable of, and responsive to, the dolphins' salient and subtle signals, not only may interspecies miscommunication occur, but our presence may likely interrupt activities (e.g., feeding, mating, and resting) vital to the wellbeing of the individual dolphin(s) as well as that of entire populations (Frohoff, 2000).

Mutually beneficial communication — or what we call, positive or collaborative cohabitation between dolphins and humans — may be exemplified when looking at human–dolphin fishing cooperatives. Scientists have documented interspecies cooperative fishing on several continents between indigenous peoples and dolphins that appears to have developed over time across many dolphin and human generations. This appears to be a form of cultural coevolution between dolphins and humans in some regions of the world. Apparently, both dolphins and fishermen benefit by mutual herding and catching of fish. Pliny the Elder (A.D. 23–79) wrote that dolphins worked with humans to drive mullet into the nets in the Rhone River in France. Later, Oppian reported cooperative fishing at Emboea and, more recently, this practice has been observed in the Mediterranean, North Africa, New Zealand, and South America (e.g., Pryor *et al.*, 1990). The Australian aborigines of Moreton Bay also used to catch fish with the aid of dolphins (Domning, 1991). The Amazon River dolphin and the Irrawaddy river dolphin have also assisted fishermen consistently (Cousteau and Diolé 1987). In Laguna, Brazil, fishermen continue to fish cooperatively with bottlenose dolphins as they have done for several generations (Domning, 1991; Pryor *et al.*, 1990). In many of these regions, families or individuals often work exclusively with one or several particular dolphins. This relationship sometimes extends to the progeny of both the dolphins and the people for several generations. Apparently, these cooperative relationships are initiated and controlled by the dolphins rather than the fishermen. Cooperative fishing between dolphins and sea birds, such as pelicans, has also been observed with some frequency and may provide insight into the nature of human–dolphin fishing cooperatives (Cousteau and Diolé, 1987).

Thus we see a continuum in the effectiveness, degree of participation and quality of consequences in interspecies communication and co-habitation. A range from "negative" miscommunication resulting in deleterious impacts — to "positive" communications being collaborative and mutually beneficial can be observed. Ideally, modes of communication between human and non-human animals and, even ecosystems, can be further identified and implemented to the betterment of individual wellbeing and population and species survival.

4. Captivity

The charisma dolphins hold for many humans has been a part of broad legislative protection (e.g., the US Federal Marine Mammal Protect Act of 1972), yet conversely, has also contributed to their exploitation by humans (e.g., captures from the wild for marine parks and circuses). Dolphins, especially the largest of the dolphin family, orcas, have been considered a key attraction to captive public display facilities around the world. In the mid 1800s in the United States, P. T. Barnum put captive dolphins and beluga whales on public display and since then, toothed dolphins and whales in captivity have been used for everything from research to captive swim-with-the dolphins programs (Frohoff, 2008).

However, recent scientific discoveries and public concerns about the ethics of capturing, and forcing dolphins to live in tanks and pens, have been burgeoning over the past few decades (White, 2007). This awareness has been reflected in legislation and policy around the world, in local and national bans on the capture or captivity of dolphins for public display. However, recent challenges to captivity are not limited to concerns about individual welfare. Capture from the wild has been identified as a serious conservation issue in various regional, national, international fora — especially when dolphins, and whales, sometimes, in large numbers, are removed from already depleted populations (Rose and Furinato, 2006).

Previously conducted primarily in captivity, one of the main goals of dolphin communication research has been to determine whether dolphins can comprehend an artificial symbolic language. The work in captivity heretofore has provided important insights into cetacean intelligence and cognition (Herman *et al.*, 1993; Reiss and McCowan 1993). Yet, captive studies are limited in external validity for a variety of reasons. These comprise the unknown and largely uncontrollable developmental-cognitive effects of living in an artificial physical, perceptual, and social environment on the generalizability of findings to wild cetaceans.

Captive studies may be confounded by the physical and psychological stress and trauma evidenced in illnesses and aberrant dolphin behavior (Frohoff, 2004). Also studies of captive dolphins likely overlook behaviors and capacities that are absent or diminished in captivity. Recently, a study of a group of wild chimpanzees revealed that their gestural repertoire was over twice the size suggested by studies of captive chimpanzees (Hobaiter and Byrne, 2011). Such studies suggest that captivity may truncate capacities under some circumstances and lead to inaccurate conclusions.

There is an arguably more compelling reason to consider adopting a new paradigm for studying cetacean biocommunication and cognition. This has to do with the essential importance of adjusting our behaviors, protocols, and paradigms to the very information revealed in our scientific endeavors with them. As Marino and Frohoff (2011: 2–3) conclude, In our view, the conclusion from decades of cumulative results of both captive and field studies is that cetaceans possess a level of intelligence, awareness and psychological, and emotional sensitivity that makes it unacceptable to continue to keep them in captivity if not necessary for their welfare, survival, or conservation. "We do not deny that captive studies have contributed substantially to this conclusion. Our point is that now that we have this knowledge about cetaceans it is incumbent upon us to revise our approaches to studying them."

5. Revisiting Research: Interspecies Collaborative Communication

The complex sentience of other animals such as dolphins must be recognized and their physical, psychological, and behavioral needs appropriately protected. Further, there is much that humanity can learn from the ethics and morals exhibited in the social behavior extended from animals of other species (referred to as "Animality" by Frohoff, (Crane, 2015). Accordingly, scientists are now faced with the task of accommodating this contemporary knowledge of cetacean neuroanatomy and behavior in ways that alter their research approaches and priorities. Frohoff and Marino (2010) have proposed developing ways to more directly assess cognition in wild individuals that may replace studies in captivity and form the basis for a more extensive cognitive ethological approach in cetaceans; one that also encompasses aspects of their behavioral ecology (Fig. 2) as well as their cultural and ethical dimensions.

There are a number of protocols available for respectfully studying cognition and other capacities of cetaceans. New and innovative approaches are being taken with free-ranging dolphins who are habituated to humans, or even initiate gregarious sociable encounters (Marino and Frohoff, 2011). Some methods used to study captive dolphins can be potentially modified and transferred to research on free-ranging dolphin groups or free-ranging individuals. One of the keys to being able to transfer cognitive tasks

Figure 2. Dolphins communicate using a vast array of behaviors, including visual and tactile interactions. © Atmoji, Wildquest, with permission.

from the captive situation to the wild is the opportunity to work with individual dolphins one-on-one. While we discuss several approaches below, our emphasis is consistent and firm on one point: that it is vital for the non-humans/dolphins to be in control of whether or not — and the degree to which — they participate and encounter humans. Only in this way can we "take advantage" of the curious or friendly ways in which some dolphins approach and interact with humans — without literally taking advantage of the dolphins themselves. In this regard, "respect" is the operative word, and the wellbeing of the dolphins or other non-human animals takes priority over the success of the research.

Individuals known as lone, sociable dolphins present extraordinary potential for cognitive and communicative study. Lone sociable dolphins are free-ranging cetacean individuals who are often solitary, yet, for one reason or another, have initiated, or participated in, sociable interactions with humans in the wild with some regularity (Frohoff *et al.*, 2006; Lockyer and Muller, 2003). Some of these individuals were orphaned and have become separated from their social group and are truly isolated from conspecifics. Others vacillate between interactions with humans and members of their own (or other) species but nevertheless fall under the category of lone sociables. There are numerous known individual cetaceans who have fit this description; mostly bottlenose dolphins in various regions (*Tursiops truncatus*), beluga whales (*Delphinapterus leucas*) in Eastern Canada and, more rarely, orcas (*Orcinus orca*) in the US Pacific Northwest. And while not all of them will be good candidates for research, many of them can be with the right circumstances and proclivities of the individual dolphin or dolphin group.

6. Changing Paradigms of Research

In fact, ongoing long-term field studies of social complexity, foraging, and culture in dolphins and whales continue to yield some of the most intriguing insights into cetacean behavior. Examples include long-term studies of bottlenose dolphins (Tursiops sp.) in

Shark Bay, Western Australia, which have led to insights into vocal learning and referential signaling in free-ranging cetaceans in the wild (Ford, 1991; Tyack and Sayigh, 1997).

Similar methods can potentially be used to engage groups — as well as individual lone sociable free-ranging dolphins in tests of language comprehension. For instance, interactive underwater keyboards containing visual symbols that dolphins could select have been used to study these skills in captive dolphins (Reiss and McCowan, 1993), providing a closer approximation to two-way communication. Denise Herzing and her colleagues piloted the use of an underwater keyboard with a habituated group of wild spotted dolphins with some success. Moreover, Herzing and collaborators from Georgia Tech in Atlanta are currently developing a cutting edge technology that will potentially provide a much more sophisticated interactive interface for human–dolphin communication in the wild (Herzing and Johnson 2015).

7. Non-Human Animal Personhood and Communication

Communication across the species barrier has potential to build relationships, mutual understanding and a broader circle of personhood. Individuals and cultures draw conceptual boundaries around those viewed as "persons", and most western developed countries extend this privilege only around humans and some companion animals (Ingold, 2000). Personhood designations align closely with a culture's matrix of values, attitudes, beliefs and with each individual's sense of personal identity (Oriel, 2014). Yet communication across species is a grail that extends these lines of commonality, respect, and even kinship.

Though legal "personhood" is often pursued through evidence of mental capacities, identified through communicative capacities, studies of biocommunication between humans, and non-human animals suffer from reductionism when they eliminate emotional bonds of sharing meanings. Philosopher Mary Midgely (2003) sees evidence of "personhood" in emotional capacities or "emotional fellowship", as the markers of mental capacities would grant machines "personhood" before dolphins or apes. Communication facilitates sharing states of mind, intention, and with this enlarged understanding, can follow empathy and inclusion into one's circle of kin, friends, or "persons".

The human/companion-animal relationship develops and deepens through interspecies communication that relies on gestures, expressions of the eye and face and active signals. The human often comes to view the companion dog or cat as a person with basic or even more advanced rights. Feminist Studies scholar Donna Haraway says of her relations with her dog companions, "We are training each other in acts of communication we barely understand. We are, constitutively, companion species. We make each other up, in the flesh." (2007: 16). Does this relationality and intimacy, built on shared messages and meanings, occur only between humans and domestic species, or is this possible with free-ranging animals? Those who have had sociable interactions initiated by habituated or, especially sociable dolphins in the wild express similar sentiments and experience.

Solitary, sociable dolphins in the wild present what may be the most potent examples of collaborative, social, interspecies communication in the wild (Fig. 3). For example, mutual swimming activities, interactions, and games between a local beach community and a solitary sociable dolphin, Moko, in New Zealand, built familiarity

Figure 3. A popular, sociable, free-ranging dolphin in Ireland makes a surprise visit. © Toni Frohoff.

and relationships over time, with some adoring him and others aggravated by his antics. When Moko rescued mother and calf pygmy whales that were beached by guiding them out to open ocean, these relationships continued to grow, evidenced by the film *Soul in the Sea*, Moko's "own" Facebook page, and personal blogs, maintained by those who knew him. Two years after the famous rescue, in 2010 Moko was found dead on a nearby island, and hundreds in the community mourned his loss with a funeral service and burial of his body in a coffin (Tornquist, 2010). The communications involved in these interactions not only influenced the direct participants, but also many people across the world who read these stories in the press. Moko has been one among many solitary sociables, such as dolphins Mara and Fungi in Ireland, various individual belugas in Eastern Canada, and the orca, Luna, in British Columbia (Frohoff *et al.*, 2006), who have been treated as valued members of an interspecies community, and raise awareness about cetacean sentience, communicative abilities, emotional complexity, leading to inclusion within personhood boundaries (Fig. 4).

Figure 4. Friendly gray whales initiate tactile contact with a boater in Baja, Mexico. Both © Toni Frohoff.

The "personhood" boundary is gradually opening and losing the anthropocentric emphasis as many nations' governments and legal frameworks are offering certain animals and other life systems inclusion (such as New Zealand, Switzerland, Bolivia, and India) (Oriel, 2014). This mutually expansive and leveling movement, which puts humans on the same moral and conceptual ground as some other species will benefit those included and perhaps others greatly, through access to legal rights and inclusion into a domain that has been reserved for humans in the West for centuries.

In certain non-western and indigenous cultures, the concept "person" may apply to animals, plants, humans, life systems, climatic events, landscapes, and spirits (Ingold, 2000: 90). Cultural studies of indigenous ontologies and work by indigenous researchers present an inclusive model of personhood and a relational approach to the world that is mirrored in the philosophical tradition of phenomenology. While some countries pursue biocentric parameters for personhood, others have chosen ecocentrism, which offers protection to whole ecosystems, including landforms.

Human cultures in the West have been colonized by certain approaches to relationality (Kohn, 2013) and communication, which finds expression in the anthropocentrism inherent in western thought and activity. Our language has been colonized by assumptions about non-human animals, as we repeat derogatory phrases, such as "dumb as a dormouse, stubborn as a mule." Anthropologists Kohn (2013), Ingold (2000), and Viveiros de Castro (2009) strip through layers of cultural assumptions and write of the perception, language, and correspondence of a more inclusive paradigm.

The "perspectivism" that Kohn (2013) draws on can form theoretical frameworks for human–animal communication studies in which *animals'* perspectives of themselves and their world are both sought and considered. As Vivieros de Castro (2009) says, these perspectives may even prove to hold meanings that are inverse from those of humans. He contends that being in thriving relationships with other animals requires adopting their perspective, which would be a noble goal for further communication research on cetaceans and other species. The ability to hold another's perspective creates an intrinsic conception of "personhood" toward other beings who are viewed as having apperception and intentionality. It appears that regarding individuals from other species for *who* they are, as opposed to *what* they are, is intrinsic to our ability to communicate with them.

8. The Grail of Positive Cohabitation with Interspecies Communication

Interspecies communication can merge identities, merge sentience, and promote a mutual thriving — or, what we call, a positive cohabitation — for humans and marine mammals (Oriel and Frohoff, 2015). In certain coastal areas of the world, humans and coastal systems and species including marine mammals, live within a general trend of thriving and even at times, mutually collaborative cohabitation. In these areas, awareness increases among fishermen, ferry, and tour boat operators, leisure boaters, farmers, and other human residents of human impacts on coastal habitats, with some pro-actively taking measures to protect marine life. Exploring such examples reveals the roots of a continuum of cohabitation, ranging from neutral, passive coexistence, to active and sometimes mutual, collaboration. These thriving scenarios are especially distinctive when counter-poised with failing cohabitation in the Indian River Lagoon, Florida, where large numbers of dolphins, manatees, pelicans, fish, and sea grasses die each year, in a so-called "ecological collapse" (Wines, 2013; NOAA, 2013). Human cultures in these small-scales regions of mutual thriving reside within interspecies communities, engaged and generated through biocommunication (Oriel and Frohoff, 2015). These communities of mutual thriving pose models for other coastal communities, and represent an under-researched arena within Human–Animal Studies.

The continuum of cohabitation provides a fascinating entry into co-cultural evolution as a characteristic of positive cohabitation that extends to other cetaceans, including baleen whales. In some regions, as in the breeding and birthing lagoons of Baja California in Mexico, gray whales of all ages, including females and their newly-born calves, often seek close, sociable contact (sometimes soliciting tactile contact,) with humans in small boats. Perhaps more surprising is that this same whale population was almost hunted to extinction — in part in these same lagoons — when calves were injured to lure their larger mothers. Not until the 1940s did the hunting in this area cease. Yet, in these lagoons that once ran red with blood, the whales have initiated contact and locals respond — a stunning example of human-whale cultural co-evolution and thriving, collaborative cohabitation.

As mentioned earlier, the fishermen and dolphins who share foraging and fishing strategies over many centuries have also been creating an intertwined culture across species. Dolphins seem to enter this foraging alliance voluntarily, which makes it unique among human/non-human animal foraging interactions. This collaboration creates a tangible mutual thriving, and provides rich terrain for interspecies communication studies. In a region of coastal Vietnam, fishermen worship whales as part of a whale-centered religion, evidently instigated by stories of whales saving community leaders. Stories about whales saving humans (or Moko saving pygmy whales) can be central to interspecies cultural and emotional bonds, as do stories about transformations between human and cetacean forms for certain indigenous people like the Noonuccal tribe in Queensland, Australia (Cressey, 1998). Models and the salient features of thriving cohabitation, which rely on interspecies communication, can guide environmental education and conservation.

9. Conclusion

The lure of dolphins has created a somewhat unparalleled curiosity for humans to communicate with other, yet very different, species from our own. This grail quest to understand what dolphins and other non-humans are feeling, thinking, and communicating develops with more inclusive and holistic paradigms (Safina,

2015). Cognitive ethology, Interspecies Collaborative Research, Interspecies Cohabitation Research (Oriel and Frohoff, 2015), and the breakthroughs along the linguistic/ethological border from scientists who build a common language across species, contribute to an equalizing research paradigm from which we transcend excessively anthropocentric models.

Yet the quest for this grail is hampered, as is the archetypal quest, by our human limitations (rather than those of our fellow species as was commonly presumed). In part, our own human colonization, as Kohn (2013) suggests, leads to assumptions that inhibit possible entry points in communication with those of other cultures, regardless of species. Therefore, studying communication in non-human animals through sociological and anthropological methodologies provides great promise (see Whitehead and Rendell, 2014; Herzing and Johnson, 2015). Yet we still need to be cautious about the cultural and species-centric reductionism that lingers not only in science, but also in our personal worldviews that impacts our cohabitation with other human cultures, let alone all life.

Drawing together these related but diverse topics — biocommunication, personhood, cohabitation, collaboration, even multispecies, and multigenerational indigenous understandings across species — they all contain multiple interfaces through which human and non-human animals connect. We have the opportunity to redefine not only animal nature in other species but also revisit and refine our own human–animal nature. Communicative connection points between species can all serve to build on a more thriving relationship between, and for, the benefit of humans and all other animals. Yet what we do with the grail of interspecies communication is up to us.

References

Barrett L and Würsig B. 2014. Why dolphins are not aquatic apes. *Animal Behavior and Cognition*, 1(1), 1–18.

Bearzi M and Stanford C. 2010. *Beautiful Minds: The Parallel Lives of Great Apes and Dolphins.* Harvard University Press, Cambridge, MA.

Brent LJN *et al.* 2015. Ecological knowledge, leadership, and the evolution of menopause in killer whales. *Current Biology*, 25(6), 746–750.

Courage M. 2010. The grandmother factor: why do only humans and whales live long past menopause? *Scientific American Blog*. Available at: http://blogs.scientificamerican.com/observations/the-grandmother-factor-why-do-only-2010-06-30/.

Cousteau J and Diole P. 1987. *Dolphins*. Arrowood Press, New York.

Crane JK. 2015. *Beastly Morality: Animals as Ethical Agents*. Columbia University Press, New York.

Cressey J. 1998. Making a splash in the pacific: Dolphin and whale myths and legends of Oceania. *Rapa Nui Journal*, 12, 75–84.

Diamond J. 2001. Unwritten knowledge. *Nature*, 410, 521. (29 March 2001).

Domning DP. 1991. A former dolphin-human fishing cooperative in Australia. *Marine Mammal Science*, 7(1), 94–96.

Dudzinski KD and Frohoff T. 2008. *Dolphin Mysteries: Unlocking the Secrets of Communication*. Yale University Press, New Haven, Connecticut.

Fagan R. 1981. *Animal Play Behavior*. Oxford University Press, New York.

Ford JKB. 1991. Vocal traditions among resident killer whales (*Orcinus orca*) in coastal water of British Columbia. *Can J Zool*, 69, 1454–1483.

Frohoff TG. 1993. Behavior of captive bottlenose dolphins (Tursiops truncatus) and humans during controlled in-water interactions. M.S. thesis, Texas A&M University.

Frohoff TG. 2000. The Dolphin's Smile. In Bekoff M (ed) *The Smile of a Dolphin: The Emotional Lives of Animals*. Discovery Books. (pp. 72–73) New York.

Frohoff TG and Peterson B. 2003. *Between Species: Celebrating the Dolphin-Human Bond*. Sierra Club Books, San Francisco.

Frohoff TG. 2004. Stress in Dolphins. In Bekoff M (ed.) *Encyclopedia of Animal Behavior* (pp. 1158–1164), Greenwood Press, Westport, Connecticut.

Frohoff T, Vail CS and Bossley M. 2006. Preliminary Proceedings of the Workshop on the Research and Management of Solitary, Sociable Odontocetes convened at the 16th Biennial Conference on the Biology of Marine Mammals. San Diego, California, December 10, 2005. International Whaling Commission Scientific Committee, SC/58/13.

Frohoff T. 2008. Marine Animal Welfare (mammals and non-mammals). In Lück M (ed.) *Encyclopedia of Tourism and Recreation in Marine Environments* Wallingford, 515–521, CABI Press, UK.

Frohoff T and Marino L. 2010. Towards a New Ethical Research Paradigm on Cetacean Cognition. Proceedings from the 12th *International Conference of the American Cetacean Society*, Monterey, California.

Haraway D. 2007. *When Species Meet*. University of Minnesota Press. Minneapolis, p. 16.

Herman LM, Kuczaj SA, and Holder MD. 1993. Responses to anomalous gestural sequences by a language-trained dolphin: evidence for processing of semantic relations and syntactic information. *J Exp Psy*, 122, 184–194.

Herzing DL. 2005. Transmission mechanisms of social learning in dolphins: underwater observations of free-ranging dolphins in the Bahamas. *Autour de L'Ethologie et de la Cognition Animale*. Presses Universitaires de Lyon, Publ, 185–193.

Herzing DL and Johnson CM. 2015. *Dolphin communication and cognition: past, present and future*. MIT Press, Cambridge, MA.

Herzing DL. 2015. Synchronous and rhythmic vocalizations and correlated underwater behavior of free-ranging Atlantic spotted dolphins (*Stenella frontalis*) and bottlenose dolphins (*Tursiops truncatus*) in the Bahamas. *Animal Behavior and Cognition*, 2(1), 14–29.

Hobaiter C and Byrne RW. 2011. The gestural repertoire of the wild chimpanzee. *Anim Cog*, 14(5), 745–767.

Ingold T. 2000. *The Perception of the Eenvironment: Essays in Livlihood, Dwelling, and Skill*. Routledge, London.

Kohn E. 2013. *How Forests Think*. University of California Press, Berkeley.

Krebs JR and Dawkins R. 1991. Animal signals: mind-reading and manipulation. In Krebs JR and Davies NB (eds.). *Behavioral Ecology: An Evolutionary Approach*, p. 380. Blackwell Scientific Publications, Oxford.

Krützen M *et al*. 2005. Cultural transmission of tool use in bottlenose dolphins. *Proc Natl Acad Sci USA*, 102, 8939–8943.

Lockyer C and Müller M. 2003. Solitary, yet sociable. In Frohoff T and Peterson B (eds.). *Between Species: Ccelebrating the Dolphin-Human Bond*, (pp. 138–150). Sierra Club Books, San Francisco.

Lusseau D *et al*. 2006. Quantifying the influence of sociality on population structure in bottlenose dolphins. *J Anim Ecol*, 75(1), 14–24.

Marino L, *et al.* 2008. A claim in search of evidence: Reply to Manger's thermogenesis hypothesis of cetacean brain structure. *Biol Rev Cambridge Phil Soc*, 83, 417–440.

Marino L and Frohoff T. 2011. Towards a new paradigm of non-captive research on cetacean cognition. *PLoS ONE*, 6(9), 1–9.

Midgley M. 2003. Is a dolphin a person? In Armstrong S and Boltzer R (eds.), *The Animal Ethics Reader* (p. 167). Routledge, London.

NOAA 2013. *Bottlenose Dolphin Unusual Mortality Event in Florida.* Available at: http://www.nmfs.noaa.gov/pr/health/mmume/floridadolphins2013.html.

Oriel E. 2014. Whom would animals designate as 'Persons'? On the avoidance of anthropocentrism and the inclusion of others. *Journal of Evolution and Technology*, 24(3), 44–59.

Oriel E and Frohoff T. 2015. Models of mutual thriving and collaborative cohabitation among humans and non-human coastal species inform conservation, education, and wellbeing. *Proceedings of the International Compassionate Conservation conference*, July 28–31, 2015.

Payne K, *et al.* 1986. Infrasonic calls of the Asian elephant (Elephas maximus). *Behavioral Ecology and Sociobiology*, 18(4), 297–301.

Philips M and Austad SN. 1990. Animal communication and social evolution. In Bekoff M and Jamieson D (eds.), *Interpretation and explanation in the study of animal behavior, Volume 1: Interpretation, intentionality, and communication* (pp. 254–268). Westview Press, Oxford.

Pliny the Elder 1940. Natural History. Vol. 3, books 8–11, trans. H. Rackham. *Loeb Classical Library*. Harvard University Press, Cambridge, MA.

Pryor K. 1990. Non-acoustic communication in small cetaceans: Glance, touch, position, gesture, and bubbles. In Thomas J and Kastelein R (eds.), *Sensory Methods of Cetaceans: Laboratory and Field Evidence* (pp. 537–543). Plenum Press, New York.

Pryor K *et al.* 1990. A dolphin-human fishing cooperative in Brazil. *Marine Mammal Science* 6(1), 77–82.

Reiss D and McCowan B. 1993. Spontaneous vocal mimicry and production by bottlenose dolphins (*Tursiops truncatus*): Evidence for vocal learning. *J Comp Psych*, 107(3), 301–312.

Rose N and Furinato R. 2006. The Case Against Marine Mammals in Captivity. Report for *The Humane Society of the United States*.

Safina C. 2015. *Beyond Words: What Animals Think and Feel*. Henry Holt, New York.

Smith J. 1990. Communication and expectations: a social process and the cognitive operations it depends upon and influences. In Bekoff M and Jamieson D (eds.), *Interpretation and explanation in the study of animal behavior, Volume I: Interpretation, Intentionality, and Communication* (pp. 234–252). Westview Press, Oxford, UK.

Tornquist B. 2010. Funeral Services for Famous New Zealand Dolphin, Moko. BBC News. [Online] Available from: http://www.bbc.co.uk/news/world-asia-pacific-10666513. (Accessed on 23 February 2015).

Tyack PL and Sayigh LS. 1997. Vocal learning in cetaceans. In Snowdon CT and Hausberger M (eds.). *Social influences on vocal development* (pp. 208–233). Cambridge University Press, Cambridge.

Viveiros de Castro E. 2009. *Metaphysiquales cannibales: Lignes d'anthroppologie post-structural.* Presses universitaires de France, Paris.

White T. 2007. *In Defense of Dolphins: The New Moral Frontier.* Blackwell Publishing, Oxford, UK.

Whitehead H and Rendell L. 2014. *The Cultural Lives of Whales and Dolphins.* University of Chicago Press, Chicago.

Wines M. 2013. Deaths of manatees, dolphins, and pelicans point to estuary at risk. New York Times (Online). Available from http://www.nytimes.com/2013/08/08/us/deaths-of-manatees-dolphins-and-pelicans-point-to-estuary-at-risk.html?_r=0. (Accessed on 22 February 2015).

Index

D

N